Mechanical Engineering Series

Series Editor

Vish Prasad, Department of Mechanical and Energy Engineering, University of North Texas, Denton, USA

The Mechanical Engineering Series presents advanced level treatment of topics on the cutting edge of mechanical engineering. Designed for use by students, researchers and practicing engineers, the series presents modern developments in mechanical engineering and its innovative applications in applied mechanics, bio-engineering, dynamic systems and control, energy, energy conversion and energy systems, fluid mechanics and fluid machinery, heat and mass transfer, manufacturing science and technology, mechanical design, mechanics of materials, micro- and nano-science technology, thermal physics, tribology, and vibration and acoustics. The series features graduate-level texts, professional books, and research monographs in key engineering science concentrations.

Mehrdaad Ghorashi

Introduction to Machine Dynamics

Springer

Mehrdaad Ghorashi
Department of Engineering
University of Southern Maine
GORHAM, ME, USA

ISSN 0941-5122 ISSN 2192-063X (electronic)
Mechanical Engineering Series
ISBN 978-3-031-94735-3 ISBN 978-3-031-94736-0 (eBook)
https://doi.org/10.1007/978-3-031-94736-0

This Springer imprint is published by the registered company Springer Nature Switzerland AG
The registered company address is: Gewerbestrasse 11, 6330 Cham, Switzerland

To the memory of my beloved mother,
Touraandokht Khoshravaan,
my dear brother, Mahdaad,
and all who selflessly strive for a better future for others.

Preface

What Is It About? This book presents an introduction to machine dynamics a good knowledge of which is necessary in any application that includes design or analysis of mechanisms and machines. Examples are everywhere around us. Designing various types of robots that are used in industry for manufacturing or materials handling, designing the landing gear of an airplane, analyzing the motion of a piston in an internal combustion engine of a vehicle, and designing a Mars rover, all are examples of where the material covered in this book is applicable. Wherever a machine is operated for force or power transmission, or a mechanism is used to provide a desired motion, the methods studied in machine dynamics present the fundamental knowledge that is needed for an optimal or at least a satisfactory design.

Why Another Book? Because of its importance, the subject of machine dynamics is offered as a required or a technical elective course in many mechanical engineering programs, including the one offered at University of Southern Maine (USM), where I am a professor of mechanical engineering. I have developed and taught the *Machines and Mechanisms Design (MEE 373)* course for sixteen years. During these years, I have used three textbooks for teaching this course and I have learned from and enjoyed reading every one of them. However, while they were all great resources and wonderful books, the feedback that I have received from my students about using them as textbooks has not been as positive as I expected. I have taught many other courses in the mechanical engineering curriculum as well, so I knew the student satisfaction level about the textbooks was below what was normally expected. Based on my students' feedback and my own analysis I could conclude that, while those books are all *comprehensive and informative*, for the reasons that will follow, there is still considerable room for *improving the way the material is being presented* to the audience, especially to students who learn the material for the first time. Teaching is only effective if it is followed by learning. These were some of my reasons (and hopes!) while writing this book.

Some Highlights? Here are some highlights of the present book. In many cases a problem has been solved using various methods to provide students with the necessary confidence in using each one of those methods. The presented solutions are in complete details. You will not see many, "*It can be shown.*" statements here! The book includes fully-explained solutions to progressively difficult problems. In this way, not only application of the discussed methods is mastered and the relative difficulty of using each method is illustrated, but also the results are compared with one another and verified. Consequently, the important habit of verifying and justifying the obtained solutions is reinforced all over the book.

In this book, I have given details of using the Newton-Raphson method for performing position analysis of mechanisms (which is usually a highly nonlinear problem). In the past, I have used the same method on subjects stretching from the limit analysis and plastic design of structures to the dynamics of composite rotating beams, see Ghorashi (1994) and Ghorashi (2016), respectively. Also, in many textbooks, calculation of the gear ratio for planetary gears is done using the superposition principle and tables. Here, this calculation is performed by implementing the general rolling contact equations that simplifies the solution considerably.

I have tried not to convert this book to another handbook on machines and mechanisms, so it is focused on the most fundamental topics. For example, while the geometrical method (that some other textbooks focus on) is an interesting method for visualizing the solution, here the focus is on the vector-loop method instead (that can be easily extended to more complex problems). The use of the method of kinematic coefficients has been illustrated here only for velocity analysis, not for acceleration. Students, in general, feel much more comfortable using more familiar concepts like velocity and acceleration compared to kinematic coefficients.

I have included static and dynamic balancing of rotors, as well as an introduction to spatial dynamics and gyroscopic motion; these are topics that I believe any graduate of a course on machine dynamics needs to know. The book includes almost 400 pictures and illustrations that assist the reader in visualizing and understanding the material. As to systems of units, both the SI (Système International) and USCS (U.S. Customary System) units have been used in this book because a graduate of such a course should feel comfortable using either of them. In the United States, this competency is also tested by the Fundamentals of Engineering (F.E.) and the Principles and Practice of Engineering (P.E.) exams, that are required for earning a Professional Engineering (P.E.) license. Use of both units of measurement in this book would help students succeed in those exams and in their career.

Secrete Sauce? This book is based on my class-tested lecture notes on machine dynamics and engineering dynamics that I have taught in Sharif University of Technology, Carleton University, and University of Southern Maine (USM). I have used an effective method for continuously updating my lecture notes before and after each lecture. Since 2011, and initially with the aim of assisting students with disabilities in their notetaking challenge, I have been using tablets for writing and projecting my class notes during lectures. That practice proved to have an important added benefit: it significantly helped me with updating my lecture notes after each class.

I continuously merged the new items that were discussed in class in my personal notes of the day and I also used this method as a means for including students' feedback in my lecture notes. The written notes on tablet have always been reviewed after class for accuracy as well, and only then the notes were posted on Brightspace (that is our learning management system). The feedback that I have received from my students about using this method has been overwhelmingly positive. They appreciate that the method allows them to focus on listening to lectures and participate in discussions, rather than just being involved in thoughtless notetaking or watching slides. The course also includes hands on exploration of real-life mechanisms in class and completing a term project. The results of an extended version of one of these projects were published in Davis and Ghorashi (2018).

Required Prior Knowledge? This book can be used for an introductory presentation of machine dynamics to students in several engineering disciplines including mechanical, mechatronics, and electro-mechanical engineering. It can also be used by practicing engineers who work in this field. While the book is at an introductory level, the reader is still expected to have a fair background of a few prerequisite materials that will only briefly be reviewed in this book. A fair background of mechanical engineering statics, dynamics, linear algebra, computer programming (especially using MATLAB), and multivariable calculus would be helpful in understanding the topics that are brought up in this book. The course that I offer at USM also includes a one-credit part focused on computer simulation of mechanisms using SolidWorks which uses Chang (2024) as a textbook. While some SolidWorks and MATLAB simulation results are presented here, the emphasis is on understanding the foundations of the theories used in machine dynamics.

What to Expect? This book consists of an introductory Chap. 1 on defining structures, kinematic chains, machines, and mechanisms, as well as two major parts. Part I is about kinematics of machines and mechanisms. This is the study of motion with no regard to the forces and moments that generate those motions. Kinematics is a purely geometrical analysis of motion. Part II is about kinetics of machines in which motion and its causes (i.e., forces and moments) are studied simultaneously. Chapter 2 is on basic characteristics of a mechanism and its components. In this chapter concepts like joints, links, and degrees of freedom (DOF) are introduced. Finding kinematic inversions of a mechanism is discussed and it is shown how Grashof's law can be used to predict the absolute motion of each one of those kinematic inversions. Finally, a frequently-used type of mechanism, i.e., the quick-return mechanism is also introduced. Chapter 3 is dedicated to the position analysis of mechanisms. Both closed-form solutions and numerical solutions using Taylor series expansion and the Newton-Raphson algorithm are presented. The latter is easily applicable to solving nonlinear, multi-degree-of-freedom systems. Chapter 4 presents velocity analysis of mechanisms using algebraic, kinematic coefficients, relative motion, and instantaneous centers methods. Acceleration analysis that is given in Chap. 5 is the major link that relates kinematic analysis to kinetic analysis using Newton's second law. In Chaps. 6, 7, and 8, various types of gears and gear trains are discussed. Spur, bevel, helical, and worm and worm gear are among the presented topics. The relation between the fundamental law of gearing and the angular-velocity-ratio theorem provides a bridge between the instantaneous centers topic and the gears section. Fixed-axis and planetary gear trains and the way their speed ratios can be calculated using the rolling contact equations are illustrated.

Design aspects are discussed throughout the book and especially in Chap. 9 that is mainly about using Freudenstein's equations for creating function-generating mechanisms.

Part 2 of the book is about kinetics and inclusion of forces and moments in the analysis. It starts with the static equilibrium analysis of mechanisms in Chap. 10 that also includes discussion of two- and three-force members as well as force analysis in spur gears. Chapter 11 is dedicated to the definition and calculation of mechanical advantage in various machines. In many of these machines the mechanical advantage is not a constant value, as we have seen, e.g., in a lever, and it changes during motion. A brief review of planar dynamics has been given in Chap. 12. The three main solution methods, i.e., Newton's second law, impulse and momentum, and work and energy are briefly reminded. In Chap. 13, the inverse planar dynamic problems are introduced in which the desired motion is known and we are looking for compatible forces and moments that can generate those motions. The connecting link between motion, forces, and moments are the acceleration terms that are used in Newton's second law. The calculated dynamic forces that are needed to generate the desired motion are compared with the corresponding static equilibrium ones and used as a means for verification of the obtained results. It is shown how static solutions can provide close approximations for dynamic solutions in cases where inertia effects are not significant (low mass and low accelerations). In Chap. 13, application of matrix algebra for solving the systems of equations of equilibrium (in statics) or equations of motion (in dynamics) is also illustrated. The topic of static and dynamic balancing of short and long rotors is discussed in Chap. 14. This is a very important topic and finds applications in designing rotating machinery. Chapter 14 also includes discussions on calculating shaking forces in mechanisms with reciprocating members, like pistons in internal combustion engines and reciprocating compressors. Finally, the book concludes in Chap. 15 with coverage of three-dimensional dynamics, including Euler's equations of motion and equations of parallel-plane motion. A discussion on gyroscope and the gyroscopic effect concludes this chapter.

Thank you! This work was completed while I was on a one-semester sabbatical leave from University of Southern Maine (USM). The book includes illustrations from the outcomes of many term projects of my students in the course, *Machines and Mechanisms Design (MEE 373)*, and several senior design capstone projects at USM during the past 16 years. The students' contributions to this book, by building all of those mechanisms and simulation models, have been substantial. Such experiences have reinforced student learning through our motto (*If you don't understand it, build or simulate it!*) and I'm very glad to see that they are now a central part of this book. I also appreciate the Springer Nature editorial and production staff for all their assistance in publishing this book. Last, but certainly not least, I appreciate the love and support that I have received from my wife, Marjaneh, my daughter Mehrnaz, my son Ali, and our fabulous feline Stormy Amber throughout the years.

Gorham, ME, USA Mehrdaad Ghorashi

References

Chang, K. H. (2024). *Motion simulation and mechanism design with SolidWorks motion 2024*. SDC Publications.

Davis, M., & Ghorashi, M. (2018). A holistic approach for deciphering the operation of planetary gear train systems. *ASEE's Journal of Engineering Technology, 35*, 16–25.

Ghorashi, M. (1994). Limit analysis of circular plates subjected to arbitrary rotational symmetric loadings. *International Journal of Mechanical Sciences, 36*(2), 87–94.

Ghorashi, M. (2016). *Statics and rotational dynamics of composite beams*. Springer International Publishing AG.

Competing Interests The author has no competing interests to declare that are relevant to the content of this manuscript.

Contents

Part II Kinetics of Machines and Mechanisms

Nomenclature

$(\ )'$	First-order kinematic coefficient of ( )
$(\ )'$	Dynamic value of force ( ) (Chap. 13)
$(\ )/[\ ]$	( ) relative to [ ]
$(\ )^{a}$	Axial component of ( )
$(\ )^{n}$	Normal component of ( )
$(\ )^{r}$	Radial component of ( )
$(\ )^{t}$	Tangential component of ( )
$(\ )^{t}$	Transpose of ( ) (Chap. 13)
$(\dot{\ })$	Time derivative of ( )
$(\ddot{\ })$	Second time derivative of ( )
$(\hat{\ })$	Unit vector ( )
a	Linear acceleration
a	Addendum
A	Coefficient matrix
B	Vector of known right-hand-side values
C	Number of cylindrical joints
c	General name for a constant quantity
C	Center distance
C	Contact point
C	Instantaneous center of zero velocity
CCW	Counter-clockwise
CW	Clockwise
d	Dedendum
d	Distance between center of percussion and G
DOF	Degrees of Freedom
e	Eccentricity
e_c	Correction mass eccentricity
e_G	Mass center eccentricity
e_l	Left-plane correction mass eccentricity
e_r	Right-plane correction mass eccentricity
F	Force

F_c	Inertia force due to correction mass
F_{ij}	Force imposed by i on j
g	Gravitational acceleration
G	Mass center
H	Number of helical joints
H	Angular momentum
I	Area moment of inertia
I	Mass moment of inertia
I_{ij}	Instantaneous center of i and j
IC	Instantaneous center
J	Jacobian matrix
j_1	Number of joints with one DOF
j_2	Number of joints with two DOF
k	Non-dimensional parameter (Chap. 9)
k	Radius of gyration
l	Lead
L	Link length
l	Pendulum length
LHS	Left-hand side
m	Mobility
m	Mass
m	Module
M	Moment
$M.A.$	Mechanical advantage
m_c	Contact ratio (Chap. 6)
m_c	Correction mass (Chap. 14)
m_{cb}	Counterbalance mass (for reciprocating masses)
m_l	Left-plane correction mass
m_r	Right-plane correction mass
n	Number of members (for mechanisms)
N	Thread number (for screws and worms)
N	Normal force
N	Number of gear teeth
O	Center point
O	Fix point
P	Number of prismatic joints
p	Pitch
P	Diametral pitch
p	Circular pitch
P	Power
P	Pitch point
P	Combustion force on piston (Chap. 10)
p	Half perimeter of a triangle
p_a	Axial pitch

p_b	Circular pitch of the base circle
p_n	Normal circular pitch
P_n	Normal diametral pitch
p_t	Transverse circular pitch
P_t	Transverse diametral pitch
Q	Advance-to-return ratio
r	Link length
R	Number of revolute joints
R	General name for radius
R	Radius of the pitch circle
r	Radius of the inscribed circle
R	Reaction force
R_b	Radius of the base circle
r_{cb}	Counterbalance mass radius (for reciprocating masses)
RHS	Right-hand side
S	Number of spherical joints
S	Area of a triangle
T	Torque
t	Wheel track
T	Tensile force
T	Kinetic energy
T_{ij}	Torque imposed by i on j
$U_{1\text{-}2}$	Work done by all external forces during 1 to 2 motion
$U'_{1\text{-}2}$	Work done by non-conservative forces during 1 to 2 motion
v	Solution of nonlinear equation (Chap. 3)
v	Velocity
V_e	Elastic potential energy
V_g	Gravitational potential energy
w	Wheel base
x	Coordinate x
x	General name for unknown quantity
X	Vector of unknowns
y	Coordinate y
z	Coordinate z
α	Angular acceleration
γ	Transmission angle
γ	Pitch cone angle
Δ	Change
ε	Small quantity
θ	Angular position
θ_c	Correction mass angular position
θ_l	Left plane correction mass angular position
θ_r	Right plane correction mass angular position
λ	Crank to connecting rod length ratio

λ	Lead angle
μ	Coefficient of friction
Π	Product
ρ	Mass density (per unit volume: Chap. 12)
ρ	Mass density (per unit length: Sect. 15.3.1)
ρ	Mass density (per unit area: Sect. 15.4)
Σ	Summation
Σ	Shaft intersection angle
φ	Pressure angle
φ	Belt and pulley contact angle
φ'	Effective pressure angle
φ_n	Normal pressure angle
φ_t	Transverse pressure angle
ψ	Helix angle
ω	Angular velocity
ω_p	Precession angular velocity
ω_s	Spin angular velocity

Chapter 1
Structures, Mechanisms, and Machines

1.1 Introduction

This chapter is an introduction to the three main subjects that will be discussed in this book, i.e., structures, mechanisms, and machines. While the focus of this book is on mechanisms and machines, there are some interesting relations between them on the one hand and structures on the other hand that provide readers with a broad understanding of the subject. We will see later how the same methods that we use for the calculation of degrees of freedom of mechanisms and machines can be used for detecting if what we have in mind is a mechanism or a stationary structure with no ability to move. Same methods can help us detect if a structure is statically determinate or it has redundant members. Later we will also see how the static analysis methods that we use for structural analysis can provide first estimations for solution of dynamic problems. Such results can also be used for verifying the accuracy of dynamic solutions.

1.2 Structures

A structure is an entity that is created so that it can withstand the applied loads on it, while it serves its purpose. A building, a car body, and a bridge all are examples of structures. Figure 1.1 illustrates a picture of a bridge structure. Structures that we deal with in this book consist of a series of rigid bodies (called links) that are connected to each other by joints. Of course, in a structure no motion or power transmission is performed. (Unless the load is so excessive that the structure fails to support it and forms what is called a collapse mechanism, see Ghorashi [2]).

The minimum subset of a truss structure, shown in Fig. 1.2, has three links 1, 2, and 3. The illustrated joints at *A*, *B*, and *C* are hinges (i.e., they allow rotation of the links with respect to each other, so no bending moment is created in the members or

M. Ghorashi, *Introduction to Machine Dynamics*, Mechanical Engineering Series,
https://doi.org/10.1007/978-3-031-94736-0_1

Fig. 1.1 Northampton Lattice Truss Bridge, Northampton, Hampshire County, MA. Public domain picture, https://garystockbridge617.getarchive.net/amp/media/boston-and-maine-railroad-northampton-lattice-truss-bridge-northampton-hampshire, accessed 12/1/2024

Fig. 1.2 A building block of a truss structure

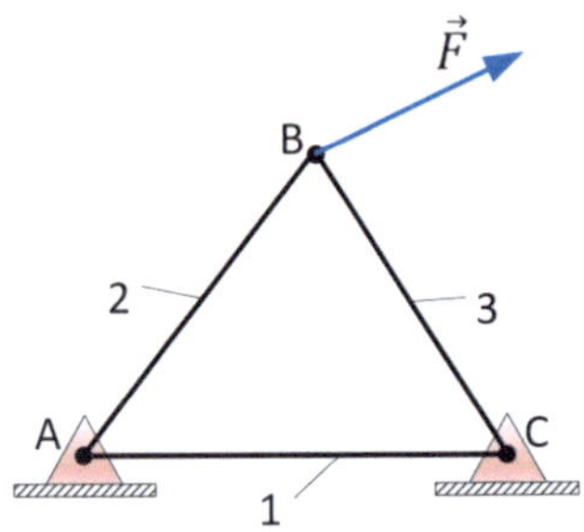

at the joints). Link 1 (or AC) is fixed to the immovable base. This structure can support the applied load $\vec{F}$ only by the axial forces that are generated in its links AB and BC (both are two-force members). These are members that can support the applied load only along the direction determined by their two ends. (It should be intuitively clear that link AB is in tension and link AC is in compression.) In a structure, the aim is to support the applied loads while the links can be assumed to remain immovable and rigid.

1.3 Kinematic Chains

If a few links are connected to each other by joints, but unlike structures the links can move with respect to each other and none of them is fixed, the set is said to form a kinematic chain. As an example, by adding a fourth link to the 3-link basic truss element shown in Fig. 1.2, and removing the connections to the ground, one can get the four-bar kinematic chain shown in Fig. 1.3. This is a closed kinematic chain. If one of the joints, e.g., the joint at D is broken, the kinematic chain would be an open kinematic chain.

An application of a kinematic chain is shown in Figs. 1.4 and 1.5 that are illustrations of a vise-grip in closed and open postures, respectively. We see that a vise-grip has four principal members that are connected by hinges and none of the members is stationary. So a vise-grip is a closed four-bar kinematic chain.

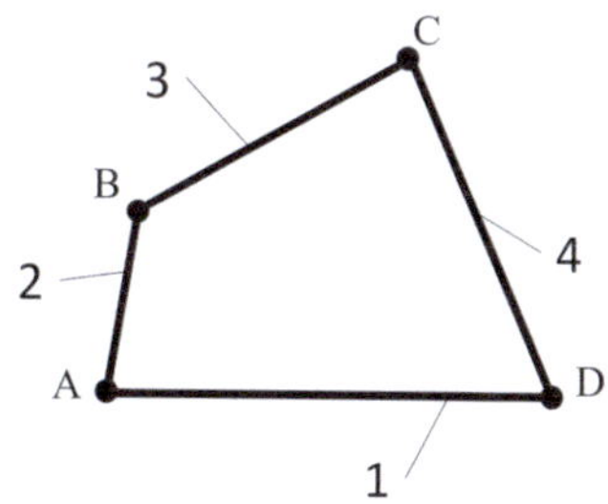

Fig. 1.3 A closed four-bar kinematic chain

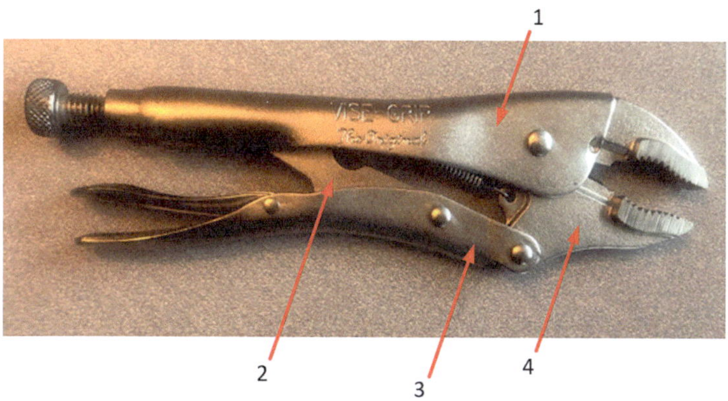

Fig. 1.4 A vise-grip in its closed (locked) posture

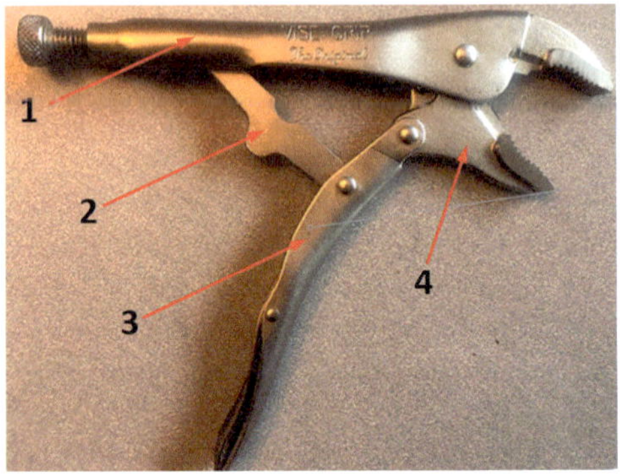

Fig. 1.5 Vise-grip in its open posture

1.4 Mechanisms and Machines

Similar to a kinematic chain, a mechanism consists of a series of links that are connected to each other by joints. However, unlike kinematic chains, a mechanism has a fixed link (that is usually numbered as 1) and motions of the other links are measured with respect to this link. The fixed link is also called the frame, or the ground member. A mechanism is principally used for motion transmission, that is a kinematical objective (i.e., motion analysis without including forces that generate it).

A machine has all the characteristics of a mechanism with the added duty that it is used for power transmission. Therefore, a machine has similar components and motions as a mechanism, but the main objective of using a machine is to do work and to transmit power, not just move in a desired way.

The door-closing mechanism shown in Fig. 1.6 has four members and its fixed link is the door frame 1. This is an example of a four-bar linkage in which, by moving any movable link (in this case, link 4, i.e., the door) the mechanism moves. This mechanism has one degree of freedom because by assigning the angular position of only one of the links (e.g., the door) the positions of the rest of the links would be fully known. (The concept of degrees of freedom will be discussed in detail in Chap. 2).

For ease of demonstration, such a mechanism is usually illustrated by a simplified (also called, skeleton) diagram. Figure 1.7 shows the top view skeleton diagram of the door-closing mechanism.

Another important four-bar linkage is the one that is used in a hand water pump and is shown in Fig. 1.8. The mechanism includes the cylindrical frame 1, link 2, handle 3, and the reciprocating piston rod 4. The latter is connected to a piston fitted with a non-return (one-way) valve. Another non-return valve is fitted at the base of cylinder 1. Right at the base of the cylinder a pipe (not shown) is connected to the cylinder that is submerged in water. When piston rod 4 is pushed down, the piston valve is open and water flows from below the piston to the top of the piston. When piston rod 4 is pulled back up, the piston valve is closed and it forces water out

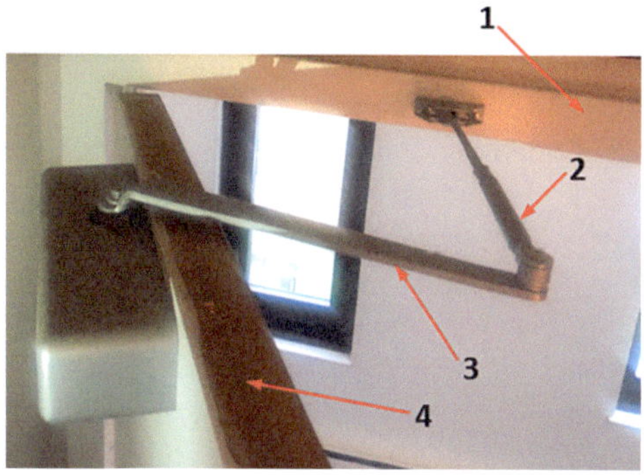

Fig. 1.6 Door-closing mechanism

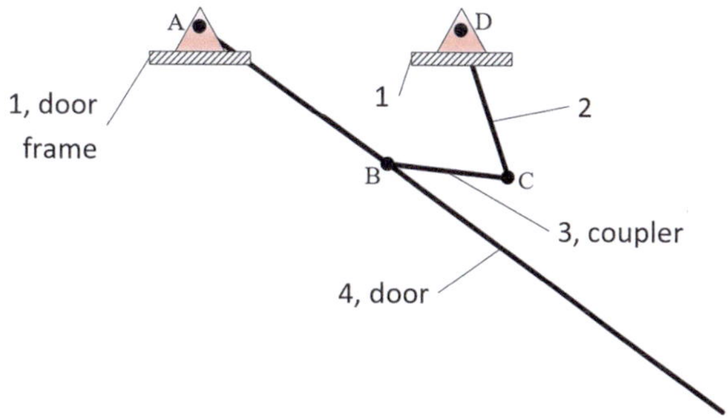

Fig. 1.7 The top view skeleton diagram of Fig. 1.6

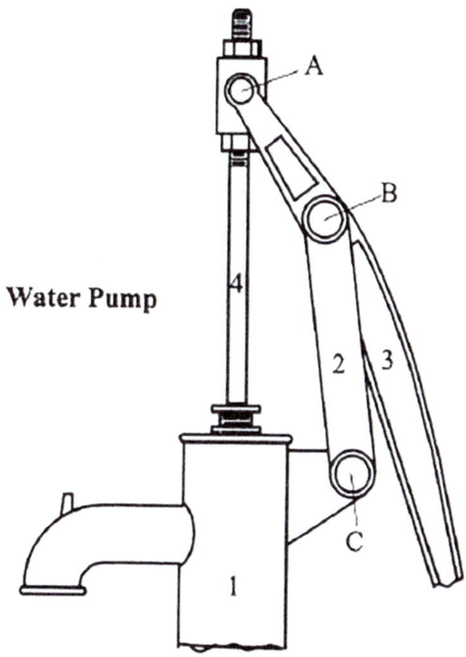

Fig. 1.8 A four-bar linkage used in a hand water pump. Page 54 of Waldron and Kinzel [4]. © 2004 John Wiley & Sons, Inc. (reproduced with permission)

of the pump, which is our aim. The opposite of these operations happen at the valve that is fitted at the base of cylinder 1.

Now, consider the four-bar linkage shown in Fig. 1.9. In the illustrated mechanism, the input link 2 can make full 360° rotations and for this reason it is called a crank. The crank can, e.g., be run by a continuously rotating shaft of an electric motor. Link 4 is the output of the mechanism and that is why we call it a follower. In Fig. 1.9, the link that connects the crank to the follower is named, the coupler.

In Fig. 1.10, the mechanism shown in Fig. 1.9 is subjected to driving and driven torques, T_{in} and T_{out}, respectively. Link 2 is subjected to torque T_{in} as the input torque. This torque is supposed to overcome the load or output torque, T_{out}. The ratio T_{out}/T_{in}, that depends on the posture of the machine, is the mechanical advantage of this machine and for static cases it will be discussed in Chap. 11.

$$\text{Mechanical Advantage (M.A.)} = \frac{T_{\text{out}}}{T_{\text{in}}} \tag{1.1}$$

If dynamic effects cannot be ignored, methods discussed in Chap. 13 should be used to provide dynamic estimations of M.A.

In an internal combustion engine shown in Fig. 1.11, the driving force originates from the pressure generated by the products of combustion inside cylinder 1 and imposed on piston 2. The output is the motion and power transmitted to the crank. Crank 4 is connected to the engine base by a bearing and it can make full 360° rotations. In the beginning of motion, and while starting the engine, the crank is rotated by the starter and the flywheel. The rotational motion of the crank is

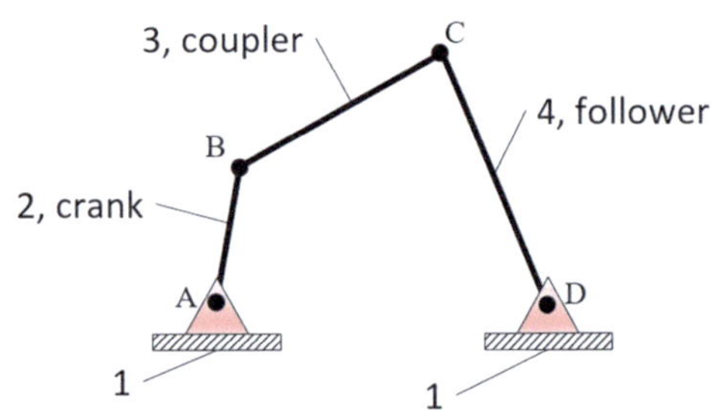

Fig. 1.9 Skeleton diagram of a four-bar linkage (mechanism)

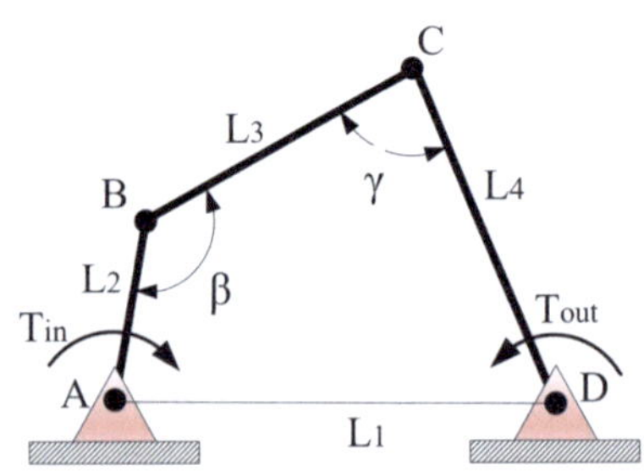

Fig. 1.10 A four-bar linkage, now operating as a machine

Fig. 1.11 Cutaway of an internal combustion engine. 1. Engine body, 2. Piston, 3. Connecting rod, and 4. Crankshaft

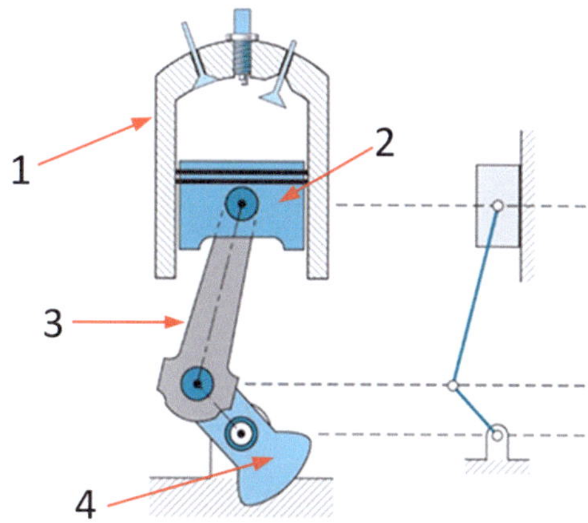

Fig. 1.12 A slider-crank mechanism (left); and its skeleton representation (right). 1. Cylinder, 2. Piston, 3. Connecting rod, and 4. Crankshaft. Page 4 of Cleghorn and Dechev (2015) © 2015 Oxford University Press (reproduced with permission)

converted to the translational motion of the piston. After the engine starts, power flows in the reverse direction from piston to the crank.

A sketch of the longitudinal cross-section of an internal combustion engine with two one-way valves, that was just described, is shown in Fig. 1.12 (left). This is a slider-crank mechanism and Fig. 1.12 (right) is its equivalent skeleton diagram and the way this mechanism is usually shown in our analyses. As mentioned before, this mechanism has a piston (member 2) that performs reciprocating motion inside a cylinder frame (member 1). This motion is then transmitted to the connecting rod (member 3). Finally, energy is transmitted to the crank (member 4) that in an internal combustion engine is a representation of the crankshaft. The crank rotates about a fixed point in the engine. In this way, the slider-crank mechanism converts the reciprocating motion of the piston to the rotational motion of the crank. The connecting rod undergoes the general planar motion; that is a mixture of rotation and translation and will be explained in Chap. 2.

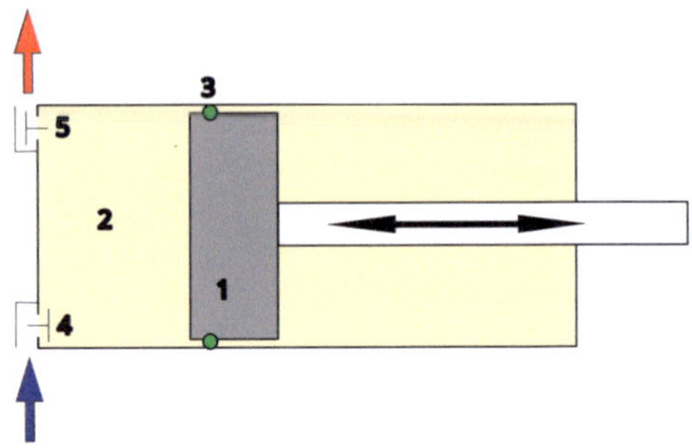

Fig. 1.13 Basic operation of a reciprocating air compressor. Public domain picture, https://commons.wikimedia.org/wiki/File:Kompresors.svg. Accessed 12/1/2024

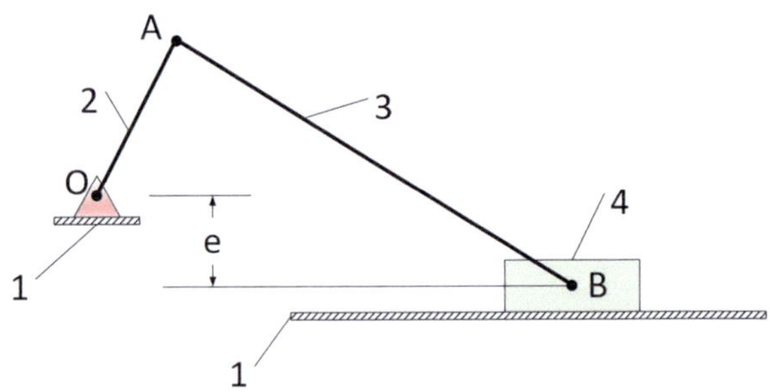

Fig. 1.14 Eccentric slider-crank mechanism

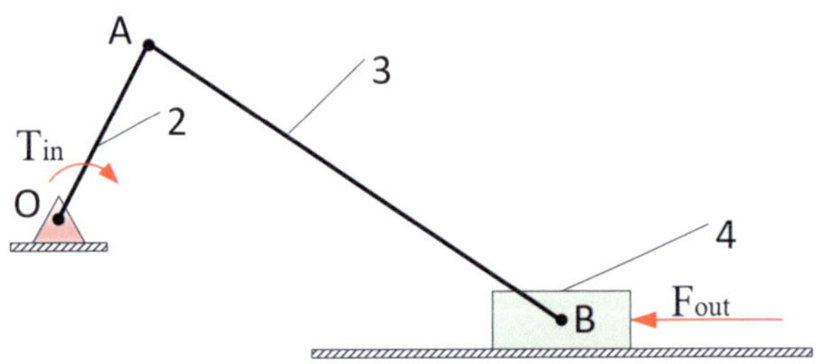

Fig. 1.15 Input torque on crank and output force on slider in an eccentric slider-crank mechanism

Recalling the slider-crank mechanism again, it is also possible that energy is provided to the crank as the input link of the mechanism by, e.g., an electric motor. In this case, the piston motion would be the output as it is seen in applications like reciprocating air compressors that again have two no-return (one-way) valves. The basic operation of this machine is shown in Fig. 1.13. As the piston moves to the right, the intake valve 4 is open and air is sucked in. During this motion, the high-pressure exit valve 5 is closed. Then, as the piston moves to the left, the reverse happens: the intake valve 4 is closed and the exit valve 5 is opened while the already sucked in air inside chamber 2 is being compressed by the motion of piston 1. Existence of O-ring 3 prevents leakage of the generated high-pressure air.

Figure 1.14 shows a skeleton diagram of a generalized version of a slider-crank mechanism; this time with eccentricity, *e*.

In Fig. 1.15, the slider-crank mechanism shown before in Fig. 1.14 is subjected to input torque T_{in}. This torque is supposed to overcome the load applied on the slider, or F_{out}. The ratio F_{out}/T_{in} that depends on the posture of the machine, is the mechanical advantage of this machine and it will be discussed for the static case in Chap. 11.

$$\text{Mechanical Advantage (M.A.)} = \frac{F_{\text{out}}}{T_{\text{in}}} \tag{1.2}$$

If dynamic effects cannot be ignored, methods discussed in Chap. 13 should be used to provide dynamic estimations of M.A.

Another interesting application of machines and mechanisms is the centrifugal governor, shown in Fig. 1.16, that is used for controlling the speed of rotation of the shaft of an engine. At steady state, the governor rotates at a speed that is proportional to the speed of rotation of the engine shaft. The figure shows the sequence of events that happen if, for example, a disturbance causes an increase in the speed of rotation of the engine shaft from the set steady-state value. We notice that as a result of the sequence of events shown in Fig. 1.16, the engine fuel control valve is closed a bit more so the velocity increase is eliminated. Therefore, this mechanism provides a mechanical negative feedback that controls the speed of an engine by adjusting the flow rate of the fuel given to the engine.

If the speed of rotation of the engine shaft and the governor fall below a certain set limit, a sequence of events happen that are in the opposite directions of what we see in Fig. 1.16. Therefore, this time the result is opening the control valve a bit more which increases the speed of rotation of the engine shaft and the governor. The outcome of all these events is that the engine will work in a tiny speed range around the set (desired) speed.

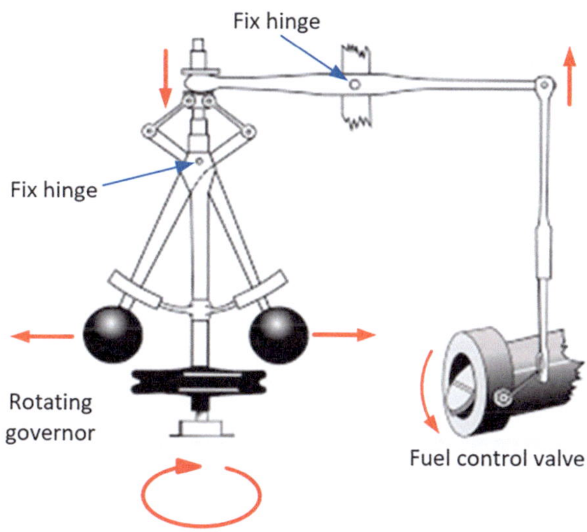

Fig. 1.16 Centrifugal governor for speed control of an engine. Public domain picture, https://commons.wikimedia.org/wiki/File:Centrifugal_governor.svg, accessed 12/1/2024

The centrifugal governor is a mechanical example of a proportional controller where the applied correction to the control valve is proportional to the difference between the set (desired) speed and the measured (actual) speed. This is a bit more complicated than the on and off control that you get by using, for example, a bi-metallic thermostat, but less complicated than the PID (Proportional, Integral and Derivative) control.

Finally, Fig. 1.17 illustrates the components of a Wankel rotary engine. A Wankel engine has an eccentric triangular rotor 6 that has convex sides. The three corners of the rotor maintain contact with the housing wall 3 as it rotates. The internal gear 7 inside the rotor is concentric with the rotor and is in mesh with a stationary external gear 5 that is concentric with the eccentric output shaft, 8. The gear mesh is not used for power transmission. It is used only for coordinating the motion of the rotor with the shaft. The ratio of internal to external gear teeth numbers is 3:2 and as a result, for every one revolution of the rotor the shaft will turn 3 times, Leemhuis and Soedel [3].

The rotor of the Wankel engine turns on the output shaft in a hula-hoop fashion seen in Fig. 1.18. Between the rotor and the housing there are three separate chambers 4 that each complete one cycle of combustion in one full turn of the rotor, as shown in Fig. 1.17. Therefore, in one revolution of the rotor, the combustion cycle has been completed three times. In a Wankel engine, there is no conversion of reciprocating to rotary motion, and no need for cam-based valves as we saw in Figs. 1.11 and 1.12. Recently, Wankel engines are being used in small aircraft and UAV (Unmanned Aerial Vehicles), since they have a high power-to-weight ratio, Cleghorn and Dechev [1].

In this chapter, we got familiar with what structures, kinematic chains, mechanisms, and machines are and we saw how they are used in real-life applications. In the rest of this book we will focus on how we can analyze and design mechanisms and machines. In Part I, the focus is on motion analysis with no regard to the forces and moments that generate those motions (i.e., kinematics). In Part II, we add forces and moments to the discussion and analyze both static and dynamic behavior of machines under applied forces and moments.

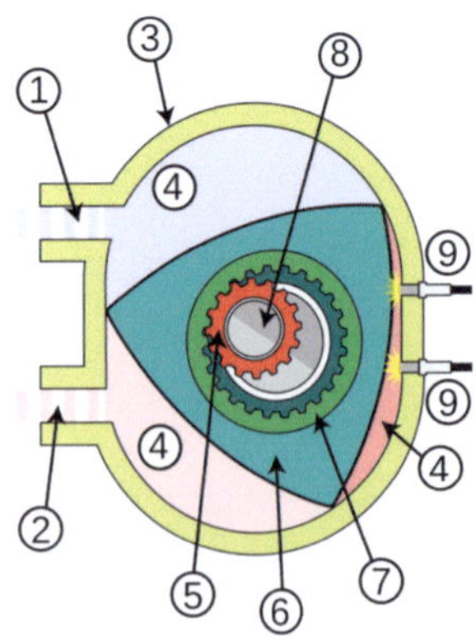

Fig. 1.17 Cross-section of a Wankel rotary engine. 1. Engine inlet, 2. Exhaust outlet, 3. Housing, 4. Combustion chambers, 5. Stationary gear, 6. Rotor, 7. Internal gear, 8. Eccentric shaft, 9. Spark plugs. Public domain picture, https://commons.wikimedia.org/wiki/File:Wankel_engine_diagram-gear-fixed.svg, accessed 12/3/2024

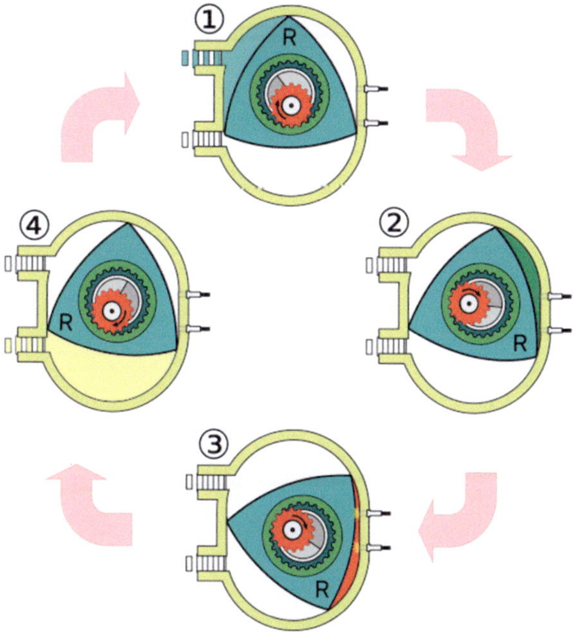

Fig. 1.18 Operation of a Wankel rotary engine. Public domain picture, https://commons.wikimedia.org/wiki/File:Wankel-cycle.svg, accessed 12/3/2024

References

1. Cleghorn, W. L., & Dechev, N. (2015). *Mechanics of machines* (2nd ed.). Oxford University Press.
2. Ghorashi, M. (2018). Closed-form solutions for the load-carrying capacity of annular plates. *International Journal of Mechanical Sciences, 136*, 313–320.
3. Leemhuis, R. S., & Soedel, W. (1976). Kinematics of Wankel compressors (or engines) by way of vector loops. *International Compressor Engineering Conference*. Paper 228.
4. Waldron, K. J., & Kinzel, G. L. (2004). *Kinematics, dynamics, and design of machinery* (2nd ed.). Wiley.

Part I
Kinematics of Machines and Mechanisms

Chapter 2
Mechanisms and Their Features

2.1 Introduction

In this chapter, we explore the various types of planar motions and the basic characteristics of a mechanism. Focus is on motion analysis rather than on the forces and moments that generate those motions. Topics like different types of joints, mobility, and degrees of freedom are discussed. Kinematic inversions of a mechanism are introduced and it is shown how Grashof's law can be used to predict the absolute motion of each one of the kinematic inversions. A frequently used type of mechanism, i.e., the quick-return mechanism, is also analyzed in detail.

2.2 Planar Motions of a Rigid Body

A rigid body in spatial motion can have three possible translations along three coordinate axes, x, y, and z, together with corresponding rotations about those axes. (In a car or an airplane, these rotations have specific names: pitch, roll, and yaw.) Considering these possible motions of a rigid body in space, it is said that a rigid body in space has six degrees of freedom:

In Fig. 2.1, the motions of a rigid body having a representative line AB are limited to plane x-y. Therefore, translations about the z axis as well as rotations about the x and y axes are no longer possible. So, the only remaining degrees of freedom of the object are translations in the x and y directions and rotation about the z axis. Therefore, a rigid body in planar motion is said to have three degrees of freedom.

These degrees of freedom in Fig. 2.1 are expressed by locating the position of point A (using x_A and y_A) and then locating the position of AB with respect to the x axis using rotation θ of AB about point A and in the z direction.

Another way of obtaining three degrees of freedom for a rigid body in planar motion is by noticing that the positions of the two ends of the representative rigid

M. Ghorashi, *Introduction to Machine Dynamics*, Mechanical Engineering Series,
https://doi.org/10.1007/978-3-031-94736-0_2

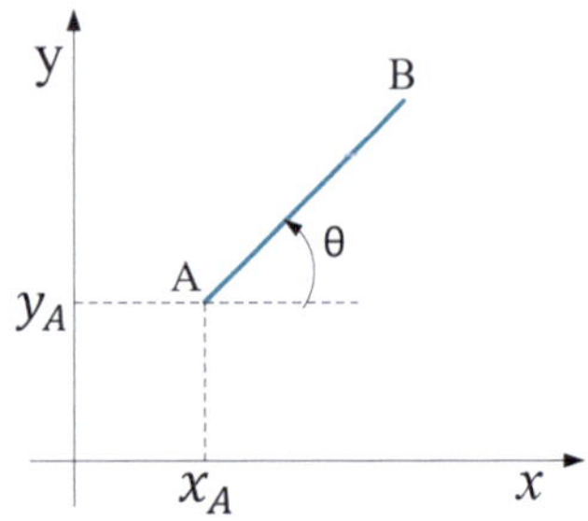

Fig. 2.1 The three degrees of freedom of a rigid link in planar motion

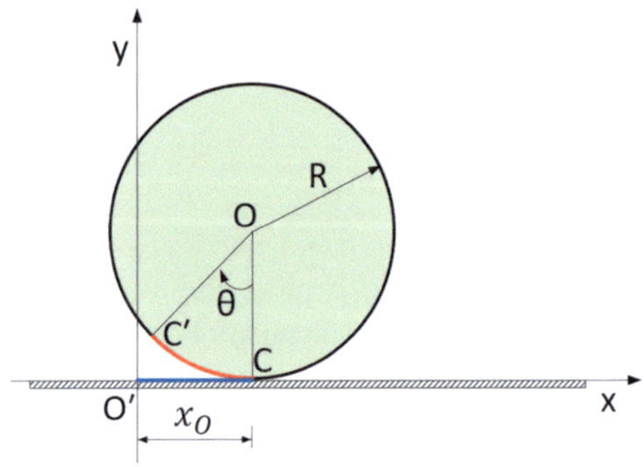

Fig. 2.2 Rolling of a disk without slipping

body AB are determined by coordinates (x_A, y_A) and (x_B, y_B). Even though there are four variables in these coordinates, they are not independent from each other. Indeed, since line AB is representative of a rigid body, its length, i.e.,

$$AB = \sqrt{(x_B - x_A)^2 + (y_B - y_A)^2} \tag{2.1}$$

is constant. Considering four variables and one constraint Eq. (2.1), we can again conclude that in planar dynamics a rigid body has three degrees of freedom. This calculation can be generalized as follows:

$$\text{\#of degrees of freedom} = \text{\#of variables} - \text{\#of constraint equations} \tag{2.2}$$

Example 2.1 Consider the pure rolling motion of a disk on the floor (i.e., without slipping) that is shown in Fig. 2.2. Determine the number of degrees of freedom of this motion.

For simplicity, and without losing generality, we can assume that the motion of the disk started when point C' (on the disk circumference) was located at origin, O' of the shown x-y coordinate system. As the disk rolls without slipping to the right, the position of the center of the disk, i.e., point O, is determined by only one variable, $O'C = x_O$. (The y coordinate of point O is a constant: $y_O = R$, as the disk is rigid.) During this motion, the disk rotates at an angle, θ.

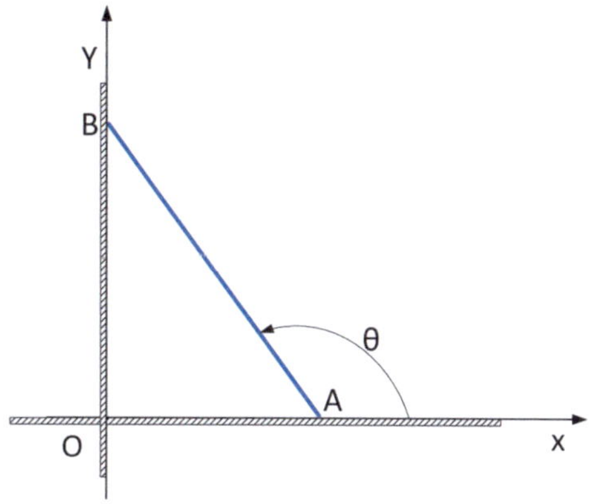

Fig. 2.3 A sliding ladder in planar motion

So, in this problem there are two variables x_O and θ, but not two degrees of freedom. Recalling Eq. (2.2), one should also look for any available constraint equations. Considering the initial condition (point C' on the disk was initially located at the origin, O') and that the disk rolls without slipping, $\overset{\frown}{CC'} = O'C = x_O$. But, $\overset{\frown}{CC'}$ is equal to $R\theta$. Therefore, x_O and θ are related by,

$$x_O = R\theta \tag{2.3}$$

Equation (2.3) is a constraint equation. Now, using Eq. (2.2) with two variables x_O and θ and one constraint equation, we conclude that the pure rolling problem has one degree of freedom.

If in the same problem rolling and sliding could happen simultaneously, the constraint Eq. (2.3) would no longer be valid. Therefore, using Eq. (2.2), the number of degrees of freedom would be 2 (as the position of the center of the disk is now independent from its rotation).

Example 2.2 Figure 2.3 illustrates a ladder that is constrained to maintain its contact with a wall and the floor. Calculate the number of its degrees of freedom.

The problem has three variables; coordinates: x_A, y_B, and the rotation angle, θ. However, these variables are related by two constraint equations. One is that the length of the ladder is a constant:

$$AB = \sqrt{x_A^2 + y_B^2} \tag{2.4}$$

And the other is a trigonometric equation for θ,

$$\tan\theta = -\frac{y_B}{x_A} \tag{2.5}$$

Fig. 2.4 Ferris wheel in Phoenix Arizona. Its frame performs rotation about a fixed point. Public domain picture, https://commons.wikimedia.org/wiki/File:Ferris_wheel_in_Phoenix,_Arizona.jpg accessed 12/4/2024

Using Eq. (2.2), with three variables and two constraint Eqs. (2.4) and (2.5), one obtains only: $3 - 2 = 1$ degree of freedom for this system.

So far the discussion was limited to locating a rigid body in a plane. Now, let's explore all possible ways a rigid body can perform planar motions.

A. **Rotation about a fixed point**

In this case, the rotational motion is the only remaining degree of freedom. The frame of a Ferris wheel shown in Fig. 2.4 performs rotation about a fixed point. The crank in the slider-crank mechanism shown in Fig. 1.12 also rotates about a point that is fixed to its base.

B. **Rectilinear translation**

As we see in Fig. 2.5, in the case of rectilinear translation all points on a representative line *AB* of an object move along straight lines (hence the motion is called rectilinear). Furthermore, all line elements on the object (like *AB*) move parallel to themselves (which is the definition of translation). The piston in the

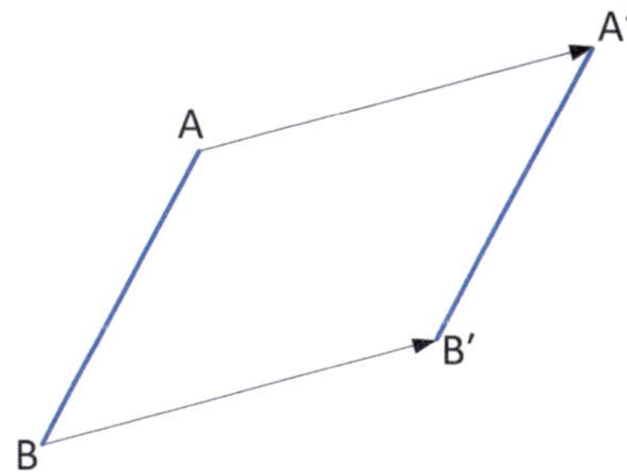

Fig. 2.5 Line element *AB* has a rectilinear translation motion to its new position, *A*′*B*′

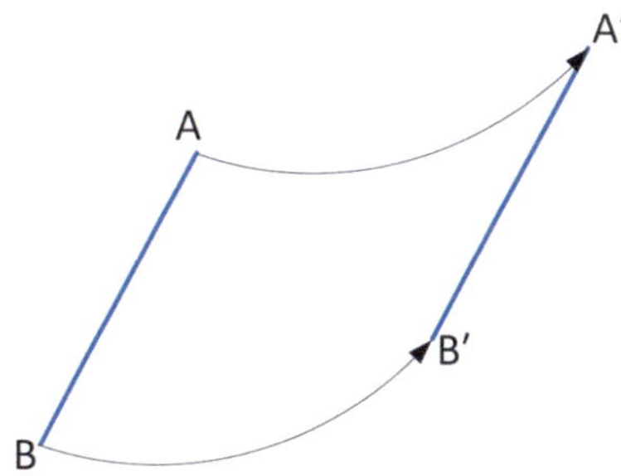

Fig. 2.6 Line element *AB* has a curvilinear translation motion to its new position *A*′ *B*′

slider-crank mechanism shown in Fig. 1.12 moves like this. This motion of a rigid body is fully defined by expressing the movement of one of its representative points (like point *A*) on a straight line (like on *AA*′); so it has one degree of freedom.

C. **Curvilinear translation**

In the curvilinear translation motion shown in Fig. 2.6, any representative line element *AB* on the object moves parallel to itself (which is the definition of translation), but all points on the object move on curved paths (see *AA*′). The difference between this case and case *B* is that in case *B* the path of motion of a representative point was a straight line. Here, however, the motion is fully defined by the curvilinear movement of one point of the object (see *AA*′).

The seats of a Ferris wheel perform curvilinear translation, as we see in Fig. 2.4. While the wheel and all points of its parts move on curved lines, each seat essentially remains parallel to itself (ignoring swinging about the equilibrium position). Therefore, each bench is essentially performing translational motion even though every single point on it is moving on curves. This is expected; after all what would happen to the people sitting on the seats if the seats were to rotate?!

In Fig. 2.7, the four-bar linkage *ABCD* forms a parallelogram and remains to be a parallelogram during its motion to the *AB*′*C*′*D* posture Consequently, link 3 experiences curvilinear translation. As a result, even though every point on link 3 moves on curves, link 3 itself always remains parallel to the fixed base link 1 (i.e. *AD*) and therefore undergoes translation (but curvilinear translation).

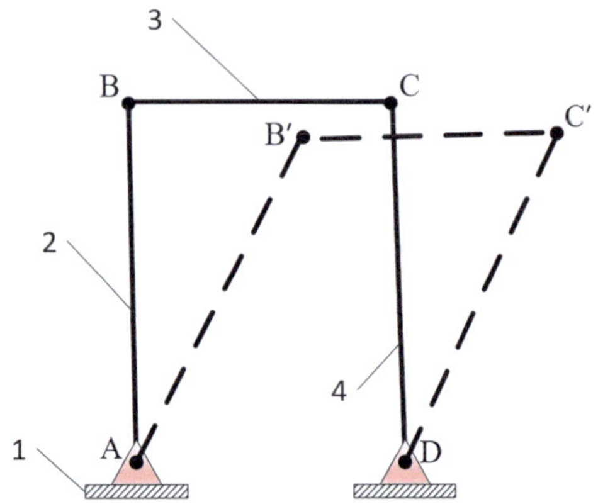

Fig. 2.7 Curvilinear translation of the coupler link *BC* in a parallelogram mechanism

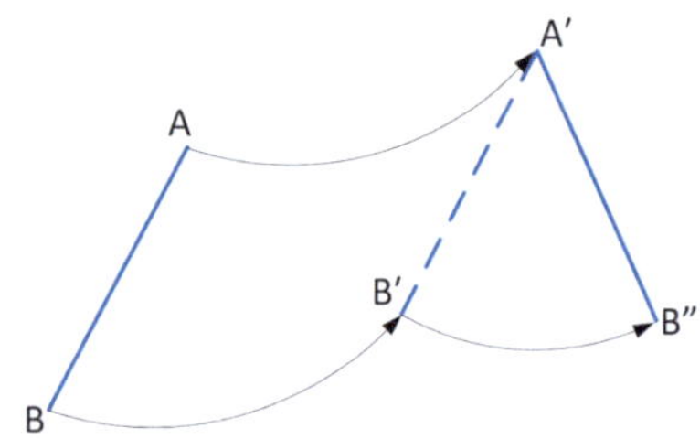

Fig. 2.8 Line element *AB* experiences general planar motion (decomposed into translation and rotation) while moving to its new position $A'B''$

D. **General plane motion**

Figure 2.8 illustrates how in general the motion of an object in a plane (e.g., *x*-*y*) can be broken down into two components. The first one is a translational component (with two degrees of freedom and motions in the *x* and *y* directions, as explained in case *C*). And the second one, that is a one degree of freedom rotational component about a fixed point (and axis *z*), as mentioned in case *A*. So the general planar motion of a rigid body represented by line *AB* in Fig. 2.8 has three degrees of freedom.

In the general plane motion, the movement of a representative line of the object, from the initial *AB* position to its final $A'B''$ position can be thought of as a combination of a curvilinear translation of the object with point *A* along the AA' path, till it reaches the $A'B'$ position, followed by rotation of the object about the fixed point A' that moves point B' to its final position, B''.

The motion of the connecting rod in the slider-crank mechanism shown in Fig. 2.9 is an example of general plane motion.

2.3 Types of Joints

In mechanisms, joints are used to provide contact between links and limit their motions. Joints permit relative motion in some directions while constraining motion in the others. Some joint are called lower-pair joints because they have a lot of

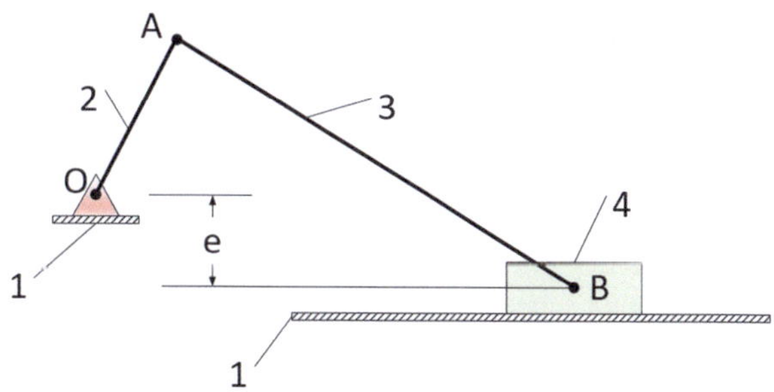

Fig. 2.9 Slider-crank mechanism: Crank 2 rotates about a fixed point, piston 4 undergoes rectilinear translation, and connecting rod 3 experiences general plane motion

contact area and therefore low stresses. Revolute joints are good examples of lower-pair joints with low stresses as they have large contact areas. Some joints are called higher-pair joints as they have only point or line contacts and therefore are associated with high stresses. Examples are gear mesh contact or pin-in-slot contact.

2.3.1 Lower-Pair Joints

There are six fundamentally different lower pair (i.e., low stress, high contact area) joints that are illustrated in Fig. 2.10. In contrast, there is an infinite number of possible higher pair (i.e., high stress, low contact area) joints including cam and follower, cylinder and plane, ball and plane, and ball in tube. Because of high contact area and low stress, in many applications, we prefer to use the lower-pair joints. Furthermore, tight clearances between surfaces in lower-pair joints provide ideal lubrication conditions.

Case *a*: Revolute Joint (Hinge)
Figure 2.10a illustrates a revolute joint. This joint is used, e.g., in a door hinge and provides one degree of freedom (that is rotation) so that one can open or close a door. Figure 2.11 illustrates a Cardan (universal) joint that has two revolute joints that connect the center cross to the two yokes. Cardan joint is a spatial mechanism that is used for transmitting power between two shafts that are positioned at different angles, as shown in Fig. 2.11.

Case *b*: Prismatic Joint
The prismatic joint shown in Fig. 2.10b is used to provide the capability of relative translational motion. Figure 2.12 illustrates an internal spline at the bore of a ring gear used in a planetary gear train of an automatic power transmission. The spline allows the gear to transmit power to its shaft and also slide longitudinally on it. The shaft (not shown) has matching external spline. This joint provides one degree of freedom for the relative motion of the gear and its shaft.

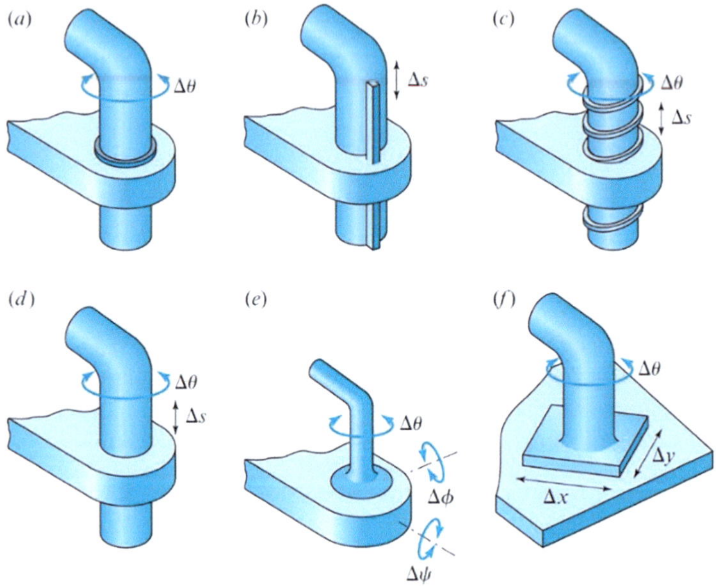

Fig. 2.10 The six lower pair (i.e., low stress, high contact area) joints. (**a**) Revolute; (**b**) prism; (**c**) screw; (**d**) cylinder; (**e**) sphere; (**f**) flat pairs. Page 9 of Uicker et al. [6]. © 2016 Oxford University Press (reproduced with permission)

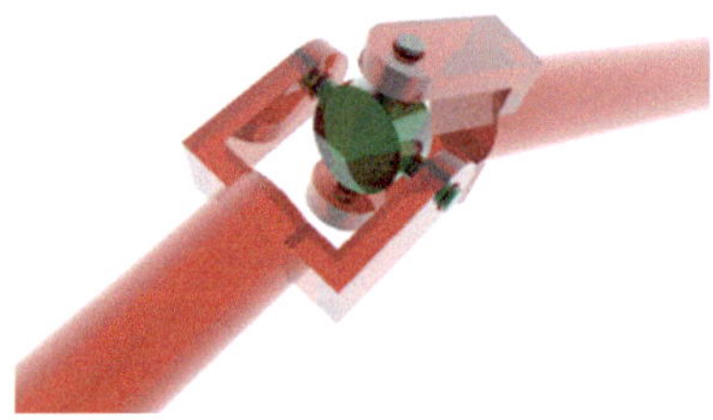

Fig. 2.11 Cardan (universal) joint; the center cross is connected to the yokes by revolute joints. Public domain picture, https://commons.wikimedia.org/wiki/File:Universal_joint.gif. Accessed 12/4/2024

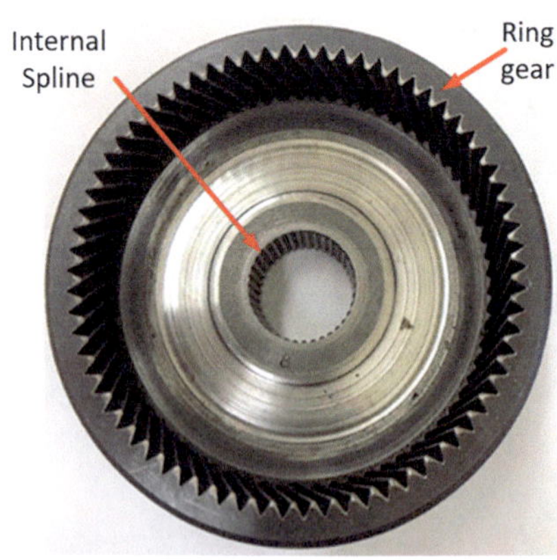

Fig. 2.12 Ring gear and internal spline for power transmission and sliding capabilities

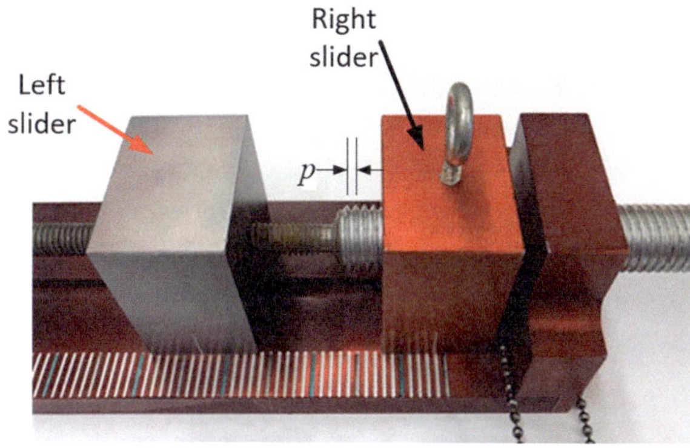

Fig. 2.13 Differential screw mechanism for fine adjustments. (USM term project)

Case *c*: Helical Joint (Screw)

This joint is shown in Fig. 2.10c and is essentially what is obtained by screwing a bolt into a nut. A helical joint allows two motions, i.e., rotation of the bolt, as well as the resulting longitudinal motion relative to the nut. We might think this joint has two degrees of freedom; however, the two motions and the corresponding variables are related. So, this joint only has one degree of freedom.

One full revolution of a bolt with pitch, p (see Fig. 2.13) results in an axial displacement of the bolt head equal to its lead l,

$$l = Np \tag{2.6}$$

where N is the number of threads (i.e., helix starters; see Sect. 8.3 for more information). For single-threaded bolts, $N = 1$, for double-threaded bolts, $N = 2$, and so on. If the bolt is rotated by angle θ, the corresponding axial motion, x of the bolt relative to the nut would be a corresponding fraction of l:

$$x = l\theta/(2\pi) \tag{2.7}$$

This is the constraint equation that relates bolt rotation angle, θ to the bolt axial displacement, x. As a result, the screw joint only has one degree of freedom, not two.

Figure 2.13 illustrates a differential screw that has two segments of a bolt with different pitch values. This device can be used for fine distance adjustments even when the bolt segments do not have fine threads. Fine threads are weak and tend to jam due to the accumulation of dirt so using coarse threads that can do the same job would be advantageous.

Consider the pitch values of the two segments are p_l and p_r for left and right segments, respectively. Rotations of the bolt do not displace the right slider at all as for this slider the two actions to the left (with the bolt) and to the right (with respect to the bolt) are equal and cancel each other out. However, one turn clockwise rotation of the bolt will move the left slider to the left with the bolt motion due to the p_r pitch and simultaneously it moves the left slider to the right due to the p_l pitch. So, the net change in the distance between the two sliders as a result of one clockwise turn of the bolt equals to:

$$\Delta l = N(p_r - p_l) \tag{2.8}$$

In Fig. 2.13, $p_r > p_l$, so, clockwise rotation of the bolt would move the left slider to the left and the distance between the sliders increases.

Now, if the bolt is rotated by an angle θ, the corresponding change in the distance between the two sliders would be a function of θ and would be equal to the corresponding fraction of what we got in Eq. (2.8):

$$\Delta l(\theta) = N(p_r - p_l)\theta/(2\pi) \tag{2.9}$$

For proper values of p_r and p_l the axial displacement Δl that is achieved can be very small, allowing the user to apply this device for fine position adjustments.

Example 2.3 In a differential screw like the one shown in Fig. 2.13, the left segment has 20 threads per inch (TPI), the right segment has 18 TPI, and they are both single threaded. If the screw is turned half a revolution clockwise, how far and in what direction will the carriage move?

The pitch values are:

$$p_l = \frac{1}{20}\text{in},$$

$$p_r = \frac{1}{18}\text{in}$$

Therefore, using Eq. (2.9) we get:

$$\Delta l(\theta) = \left(\frac{1}{18} - \frac{1}{20}\right) \times 0.5 = \frac{1}{180} \times 0.5 = 0.0028\ \text{in}$$

Since the calculated change in the length is positive, the left slider will move to the left-hand side and the gap between the sliders increases. It can also be seen that the equivalent fine thread for this differential screw has 180 TPI that is much finer than either of the two threads that are used in it.

Case *d*: Cylindrical Joint

A cylindrical joint, shown in Fig. 2.10d, allows both rotation about an axis and longitudinal translation along the same axis. The two motions are independent from each other and therefore this joint has two degrees of freedom.

Case *e*: Spherical, or Ball-and-Socket Joint

Figure 2.10e illustrates a spherical joint. A spherical joint as shown in Fig. 2.14 in a hip joint, allows only rotational motions about each one of the three axes in space. Therefore, it has three degrees of freedom. This joint is also frequently used for

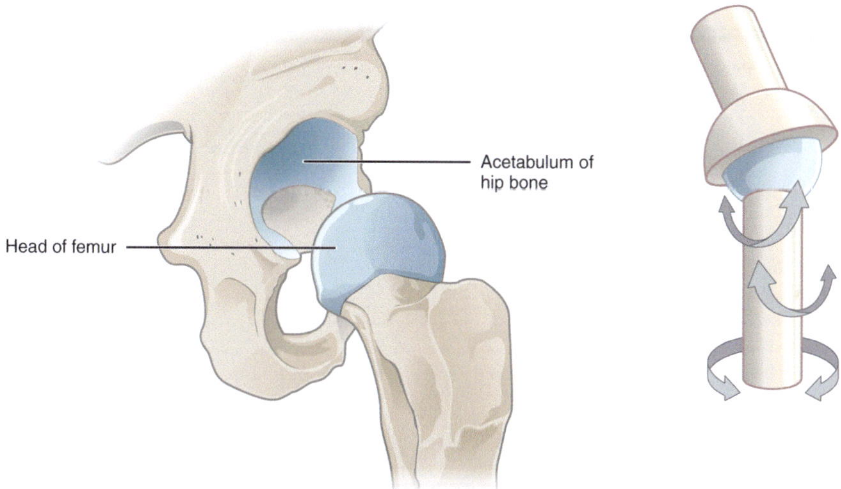

Fig. 2.14 Hip joint and its degrees of freedom. Public domain picture, https://commons.wikimedia.org/wiki/File:903_Multiaxial_Joint.jpg. Accessed 12/5/2024

holding the rear-view mirrors in vehicles hence allowing proper positioning of this mirror by a driver.

Case *f*: Flat Joint (Planar)
As shown in Fig. 2.10f, a flat joint allows a flat surface to translate (along two axes) and rotate (along an axis perpendicular to the flat surface) with respect to another flat surface while the two surfaces remain in contact. Therefore, this joint allows three degrees of freedom.

2.3.2 Higher-Pair Joints

A cam and follower, a pair of meshed gears, or a disk that is rolling and sliding on a flat surface are all examples of joints with limited contact areas. These are therefore higher-stress joints, also called higher-pair joints. To minimize wear that happens under combined high stresses and relative sliding motions, proper lubrication is necessary. That is why when needed, gears are designed to work in gearboxes where they are subjected to continuous lubrication. The same is true regarding cams and followers that are ideal for operating valves.

Figure 2.15 (left) illustrates the pin-in-the-slot higher-pair joint. This joint imposes a kinematic constraint on the relative motion of the pin and the slot in the common normal direction. However, it allows both relative rotational and relative translational motions. This higher-stress, higher-pair joint is kinematically

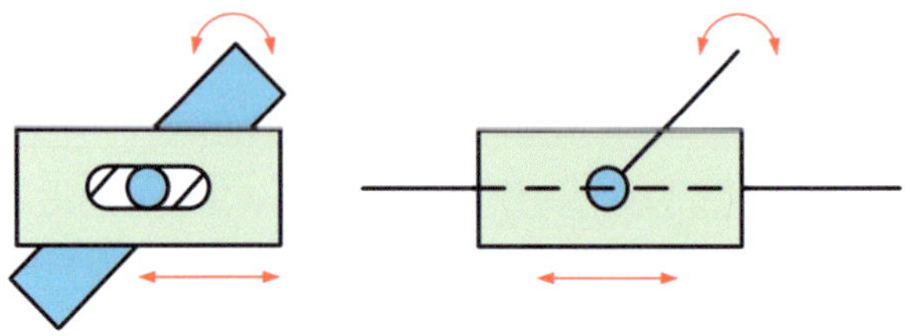

Fig. 2.15 Pin-in-the-slot higher-pair joint (left); Equivalent one, with two lower-pair joints (right)

Fig. 2.16 Geneva mechanism; the driver runs freely since the driver pin has not yet engaged with the driven wheel slot (locked position)

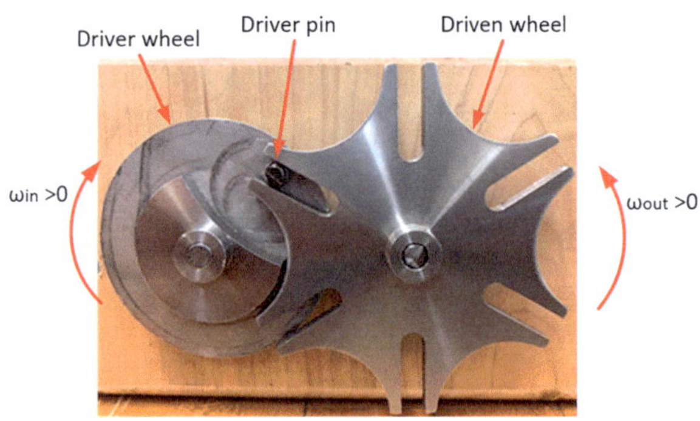

Fig. 2.17 Geneva mechanism; the driver pin engages with the driven wheel slot and starts turning the driven wheel

equivalent to the joint shown in Fig. 2.15 (right). That one is a combination of a revolute and a prismatic or sliding joint (i.e., two lower-stress, lower-pair joints). Both the original and the equivalent mechanisms have two degrees of freedom.

An application of higher-pair joints can be seen in the Geneva mechanism shown in Figs. 2.16 and 2.17.

Geneva mechanism is an indexing mechanism meaning that the driver wheel rotates the driven wheel (aka the star wheel) intermittently (in periods of motion and rest) and through a designed angle. The pin shown in Figs. 2.16 and 2.17 drives the driven wheel by 60° each time it engages with and then exits a slot in the driven wheel. After one engagement, the pin then engages with the next five slots one after the other to drive the star wheel one complete CCW turn.

The number of slots of the driven wheel determines the ratio between dwell and motion periods of the driven shaft. The lowest possible number of slots is three. In practice, the three slot Geneva mechanism is seldom used because of extremely high acceleration values encountered. Geneva mechanisms with more than 18 slots are also infrequently used because they require wheels with comparatively large diameters, Sclater and Chironis [4], page 190. In the external Geneva mechanism shown in Figs. 2.16 and 2.17, the dwell period always exceeds the motion period. The opposite is true of the internal Geneva mechanism. Thus more time is left for the star wheel to reach maximum velocity and therefore acceleration is lower, Sclater and Chironis [4], page 193.

2.4 Mobility of a Mechanism

Mobility (or the number of degrees of freedom of a mechanism) is the minimum number of position variables (of all members) that should be stated relative to the fix member so that the positions of all members in the mechanism are fully specified.

Mobility can also be thought of as the number of actuators that should exist in a mechanism or machine so that it can operate in a controlled way. A loader with two hydraulic jacks needs two inputs to operate and that fact implies that it has two degrees of freedom!

2.4.1 Mobility in Planar Mechanisms

Since a mechanism is an assembly of members and joints, the mobility of a mechanism depends on the number of degrees of freedom of its members and joints. We consider their effects separately and then combine them together to calculate the mobility of the whole mechanism.

In Sect. 2.2, we realized that a rigid body that undergoes a planar motion has three degrees of freedom. If a mechanism consists of n rigid bodies, it is already known that one of them is the stationary ground member 1 (which has no degrees of freedom). The total number of degrees of freedom for the remaining n-1 links, in planar motion, would be: $3(n-1)$. However, in a mechanism, these members are not free to move independently as they are attached together. Because of those constraints, the total number of degrees of freedom of the mechanism will be fewer than $3(n-1)$.

In planar mechanisms, the lower-stress, lower-pair joints are either revolute or prismatic joints that both have only one DOF. If the total number of joints with one degree of freedom (like rolling without slipping, or the sliding joint in a slider) is j_1, since every joint allows one motion and restrains two others, the total number of constraints because of the j_1 number of constraints would be $2j_1$. Therefore, the mobility or the total number of degrees of freedom of a planar mechanism

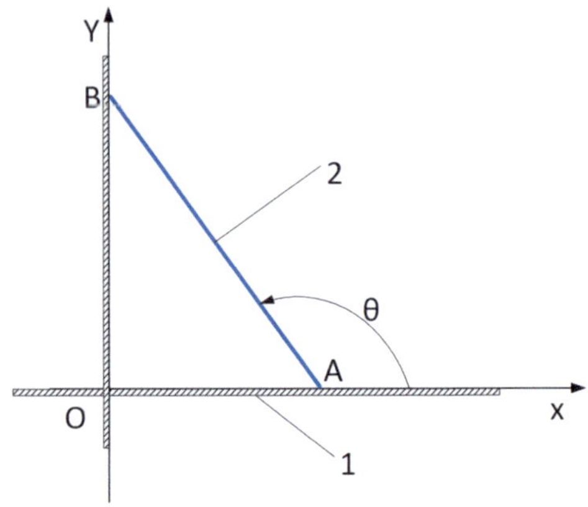

Fig. 2.18 A sliding ladder in planar motion

considering only the number of members and the number of one degree of freedom joints is:

$$m = 3(n - 1) - 2j_1$$

Now, if the total number of joints with two degrees of freedom (like rolling and sliding of a disk, or a pin sliding in a guide) is j_2, since every joint allows two motions and restrains one other, the total number of constraints because of the j_2 number of joints would be $1 \times j_2$. Therefore, the mobility or the total number of degrees of freedom of a planar mechanism considering both the number of members and the number of one and two degrees of freedom joints is, Kutzbach [3]:

$$m = 3(n - 1) - 2j_1 - j_2 \tag{2.10}$$

This is called Kutzbach's equation.

Example 2.4 Consider the sliding ladder shown in Fig. 2.18 and use Eq. (2.10) to calculate its mobility.

There are two members, i.e., the wall/ground combination as well as the ladder itself; so $n = 2$. Now, sliding at points A and B can happen with one DOF (along axes x and y) so $j_1 = 2$. Substitution of these findings in Eq. (2.10) results in:

$$m = 3(n - 1) - 2j_1 - j_2 = 3(2 - 1) - 2 \times 1 - 0 = 1.$$

This result is compatible with what we obtained in Sect. 2.2.

Example 2.5 Consider the ball in rolling without slipping motion shown in Fig. 2.19 and use Eq. (2.10) to calculate its mobility. Repeat the calculation for when the ball rolls with slipping.

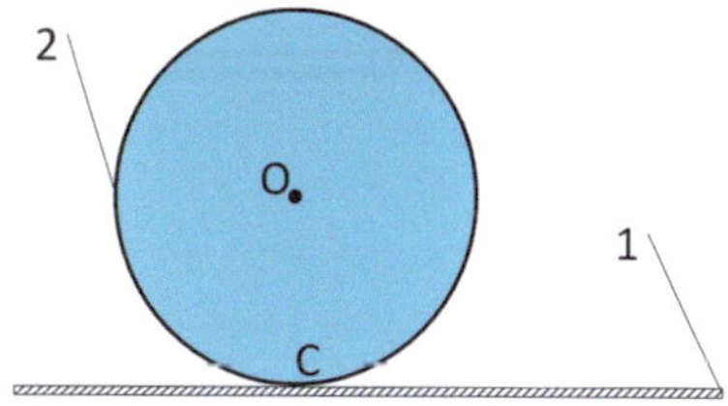

Fig. 2.19 A ball in planar motion

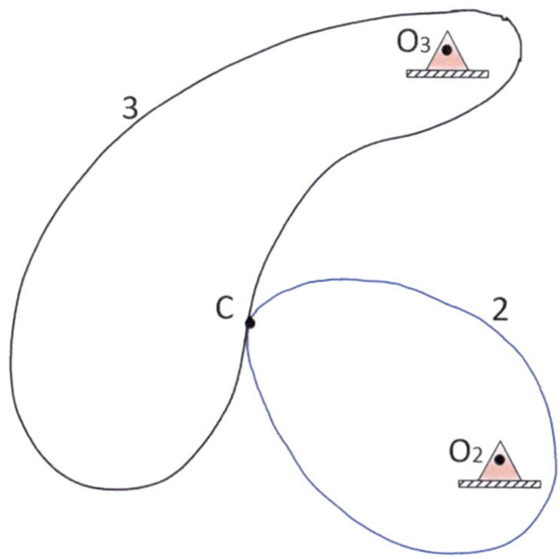

Fig. 2.20 Cam 2 drives follower 3

In this case there are two members so: $n = 2$, and at the pure rolling contact point C, we have a 1DOF joint, so $j_1 = 1$. Substitution of these findings in Eq. (2.10) results in:

$$m = 3(n - 1) - 2j_1 - j_2 = 3(2 - 1) - 2 \times 1 - 0 = 1$$

This result is compatible with what we got in Sect. 2.2 using a different approach. If the ball rolls with sliding, while still $n = 2$, due to the sliding contact at point C we have $j_1 = 0$ and $j_2 = 1$. Therefore Eq. (2.10) gives,

$$m = 3(n - 1) - 2j_1 - j_2 = 3(2 - 1) - 2 \times 0 - 1 = 2$$

This result is also compatible with what we got in Sect. 2.2 using a different approach.

Example 2.6 In Fig. 2.20, a cam is illustrated that drives a follower. Calculate the mobility of this mechanism using Eq. (2.10).

This mechanism has three members, i.e., base 1, cam 2, and follower 3, so $n = 3$. There are two revolute joints at O_2 and O_3, so $j_1 = 2$, and there is a rolling and sliding joint at contact point C. Therefore, we have $j_2 = 1$ and using Eq. (2.10) we have:

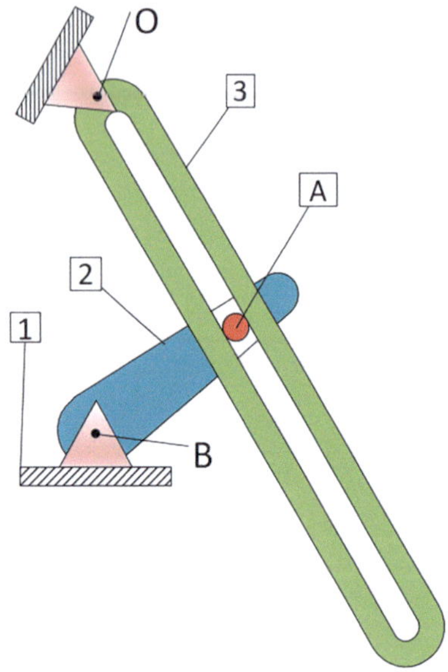

Fig. 2.21 A pin-in-the-slot mechanism

$$m = 3(n-1) - 2j_1 - j_2 = 3(3-1) - 2 \times 2 - 1 = 1$$

So, this cam and follower mechanism has one degree of freedom which simply means it can be run using one input that is provided to the cam by a rotary motor.

If in this cam and follower problem, one assumes pure rolling (instead of sliding) at the contact point, then $j_1 = 3$, $j_2 = 0$, and the new mobility would be:

$$m = 3(n-1) - 2j_1 - j_2 = 3(3-1) - 2 \times 3 - 0 = 0$$

Therefore, under the mentioned conditions, the mechanism would be immovable! This is a result of the incorrect rolling assumption at the contact point.

Example 2.7 Consider the pin-inside-slot mechanism shown in Fig. 2.21 and calculate its mobility.

Pin *A* is rigidly attached to link 2 and can slide along the slot in link 3. The joint at *A* is the pin-inside-slot joint that was shown in Fig. 2.15 and has two DOF. The slot imposes only one constraint on the motion of the pin and that is: no motion perpendicular to the slot (i.e., the common normal direction of pin and slot) is possible. So, this joint still allows rotation and translation along the slot and is a j_2 joint. The number of DOF in this mechanism is thus,

$$m = 3(n-1) - 2j_1 - j_2 = 3(3-1) - 2 \times 2 - 1 = 1$$

So, the three-member mechanism has 1 DOF.

Equation (2.10) is also valid for structures, as we will see in the coming example.

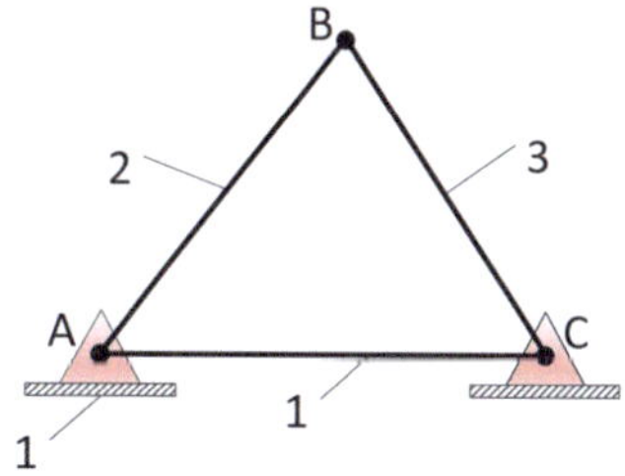

Fig. 2.22 A basic truss element

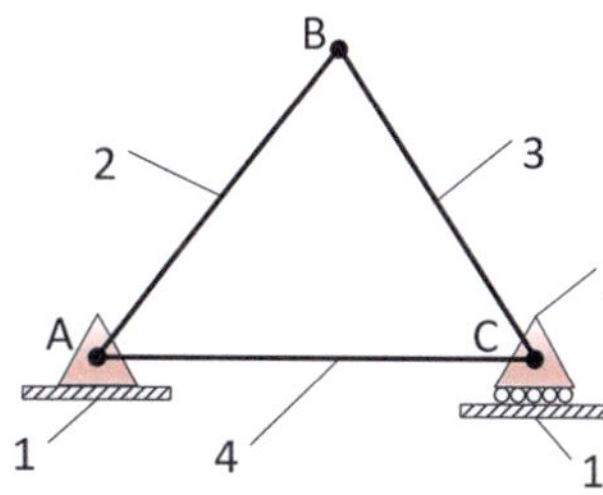

Fig. 2.23 A basic truss element with a sliding member

Example 2.8 Consider the truss element shown in Fig. 2.22 that has three revolute joints. Calculate its mobility.

In this case, there is no difference between member AC and frame #1 as the two are rigidly attached to each other. Using Eq. (2.10) we get:

$$m = 3(n-1) - 2j_1 - j_2 = 3(3-1) - 2 \times 3 - 0 = 0$$

And of course in a structure the expectation is to have no motion and zero DOF! A mobility of zero or negative means we have a structure. The zero value for mobility refers to a statically determinate structure where static equilibrium equations can be solved for all of the unknown forces and moments.

So far, each joint was used for connecting only two members together. If three members are connected to one another, that connection would be counted as two single joints of the same type. In general, if a connection is used for connecting i members to one another, this shared joint will be counted as $i - 1$ single joints of the same type. Let's solve a problem on this subject.

Example 2.9 Consider the alternative structure model shown in Fig. 2.23 in which the right-hand support is free to slide horizontally. Calculate its mobility.

In this case, member AC is not rigidly attached to frame #1 so it is counted as a separate member and we have a total of five members. Joints are numbered in Fig. 2.24. Point A is connecting members 1, 2, and 4 together; however, point A is indeed two collocated joints as it is connecting member 1 to 2, and member 2 to 4 with two one-DOF joints. The connection of member 1 to 4 is assured by the other two joints and therefore that is not counted as a separate joint. The same is true at

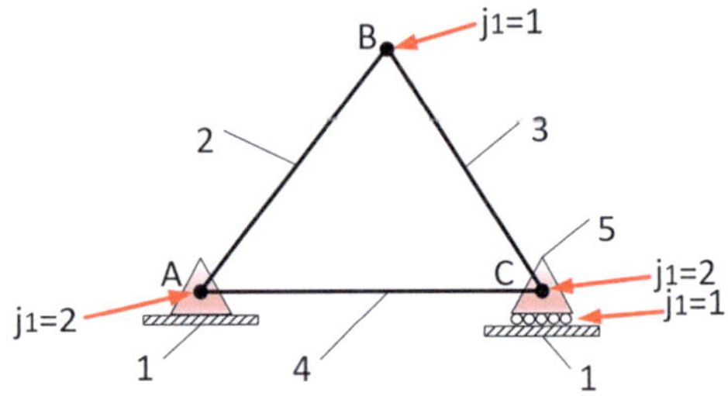

Fig. 2.24 Counting joints in Fig. 2.23

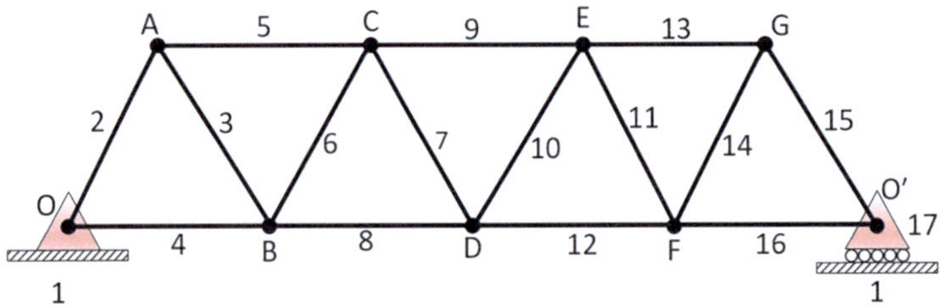

Fig. 2.25 A truss model

point C. So, there are two 1DOF joints at connection points A and C (that connect three members to one another).

Using Eq. (2.10) we get:

$$m = 3(n-1) - 2j_1 - j_2 = 3(5-1) - 2\times 6 - 0 = 0$$

Again, the zero value for mobility refers to a statically determinate structure. Now, let's consider a more complex case.

Example 2.10 Calculate the mobility of the truss model shown in Fig. 2.25.

Counting the number of joints in this case is more challenging as there are a few cases where several members share the same connection point. We recall that in general if a connection point is used for connecting i members to one another, this shared point is to be counted as i-1 single joints of the same type.

In Fig. 2.26, we see that point A is connecting members 2, 3, and 5 together, however, point A is indeed two collocated joints as it is connecting member 2 to 3, and member 3 to 5 with two one-DOF joints. The connection of member 2 to 5 is assured by the other two and therefore that is not counted as a separate joint. The same is true at point B that is connecting members 4, 3, 6, and 8 and is counted as three (i.e., 4 − 1) one-DOF joints.

Connection points O, G, and O' are like A, and each count as two one-DOF joints. Connection points at C, D, E, and F are like B, and each count as three one-DOF joints. Finally, we have a single rolling joint at the right-hand support between member 17 and the frame. So the total number of one-DOF joints is:

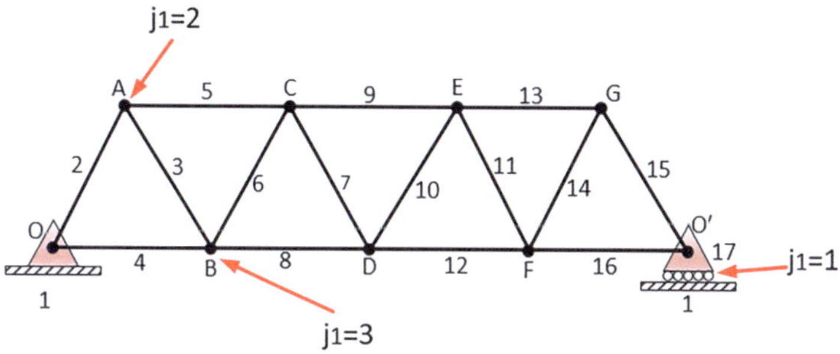

Fig. 2.26 Counting joints in Fig. 2.25

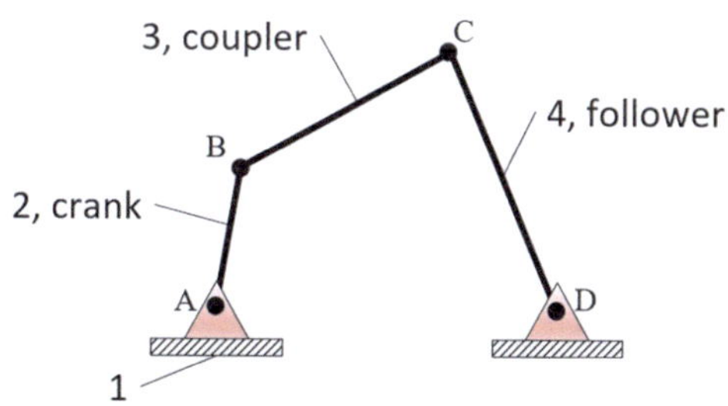

Fig. 2.27 A four-bar linkage

$$j_1 = 4 \times 2 + 5 \times 3 + 1 = 24$$

Therefore, using Eq. (2.10), the mobility of this truss structure is:

$$m = 3(n - 1) - 2j_1 - j_2 = 3(17 - 1) - 2 \times 24 - 0 = 0$$

Again, zero mobility is an indication of the fact that this is a structure that is statically determinate.

One of the most important mechanisms is the four-bar linkage that is the subject of the next example. This mechanism is obtained by adding one link to the basic truss element shown in Fig. 2.22. A four-bar linkage has four revolute joints and that is why it is also called a 4R linkage.

Example 2.11 Calculate the mobility of the four-bar linkage shown in Fig. 2.27.

Using Eq. (2.10) we get:

$$m = 3(n - 1) - 2j_1 - j_2 = 3(4 - 1) - 2 \times 4 - 0 = 1$$

So this mechanism has one DOF and can be run by one input given to, e.g., member 2 using a rotary motor. If link 2 can make full rotations, it is called a crank. Link 4 is usually the output of this 4R mechanism and it is called a follower. If link 4 cannot complete full rotations and just rocks back and forth, it is called a rocker. The link that connects these two together is link 3 and it is called the coupler. Points

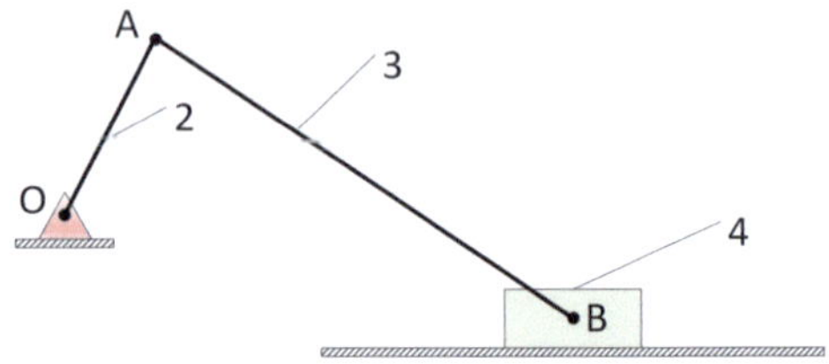

Fig. 2.28 A slider-crank mechanism

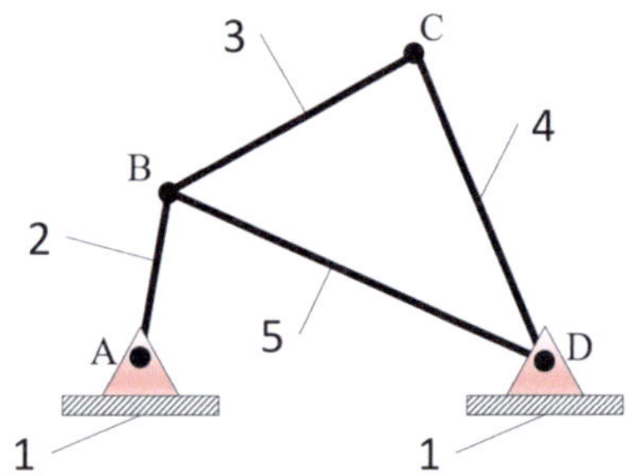

Fig. 2.29 A practically 3-link structure

on the coupler can sketch complicated curves that are called coupler curves. Points on the crank and follower only draw large and small circles.

Another important mechanism, shown in Fig. 2.28, is the slider-crank mechanism which is the subject of the next example.

Example 2.12 Calculate the mobility of the slider-crank mechanism shown in Fig. 2.28.

The slider-crank mechanism has three revolute joints and one slider (prismatic) joint. So, it is a 3R-1P mechanism. The total number of one-DOF joints or j_1 is 4; so using Eq. (2.10) we get:

$$m = 3(n-1) - 2j_1 - j_2 = 3(4-1) - 2 \times 4 - 0 = 1$$

So, this mechanism has one DOF. This is a fundamental mechanism that is used in internal combustion engines where the translational motion of piston 4 is converted to the rotational motion of crankshaft 2 that ultimately provides the necessary traction force, e.g., for driving a vehicle.

In the next example we add another link to the four-bar linkage.

Example 2.13 Calculate the mobility of the linkage shown in Fig. 2.29.

In this case, connection points B and D each connect three members to one another. Therefore, each one of them is counted as two one-DOF joints and for the whole system, $j_1 = 6$. So, using Eq. (2.10) we get:

$$m = 3(n - 1) - 2j_1 - j_2 = 3(5 - 1) - 2 \times 6 - 0 = 0$$

So, adding an extra member to the four-bar linkage has converted it to a structure. Now, if due to, e.g., an excessive loading member 5 fails, the structure again converts to a mechanism that is called a collapse mechanism, Ghorashi [1].

The same result could have been obtained by noticing that members 3, 4, and 5 form a larger rigid triangular link, let's call it link 6. Then link 6 together with links 1 and 2 form another three-member structure that, like the basic truss element considered before, has no mobility, as expected.

The following example explores the case of statically indeterminate structures (see Chap. 10 for more discussion).

Example 2.14 Calculate the mobility of the structure shown in Fig. 2.30.

The structure shown in Fig. 2.30 has been obtained by adding one more link to the structure shown in Fig. 2.29. In this case, each of the connection points *A*, *B*, *C*, and *D* connect three members. Therefore, each of them is counted as a two one-DOF joint and for the whole system, $j_1 = 8$. So, using Eq. (2.10) we get:

$$m = 3(n - 1) - 2j_1 - j_2 = 3(6 - 1) - 2 \times 8 - 0 = -1$$

A mobility value of zero or negative corresponds to a structure. The zero value for mobility refers to a statically determinate structure. Negative mobility happens for a statically indeterminate structure in which the number of static equilibrium equations is fewer than the number of unknowns. In the present case, the number of static equilibrium equations is one fewer than the number of unknown forces. Therefore, the static problem cannot be solved without additional information. If the links are modeled as elastic members, compatibility of deflections provides the necessary added equation.

The -1 DOF is also an indication that the structure in Fig. 2.30 has one redundant member which imposes extra constraints (and makes the problem statically indeterminate). Having redundant constraints in structures increases stiffness and strength and therefore improves safety. Indeed after failing one member, e.g. link 6, the structure may still be able to support forces as it is now back to the five-link structure we studied in Fig. 2.29. The collapse mechanism mentioned before will only form if

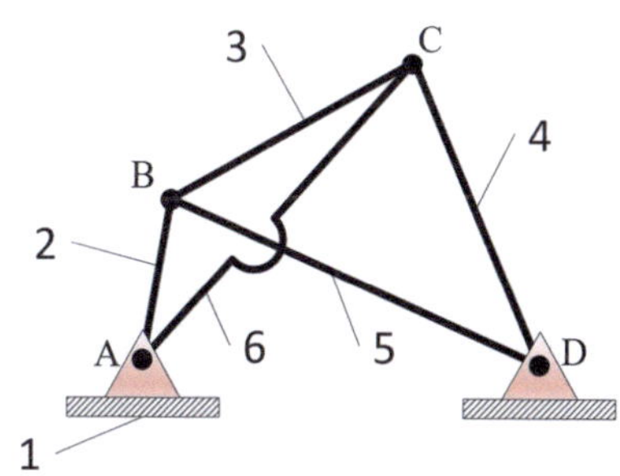

Fig. 2.30 A statically indeterminate structure

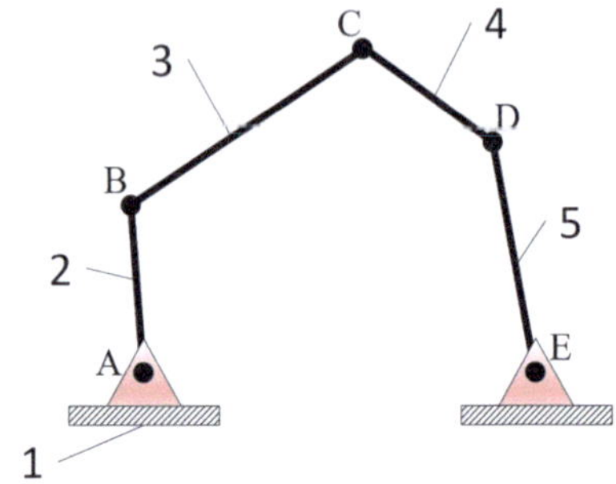

Fig. 2.31 A five-bar linkage

a second member, e.g. number 5 also fails and the structure becomes a four-bar linkage. By adding redundant members we gain some time to react to a potential collapse (by evacuations or applying reinforcements).

Now, let's look at the other side of the spectrum, i.e., mechanisms that have more than one DOF.

Example 2.15 Calculate the mobility of the five-bar linkage shown in Fig. 2.31.

In this case, each connection point is a revolute joint so, $j_1 = 5$. Therefore, using Eq. (2.10) results in:

$$m = 3(n - 1) - 2j_1 - j_2 = 3(5 - 1) - 2 \times 5 - 0 = 2$$

Having two DOF means that two inputs (given, e.g., by actuators) should be provided to operate this mechanism.

Next, we consider a more complex mechanism that draws straight lines.

Example 2.16 Figure 2.32 illustrates the Peaucillier inversor that is a straight line generator that converts the rotary motion of point A to the straight line motion of point C (and vice versa). Both motions are shown in dashed lines. Calculate the mobility of this linkage. Hinges O_1 and O_2 are connected to the base and $ABCD$ is a rhombus. Here are some specifications:

$$O_1O_2 = O_1A = R, \quad AB = BC = CD = DA = a, \quad O_2B = O_2D = b$$

The inverse of a circle is a straight line, meaning the Peaucillier inversor mechanism converts circular motion to straight line motion (and vice versa) and for this purpose it uses only rods and pins. An early mathematical analysis of this mechanism can be found in, Woolsey Johnson [8]. A physical realization of this mechanism that includes the degrees of freedom at each joint is shown in Fig. 2.33.

Using Eq. (2.10) for this mechanism results in:

$$m = 3(n - 1) - 2j_1 - j_2 = 3(8 - 1) - 2 \times 10 = 1$$

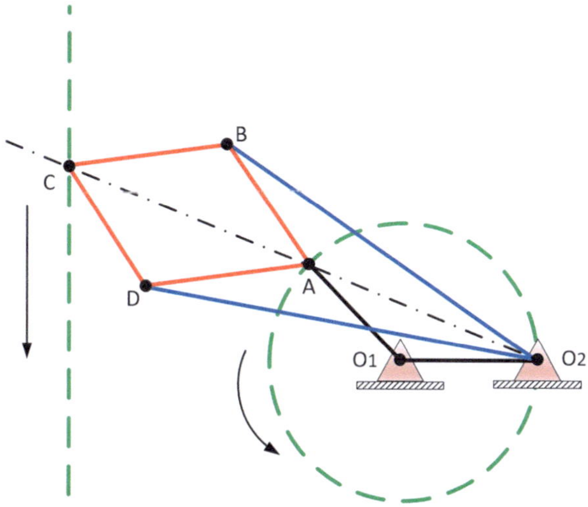

Fig. 2.32 Peaucillier inversor straight line generator

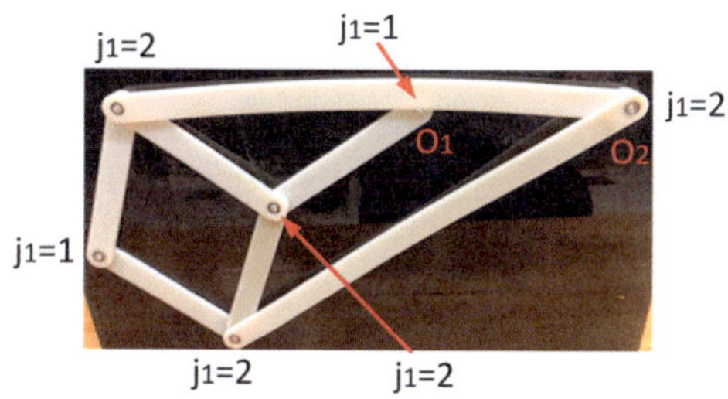

Fig. 2.33 Peaucillier inversor straight-line generator. (USM term project)

So, this mechanism has 1 DOF. An even more complicated machine is given in the next example.

Example 2.17 Calculate the mobility of the excavator given in Fig. 2.34 in planar motion.

First, let us determine the number of members. In Fig. 2.35, seven members that include all of the members of the machine other than the hydraulic jacks are numbered. Also, there are four hydraulic jacks (labeled Jack *A* to *D*) each one of them has two parts that make up a prismatic joint. So the hydraulic jacks are counted as $4 \times 2 = 8$ members. Therefore, the total number of members in this machine is: $n = 7 + 8 = 15$.

Next, we count the number of joints. There are seven structural revolute joints *R*1 to *R*7 that are marked in Fig. 2.35. Also, each hydraulic jack has two revolute joints at its two ends and a prismatic joint that represents its operation. Therefore, the hydraulic jacks contribute to a total of $4 \times 3 = 12$ one-DOF joints. Adding these to the seven structural revolute joints, we get $j_1 = 7 + 12 = 19$ one-DOF joints.

Therefore, using Eq. (2.10) for planar motion of this machine results in:

Fig. 2.34 Man and excavator. Public domain picture, https://www.pickpik.com/excavators-blade-construction-machine-backhoe-bucket-machine-wireframe-134952. Accessed 12/6/2024

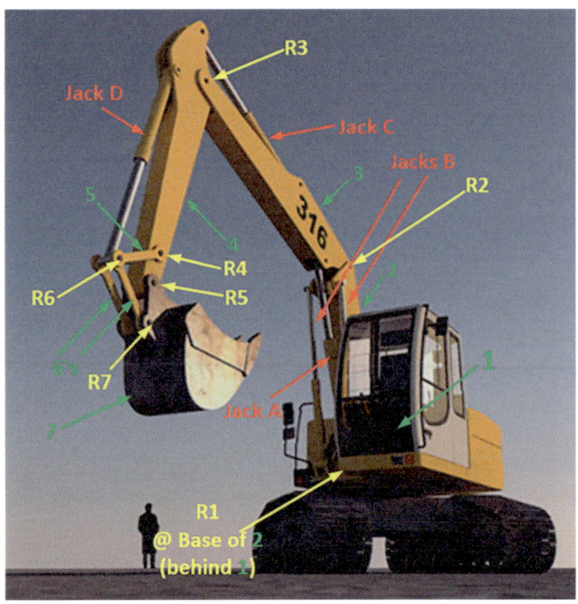

Fig. 2.35 Labeled version of Fig. 2.34

$$m = 3(n - 1) - 2j_1 - j_2 = 3(15 - 1) - 2 \times 19 - 0 = 4$$

Interestingly, for the planar motion of this machine we have four hydraulic jacks too! That is, each jack is used for controlling one of the degrees of freedom.

Notice that hydraulic jacks *B* in the figure are parallel and perform the same motions so as far as it is related to mobility calculations, they are counted as just one jack. The same is true for members 6 that are parallel and are counted as just one member.

2.4.2 Exceptional Cases

There are two cases where Kutzbach's Eq. (2.10) cannot produce correct results, one is the rolling contact case and the other is when the mechanism has parallel members.

Case 1. Rolling Members

Consider rolling contact between two friction disks that is illustrated in Fig. 2.36.

Pure rolling contact is a one-DOF joint, the two revolute joints at the hinges are also one-DOF joints. Therefore, the mechanism in Fig. 2.36 has three one-DOF joints so $j_1 = 3$ and using Eq. (2.10) we get:

$$m = 3(n - 1) - 2j_1 - j_2 = 3(3 - 1) - 2 \times 3 - 0 = 0$$

So, calculation results in zero mobility and therefore no motion in this system is expected! This is an obviously incorrect answer as we know that pure rolling motion in this mechanism *is* happening. The contradiction originates from the fact that the Kutzbach's equation is an accounting equation that has no way of including the special geometry of the members of the mechanism into consideration. In this case, the special form of the disks that are circular allows the pure rolling motion to happen. In the general case of a non-circular cam and follower pure rolling motion is indeed impossible as Eq. (2.10) predicts!

The discussion on rolling contact of friction disks can be extended to gears in mesh (see Fig. 2.37). To this end, the gears are replaced by rolling disks having radii equal to the pitch circle radii of the gears (see Chap. 6).

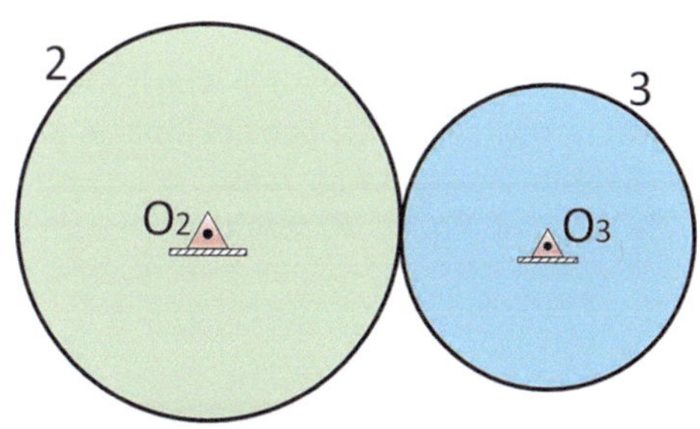

Fig. 2.36 Rolling contact between two friction disks

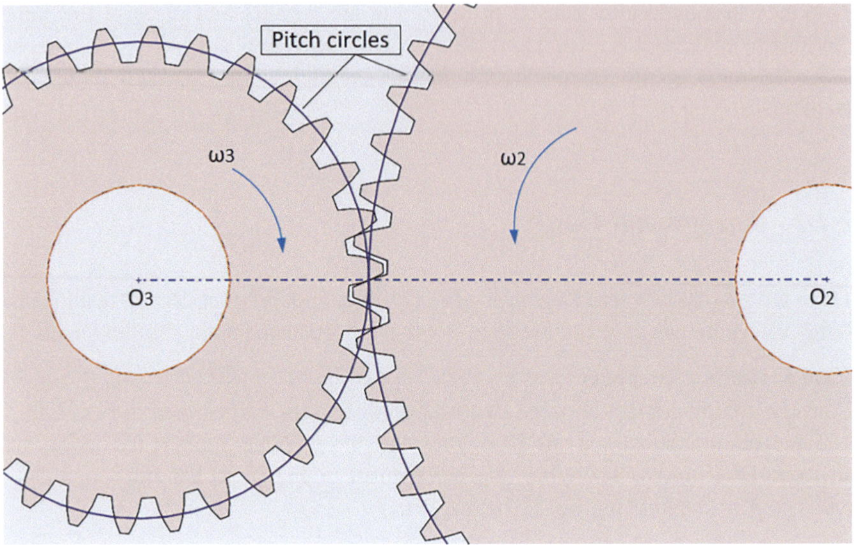

Fig. 2.37 Gear mesh replaced by pitch circles in rolling contact

Fig. 2.38 A parallelogram mechanism

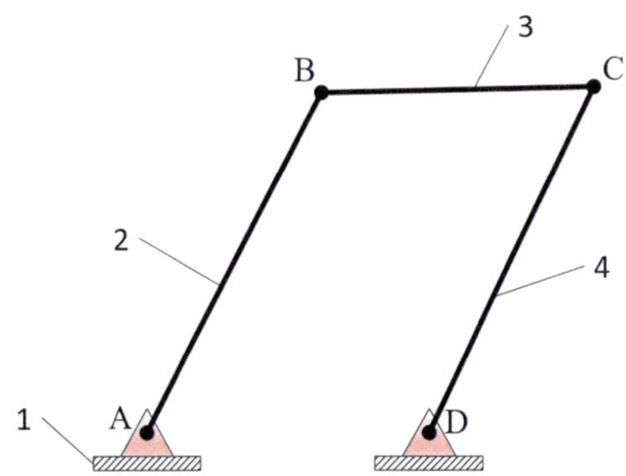

Case 2. Parallel Members

The second exceptional case in which Kutzbach's Eq. (2.10) cannot produce correct results is when there are parallel links in a mechanism. Consider the parallelogram mechanism shown in Fig. 2.38.

In this case Eq. (2.10) results in,

$$m = 3(n - 1) - 2j_1 - j_2 = 3(4 - 1) - 2 \times 4 - 0 = 1$$

Now, that mechanism is modified as we see in Fig. 2.39 where a rigid piece has been welded to the coupler link *BC* and link *EF* is hinged at point *F* to that piece.

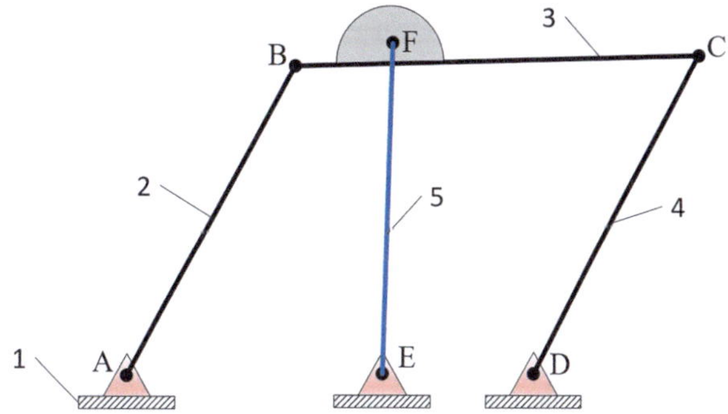

Fig. 2.39 The modified parallelogram mechanism is now a structure

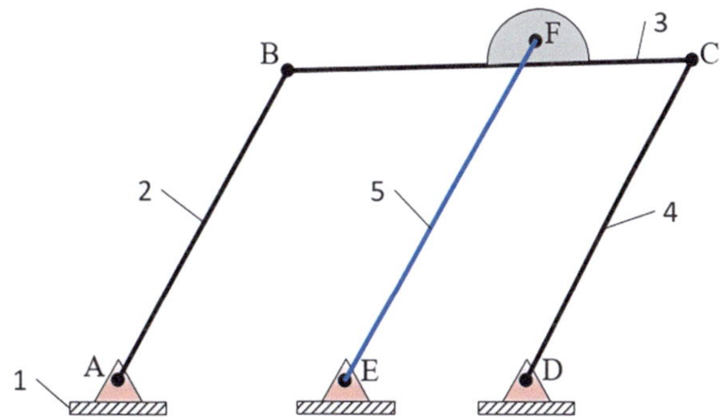

Fig. 2.40 The parallelogram mechanism with one extra link

Using Eq. (2.10) we get:

$$m = 3(n - 1) - 2j_1 - j_2 = 3(5 - 1) - 2 \times 6 - 0 = 0$$

Therefore, addition of link *EF* converted the mechanism shown in Fig. 2.38 to the structure we see in Fig. 2.39 that can remain stationary under applied loads.

Now, looking back at Fig. 2.38, consider adding link 5 that is parallel to links 2 and 4 using a small rigid piece welded to link 3. The result is shown in Fig. 2.40.

This mechanism has the same motion as the one shown in Fig. 2.38 with $m = 1$ (links 2, 4, and 5 are parallel and will stay parallel during motion). However, Eq. (2.10) still gives:

$$m = 3(n - 1) - 2j_1 - j_2 = 3(5 - 1) - 2 \times 6 - 0 = 0$$

Like in the previous exceptional case, this incorrect result is obtained because Kutzbach's equation is an accounting equation that has no way of including the special geometry of the members of the mechanism into consideration. In this case, the links are parallel and that is the reason the mechanism shown in Fig. 2.40 can move. In the general case of non-parallel members, as we saw in Fig. 2.39, the motion is indeed impossible, as Eq. (2.10) correctly predicts! Therefore, mechanisms that include parallel members should be treated as exceptional cases.

2.4.3 Extension to Spatial Mechanisms

The purpose of this section is to derive a three-dimensional version of the mobility equation, i.e., Eq. (2.10). A link (or any rigid body) in three-dimensional motion has three translational and three rotational degrees of freedom. So, if a mechanism has n members, considering one of them to be the fixed base, the number of DOF would be: $6(n - 1)$.

Similar to the two-dimensional case, the presence of joints should also be included in the mobility calculation. Recall the six lower-stress lower-pair joints that were discussed in Sect. 2.3. We now reconsider their effects on mobility of a mechanism in three-dimensional motion. Table 2.1 includes the number of DOF that each one of these joints allows, the number of DOF that each joint constrains, and finally the effect that each particular joint type will have on mobility of the mechanism.

The impact of all joints on mobility is the summation of all of the terms in the last column of Table 2.1. Subtracting it from $6(n - 1)$, results in the following equation for the mobility of a mechanism in three-dimensional motion:

$$m = 6(n - 1) - 5(R + P + H) - 4C - 3(S + F) \tag{2.11}$$

Parallel links are frequently used in spatial mechanisms as well; and like the 2D case there are exceptions in which Eq. (2.11) does not work. One famous example is the spatial four-bar Revolute-Cylindrical-Cylindrical-Revolute (*R-C-C-R*) linkage in which the axes of one revolute and one cylindrical joint in each *R-C* pair are parallel with each other. Using Eq. (2.11) we get:

$$m = 6(4 - 1) - 5(2) - 4(2) = 0$$

But this is a special case of parallel members and the mechanism has indeed 1DOF, not zero.

Table 2.1 The six lower-stress lower-pair joints and their effects on reduction of the number of DOF in a spatial mechanism

	# of DOF kept per joint	# of DOF constrains per joint	# of DOF reduced, if more joints are used
R: Revolute	1	5	5R (for R joints)
P: Prismatic	1	5	5P (for P joints)
H: Helical	1	5	5H (for H joints)
C: Cylindrical	2	4	4C (for C joints)
S: Spherical	3	3	3S (for S joints)
F: Flat (planar)	3	3	3F (for F joints)

2.5 Quick-Return Mechanisms

In many manufacturing processes, like shaping and sawing, work is done in the so-called advance stroke. However, no work is performed during the return stroke. The return stroke is just needed so that the advance stroke operation could be repeated again.

The slider-crank mechanism shown in Fig. 2.41 is an example of a quick-return mechanism used as a shaper. The crank OA rotates in the CW direction and starting from position OA_1 to OA_2 the crank pushes piston rod AB and moves slider 4 from position B_1 to B_2. Arc α, or the advance angle, is the crank angular displacement in this working stroke during which a cutting force F is applied on the slider (that represents the cutting tool).

In the return stroke, while crank OA rotates from position OA_2 back to position OA_1, the slider returns from B_2 to B_1. Arc β, or the return angle is the crank angular displacement in the return stroke during which no cutting force is applied on the slider. The slider-crank mechanism has two limit positions shown by subscripts 1 and 2. Subscript 1 refers to the limit position at the start of the advance stroke and subscript 2 refers to the limit position at the end of the advance stroke. At the limit position 1 the crank is collinear with the piston rod AB, but in the opposite direction. At the limit position 2 the crank is collinear with the piston rod AB and in the same direction.

Therefore, motion from B_1 to B_2 is the advance (working) stroke and motion from B_2 back to B_1 is the return stroke.

In quick-return mechanisms in order to make the process more economical one needs to maximize the time needed to perform the advance (working) stroke, t_a, compared to the time needed to perform the return stroke t_r. In other words, the ratio of the time for the advance stroke over that of the return stroke should be maximized. This ratio is called the advance-to-return ratio and it is defined as follows:

$$Q = \frac{\text{time of the advance stroke}}{\text{time of the return stroke}} = \frac{t_a}{t_r} \tag{2.12}$$

Fig. 2.41 The slider-crank mechanism used as a quick-return mechanism in a shaper

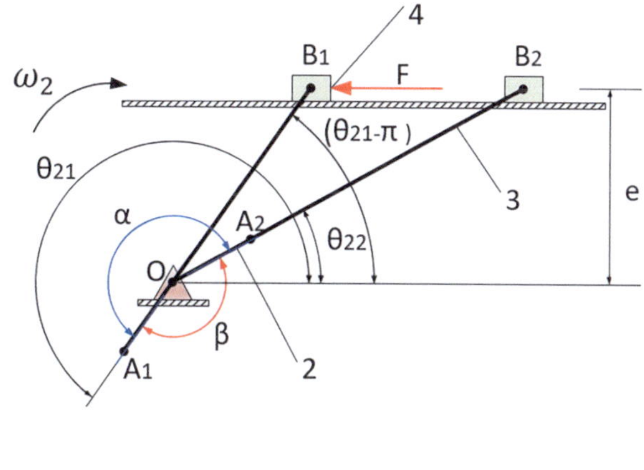

If the angular velocity of the crank is ω (CW) and it is constant, for the mechanism shown in Fig. 2.41 we have:

$$Q_{CW} = \frac{\omega t_a}{\omega t_r} = \frac{\alpha}{\beta} > 1 \tag{2.13}$$

So, the advance-to-return ratio is independent of the crank angular velocity and it is equal to the ratio of the advance angle over the return angle. This ratio is a function of the geometry of the mechanism and can be adjusted by changing the lengths of the links. In the mentioned mechanism $\alpha > \beta$, so for CW motion of the crank, return happens faster than advance, and hence it is a quick-return mechanism. This faster return is obtained even though the crank angular velocity during the two strokes remains constant! If the crank rotates in the CCW direction, α would be the return angle and β would be the advance angle. So, the new Q would be:

$$Q_{CCW} = \frac{\beta}{\alpha} < 1$$

And the mechanism will no longer be a quick-return mechanism. Therefore, proper choice of the direction of rotation of the crank is also critical and determines if the mechanism has quick-return property.

Figure 2.42 illustrates the Scotch-yoke mechanism that converts the rotational motion of crank 2 to the oscillatory translational motion of slider 4. This mechanism is 2R-2P, meaning it has two revolute joints and two prismatic joints. Both of the limit positions of this mechanism happen when the crank is horizontal and collinear with the direction of motion of the slider 4. As a result, the advance and the return angles that the crank would traverse are both 180° and the advance-to-return ratio, Q, is 1.

Figure 2.43 illustrates a manufactured Scotch-yoke mechanism that uses a pin-in-the slot joint (higher pair) in place of a revolute and a slider joint (lower-pair) seen in Fig. 2.42.

The pin-in-the slot mechanism shown in Fig. 2.44 is also capable of illustrating the quick-return response. The two parts of Fig. 2.44 illustrate the two limit positions of this mechanism.

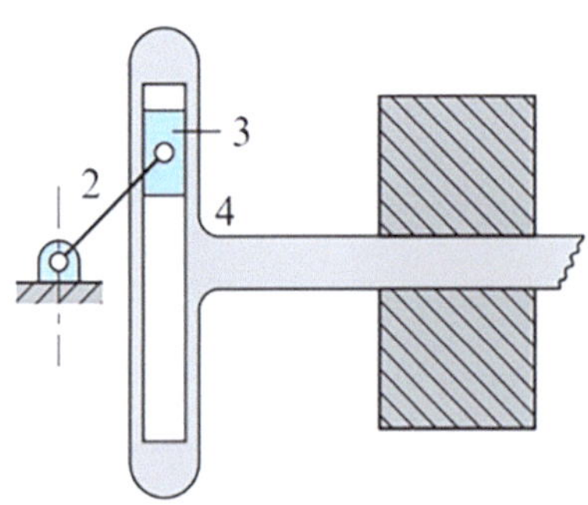

Fig. 2.42 The Scotch-yoke mechanism. Page 22 of Uicker et al. [6] © 2016 Oxford University Press (reproduced with permission)

Fig. 2.43 Manufactured Scotch-yoke mechanism with a pin-in-the slot joint. (USM term project)

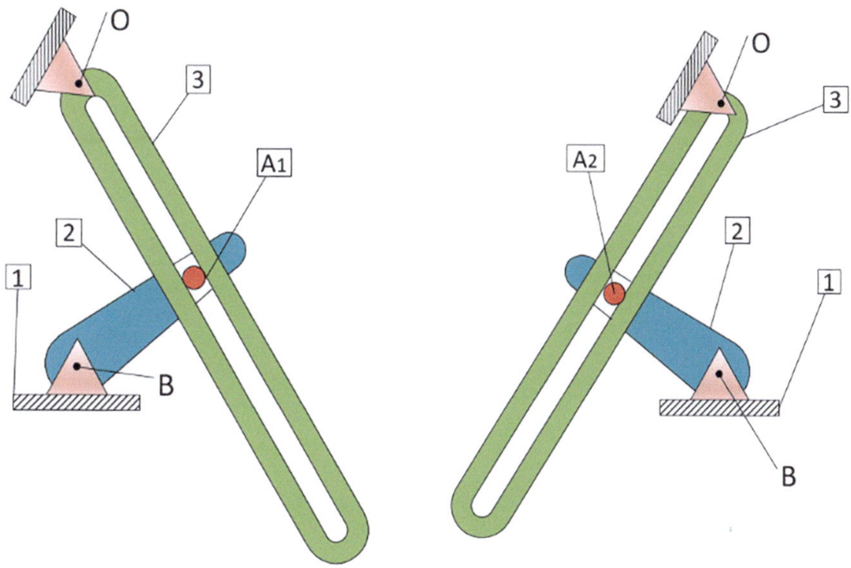

Fig. 2.44 The two limit positions of a pin-in-the slot mechanism

In Fig. 2.44 (left) as crank 2 rotates CW, link 3 goes CW as well until the crank traverses angle α (see Fig. 2.45) and reaches the position shown in Fig. 2.44 (right). If the crank rotates any further, link 3 will now go CCW and after the crank traverses an angle β (see Fig. 2.45), it returns back to the configuration shown in Fig. 2.44 (left). So, the crank makes full turns while link 3 experiences a rocking motion that happens between the two limit positions shown in Fig. 2.44. These limit positions happen when rod 2 is perpendicular to the slot in rod 3, or, alternatively, the circular path of pin *A* is tangent to the slot in rod 3. If the CW motion of link 3 is defined as the advance motion and its CCW rotation defined as the return motion, this mechanism would also be a quick-return mechanism (if the crank rotates CW). The corresponding angles and a skeleton diagram of the limit positions are illustrated in Fig. 2.45, where $r_1 = OB$ and $r_2 = AB$.

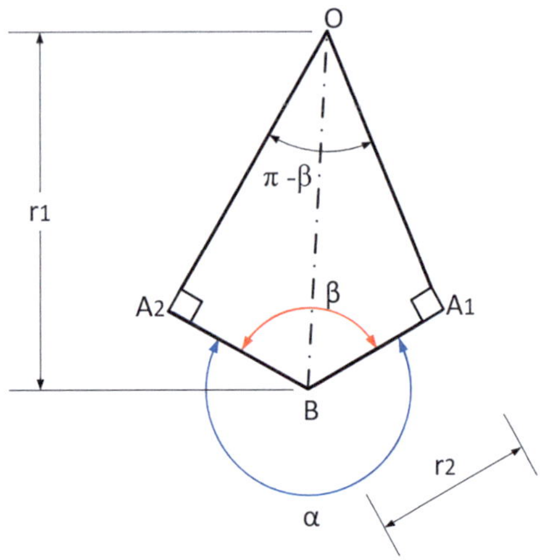

Fig. 2.45 Skeleton diagram of the two limit positions of the mechanism shown in Fig. 2.44

Based on the foregoing discussion, the advance-to-return ratio in this case can be expressed as,

$$Q_{CW} = \frac{\alpha}{\beta} > 1 \tag{2.14}$$

Referring to Fig. 2.45 we have,

$$\cos(\beta/2) = \frac{r_2}{r_1} \tag{2.15}$$

And,

$$\alpha + \beta = 2\pi \tag{2.16}$$

Substituting Eq. (2.16) into Eq. (2.14) results in,

$$Q_{CW} = \frac{2\pi - \beta}{\beta} \tag{2.17}$$

Substituting Eq. (2.15) into Eq. (2.17) gives,

$$Q_{CW} = \frac{\pi}{\cos^{-1}\left(\frac{r_2}{r_1}\right)} - 1 \tag{2.18}$$

Example 2.18 Considering the quick-return mechanism shown in Fig. 2.44, calculate the advance-to-return ratio if, $r_1 = 220$ mm and $r_2 = 80$ mm.

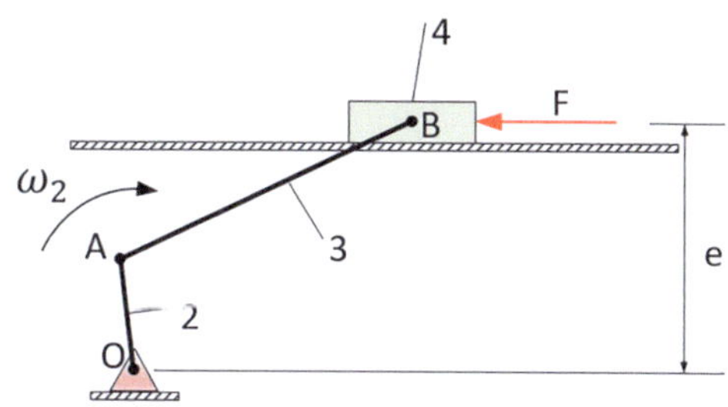

Fig. 2.46 The slider-crank mechanism used as a quick-return mechanism in a shaper

By substituting the given numerical values in Eq. (2.18) the resulting advance-to-return ratio would be:

$$Q_{CW} = \frac{\pi}{\cos^{-1}\left(\frac{80}{220}\right)} - 1 = 1.621$$

Also, using Eq. (2.15) we get, $\beta = 137.35°$. The total rocking angle of OA is shown in Fig. 2.45 and it is,

$$\pi - \beta = 180° - 137.35° = 42.65°$$

Now let us go back to the calculation of the advance-to-return ratio of a shaper slider-crank mechanism explained in Fig. 2.41, with the original mechanism reproduced in Fig. 2.46. The mechanism has eccentricity e (shown in the figures). The advance direction of motion of the slider is its motion to the right; during which piston rod 3 pushes slider 4 to the right. During the advance motion, the slider is subjected to the machining force F exerted on it by the workpiece. In the return motion, piston rod 3 pulls the slider back to the left. (Of course, in general the advance motion direction is determined by what the machine or the mechanism is supposed to do.)

In Fig. 2.41 crank 2 of length r_2 is rotating clockwise. Let us consider position OA_1 as the start of motion of the crank. This position is one of the two limit positions of the mechanism at which the slider is located at B_1 (the left limit position of the slider). Motion of the crank is transmitted to the slider by the piston rod AB of length r_3. The crank should now rotate by angle α in order for the slider to reach its right limit position, B_2. The B_1B_2 interval is the advance part of the motion (aka, the working stroke of the shaper). If the crank rotates any further, the slider will move back to the left until it returns all the way back to its left limit position, B_1. The B_2B_1 interval is the return segment of the motion (aka, the return stroke of the shaper).

Considering the discussed motion and Fig. 2.41, the angular position of the crank at the right limit position of the slider, θ_{22} (measured from the slider direction of motion) can be written as:

$$\sin\theta_{22} = \frac{e}{OA_2 + A_2B_2}$$

Using,

$$OA_2 = r_2, \qquad A_2B_2 = r_3$$

One obtains,

$$\sin\theta_{22} = \frac{e}{r_2 + r_3} \tag{2.19}$$

Or,

$$\theta_{22} = \sin^{-1}\left(\frac{e}{r_2 + r_3}\right) \tag{2.20}$$

Similarly, the angular position of the crank at the left limit position of the slider, θ_{21} (measured from the slider direction of motion) can be written as:

$$\sin(\theta_{21} - \pi) = \frac{e}{A_1B_1 - OA_1}$$

Using,

$$OA_1 = r_2, \qquad A_1B_1 = r_3$$

We get,

$$\sin(\theta_{21} - \pi) = \frac{e}{r_3 - r_2} \tag{2.21}$$

Or,

$$\theta_{21} = \pi + \sin^{-1}\left(\frac{e}{r_3 - r_2}\right) \tag{2.22}$$

Referring to Fig. 2.41, the advance angle, α, and the return angle, β, can be expressed in terms of crank locations at the limit positions, θ_{22} and θ_{21}, as follows:

$$\alpha = \theta_{21} - \theta_{22}, \beta = 2\pi - \alpha \tag{2.23}$$

Substituting Eqs. (2.20) and (2.22) into Eq. (2.23) we get:

$$\alpha = \pi + \sin^{-1}\left(\frac{e}{r_3 - r_2}\right) - \sin^{-1}\left(\frac{e}{r_2 + r_3}\right) \tag{2.24}$$

$$\beta = \pi + \sin^{-1}\left(\frac{e}{r_2 + r_3}\right) - \sin^{-1}\left(\frac{e}{r_3 - r_2}\right) \tag{2.25}$$

Substituting Eqs. (2.24) and (2.25) into Eq. (2.13), the advance-to-return ratio for the mechanism shown in Fig. 2.46 is obtained as,

$$Q_{CW} = \frac{\pi + \sin^{-1}\left(\frac{e}{r_3 - r_2}\right) - \sin^{-1}\left(\frac{e}{r_2+r_3}\right)}{\pi + \sin^{-1}\left(\frac{e}{r_2+r_3}\right) - \sin^{-1}\left(\frac{e}{r_3 - r_2}\right)} > 1 \tag{2.26}$$

The quick-return response (considering the defined advance direction of the slider) is materialized if the crank rotates in the CW direction. If the direction of rotation of crank is changed to CCW, the advance would happen as the crank traverses the β angle and the return would happen as it covers the α angle. Therefore,

$$Q_{CCW} = \frac{\beta}{\alpha}$$

Or,

$$Q_{CCW} = \frac{\pi + \sin^{-1}\left(\frac{e}{r_2+r_3}\right) - \sin^{-1}\left(\frac{e}{r_3 - r_2}\right)}{\pi + \sin^{-1}\left(\frac{e}{r_3 - r_2}\right) - \sin^{-1}\left(\frac{e}{r_2+r_3}\right)} < 1 \tag{2.27}$$

And the mechanism will no longer be a quick-return mechanism. Therefore, proper choice of the direction of rotation of the crank is critical and determines if the mechanism has quick-return property.

One obvious requirement for the validity of these results is that $r_3 - r_2 > e$, and the other is, $r_2 + r_3 > e$. (Since sine of an angle cannot exceed one!). Clearly, if the first condition is satisfied, the second one will be met as well. Focusing on $r_3 - r_2 > e$, this condition has a simple geometrical interpretation. Referring to Fig. 2.41, at the left limit position, it should be clear that if $r_3 - r_2 = e$, the crank at the left limit position is vertical. Now if, $r_3 - r_2 < e$, the assembly at the left limit position simply cannot be constructed!

Now, let us consider a numerical example.

Example 2.19 Calculate the advance-to-return ratio of the mechanism shown in Fig. 2.46 if, $e = 1$ in, $r_2 = 3$ in, and $r_3 = 5$ in (for the CW direction of rotation of the crank).

For a CW direction of rotation of the crank, and using the given numerical values, Eq. (2.26) results in:

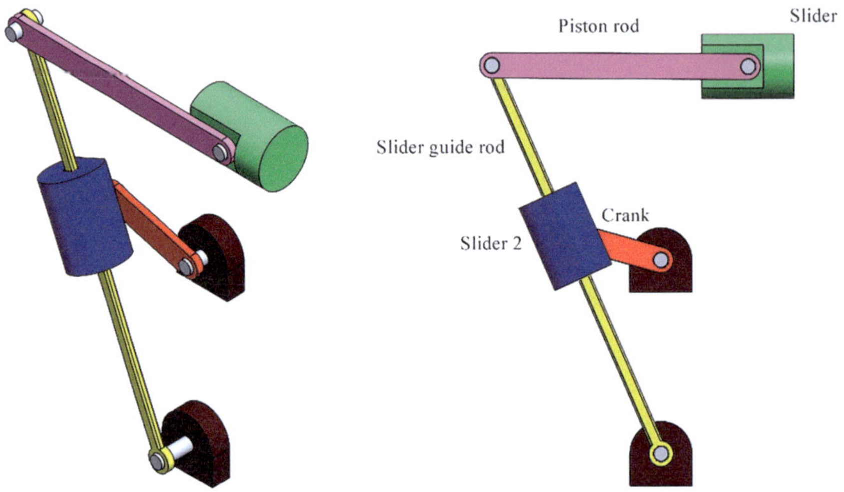

Fig. 2.47 A six-member quick-return mechanism modeled in SolidWorks, Stevenson [5]

$$Q_{CW} = \frac{\pi + \sin^{-1}\left(\frac{e}{r_3 - r_2}\right) - \sin^{-1}\left(\frac{e}{r_2 + r_3}\right)}{\pi + \sin^{-1}\left(\frac{e}{r_2 + r_3}\right) - \sin^{-1}\left(\frac{e}{r_3 - r_2}\right)} = \frac{180^\circ + 30^\circ - 7.18^\circ}{180^\circ + 7.18^\circ - 30^\circ} = \frac{202.8^\circ}{157.2^\circ} = 1.29 > 1$$

Figure 2.47 illustrates a six-member quick-return mechanism. This mechanism was modeled using the SolidWorks Motion software as part of a capstone project of one of my students at USM, see Stevenson [5]. The mechanism has one crank, two sliders, a guide rod that directs slider 2, and a piston rod that pushes slider 1 to the right (the advance direction) or pulls it to the left (the return direction).

Considering the direction of the advance motion of slider 1, the quick-return characteristic of the mechanism will be experienced if the crank rotates CW. The two limit positions of the guide rod are achieved when the crank is perpendicular to the guide rod, as shown in Fig. 2.48. The advance part of the motion of slider 1 starts when the mechanism reaches the position shown in the left limit position. Then, as the crank rotates CW, the guide rod rotates CW as well and as a result the piston rod pushes slider 1 to the right, i.e., in the advance direction. This segment of the motion concludes when the mechanism reaches the right limit position. As the crank passes this position, the quick-return segment of the motion starts in which while the crank rotates CW, the guide rod rocks CCW and moves back toward the left limit position, completing one cycle.

Calculations in SolidWorks show that for this mechanism the advance angle, α, of the crank is 252.4° and the corresponding return angle, β is 107.6°. Therefore, the advance-to-return ratio is,

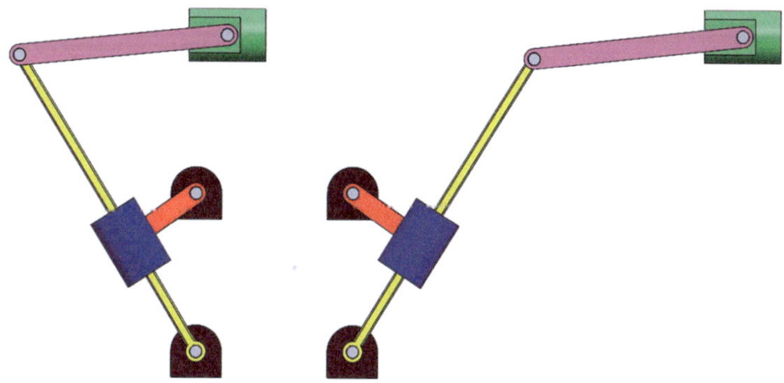

Fig. 2.48 Limit positions of the six-member quick-return mechanism, Stevenson [5]

$$Q_{CW} = \frac{\alpha}{\beta} = \frac{252.4\,^\circ}{107.6\,^\circ} = 2.35$$

2.6 Grashof's Law and Kinematic Inversions

One major practical question in a four-bar linkage design is: *Under what conditions can a link be connected to an electric motor and make complete revolutions?* The answer to this question is given by the Grashof's law to be discussed in this section. We will see that when the Grashof's law condition is satisfied and making such complete revolutions is possible, still two mechanisms with the same lengths of links can behave very differently (relative to a stationary observer) depending on which one of the links is held fixed (see Fig. 2.49). The possible types of behavior are: crank-rocker, rocker-rocker, and crank-crank (also known as drag-link). In each case, the first term refers to the way the input member behaves, and the second term refers to the response of the output member.

Mechanisms that have the same links and their only difference is in which link has been held fixed are called kinematic inversions of one another. The motion that can be viewed relative to the fixed member in one inversion will be different from the one viewed in another inversion. Figure 2.49 shows four kinematic inversions of a four-bar linkage. In general, in an n-link kinematic chain, each of the links can be chosen as the fixed member. Since there are n ways for choosing the fixed link, the n-member kinematic chain has n kinematic inversions, each with a different input–output relation. While the absolute motion of the members as seen by a stationary observer in one of these inversions would be different from the others, the relative motion of the links with respect to each other in different kinematic inversions are identical and that is what Grashof's law refers to.

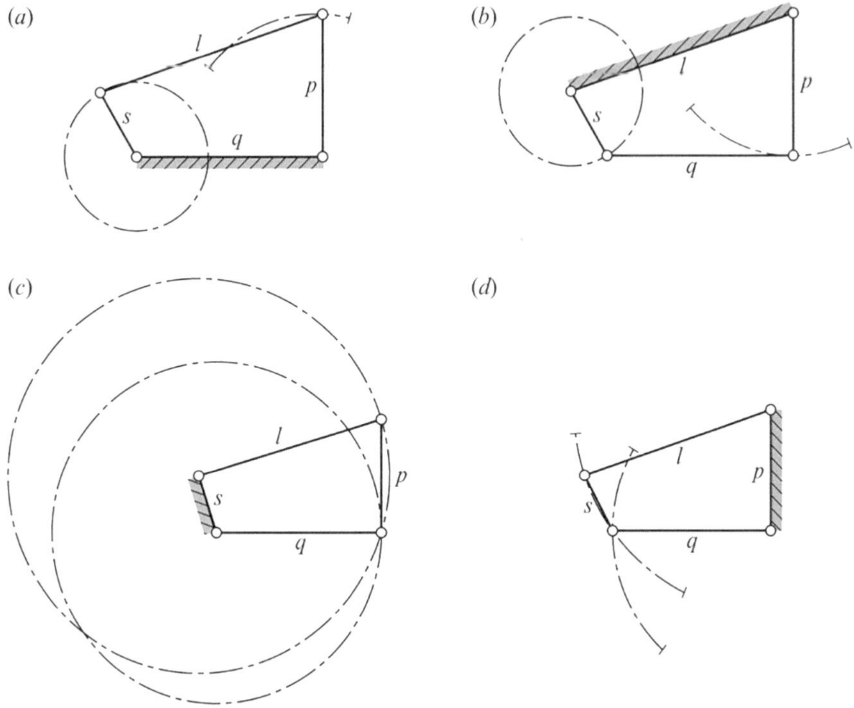

Fig. 2.49 Kinematic inversions of a four-bar linkage (that satisfies Grashof's law) and their motions with respect to a stationary observer. (**a**, **b**) crank-rocker linkages, (**c**) drag-link linkage, (**d**) rocker-rocker linkage. Page 35 of Uicker et al. [6] © 2016 Oxford University Press (reproduced with permission)

Consider the four-bar linkage shown in Fig. 2.50 in which link 2 is the input and link 4 is the output. The trajectories of the end points of links 2 and 4 show that this is a crank-rocker mechanism.

Consider the following notations for the longest and the shortest links,

$$L_{min} = Min\{L_1, L_2, L_3, L_4\}; L_{max} = Max\{L_1, L_2, L_3, L_4\} \tag{2.28}$$

The lengths of the remaining two links are named, L_a and L_b.

Grashof's law states that if in a four-bar linkage we have,

$$L_{min} + L_{max} < L_a + L_b \tag{2.29}$$

in all kinematic inversions, there are two joints that perform complete rotations and two that oscillate between motion limits. The two fully rotatable joints are those on either end of the shortest link. Links that do not obey Grashof's law have no fully rotatable joints. All four joints then oscillate between motion limits, Waldron and Kinzel [7], page 33.

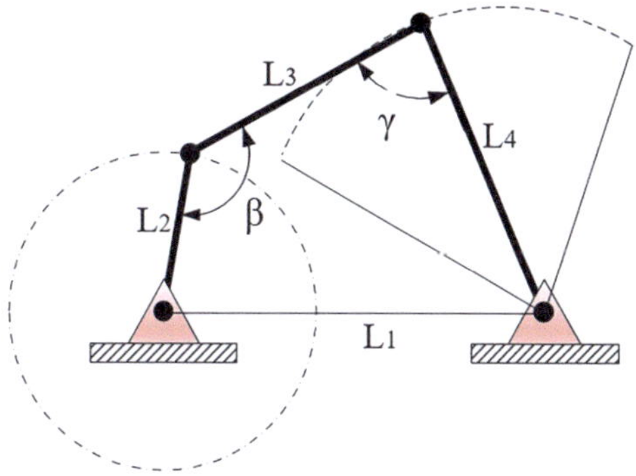

Fig. 2.50 A four-bar linkage operating as a crank (2)-rocker (4) mechanism

In order to prove this law, let us first develop the conditions that allow us assemble a mechanism using links of lengths L_1, L_2, L_3, and L_4. Since the shortest distance between two points is a straight line, in order to form a mechanism, the length of each link should be less than the summation of the other three. So,

$$\begin{aligned} &L_1 < L_2 + L_3 + L_4; L_2 < L_1 + L_3 + L_4 \\ &L_3 < L_1 + L_2 + L_4; L_4 < L_1 + L_2 + L_3 \end{aligned} \tag{2.30}$$

Recalling the (2.28) definitions, conditions (2.30) can be rewritten as follows:

$$\begin{aligned} &L_{max} < L_{min} + L_a + L_b; L_{min} < L_{max} + L_a + L_b \\ &L_a < L_{max} + L_{min} + L_b; L_b < L_{max} + L_{min} + L_a \end{aligned} \tag{2.31}$$

It should be clear that all of these conditions for forming a mechanism will be satisfied only if the first of conditions (2.31) is valid, i.e.,

$$L_{max} < L_{min} + L_a + L_b \tag{2.32}$$

Considering Fig. 2.50, now the aim is to find conditions that guarantee the assembled mechanism has two joints that can perform complete rotations and two that oscillate between motion limits. Assume that link 2 is the one with two joints that perform complete rotations and link 4 is the one that has two joints that oscillate between motion limits. For this matter to happen, the mechanism should be able to pass through two limit positions that correspond to $\beta = 180°$ (where links 2 and 3 are collinear and they are in the same direction, as seen in Fig. 2.51) and $\beta = 0°$ (where links 2 and 3 are collinear, but they are in opposite directions, as seen in Fig. 2.52). These are the two limit positions of the rocking motion of link 4.

Referring to Fig. 2.51, since the shortest distance between two points is a straight line, in any triangle the length of each side is less than the summation of the lengths of the other two sides, therefore:

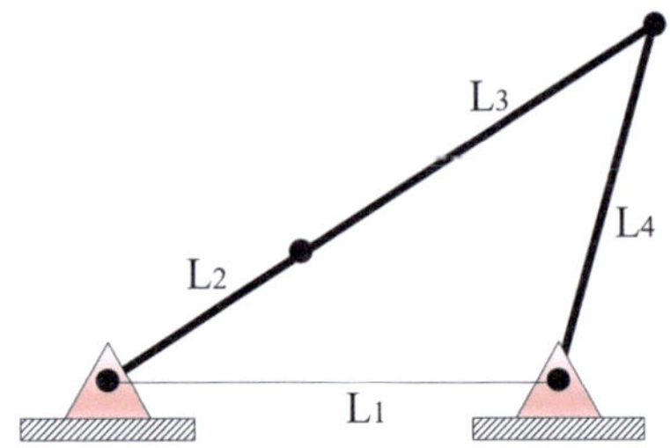

Fig. 2.51 First limit position of rocker 4 at $\beta = 180°$

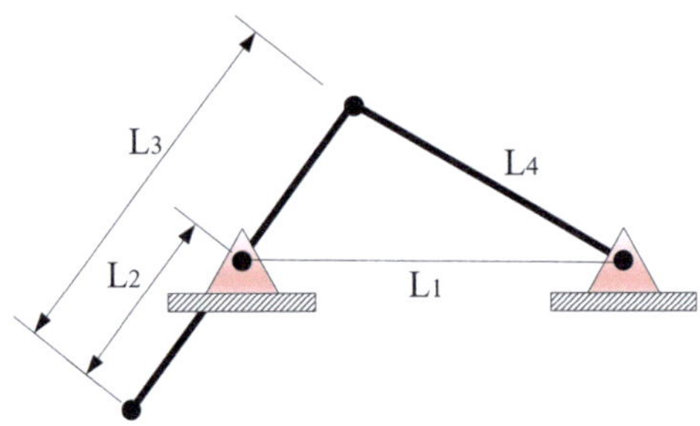

Fig. 2.52 Second limit position of rocker 4 at $\beta = 0$

$$L_2 + L_3 < L_1 + L_4 \tag{2.33}$$

Similarly, for the triangle that corresponds to the second limit position shown in Fig. 2.52 one can write,

$$L_1 < (L_3 - L_2) + L_4 \tag{2.34}$$

$$L_4 < (L_3 - L_2) + L_1 \tag{2.35}$$

The three conditions in (2.33), (2.34), and (2.35) are now added two by two at a time. From the first two we get,

$$L_2 < L_4$$

From (2.34) and (2.35) we get:

$$L_2 < L_3$$

From (2.33) and (2.35) we get:

$$L_2 < L_1$$

By comparing the obtained results we conclude that link 2 (that has two joints that perform complete rotations) has to be the shortest link:

$$L_2 = L_{min} \tag{2.36}$$

Now, back to Eqs. (2.33), (2.34), and (2.35), they can be rewritten as:

$$L_2 + L_1 < L_3 + L_4 \tag{2.37}$$

$$L_2 + L_4 < L_1 + L_3 \tag{2.38}$$

It was proven in Eq. (2.36) that, $L_2 = L_{min}$. Among links 1, 3, and 4 one of them is the longest; therefore, using (2.28) the above three inequalities can be rewritten as follows:

$$L_{min} + L_{max} < L_a + L_b \tag{2.39}$$

$$L_{min} + L_a < L_{max} + L_b \tag{2.40}$$

$$L_{min} + L_b < L_{max} + L_a \tag{2.41}$$

Among these conditions, the first one is the strongest inequality, meaning that if Eq. (2.39) is satisfied, the other two will also be satisfied. (For the rest of them the LHS is smaller and the RHS is greater!) Let us formally prove that if Eq. (2.39) is valid, then Eq. (2.40) will be automatically satisfied. Starting from Eq. (2.39), since,

$$L_a < L_{max}; L_{min} < L_b,$$

one can write the following inequality sequence:

$$L_{min} + L_a < L_{min} + L_{max} < L_b + L_{max}$$

Therefore, Eq. (2.40) is proven.

So, for having a mechanism that has two joints that perform complete rotations and two that oscillate between motion limits the following inequality should be satisfied,

$$L_{min} + L_{max} < L_a + L_b \tag{2.39}$$

(In words, the summation of the lengths of the shortest and the longest links should be less than the summation of the lengths of the other two links.) Considering Eq. (2.36), the two fully rotatable joints are those on either ends of the shortest link. Links that do not obey Grashof's law have no fully rotatable joints. All four joints then oscillate between motion limits, Waldron and Kinzel [7], page 33. This result is an expression of the Grashof's law.

Stated differently, in order that a link can make full turns *with respect to the other links*, it should be the shortest link in the mechanism and also the summation of the lengths of the shortest and the longest links should be less than the summation of the lengths of the other two links.

For a four-bar linkage that satisfies Grashof's law, the absolute motion that we see depends on the particular link in each of the four kinematic inversion that is held fixed (see Fig. 2.49). Here are the four possibilities:

Inversions *a* and *b*: If the shortest link is joined to the fixed link, the mechanism is a crank-rocker one. The shortest link is the crank. The other link connected to the fixed link has joints that are not fully rotatable. Hence the other link connected to the fixed link is a rocker. The crank can be operated by either of its two fully rotatable joints.

Inversion *c*: The shortest link is the fixed link, the other links connected to the fixed link are cranks and the mechanism is a crank-crank, double-crank, or drag-link mechanism. It can be actuated at either of its fully rotatable joints.

Inversion *d*: The shortest link is the coupler (opposite to the fixed link) then the mechanism is a rocker-rocker mechanism. Since the mechanism satisfies the Grashof's law, the coupler's joints are both fully rotatable and the coupler can perform complete rotations with respect to a stationary observer. In this case, the shortest link completes full rotations while the other moving links only perform rocking (oscillating) motions. The coupler joints can be actuated by a rotary motor. This inversion has applications is oscillating fans, Waldron and Kinzel [7], page 254. Mechanisms in which the coupler is the input link exist and are used, for example, for trunk lid opening in cars.

Example 2.20 In the mechanism shown in Fig. 2.53 we know that:

$$L_1 = 11.5 \text{ in}, L_2 = 4.5 \text{ in}, L_3 = 8 \text{ in}, L_4 = 9.5 \text{ in}$$

A. Does this mechanism satisfy Grashof's law?
B. What kind of response do you expect to see from the inversion of this mechanism drawn in Fig. 2.53?
C. What kind of response do you expect to see from another inversion of this mechanism in which link 2 is fixed instead of link 1?

A. We check to see if condition (2.39) is satisfied. In this case,

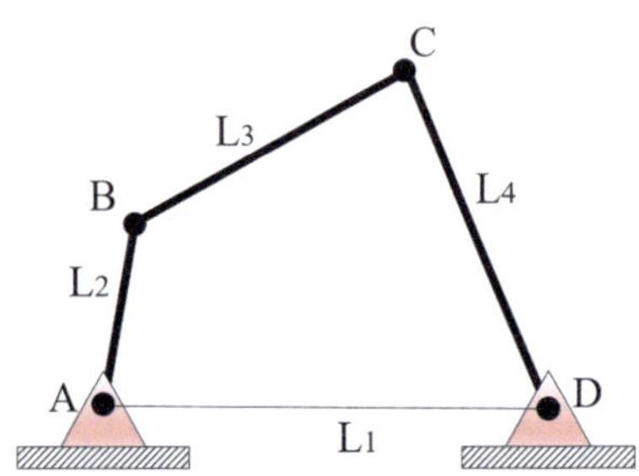

Fig. 2.53 A four-bar linkage

$$L_{max} = 11.5 \text{ in}, L_{min} = 4.5 \text{ in}, L_a = 8 \text{ in}, L_b = 9.5 \text{ in}$$

Therefore, condition (2.39) results in: $11.5 + 4.5 < 8 + 9.5$, that is valid. Therefore, this mechanism satisfies the Grashof's law condition.

B. Since the Grashof's law condition is satisfied and the shortest link is adjacent to the base (fixed member), we conclude that this is a crank-rocker mechanism (inversions *a* and *b*) where link 2 is the crank and link 4 is the rocker.
C. In another inversion of this mechanism in which link 2 is fixed instead of link 1, the shortest link is fixed (inversion *c*). Therefore, we have a double-crank or drag-link case where links 1 and 3 are the cranks.

Design/Manufacturing Tip When designing/manufacturing mechanisms, to allow for full rotation, it is important that links do not interfere with each other. Such an interference is not detectable in a planar view so examining other views is needed! What should happen is that the members should move in parallel planes (see Fig. 2.54) so they do not bump into one another.

It should be clear that if Grashof's condition (2.39) is satisfied, then the condition for forming a mechanism, (2.32), is also satisfied.

$$L_{max} < L_{min} + L_a + L_b \tag{2.32}$$

So, Grashof's law encompasses the condition for existence of a mechanism.

What if Grashof's condition is not satisfied?

If:

$$L_{min} + L_{max} > L_a + L_b \tag{2.42}$$

the response will always be double-rocker. However, in such a case, the coupler can no longer complete full rotations (unlike inversion *d* discussed earlier).

If:

$$L_{min} + L_{max} = L_a + L_b \tag{2.43}$$

we have a transition, change point, or toggle point mechanism; see Fig. 2.55. Once the mechanism reaches the flattened position (where all links are collinear), it can continue in either of the two configurations shown in Fig. 2.56. In practice, such behavior is not desirable as it can happen with no warning and create excessive loads and stresses.

Example 2.21 State the different inversions of a 2R-2P mechanism.

The oscillating Scotch-yoke mechanism shown in Fig. 2.42 is one of the 2R-2P inversions in which a revolute joint and a prismatic joint are attached to the frame.

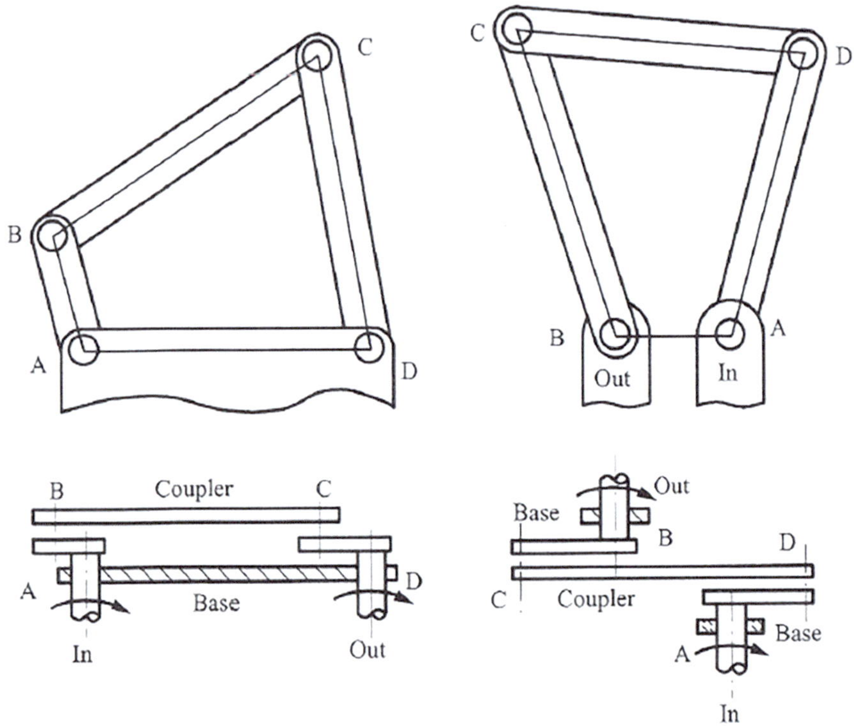

Fig. 2.54 Shaft drive of crank-rocker (left) and drag-link linkages (right) Page 41 of Waldron and Kinzel (2004) © 2004 John Wiley & Sons, Inc. (reproduced with permission)

Fig. 2.55 A change point or toggle point mechanism

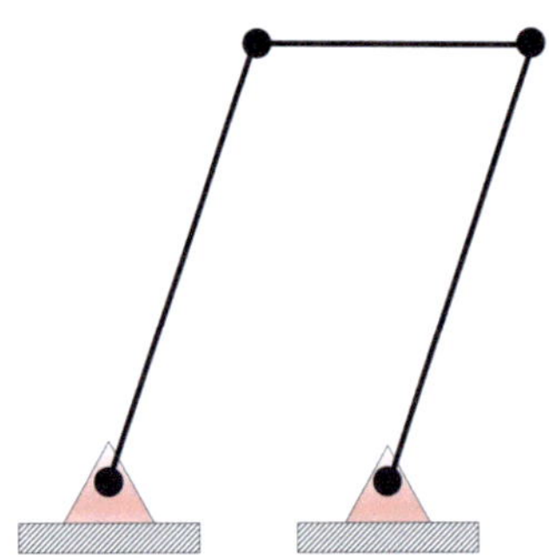

Another inversion is an R-2P-R mechanism that is realized in an Oldham coupling shown in Fig. 2.57. In this mechanism both revolute joints are attached to the frame. In Oldham coupling, the input and output shafts must be parallel, but they can be offset from each other.

Finally, the third inversion is the double-slider shown in Fig. 2.58. This is a P-2R-P mechanism where both of the prismatic joints are attached to the frame. This

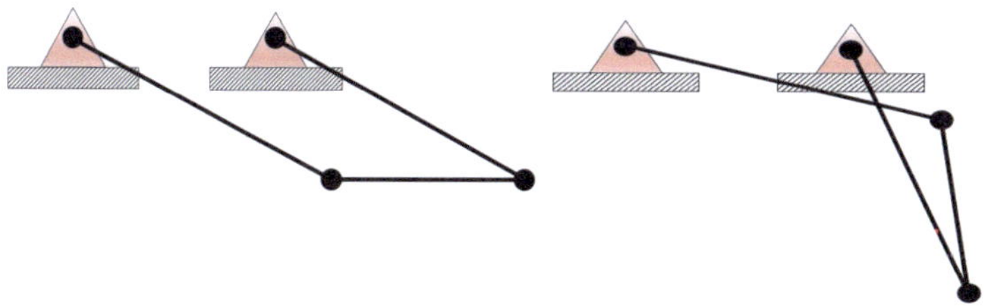

Fig. 2.56 The two possible configurations of a change point or toggle point mechanism

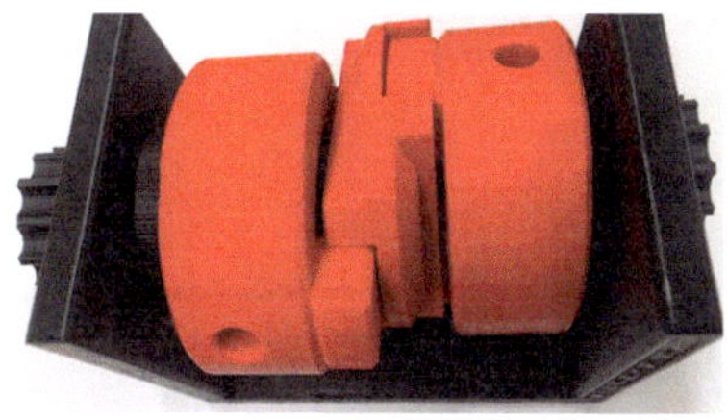

Fig. 2.57 A 3D-printed Oldham coupling model (USM term project)

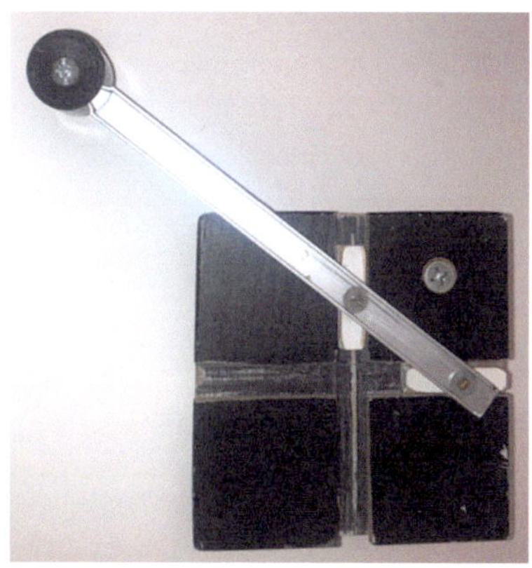

Fig. 2.58 Trammel of Archimedes (USM term project)

is the trammel of Archimedes that is an ellipsograph (i.e., it can be used for drawing ellipses). Here, the center point between the hinges on the link draws a circle. All other points on the link draw ellipses, Hartenberg and Denavit [2], page 57. The fourth inversion is another Scotch-yoke mechanism that is different from the first one.

References

1. Ghorashi, M. (2018). Closed-form solutions for the load-carrying capacity of annular plates. *International Journal of Mechanical Sciences, 136*, 313–320.

2. Hartenberg, R. S., & Denavit, J. (1964). *Kinematic synthesis of linkages*. McGraw-Hill.
3. Kutzbach, K. (1929). Mechanische leitungsverzweigung, ihre Gesetze und Anwendungen. *Maschinenbau, 8*, 710–716.
4. Sclater, N., & Chironis, N. P. (2007). *Mechanisms and mechanical devices* (4th ed.). McGraw-Hill.
5. Stevenson, B. (2018). *Design and manufacture of several mechanisms to demonstrate the concept of mobility*. Capstone Project Report, Department of Engineering, University of Southern Maine.
6. Uicker, J. R., et al. (2016). *Theory of machines and mechanisms* (5th ed.). Oxford University Press.
7. Waldron, K. J., & Kinzel, G. L. (2004). *Kinematics, dynamics, and design of machinery* (2nd ed.). Wiley.
8. Woolsey Johnson, W. M. (1875). The Peauceller machine and other linkages. *The Analyst, 2*(2), 41–45.

Chapter 3
Position Analysis

3.1 Introduction

This chapter is dedicated to the position analysis of mechanisms. Both closed-form and numerical solutions are presented. Position analysis is the most fundamental type of kinematic analyses. Once position of a point or orientation of a rigid body is obtained as a function of time, finding velocity or acceleration functions would be easy; as they just need differentiation of the position function with respect to time. One way for performing position analysis is to use the vector-loop method. This method results in simultaneous nonlinear equations that have multiple solutions.

Let us first clarify what nonlinear equations are and how they may have multiple solutions. Linear problems have variables that are raised to the power of one. (No sines, squares, or logarithms of the unknown variables, for example, appear in linear equations). Here is an example:

$$\begin{cases} x + y = 4 \\ x - y = 1 \end{cases}$$

One can easily find its unique answer as: $x = 2.5$, $y = 1.5$. The following system, however, is not linear,

$$\begin{cases} \cos x + \cos y = 0.7 \\ \sin x - \sin y = 0.1 \end{cases}$$

Finding solutions of nonlinear problems is usually not easy and requires application of numerical methods. The following system is also nonlinear:

$$\begin{cases} x^2 - y^2 = 0 \\ x^2 + y^2 = 50 \end{cases}$$

M. Ghorashi, *Introduction to Machine Dynamics*, Mechanical Engineering Series,
https://doi.org/10.1007/978-3-031-94736-0_3

This one, however, has simple analytical solutions:

$$x = \pm 5;\ y = \pm 5$$

In fact, the acceptable solution set is: (5,5), (5,−5), (−5,5), (−5,−5). So this nonlinear problem has multiple (four) answers even though it was a set of two equations for two unknowns.

In general, solutions of nonlinear equations should be calculated numerically using methods like the Newton-Raphson algorithm which is easily applicable to solving nonlinear, multi-degree-of-freedom systems. In contrast to the position problem, and as we will see in Chaps. 4 and 5, the velocity and the acceleration problems are linear and can be solved easier.

3.2 Vector-Loop Equations of a Mechanism

The vector-loop method is a powerful way of formulating the geometry of a mechanism. The formulated geometry can then be used for solving position, velocity, and acceleration at points of interest in a mechanism. In order to demonstrate the application of the vector-loop method in position analysis, let us consider the four-bar linkage shown in Fig. 3.1.

In Chap. 2 we learned that the posture of this mechanism can be controlled (and fully defined) by setting the orientation of only one of its links (and that is why it has one DOF). To show the orientations of links, one may replace each link by a vector defined by its length and angular position, with the latter measured from the positive x axis. The x axis is drawn aligned with line AD (i.e., the base member defined by the two fixed hinges) and with a positive direction from A to D. The corresponding y axis is obtained by rotating the x axis in the CCW direction by 90°. In this way, we obtain the vector loop shown in Fig. 3.2. The corresponding vector-loop equation is:

$$\vec{r}_2 + \vec{r}_3 = \vec{r}_1 + \vec{r}_4 \tag{3.1}$$

Projecting Eq. (3.1) in the $\vec{r}_1$ direction (i.e., the x direction) results in:

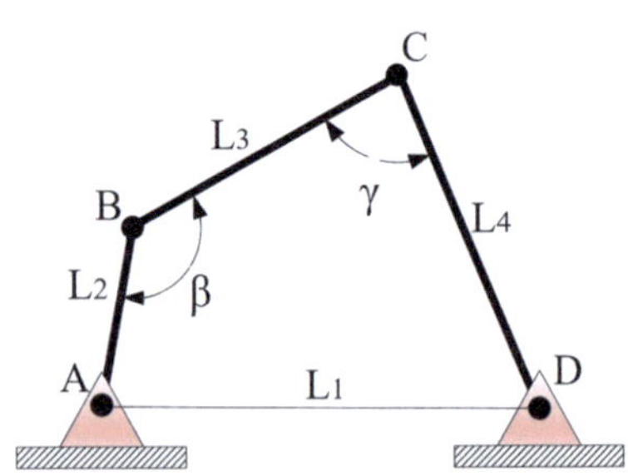

Fig. 3.1 A four-bar linkage

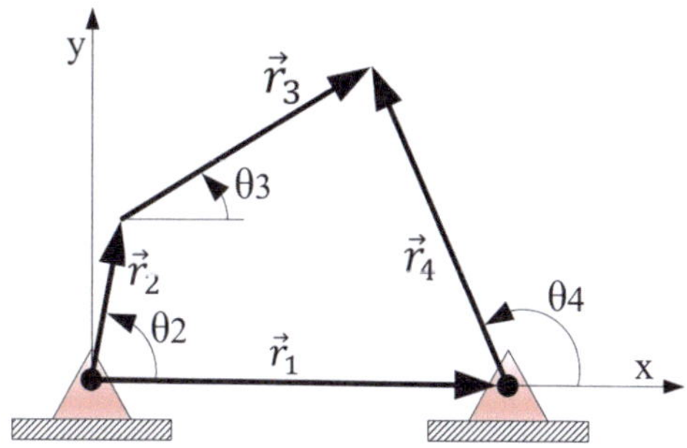

Fig. 3.2 The vector loop corresponding to the mechanism shown in Fig. 3.1

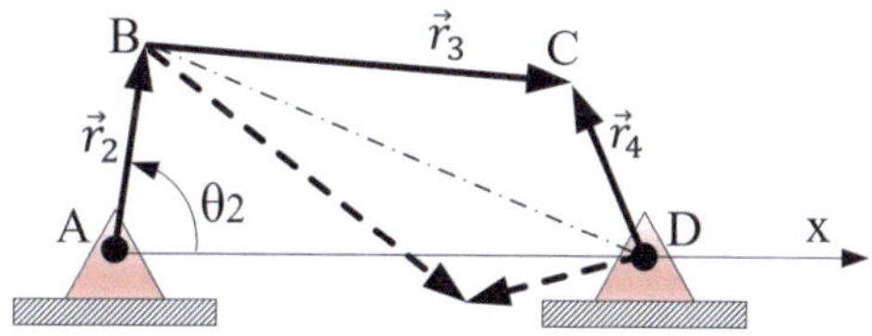

Fig. 3.3 Open posture (solid line) and crossed posture (dashed line) solutions for the same input angle θ_2

$$r_2 \cos \theta_2 + r_3 \cos \theta_3 = r_1 + r_4 \cos \theta_4 \tag{3.2}$$

Similarly, projecting it in the direction perpendicular to $\vec{r}_1$ (i.e., the y direction) gives:

$$r_2 \sin \theta_2 + r_3 \sin \theta_3 = r_4 \sin \theta_4 \tag{3.3}$$

Equations (3.2) and (3.3) are the projected vector-loop equations of the mechanism. These two equations have three variables, θ_2, θ_3, and θ_4. We take link 2 as the input link (i.e., the one whose orientation is being controlled) and therefore consider θ_2 as the input angle. To show this matter, let us rewrite Eqs. (3.2) and (3.3) as follows:

$$r_3 \cos \theta_3 - r_4 \cos \theta_4 = r_1 - r_2 \cos \theta_2 \tag{3.4}$$

$$r_3 \sin \theta_3 - r_4 \sin \theta_4 = -r_2 \sin \theta_2 \tag{3.5}$$

Once θ_2 is known, the system of nonlinear trigonometric Eqs. (3.4) and (3.5) can be solved for the remaining two unknown angles θ_3 and θ_4. In this way, the posture of the mechanism at the chosen value of the input angle θ_2 will be fully known. Hence, it is verified that the mechanism has indeed one DOF.

The obtained system of Eqs. (3.4) and (3.5) for unknown angles θ_3 and θ_4 is nonlinear and therefore it is a bit hard to solve. One should also remember that for such a nonlinear system there is always the possibility of having multiple solutions. Figure 3.3 clearly shows such a possibility. Here we see that for the same input angle θ_2, and the same set of link lengths, one can find two solutions: an open posture (solid line) and a crossed posture (dashed line). These solutions correspond to different values of θ_3 and θ_4. The open and the crossed postures are in such a way that links 3 and 4 in the open posture are mirror images of links 3 and 4 in the crossed posture with respect to the symmetry line *BD*. The way the mechanism has been initially assembled determines which one of these possibilities will actually take place. In what follows, these matters will be explored in more detail.

3.3 Closed-Form Solutions for Position

3.3.1 *Four-Bar Linkage*

In Sect. 3.2, the system of Eqs. (3.4) and (3.5) was developed for solving the position and posture problem of a four-bar linkage. In the present section, closed-form solutions for this problem are developed. This system will later be solved numerically as well. In fact, for more complex mechanisms, a numerical method is the only practical way for solving the position problem.

Recall that in the vector loop of a four-bar linkage, shown in Fig. 3.2, the variable quantities are θ_2, θ_3, and θ_4, among which θ_2 is the given input angle. For a certain value of θ_2 there are two solutions that are shown in Fig. 3.4. The solutions are *ABCD* and *ABC′D*, and they are called the open and the crossed postures, respectively. The *BC′D* segment of the crossed posture is a mirror image of the *BCD* segment of the open posture with respect to line *BD*. In this section, for each of these postures, we will obtain closed-form solutions for θ_3, θ_4 and the transmission angle γ (its significance will be discussed in Chap. 11) in terms of the input variable, θ_2.

First, let us calculate length of line *BD* shown by s. Using law of cosines (see the Appendix for a review) in triangle *ABD* one obtains:

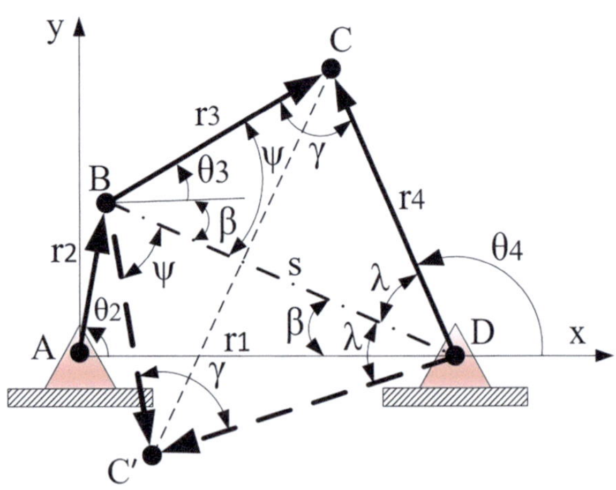

Fig. 3.4 Geometrical analysis of open (solid line) and crossed posture (dashed line) solutions ($0 \le \theta_2 \le \pi$)

$$s^2 = r_1^2 + r_2^2 - 2r_1r_2\cos\theta_2 \tag{3.6}$$

Having calculated s, we can write three more law of cosines equations in order to calculate β, ψ, and λ. Considering triangle ABD again we obtain an equation for β:

$$r_2^2 = s^2 + r_1^2 - 2sr_1\cos\beta \tag{3.7}$$

Similarly, analyzing triangle BCD provides equations for ψ and λ:

$$r_4^2 = s^2 + r_3^2 - 2sr_3\cos\psi \tag{3.8}$$

$$r_3^2 = s^2 + r_4^2 - 2sr_4\cos\lambda \tag{3.9}$$

Referring to Fig. 3.4 for $0 \leq \theta_2 \leq \pi$, and considering positive values for β, ψ, and λ, one can calculate θ_3 and θ_4 as follows:

For open posture:

$$\theta_3 = \psi - \beta, \theta_4 = \pi - \beta - \lambda \tag{3.10}$$

For crossed posture:

$$\theta_3 = -\psi - \beta, \theta_4 = \pi - \beta + \lambda \tag{3.11}$$

Similarly, considering Fig. 3.5 for $\pi \leq \theta_2 \leq 2\pi$ we have:

For open posture:

$$\theta_3 = -\psi + \beta, \theta_4 = \pi + \beta + \lambda \tag{3.12}$$

For crossed posture:

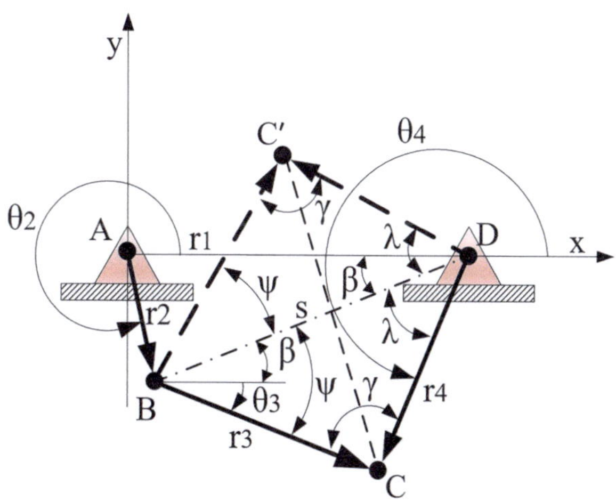

Fig. 3.5 Geometrical analysis of open (solid line) and crossed posture (dashed line) solutions ($\pi \leq \theta_2 \leq 2\pi$)

$$\theta_3 = \psi + \beta, \theta_4 = \pi + \beta - \lambda \tag{3.13}$$

The transmission angle γ can now be calculated using the law of cosines in triangle *BCD*:

$$s^2 = r_3^2 + r_4^2 - 2r_3r_4\cos\gamma \tag{3.14}$$

Substituting Eq. (3.6) into (3.14) we get,

$$r_1^2 + r_2^2 - 2r_1r_2\cos\theta_2 = r_3^2 + r_4^2 - 2r_3r_4\cos\gamma$$

Or,

$$\cos\gamma = \frac{r_3^2 + r_4^2 - r_1^2 - r_2^2 + 2r_1r_2\cos\theta_2}{2r_3r_4} \tag{3.15}$$

This function presents the way the transmission angle γ varies with the input angle θ_2. To get the maximum and the minimum values of γ, one can explore the maximum and the minimum values of $\cos\theta_2$, i.e.:

For $\theta_2 = 0$, $\cos\theta_2 = 1$ and substitution in Eq. (3.15) gives the maximum value of $\cos\gamma$ as:

$$\cos\gamma|_{max} = \frac{r_3^2 + r_4^2 - (r_1 - r_2)^2}{2r_3r_4}$$

Angle γ is an angle of a triangle and therefore, it is less than 180°. Since in the [0, 180°] interval cosine is a decreasing function, the maximum $\cos\gamma$ happens at the minimum value of γ. Therefore, the minimum transmission angle is:

$$\gamma_{min} = \cos^{-1}\left(\frac{r_3^2 + r_4^2 - (r_1 - r_2)^2}{2r_3r_4}\right) \tag{3.16}$$

And as we saw it happens at $\theta_2 = 0$. See Fig. 3.6.

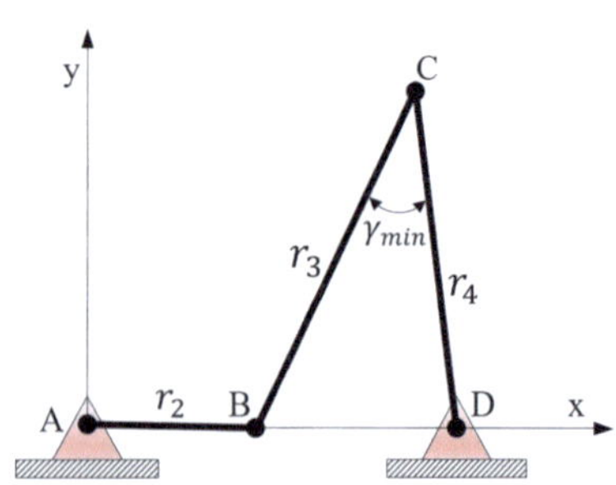

Fig. 3.6 A four-bar linkage at minimum transmission angle γ_{min}

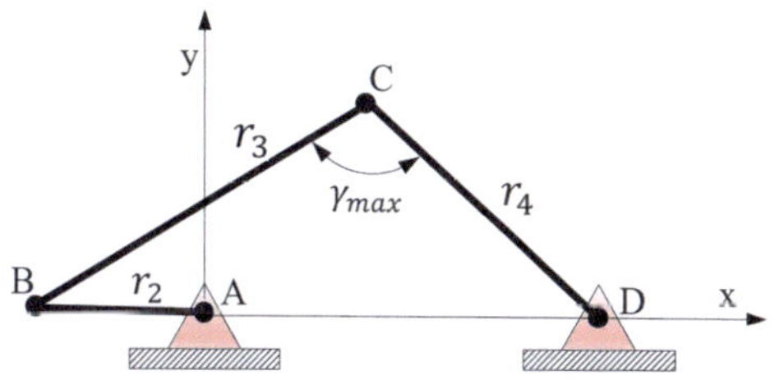

Fig. 3.7 A four-bar linkage at maximum transmission angle γ_{max}

For $\theta_2 = \pi$, $\cos\theta_2 = -1$ and substitution in Eq. (3.15) gives the minimum value of $\cos\gamma$ as,

$$\cos\gamma|_{min} = \frac{r_3^2 + r_4^2 - (r_1 + r_2)^2}{2r_3 r_4} \tag{3.17}$$

Angle γ is an angle of a triangle and therefore, it is less than 180°. Since in the [0, 180°] interval cosine is a decreasing function, the minimum $\cos\gamma$ happens at the maximum value of γ. Therefore, the maximum transmission angle is:

$$\gamma_{max} = \cos^{-1}\left(\frac{r_3^2 + r_4^2 - (r_1 + r_2)^2}{2r_3 r_4}\right) \tag{3.18}$$

And as we saw it happens at $\theta_2 = 180°$. See Fig. 3.7.

Example 3.1 Considering the mechanism shown in Fig. 3.2 and for the following data: $r_1 = 14$ in, $r_2 = 7$ in, $r_3 = 10$ in, and $r_4 = 8$ in, calculate θ_3, θ_4, and γ. Use an input angle of $\theta_2 = 70°$.

Substituting the given data into Eq. (3.6) we get: $s = 13.34$ in. Then using Eqs. (3.7), (3.8), and (3.9) one obtains:

$$\beta = 29.54\,°, \psi = 36.68\,°, \lambda = 48.31\,°$$

For open posture, since $0 \le \theta_2 \le \pi$, we use Eq. (3.10) and get:

$$\theta_3 = 7.14\,°, \theta_4 = 102.15\,°$$

For crossed posture, using Eq. (3.11) results in:

$$\theta_3 = -66.22\,°, \theta_4 = 198.77\,°$$

Finally, using Eq. (3.15) gives,

$$\gamma = 95\,°.$$

3.3.2 Offset Slider-Crank Mechanism

Figure 3.8 illustrates a slider-crank mechanism in which continuous rotations of crank 2 generate reciprocal motion of slider 4 that is hinged to connecting rod 3 at point B. The crank rotates about hinge O that has an offset from the line of action of the slider by an amount, e.

In Chap. 2, it was shown that a slider-crank mechanism has one DOF. So, its motion can be controlled by changing one of its position variables, for example, the crank angle. Those position variables are shown in Fig. 3.9 that is the vector-loop diagram of the mechanism. The variable quantities in this figure are r_1, θ_2, and θ_3, among which θ_2 is the given input angle. Orientations of the links are measured with respect to the x axis (that is collinear with the direction of motion of the slider) and the corresponding y axis (obtained by rotating the x axis in the CCW direction by 90°).

For the slider to maintain its contact with the stationary surface during motion, the following requirement should be satisfied:

$$r_2 + e < r_3. \tag{3.19}$$

The vector-loop diagram shown in Fig. 3.9 can be used for locating the slider with respect to the position of the crank hinge. The corresponding vector-loop equation is:

$$\vec{r}_2 + \vec{r}_3 + \vec{r}_4 = \vec{r}_1 \tag{3.20}$$

Projecting this vector equation in the direction of $\vec{r}_1$ (i.e., the x axis) results in,

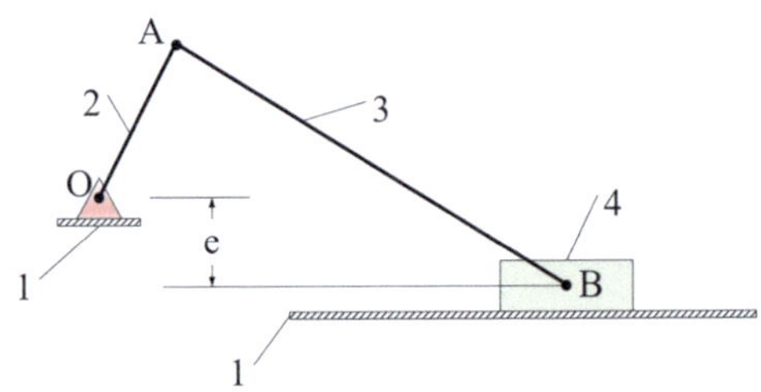

Fig. 3.8 An offset slider-crank mechanism

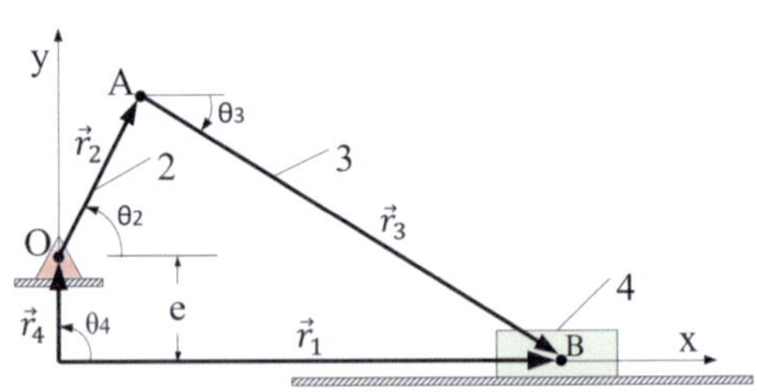

Fig. 3.9 The vector-loop diagram of the slider-crank mechanism shown in Fig. 3.8

$$r_2 \cos\theta_2 + r_3 \cos\theta_3 + r_4 \cos\theta_4 = r_1 \cos\theta_1 \tag{3.21}$$

Since angles are measured from the positive direction of the x-axis, there are two possibilities for θ_1, i.e., $\theta_1 = 0$ or $\theta_1 = \pi$ (depending on the position of the slider) and we know that, $\theta_4 = 90°$. So, Eq. (3.21) reduces to the following,

$$r_2 \cos\theta_2 + r_3 \cos\theta_3 = r_1 \cos\theta_1 \tag{3.22}$$

Similarly, projecting the vector-loop equation (3.20) in the y direction results in:

$$r_2 \sin\theta_2 + r_3 \sin\theta_3 + r_4 \sin\theta_4 = 0$$

Or, considering $r_4 = e$, one obtains,

$$r_2 \sin\theta_2 + r_3 \sin\theta_3 + e = 0 \tag{3.23}$$

Equations (3.22) and (3.23) are the projected vector-loop equations of the mechanism. If the lengths of the crank and the connecting rod, as well as the eccentricity value are known (so, r_2, r_3, and $e = r_4$ are given), the two Eqs. (3.22) and (3.23) would have three unknowns, i.e., θ_2, θ_3, and r_1. Therefore, one can choose any one of these variables: θ_2, θ_3, and r_1 (usually θ_2) as the input variable and then solve the above-mentioned system of two nonlinear trigonometric equations for the remaining two unknowns (i.e., θ_3, and r_1) as the output variables. This is another way of reaching to the conclusion that the slider-crank mechanism has one degree of freedom.

Now, the horizontal position of the slider with respect to hinge point O can be stated by,

$$x_B = r_1 \cos\theta_1 \tag{3.24}$$

We would like to express x_B and θ_3 in terms of the only degree of freedom of the mechanism, i.e., the crank angle, θ_2. Eq. (3.23) can be solved for θ_3 in terms of θ_2:

$$\sin\theta_3 = -\frac{1}{r_3}(e + r_2 \sin\theta_2) \tag{3.25}$$

In order to calculate x_B in terms of θ_2, and considering Eqs. (3.22) and (3.24), one needs to express $\cos\theta_3$ in terms of θ_2. We already have $\sin\theta_3$ given by Eq. (3.25). So, we use the trigonometric identity,

$$\sin^2\theta_3 + \cos^2\theta_3 = 1$$

That gives,

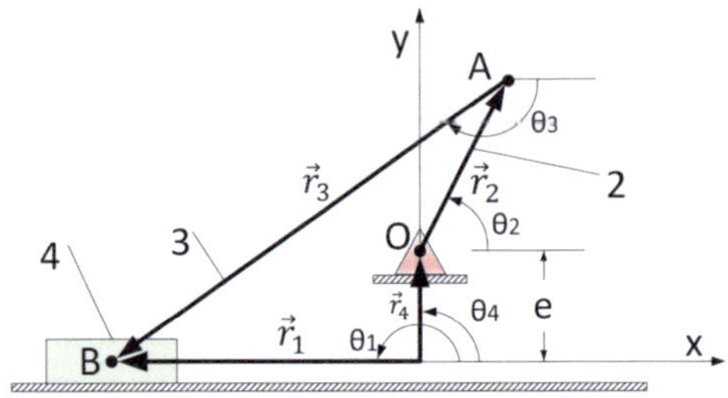

Fig. 3.10 Result of Eq. (3.28) using the negative sign. (Fig. 3.9 is the solution of Eq. (3.28) using the positive sign)

$$\cos\theta_3 = \pm\sqrt{1 - \sin^2\theta_3} \tag{3.26}$$

Substituting Eq. (3.25) into Eq. (3.26) gives,

$$\cos\theta_3 = \pm\sqrt{1 - \frac{1}{r_3^2}(e + r_2\sin\theta_2)^2} \tag{3.27}$$

Finally, substituting Eq. (3.27) into Eq. (3.22) and using Eq. (3.24) eliminate θ_3 and result in,

$$x_B = r_1\cos\theta_1 = r_2\cos\theta_2 \pm \sqrt{r_3^2 - (e + r_2\sin\theta_2)^2} \tag{3.28}$$

Equation (3.28) is a closed-form solution for the position of the slider in terms of the crank angle, θ_2. The appearance of the ± sign in this solution is a reminder that the projected vector-loop Eqs. (3.22) and (3.23) are nonlinear in terms of θ_3 and r_1, and therefore having multiple solutions is expected. This ± sign appears not only as a mathematical necessity, but also as a physical necessity. In fact, two slider-crank mechanisms with the same members and at the same input angle θ_2, can respond the way we saw in Fig. 3.9 [using the positive sign in Eq. (3.28)], or the way it is shown in Fig. 3.10 [using the negative sign in Eq. (3.28)]. The way the mechanism has been initially assembled determines which one of these possibilities will actually take place.

Example 3.2 Calculate the possible positions of a slider in a slider-crank mechanism at $\theta_2 = 30°$ if $e = 0.5$ in, $r_2 = 1$ in, and $r_3 = 2$ in.

Using the given data and the positive sign in Eq. (3.28) we get: $x_B = 2.5981$ in. If the negative sign is used, we get: $x_B = -0.8660$ in.

The answer to the question, "*Which one of these values will actually happen*?" completely depends on the initial assembly of the mechanism.

Example 3.3 Consider a slider-crank mechanism with $e = 0$ in, $r_2 = 1$ in, and $r_3 = 2$ in. Write a MATLAB code that uses Eq. (3.28) to calculate and plot the position of the slider, x_B versus the crank angle, θ_2 as a result of one complete revolution of the crank.

Here is the MATLAB code:

```
clear all
r2=1, r3=2, e=0
for i=1:1:360
x(i)=(i-1)*pi/180;
xb1(i)=r2*cos(x(i))+sqrt(r3^2-(e+r2*sin(x(i)))^2);
xb2(i)=r2*cos(x(i))-sqrt(r3^2-(e+r2*sin(x(i)))^2);
end
xdeg=x*180/pi;
figure
plot(xdeg,xb1,xdeg,xb2,'-.')
xlabel('\theta_2')
ylabel('x_B')
legend('+ sign','- sign')
grid on
```

The output of the code is shown in Fig. 3.11. We see that by using the positive sign in Eq. (3.28) the slider will oscillate between the +1 in and +3 in positions on the x axis. In contrast, if the negative sign in Eq. (3.28) is used, the slider will oscillate between the -3 in and -1 in positions on the x axis. All of these results have simple explanations that we will see in the next example.

Furthermore, the symmetry of the curves in Fig. 3.11 with respect to the 180° line is due to having no eccentricity in the mechanism ($e = 0$). As a result, the advance-to-return time ratio is 1 (i.e., the two times are equal). It is also seen that at any value of the crank angle, θ_2, there are two possible answers: one for the positive sign and the other for the negative sign used in Eq. (3.28).

Example 3.4 Referring to the data in Example 3.3 for a slider-crank mechanism with $e = 0$ in, $r_2 = 1$ in, and $r_3 = 2$ in, and using no equations, draw possible limit positions of the slider and compare them with the corresponding results obtained in Example 3.3.

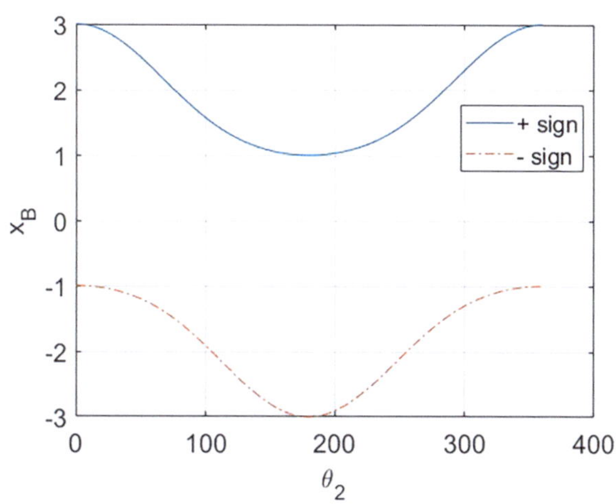

Fig. 3.11 MATLAB output for the possible positions of the slider in a slider-crank mechanism with: $e = 0$ in, $r_2 = 1$ in, and $r_3 = 2$ in, during one complete revolution of the crank

The First Mechanism

Using the first type of assembly at $\theta_2 = 0$, the mechanism posture is shown in Fig. 3.12.

The corresponding location of the slider is:

$$x_B = r_3 + r_2 = 2 + 1 = 3 \text{ in}$$

For the same mechanism at $\theta_2 = 180°$ the solution is shown in Fig. 3.13.

The corresponding location of the slider is:

$$x_B = r_3 - r_2 = 2 - 1 = 1 \text{ in}$$

These results are compatible with the limit positions obtained in Example 3.3 using the plus sign in Eq. (3.28).

The Second Mechanism

Using the second type of assembly at $\theta_2 = 0$, the mechanism posture is shown in Fig. 3.14.

The corresponding location of the slider is:

$$x_B = r_2 - r_3 = 1 - 2 = -1 \text{ in}$$

For the same mechanism at $\theta_2 = 180°$ the solution is shown in Fig. 3.15.

The corresponding location of the slider is:

$$x_B = -r_3 - r_2 = -2 - 1 = -3 \text{ in}$$

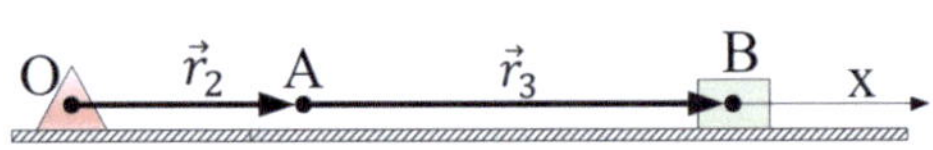

Fig. 3.12 Solution at $\theta_2 = 0$ for the first mechanism; top dead center

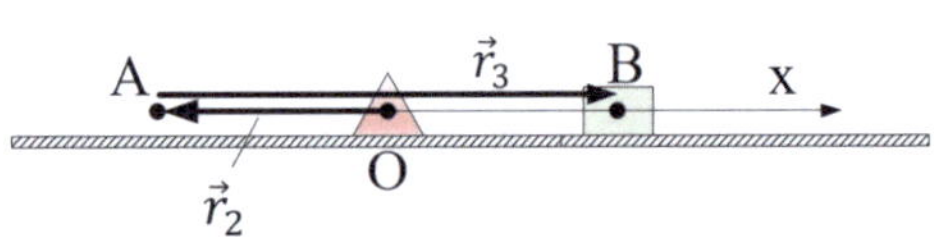

Fig. 3.13 Solution at $\theta_2 = 180°$ for the first mechanism; bottom dead center

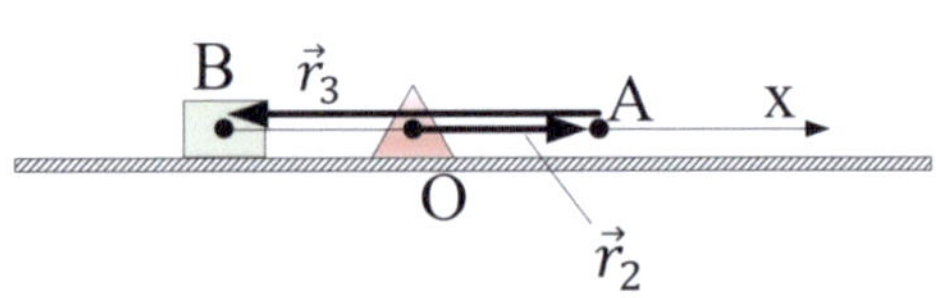

Fig. 3.14 Solution at $\theta_2 = 0$ for the second mechanism; bottom dead center

Fig. 3.15 Solution at $\theta_2 = 180°$ for the second mechanism; top dead center

These results are compatible with the limit positions obtained in Example 3.3 using the minus sign in Eq. (3.28).

Example 3.5
Consider a slider-crank mechanism with $e = 0.5$ in, $r_2 = 1$ in, and $r_3 = 2$ in studied in Example 3.2. Rerun the MATLAB code given in Example 3.3 that uses Eq. (3.28) to calculate and plot the position of the slider, x_B versus the crank angle, θ_2 as a result of one complete revolution of the crank.

The results of running the code using the new data are shown in Fig. 3.16. We see that unlike Fig. 3.11 the curves are not symmetrical with respect to the 180° line. The reason for this observation is because the eccentricity e is no longer zero. Also, in this case, the advance-to-return time ratio is no longer equal to 1 (i.e., the two time intervals are not equal). Finally, two data points that correspond to Example 3.2 are shown in Fig. 3.16. These data points are the same as those calculated in Example 3.2.

The Transmission Angle
The transmission angle γ of a slider-crank mechanism is shown in Fig. 3.17. The significance of transmission angle will be discussed more in Chap. 11; here an introduction is given. We know that the crank angle θ_2 is the input to the mechanism and the connecting rod 3 is a two-force member. Therefore, the force transmitted by the connecting rod 3 to the slider 4 acts along member 3. If γ is small, i.e., member 3 is close to being perpendicular to the direction of motion of the slider and there will be only a small amount of force that can drive the slider along its path. The larger the transmission angle is, the larger the magnitude of the transmitted force to the slider will be. That is why this angle is called the transmission angle! Therefore, understanding the variation of γ with changes of the crank angle θ_2 is important.

In order to derive an equation for calculating the transmission angle γ in terms of the crank angle, θ_2, we only need to project the geometry of the mechanism shown in Fig. 3.17 on the y axis which results in:

$$r_3 \cos\gamma = e + r_2 \sin\theta_2 \tag{3.29}$$

Now, a follow-up question is, "*At what crank angles do the maximum and the minimum values of γ occur*?" Differentiating Eq. (3.29) gives:

$$-r_3 \sin\gamma d\gamma = r_2 \cos\theta_2 d\theta_2$$

Or,

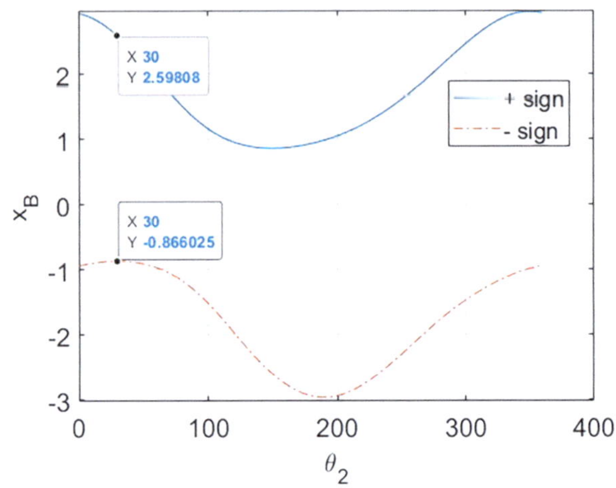

Fig. 3.16 MATLAB output for the possible positions of the slider in a slider-crank mechanism with: $e = 0.5$ in, $r_2 = 1$ in, and $r_3 = 2$ in during one complete revolution of the crank

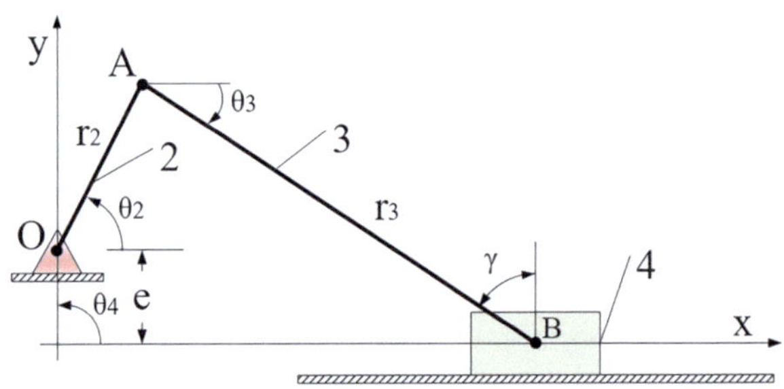

Fig. 3.17 Showing the transmission angle, γ

$$\frac{d\gamma}{d\theta_2} = \frac{-r_2 \cos\theta_2}{r_3 \sin\gamma} \tag{3.30}$$

The maximum and the minimum values of the transmission angle γ happen when this derivative is zero. Therefore γ has extreme values at $\theta_2 = 90°$ and $\theta_2 = 270°$, as shown in Figs. 3.18 and 3.19.

The minimum and the maximum values of the transmission angle γ can now be calculated by substituting $\theta_2 = 90°$ and $\theta_2 = 270°$ (respectively) into Eq. (3.29). Therefore we get:

$$r_3 \cos\gamma_{min} = e + r_2 \tag{3.31}$$

$$r_3 \cos\gamma_{max} = e - r_2 \tag{3.32}$$

Example 3.6

Referring to Fig. 3.17 calculate the maximum and the minimum values of the transmission angle γ in a slider-crank mechanism in which: $e = 2$ in, $r_2 = 4$ in, and $r_3 = 12$ in.

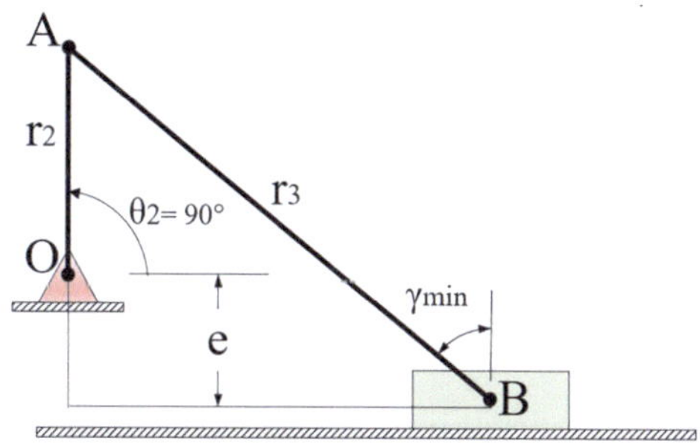

Fig. 3.18 The minimum transmission angle position at $\theta_2 = 90°$

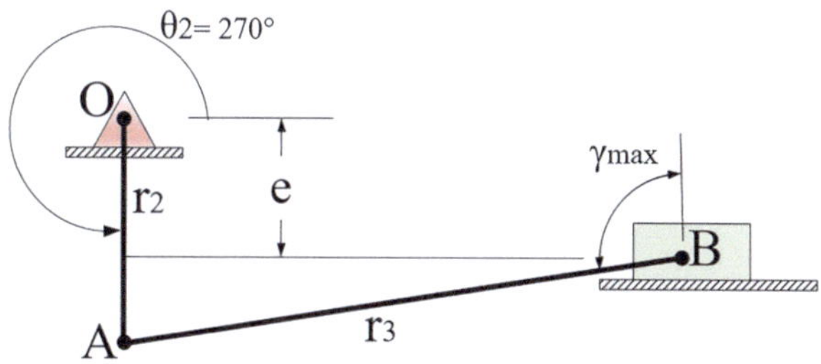

Fig. 3.19 The maximum transmission angle position at $\theta_2 = 270°$

Using the provided data in Eqs. (3.31) and (3.32) we get,

$$12 \cos \gamma_{min} = 2 + 4, \quad 12 \cos \gamma_{max} = 2 - 4$$

Therefore,

$$\gamma_{min} = 60\,°, \quad \gamma_{max} = 99.59\,°$$

3.4 Numerical Solutions for Position

In Sect. 3.3, a few closed-form solutions for the position problem in simple mechanisms were presented. For complex mechanisms, however, solving the nonlinear position problems requires application of numerical methods. In this section, a brief introduction to the Taylor series expansion for single-variable and multivariable functions is given. This introduction forms the necessary foundation for an iterative method that is used for solving complex nonlinear position problems (i.e., the Newton-Raphson algorithm).

3.4.1 Single-Variable Taylor Series Expansion

Consider a differentiable function $f(x)$ that its value is known at a certain point x_0. An approximate value for this function at a neighboring point x_1 is sought. The neighboring point is away from x_0 by an incremental amount, Δx so that we have:

$$x_1 = x_0 + \Delta x \tag{3.33}$$

The function value at the neighboring point x_1 can be written in terms of the increment Δx as well as the value of this function and its derivatives at the original point x_0 using the following Taylor series expansion of a function of one variable:

$$f(x_1) = f(x_0 + \Delta x) = f(x_0) + \Delta x f'(x_0) + (\Delta x)^2 \frac{f''(x_0)}{2!} + \ldots = \sum_{n=0}^{\infty} (\Delta x)^n \frac{f^{(n)}(x_0)}{n!} \tag{3.34}$$

One may rewrite Eq. (3.34) as follows:

$$f(x_1) = f(x_0 + \Delta x) = f(x_0) + \Delta x f'(x_0) + H.O.T.$$

In which *H.O.T.* refers to *higher order terms* (that include nonlinear powers of Δx). For small increments, the effect of higher order terms compared to the remaining linear ones is negligible. In such a case, Eq. (3.34) can be approximated as,

$$f(x_1) = f(x_0 + \Delta x) \approx f(x_0) + \Delta x f'(x_0) \tag{3.35}$$

Equation (3.35) is the linear Taylor series expansion of function $f(x)$. Figure 3.20 provides a graphical interpretation of this approximation. Starting from point $A(x_0,$

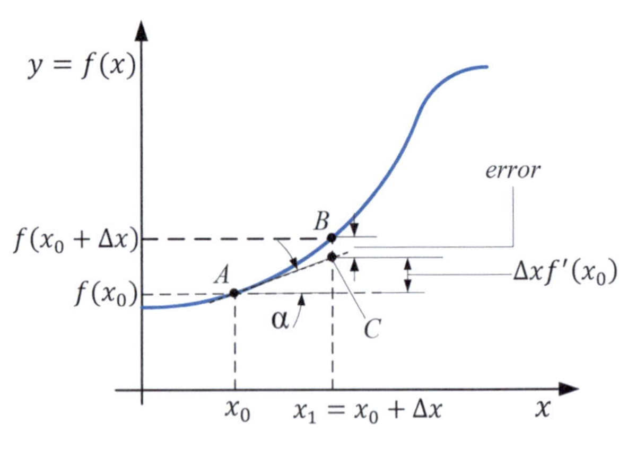

Fig. 3.20 Geometrical interpretation of the linear Taylor series approximation, Eq. (3.35)

$f(x_0)$), we are looking for an estimation of $f(x_1) = f(x_0 + \Delta x)$. Point B on the $f(x)$ curve corresponds to the exact value of $f(x_0 + \Delta x)$. The graphical interpretation of Eq. (3.35), that is a linear estimation of $f(x_0 + \Delta x)$, is obtained by moving along the line drawn tangent to the $f(x)$ curve at point A by a Δx amount and achieving at point C. The coordinates of point C correspond to the approximation given by Eq. (3.35). Notice that $f'(x_0) = \tan\alpha$ is the slope of the tangent line drawn at x_0. The resulting linearization error, BC, can be seen in the figure. Figure 3.20 also reveals that large increment Δx choices result in large estimation errors. So, for obtaining more accurate results using Taylor series, small increment values should be used in Eq. (3.35).

Example 3.7

Using Eq. (3.35) obtain a linear Taylor series approximation for $f(x) = \sin(x)$, and then calculate first-order estimations for sin(31°) and sin(35°).

Using $f(x) = \sin(x)$ in Eq. (3.35) we get:

$$\sin(x_0 + \Delta x) \approx \sin(x_0) + \Delta x \cos(x_0) \tag{3.36}$$

where incremental angle Δx is measured in radians.

In order to estimate sin(31°), we substitute $x_0 = 30° = \pi/6$ rad, and $\Delta x = 1° = 1 \times \frac{\pi}{180}$ rad in Eq. (3.36) to get:

$$\sin(31°) \approx \sin(30°) + \frac{\pi}{180}\cos(30°) = 0.5 + \frac{\pi}{180} \times \frac{\sqrt{3}}{2} = 0.51511$$

This result compares well with what could be obtained by a calculator, i.e., sin (31°) = 0.51504. The error percentage is,

$$\frac{0.51511 - 0.51504}{0.51504} \times 100 = 0.019\%$$

For estimating sin(35°), one can again use $x_0 = 30° = \pi/6$ rad, but now $\Delta x = 5° = 5 \times \frac{\pi}{180}$ rad. Since Δx is now larger, one can anticipate a larger estimation error as well. Using Eq. (3.36) with this new data we get:

$$\sin(35°) \approx \sin(30°) + 5 \times \frac{\pi}{180}\cos(30°) = 0.5 + 5 \times \frac{\pi}{180} \times \frac{\sqrt{3}}{2} = 0.5756$$

This result has 0.3349% error when compared with the exact value of sin (35°) = 0.5736. Therefore, the error percentage in calculating sin(35°) is about 18 times that of sin(31°). So it is confirmed that to get relatively accurate estimations for function values using a linearized Taylor series, the increment size Δx should be small.

3.4.2 Newton-Raphson Method for Solving Single-Variable Equations

Position analysis problems in mechanisms usually require solution of nonlinear equations. In the slider-crank problem finding, the location of the slider x_B at any crank angle θ_2 is easily done and just requires calculating x_B in Eq. (3.28), repeated below:

$$x_B = r_1 \cos\theta_1 = r_2 \cos\theta_2 \pm \sqrt{r_3^2 - (e + r_2 \sin\theta_2)^2} \tag{3.28}$$

The inverse question, i.e., finding the crank angle θ_2 that can result in a given position of the slider x_B requires solution of the following nonlinear equation for the single variable, θ_2.

$$r_2 \cos\theta_2 \pm \sqrt{r_3^2 - (e + r_2 \sin\theta_2)^2} - x_B = 0 \tag{3.37}$$

In this section we will learn how to use the Newton-Raphson method for solving equations like Eq. (3.37). This is a powerful algorithm for numerical solution of nonlinear equations. The author has used this method on various topics stretching from the limit analysis and plastic design of structures to the dynamics of composite rotating beams, see Ghorashi [4, 5], respectively. In this book we use the Newton-Raphson method for position analysis of mechanisms.

Consider a nonlinear equation for one variable, x, as follows:

$$f(x) = 0 \tag{3.38}$$

The aim is to find estimations for x that satisfy Eq. (3.38). One can start by guessing an answer. Let us show this initial guess by x_0. Considering the function,

$$y = f(x) \tag{3.39}$$

If $f(x_0)$ is zero, the problem is solved and x_0 is the answer. Otherwise, we need to improve the initial guess and change it to $x_1$instead, that is related to x_0 by:

$$x_1 = x_0 + \Delta x \tag{3.40}$$

Now we should find the best improvement on x_0, i.e., the best increment Δx. Using the linearized Taylor series expansion of $f(x)$ in the vicinity of the initial guess x_0 shown by Eq. (3.35), one can write the following approximation at the neighboring point, x_1,

$$f(x_1) = f(x_0 + \Delta x) \approx f(x_0) + \Delta x f'(x_0) \tag{3.35}$$

In order to solve Eq. (3.38) and by using Eq. (3.35), the best increment Δx is the one that can result in:

$$f(x_0) + \Delta x f'(x_0) = 0 \tag{3.41}$$

Since our aim is to satisfy Eq. (3.38), we can think of $f(x_0)$ as a residue or solution estimation error that we like it to be zero. If it was zero, Eq. (3.41) would have resulted in zero value for Δx since no corrections would have been needed! Otherwise, solving for increment Δx gives,

$$\Delta x = -\frac{f(x_0)}{f'(x_0)} \tag{3.42}$$

By combining Eqs. (3.40) and (3.42), we get the following improved solution after the first iteration:

$$x_1 = x_0 + \Delta x = x_0 - \frac{f(x_0)}{f'(x_0)} \tag{3.43}$$

One can now calculate $f(x_1)$. If $f(x_1)$ is zero, the problem has been solved and x_1 is the answer. Otherwise, $f(x_1)$ is a measure of the solution estimation error and in the second iteration the x_1 approximation for the solution is improved even further by using Eq. (3.43) again, this time replacing x_0 by the improved guess x_1,

$$x_2 = x_1 - \frac{f(x_1)}{f'(x_1)} \tag{3.44}$$

At the i-th iteration one can write a generalized version of Eq. (3.44) as follows:

$$x_i = x_{i-1} - \frac{f(x_{i-1})}{f'(x_{i-1})} \tag{3.45}$$

The procedure will be repeated till $f(x_i)$ is found to be very close to zero, or equivalently, the amount of improvement in the obtained solution estimation is smaller than a predefined small quantity, ε. That is:

$$\left| \frac{x_i - x_{i-1}}{x_{i-1}} \right| < \varepsilon \tag{3.46}$$

In such a case, the solution has converged and x_i would be an acceptable solution. If at any iteration it happens that,

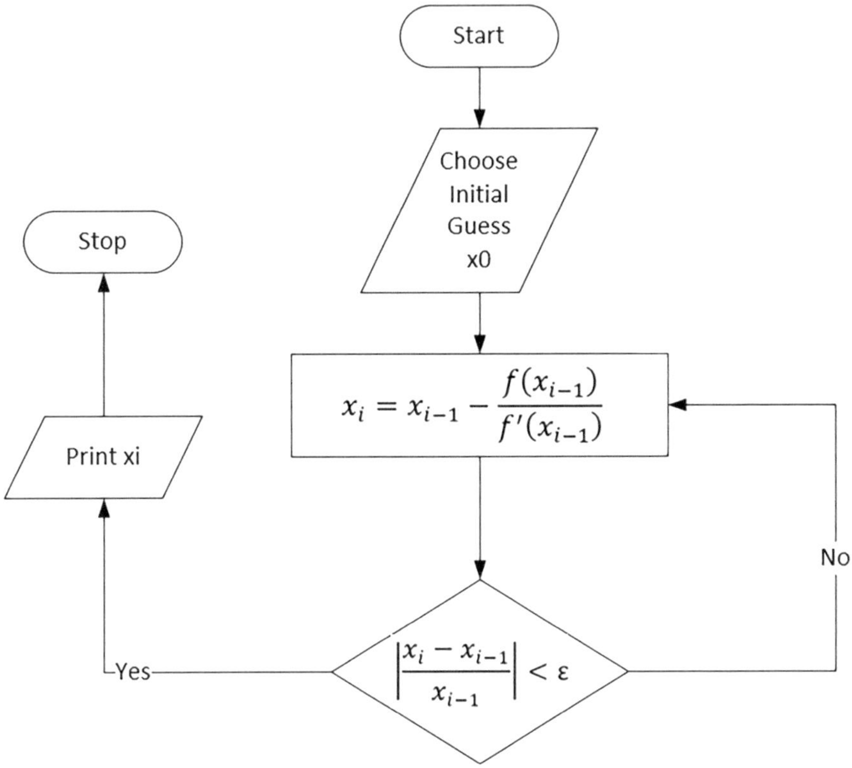

Fig. 3.21 Flow chart of the Newton-Raphson algorithm for solving Eq. (3.38)

$$f(x_i) = 0 \tag{3.47}$$

the next iteration obtained by reusing Eq. (3.45), i.e.,

$$x_{i+1} = x_i - \frac{f(x_i)}{f'(x_i)}$$

would not result in any further improvement, as we get $x_{i+1} = x_i$. Therefore x_i would be the solution of Eq. (3.38) [and it should, since Eq. (3.47) just means that!]. The mentioned iterative procedure that starts with an initial guess and improves this guess incrementally is called the Newton-Raphson algorithm. The flow chart of this algorithm is shown in Fig. 3.21.

A pictorial interpretation of the Newton-Raphson algorithm is shown in Fig. 3.22. Considering function $y = f(x)$, the aim is to find the root of $f(x) = 0$, that is where $f(x)$ intersects the x axis (point x^* in Fig. 3.22). Starting by choosing an initial guess x_0, one can find the corresponding function value $f(x_0)$ and point A in the figure. Then, applying Eq. (3.43) is equivalent to drawing a tangent line on the $f(x)$ curve at point

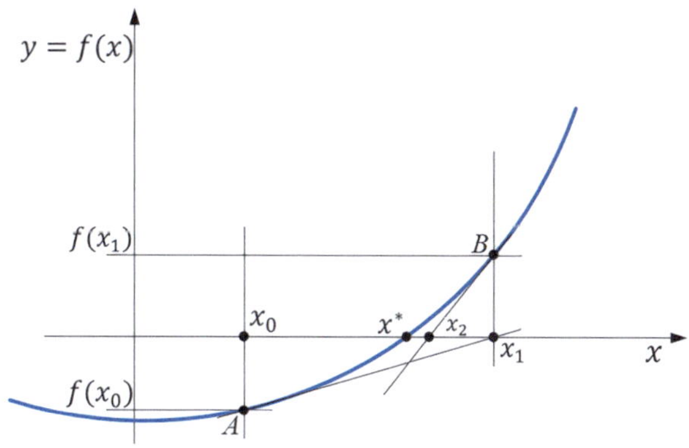

Fig. 3.22 Graphical interpretation of the Newton-Raphson algorithm

A and finding the intersection point of this line with the x axis. After finding x_1, one can calculate $f(x_1)$ and locate point B. The procedure of calculating the next iteration, x_2, using Eq. (3.44) is equivalent to drawing a tangent line on the $f(x)$ curve at point B and then finding the intersection point of this line with the x axis. This intersection point, i.e., x_2 is very close to the exact solution x^*. Of course, one can continue such iterations even further, if higher accuracy is needed.

Example 3.8
Calculate the real root of the following equation using the Newton-Raphson method.

$$x^3 - 1 = 0 \tag{3.48}$$

The obvious answer is $x^* = 1$. Let us see if the Newton-Raphson method can obtain it. In this case,

$$f(x) = x^3 - 1 \tag{3.49}$$

Substituting this function into Eq. (3.45) results in:

$$x_i = x_{i-1} - \frac{x_{i-1}^3 - 1}{3x_{i-1}^2} \tag{3.50}$$

For the first iteration we choose $x_0 = 0.5$ (the initial guess). Substituting $x_0 = 0.5$ for x_{i-1} into Eq. (3.50) results in: $x_1 = 1.6667$. Repeating the calculation, this time using $x_1 = 1.6667$ for x_{i-1} in Eq. (3.50) we get: $x_2 = 1.2311$. Table 3.1 includes the results obtained by the foregoing iterative calculations. It can be seen that after an initial overshoot to 1.6667, the Newton-Raphson algorithm has been able to converge the solution estimation to the correct answer in a total of four iterations.

Table 3.1 Results of using the Newton-Raphson method for solving Eq. (3.48)

Iteration number i	x_i	$f(x_i)$
0	0.5	−0.875
1	1.6667	3.6299
2	1.2311	0.8659
3	1.0407	0.1271
4	1.0019	0.00571

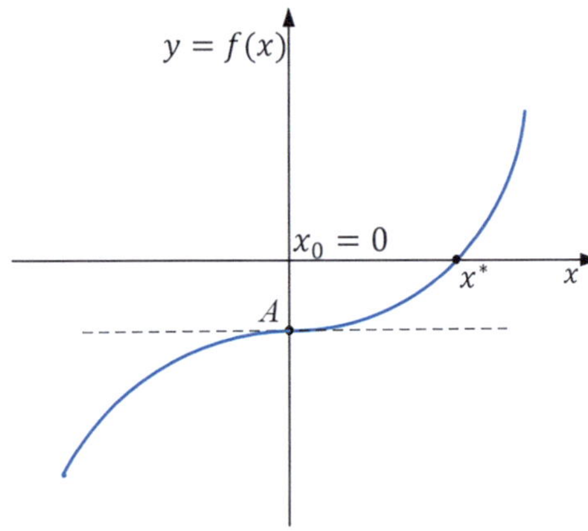

Fig. 3.23 Failure of the Newton-Raphson method due to a bad choice of initial guess in Example 3.8

Does the Newton-Raphson Method Always Work?

The method is not guaranteed to converge all the time and its convergence is heavily dependent on the proper choice of the initial guess. If in Example 3.8 for solving Eq. (3.48), $x_0 = 0$ was chosen as the initial guess, then in the first iteration (and since $f'(0) = 0$) Eq. (3.50) would have a division by zero error and the algorithm would fail.

Figure 3.23 provides a pictorial explanation for this failure. The initial guess $x_0 = 0$ corresponds to the inflection point A, of Eq. (3.49) (*i. e.* , $f''(x_0) = 0$). One can see that a line tangent to $f(x)$ at point A is horizontal and it never intersects the x axis. Therefore, no values for the improved solution x_1 is obtained and the algorithm cannot continue the way it did in Fig. 3.22.

MATLAB Implementation The following MATLAB script is based on the Newton-Raphson algorithm for solving Example 3.8 and uses $x_0 = 0.5$.

```
clear all
xinitial=0.5
x=xinitial;
for i=1:10
count(i)=i;
xx(i)=x;
f(i)=x^3-1
fp(i)-3*x^2
x=x-f(i)./fp(i)
end
figure
plot(count,f,'--o')
xlabel('iteration')
ylabel('error')
grid on
figure
plot(count,xx,'--o')
xlabel('iteration number')
ylabel('x')
grid on
```

Figure 3.24 illustrates the output of the MATLAB script (x_i values) for Example 3.8 in each iteration. This figure is compatible with Table 3.1 and shows convergence of the results to the exact solution, $x^* = 1$.

Figure 3.25 shows the solution estimation error (i.e., the $f(x_i)$ values) obtained by the MATLAB script in each iteration for Example 3.8. We notice the rapid decrease of the calculated error and its convergence to zero after an initial overshoot.

Alternatively, one can use the roots([1 0 0 −1]) command in MATLAB to get all of the roots of this equation, including the complex roots, as follows:

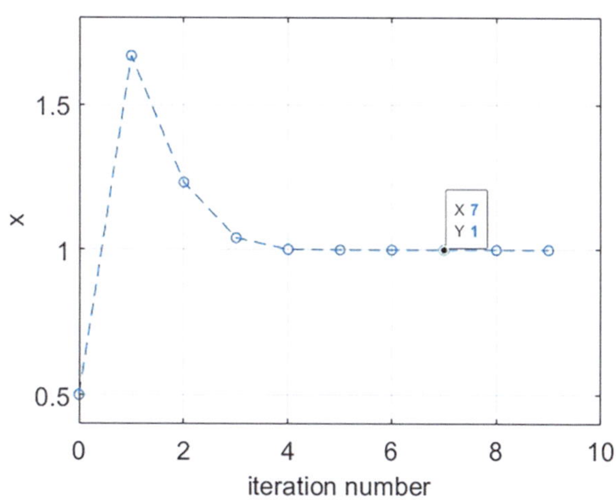

Fig. 3.24 Convergence of solution estimates to the exact solution $x^* = 1$ (obtained by MATLAB; for Example 3.8)

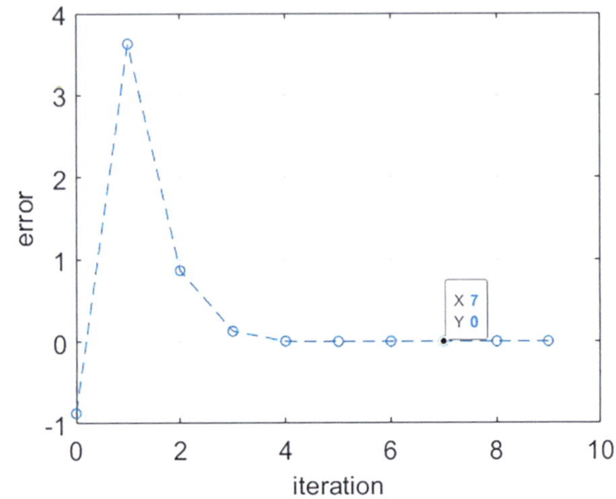

Fig. 3.25 Convergence of solution error to zero (obtained by MATLAB; for Example 3.8)

–0.5000 + 0.8660i; –0.5000–0.8660i; 1.0000 + 0.0000i

where $i = \sqrt{-1}$.

3.4.3 Newton-Raphson Method for Solving the Slider-Crank Problem

We now turn our attention back to the question that was posed in the beginning of Sect. 3.4.2. In the slider-crank mechanism shown in Fig. 3.9, we would like to solve the inverse problem of finding the crank angle θ_2 if the position of the slider x_B is known. Therefore, we need to solve the following nonlinear equation for the single variable, θ_2.

$$r_2 \cos\theta_2 \pm \sqrt{r_3^2 - (e + r_2 \sin\theta_2)^2} - x_B = 0 \tag{3.37}$$

In order to solve this problem, we can apply what we learned about using the Newton-Raphson method for problems with one unknown. In what follows, only the positive sign in Eq. (3.37) is considered. The method can also be used when the sign is negative.

To simplify the notation we choose, $x = \theta_2$. Therefore, Eq. (3.37) can be rewritten as follows:

$$r_2 \cos x + \sqrt{r_3^2 - (e + r_2 \sin x)^2} - x_B = 0 \tag{3.51}$$

Now, recall the Newton-Raphson iterative process:

$$x_i = x_{i-1} - \frac{f(x_{i-1})}{f'(x_{i-1})} \tag{3.45}$$

Considering Eq. (3.51) we can define,

$$f(x) = r_2 \cos x + \sqrt{r_3^2 - (e + r_2 \sin x)^2} - x_B \tag{3.52}$$

Substituting Eq. (3.52) into Eq. (3.45) results in,

$$x_i = x_{i-1} - \frac{r_2 \cos(x_{i-1}) + \sqrt{r_3^2 - [e + r_2 \sin(x_{i-1})]^2} - x_B}{-r_2 \sin(x_{i-1}) - \frac{[e + r_2 \sin(x_{i-1})] r_2 \cos(x_{i-1})}{\sqrt{r_3^2 - [e + r_2 \sin(x_{i-1})]^2}}} \tag{3.53}$$

Example 3.9
In Example 3.5 it was shown that a slider-crank mechanism with $e = 0.5$ in, $r_2 = 1$ in, and $r_3 = 2$ in at crank angle $\theta_2 = 30°$ will have a slider position of: $x_B = 2.5981$ in if we use the positive sign in Eq. (3.28). Solve the inverse problem and use the Newton-Raphson algorithm to verify that at $x_B = 2.5981$ in, the corresponding crank angle is, in fact $x = \theta_2 = 30°$.

The iterative Eq. (3.53) was derived using the positive sign in Eq. (3.37) and it is applicable here. The following MATLAB script uses Eq. (3.53) and x_0=10° as the initial guess and iteratively calculates the crank angle that is compatible with $x_B = 2.5981$ in:

```
clear all
r2=1, r3=2, e=0.5
xb=2.5981;
xinitial=10;
x=xinitial*pi/180;
for i=1:10
count(i)=i;
xx(i)=x;
f(i)=r2*cos(x)+sqrt(r3^2-(e+r2*sin(x))^2)-xb;
fp(i)=-r2*sin(x)-((e+r2*sin(x))*r2*cos(x))./(sqrt(r3^2-(e+r2*sin
(x))^2));
x=x-f(i)./fp(i);
end
```

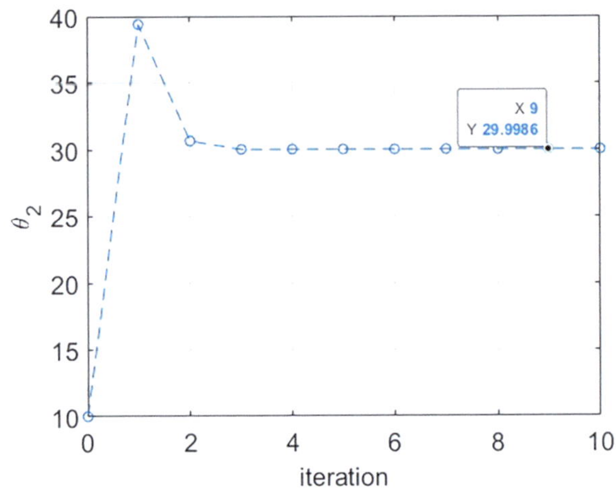

Fig. 3.26 Convergence of the estimated crank angle to $x^* = \theta_2 - 30°$, if initial guess is 10°

```
xdegrees=xx*(180/pi);
figure
plot(count,xdegrees,'--o')
xlabel('iteration')
ylabel('\theta_2')
grid on
```

The corresponding output of this script is shown in Fig. 3.26 and it can be seen that the solution quickly converges to the expected $x = \theta_2 = 30°$ answer.

If we keep everything unchanged except for the initial guess of the crank angle and choose it to be equal to $-20°$ instead, we will get Fig. 3.27 that shows convergence to $-51.7852°$, instead of 30°. But, this answer is not wrong!

The $-51.7852°$ solution is equivalent to 308.2148° that is the other possible crank angle at the slider position of $x_B = 2.5981$, as can be seen in Fig. 3.28. So, changing the initial guess can alter the solution to which the Newton-Raphson algorithm converges.

3.4.4 Multivariable Taylor Series Expansion

We noticed that the inverse slider-crank problem shown by Eq. (3.37) has only one unknown, i.e., the crank angle, θ_2. And therefore, we could use the single variable form of the Newton-Raphson method, i.e., Eq. (3.45), for solving the problem. In the case of a four-bar linkage shown in Fig. 3.1 and for the vector loop shown in Fig. 3.29 the projected vector-loop equations are given as follows:

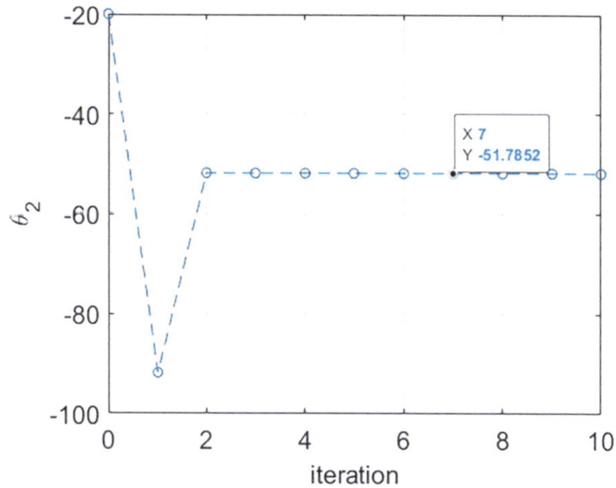

Fig. 3.27 Convergence of the estimated crank angle to $x^* = \theta_2 = -51.7852°$, if initial guess is $-20°$

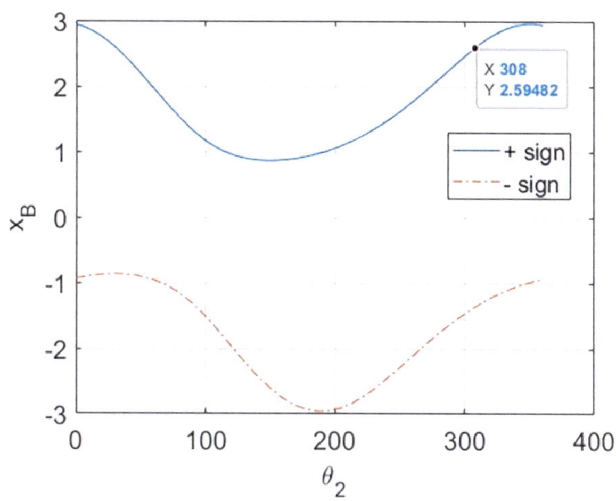

Fig. 3.28 MATLAB output for the possible positions of the slider during one complete revolution of the crank. (The same as Fig. 3.16, with a different data point highlighted)

$$\begin{cases} r_3 \cos \theta_3 + r_4 \cos \theta_4 = r_1 - r_2 \cos \theta_2 \\ r_3 \sin \theta_3 + r_4 \sin \theta_4 = -r_2 \sin \theta_2 \end{cases} \tag{3.54}$$

Therefore, for any input angle, θ_2, we need to solve a system of two nonlinear equations for two unknown angles θ_3 and θ_4. A trigonometric solution for this problem was given in Sect. 3.3.1. In what follows, we will see how the Newton-Raphson method can be expanded for solving a multivariable problem. This solution requires use of the Taylor series for multivariable functions as well as matrix algebra.

Consider a differentiable function $f(x, y)$ that its value is known at a certain point (x_0, y_0). The value of this function at a neighboring point is sought. The neighboring point (x_1, y_1) is away from (x_0, y_0) by incremental amounts, Δx and Δy:

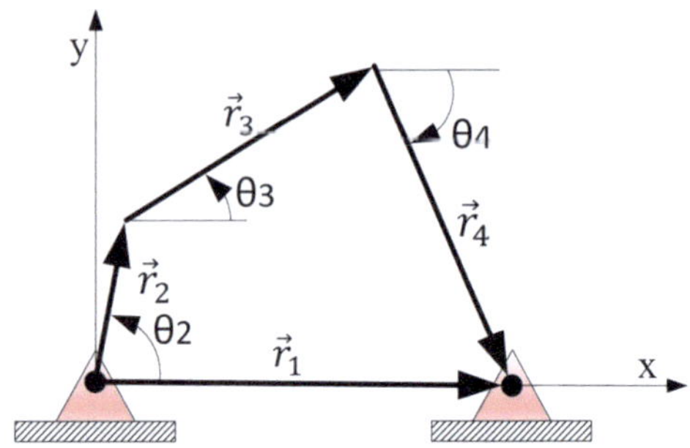

Fig. 3.29 A vector loop for the four-bar linkage shown in Fig. 3.1

$$(x_1, y_1) = (x_0, y_0) + (\Delta x, \Delta y) = (x_0 + \Delta x, y_0 + \Delta y) \tag{3.55}$$

With similar arguments as were given in the one-variable case and for small increments, Δx and Δy, using the linearized Taylor series approximation, $f(x_1, y_1)$ can be related to $f(x_0, y_0)$ as follows:

$$f(x_1, y_1) = f(x_0 + \Delta x, y_0 + \Delta y) \approx f(x_0, y_0) + \Delta x \frac{\partial f}{\partial x}(x_0, y_0) + \Delta \mathrm{y} \frac{\partial f}{\partial y}(x_0, y_0) \tag{3.56}$$

Example 3.10

Considering $f(x, y) = x \sin y$, find a first-order estimation for $f\left(1.1, \frac{\pi}{6} + 0.01\right)$ using Taylor series expansion around point $\left(1, \frac{\pi}{6}\right)$.

In this case, $x_0 = 1$, $y_0 = \frac{\pi}{6}$, $\Delta x = 0.1$, and $\Delta \mathrm{y} = 0.01$. Therefore, using Eq. (3.56) we get:

$$f\left(1.1, \frac{\pi}{6} + 0.01\right) \approx f\left(1, \frac{\pi}{6}\right) + 0.1 \frac{\partial f}{\partial x}\left(1, \frac{\pi}{6}\right) + 0.01 \frac{\partial f}{\partial y}\left(1, \frac{\pi}{6}\right)$$

or,

$$f\left(1.1, \frac{\pi}{6} + 0.01\right) \approx 1 \times \sin 30^\circ + 0.1 \sin 30^\circ + 0.01 \times 1 \cos 30^\circ = 0.5587$$

This estimation can be compared with the exact value of the function at the neighboring point, i.e.,

$$f\left(1.1, \frac{\pi}{6} + 0.01\right) = 1.1 \times \sin\left(\frac{\pi}{6} + 0.01\right) = 0.5502$$

This result is only 0.14% different from the estimated value.

3.4.5 Newton-Raphson Method for Solving a System of Equations

Now, we use the two-variable Taylor series expansion given in Eq. (3.56) in order to solve the following system of two nonlinear equations for two unknowns:

$$\begin{cases} f(x,y)=0 \\ g(x,y)=0 \end{cases} \tag{3.57}$$

We are looking for estimations for variables x and y that satisfy Eqs. (3.57). Let us start by guessing an answer, i.e., the initial guess, as $\vec{v}_0=(x_0,y_0)$. Considering functions $f(x, y)$ and $g(x, y)$ one may calculate $f(x_0, y_0)$ and $g(x_0, y_0)$. If these values are zero, the problem is solved and (x_0, y_0) is the answer. Otherwise, we need to improve our guess and choose (x_1, y_1) instead, that is related to (x_0, y_0) by:

$$\vec{v}_1=(x_1,y_1)=(x_0,y_0)+(\Delta x,\Delta y)=(x_0+\Delta x,y_0+\Delta y) \tag{3.58}$$

We want to find the best improvement on (x_0, y_0), i.e., the best increment $(\Delta x, \Delta y)$. Using Taylor series expansion (3.56) for $f(x, y)$ and $g(x, y)$ in the vicinity of the initial guess (x_0, y_0) we have the following approximations:

$$\begin{cases} f(x_1,y_1)=f(x_0+\Delta x,y_0+\Delta y)\approx f(x_0,y_0)+\Delta x\dfrac{\partial f}{\partial x}(x_0,y_0)+\Delta y\dfrac{\partial f}{\partial y}(x_0,y_0) \\ g(x_1,y_1)=g(x_0+\Delta x,y_0+\Delta y)\approx g(x_0,y_0)+\Delta x\dfrac{\partial g}{\partial x}(x_0,y_0)+\Delta y\dfrac{\partial g}{\partial y}(x_0,y_0) \end{cases} \tag{3.59}$$

In order to solve Eqs. (3.57) and considering Eqs. (3.59), the best increment $(\Delta x, \Delta y)$ is the one that can result in:

$$\begin{cases} f(x_0,y_0)+\Delta x\dfrac{\partial f}{\partial x}(x_0,y_0)+\Delta y\dfrac{\partial f}{\partial y}(x_0,y_0)=0 \\ g(x_0,y_0)+\Delta x\dfrac{\partial g}{\partial x}(x_0,y_0)+\Delta y\dfrac{\partial g}{\partial y}(x_0,y_0)=0 \end{cases} \tag{3.60}$$

Therefore, we now have two linear equations for two unknowns, i.e., Δx and Δy. At this step all other terms are known and are functions that have been calculated at the initial guess point (x_0, y_0). Moving the known terms $f(x_0, y_0)$ and $g(x_0, y_0)$ to the right-hand side of Eqs. (3.60) gives:

$$\begin{cases} \Delta x\dfrac{\partial f}{\partial x}(x_0,y_0)+\Delta y\dfrac{\partial f}{\partial y}(x_0,y_0)=-f(x_0,y_0) \\ \Delta x\dfrac{\partial g}{\partial x}(x_0,y_0)+\Delta y\dfrac{\partial g}{\partial y}(x_0,y_0)=-g(x_0,y_0) \end{cases} \tag{3.61}$$

Since our aim is to satisfy Eqs. (3.57), we can think of $f(x_0, y_0)$ and $g(x_0, y_0)$ as residues or solution estimation errors that we like them to be zero. If they were zero, Eqs. (3.61) would result in zero values for Δx and Δy, since no corrections would be needed!

In general, the mentioned residues are not zeros and Eqs. (3.61) should be solved for Δx and Δy. In order to do so, we first rewrite Eqs. (3.61) in matrix form. The matrix form can be mathematically manipulated much easier and in the general case of more than two variables it would be more convenient to use. The matrix form of Eqs. (3.61) is as follows:

$$\begin{bmatrix} \frac{\partial f}{\partial x} & \frac{\partial f}{\partial y} \\ \frac{\partial g}{\partial x} & \frac{\partial g}{\partial y} \end{bmatrix}_{(x_0,y_0)} \times \begin{Bmatrix} \Delta x \\ \Delta y \end{Bmatrix} = -\begin{Bmatrix} f(x_0, y_0) \\ g(x_0, y_0) \end{Bmatrix} \tag{3.62}$$

The two by two matrix in Eq. (3.62) is called the Jacobian matrix, J, and in Eq. (3.62), it is calculated at the initial guess point, (x_0, y_0). That is why this matrix can be shown by $J(x_0, y_0)$, i.e.,

$$J(x,y) = \begin{bmatrix} \frac{\partial f}{\partial x} & \frac{\partial f}{\partial y} \\ \frac{\partial g}{\partial x} & \frac{\partial g}{\partial y} \end{bmatrix}; J(x_0, y_0) = \begin{bmatrix} \frac{\partial f}{\partial x} & \frac{\partial f}{\partial y} \\ \frac{\partial g}{\partial x} & \frac{\partial g}{\partial y} \end{bmatrix}_{(x_0,y_0)} \tag{3.63}$$

In the Jacobian matrix, the first row refers to the first equation (i.e., for f) and the second row refers to the second equation (i.e., for g). Furthermore, the first column includes derivatives with respect to the first variable (i.e., x) and the second column includes derivatives with respect to the second variable (i.e., y). All derivatives are calculated at the initial guess point $\vec{v}_0 = (x_0, y_0)$.

Solving Eq. (3.62) for the proper unknown increments requires left multiplication of that equation by the inverse of the Jacobian matrix that results in:

$$\begin{Bmatrix} \Delta x \\ \Delta y \end{Bmatrix} = -\begin{bmatrix} \frac{\partial f}{\partial x} & \frac{\partial f}{\partial y} \\ \frac{\partial g}{\partial x} & \frac{\partial g}{\partial y} \end{bmatrix}^{-1}_{(x_0,y_0)} \times \begin{Bmatrix} f(x_0, y_0) \\ g(x_0, y_0) \end{Bmatrix} = -J^{-1}_{(x_0,y_0)} \times \begin{Bmatrix} f(x_0, y_0) \\ g(x_0, y_0) \end{Bmatrix} \tag{3.64}$$

Equation (3.64) is the two-variable equivalent of Eq. (3.42) (for the single-variable case) that is repeated below:

$$\Delta x = -\frac{f(x_0)}{f'(x_0)} = -[f'(x_0)]^{-1} f(x_0) \tag{3.42}$$

So in Eq. (3.64) for the two-variable case, the Jacobian matrix replaces the simple derivative that was used in the single-variable case. Therefore, the updated answer estimation after the first iteration can be written as follows:

$$\vec{v}_1 = (x_1, y_1) = \begin{Bmatrix} x_1 \\ y_1 \end{Bmatrix} = \begin{Bmatrix} x_0 \\ y_0 \end{Bmatrix} + \begin{Bmatrix} \Delta x \\ \Delta y \end{Bmatrix} = \begin{Bmatrix} x_0 \\ y_0 \end{Bmatrix} - J^{-1}_{(x_0, y_0)} \times \begin{Bmatrix} f(x_0, y_0) \\ g(x_0, y_0) \end{Bmatrix} \tag{3.65}$$

One can now calculate function values (aka the vector of residues) at the next iteration:

$$\begin{Bmatrix} f(x_1, y_1) \\ g(x_1, y_1) \end{Bmatrix} \tag{3.66}$$

If this residue vector is zero, the problem has been solved and (x_1, y_1) is the answer. Otherwise, we can improve the answer estimation (x_1, y_1) even further by using it in Eq. (3.65) to replace (x_0, y_0) in the second iteration:

$$\vec{v}_2 = (x_2, y_2) = \begin{Bmatrix} x_2 \\ y_2 \end{Bmatrix} = \begin{Bmatrix} x_1 \\ y_1 \end{Bmatrix} - J^{-1}_{(x_1, y_1)} \times \begin{Bmatrix} f(x_1, y_1) \\ g(x_1, y_1) \end{Bmatrix} \tag{3.67}$$

At the i-th iteration, we can write a generalized version of Eqs. (3.65) and (3.67) as follows:

$$\vec{v}_i = (x_i, y_i) = \begin{Bmatrix} x_i \\ y_i \end{Bmatrix} = \begin{Bmatrix} x_{i-1} \\ y_{i-1} \end{Bmatrix} - J^{-1}_{(x_{i-1}, y_{i-1})} \times \begin{Bmatrix} f(x_{i-1}, y_{i-1}) \\ g(x_{i-1}, y_{i-1}) \end{Bmatrix} \tag{3.68}$$

Equivalently, using vectors, we have,

$$\vec{v}_i = \vec{v}_{i-1} - J^{-1}_{\left(\vec{v}_{i-1}\right)} \times \begin{Bmatrix} f\left(\vec{v}_{i-1}\right) \\ g\left(\vec{v}_{i-1}\right) \end{Bmatrix} \tag{3.69}$$

The procedure continues till solution convergence is detected, i.e., elements in the following residue vector are very small, i.e.,

$$\begin{Bmatrix} f\left(\vec{v}_i\right) \\ g\left(\vec{v}_i\right) \end{Bmatrix} \approx 0 \tag{3.70}$$

Obviously, if Eq. (3.70) is satisfied, $\vec{v}_i = (x_i, y_i)$ would be a good approximate answer for Eqs. (3.57). More formally, we can use a solution convergence criterion that is based on the relative size (norm) of improvement in the obtained answer estimation being smaller than a predefined small positive quantity, ε. In the two-variable case that means:

$$\frac{\sqrt{(x_i - x_{i-1})^2 + (y_i - y_{i-1})^2}}{\sqrt{x_{i-1}^2 + y_{i-1}^2}} < \varepsilon \tag{3.71}$$

Recalling Eq. (3.68) the convergence criterion can also be stated as follows, using norms of vectors:

$$\frac{\left|\vec{v}_i - \vec{v}_{i-1}\right|}{\left|\vec{v}_{i-1}\right|} < \varepsilon \tag{3.72}$$

This is the two-variable version of the Newton-Raphson algorithm that solves Eqs. (3.57) using an iterative procedure. It starts with an initial guess and improves this guess incrementally till convergence is reached. A flow chart for this algorithm, that is an expanded version of the flow chart given in Fig. 3.21 for the single-variable case, is given in Fig. 3.30.

If there are more than two (in general, n) equations to be solved for n variables, the additional equations will be just added rows and the additional variables will be just added columns to Eq. (3.68):

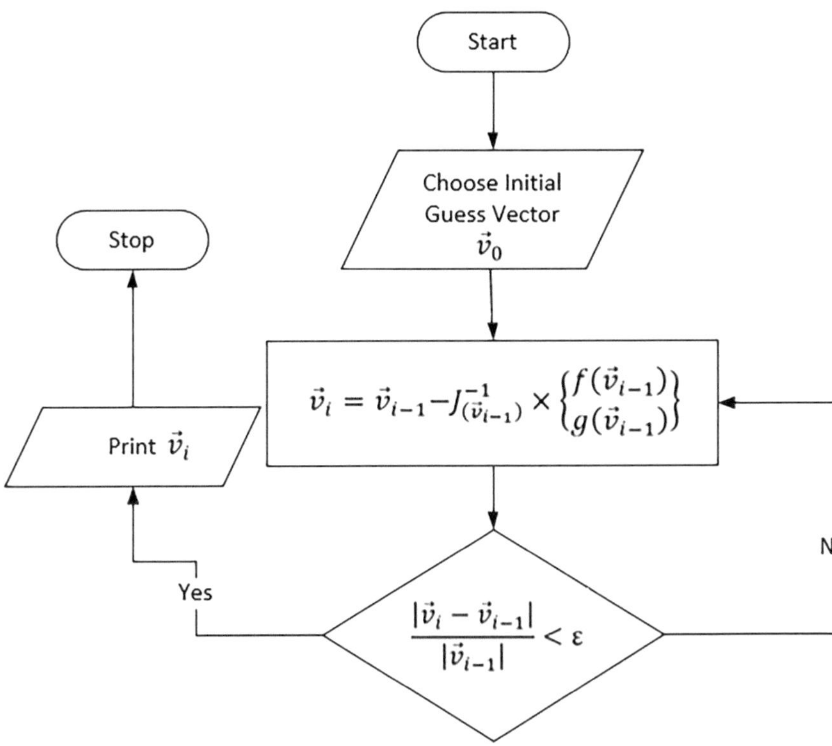

Fig. 3.30 Flow chart of the Newton-Raphson algorithm for solving Eqs. (3.57)

$$\vec{v}_i = (x_i, y_i) = \begin{Bmatrix} x_i \\ y_i \end{Bmatrix} = \begin{Bmatrix} x_{i-1} \\ y_{i-1} \end{Bmatrix} - J^{-1}_{(x_{i-1}, y_{i-1})} \times \begin{Bmatrix} f(x_{i-1}, y_{i-1}) \\ g(x_{i-1}, y_{i-1}) \end{Bmatrix} \tag{3.68}$$

The Jacobian matrix originally defined in Eq. (3.63) will then have n rows (one for each equation) and n columns (one for each variable).

Similar to the single-variable case, the method is not guaranteed to converge all the time and its convergence is heavily dependent on the proper choice of the initial guess vector, $\vec{v}_0$.

Example 3.11

Considering the system of two equations for two unknowns given in Eqs. (3.73) and starting from the origin: (0, 0) as the initial guess, write a MATLAB script that can solve Eqs. (3.73) using the presented Newton-Raphson method.

$$\begin{cases} x^2 + y = 3 \\ x + 4y^2 = 6 \end{cases} \tag{3.73}$$

Recalling the Jacobian matrix as:

$$J(x, y) = \begin{bmatrix} \frac{\partial f}{\partial x} & \frac{\partial f}{\partial y} \\ \frac{\partial g}{\partial x} & \frac{\partial g}{\partial y} \end{bmatrix}; J(x_0, y_0) = \begin{bmatrix} \frac{\partial f}{\partial x} & \frac{\partial f}{\partial y} \\ \frac{\partial g}{\partial x} & \frac{\partial g}{\partial y} \end{bmatrix}_{(x_0, y_0)} \tag{3.63}$$

in this problem we have:

$$J = \begin{bmatrix} 2x & 1 \\ 1 & 8y \end{bmatrix} \tag{3.74}$$

Using Eq. (3.74), a MATLAB script for solving Eq. (3.73) can be written as follows:

```
clear all;
x=0;
y=0;
v=[x;y];
S(1)=0;
S1(1)=x;
S2(1)=y;
for i=2:8
f=x^2+y-3;
g=x+4*y^2-6;
F=[f;g];
J= [2*x 1; 1 8*y];
v=v-inv(J)*F;
```

```
x=v(1,1);
y=v(2,1);
S(i)=i;
S1(i)=x;
S2(i)=y;
end
subplot(2,1,1);
plot(S,S1,'--o')
xlabel('iterations');
ylabel('x');
subplot(2,1,2);
plot(S,S2,'--o')
xlabel('iterations');
ylabel('y');
```

After running this script we get the solution estimation results given in Fig. 3.31. It is seen that after a brief overshoot, the solution quickly converges to the exact result of $x = 1.3879$ and $y = 1.0738$.

If in the same problem the initial guess was changed to (2,-2), the algorithm would converge to the other answer, i.e., $x = 2$ and $y = -1$, as can be seen in Fig. 3.32:

3.4.6 Newton-Raphson Method for Solving Four-Bar Linkage Problem

Now we turn our attention to the numerical solution of the four-bar linkage problem. For the vector loop shown in Fig. 3.29, we derived the following projected vector-loop equations:

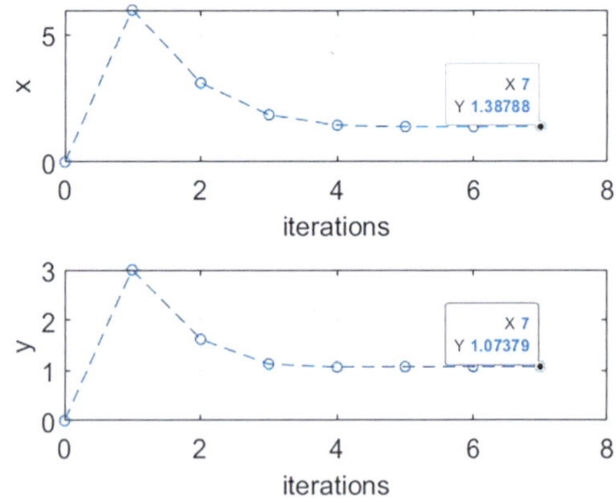

Fig. 3.31 Convergence of answer estimates for Eq. (3.73) using $(0, 0)$ as the initial guess

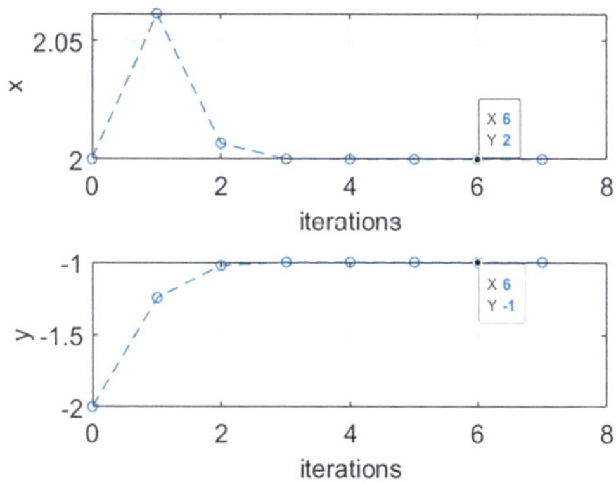

Fig. 3.32 Convergence of answer estimates for Eq. (3.73) using (2,-2) as the initial guess

$$\begin{cases} r_3 \cos\theta_3 + r_4 \cos\theta_4 = r_1 - r_2 \cos\theta_2 \\ r_3 \sin\theta_3 + r_4 \sin\theta_4 = -r_2 \sin\theta_2 \end{cases} \tag{3.54}$$

Therefore, for any input angle, θ_2, we need to solve a system of the two nonlinear equations for two unknown angles θ_3 and θ_4. A closed-form solution for this problem was given in Sect. 3.3.1. In what follows, we apply the multivariable Newton-Raphson method studied in Sect. 3.4.5 for solving the four-bar linkage problem.

Example 3.12

Consider the vector loop shown in Fig. 3.33 that corresponds to a mechanism in which $r_1 = 4$ cm, $r_2 = 1$ cm, $r_3 = 5$ cm, $r_4 = 4$ cm. The crank angle $\theta_2 = 90°$ is the input to the mechanism. Calculate output angles θ_3 and θ_4 using the following methods:

A. The Newton-Raphson algorithm;
B. The closed-form solution.

A very simple geometrical solution method for this problem results in the following exact open posture answer:

$$\theta_3 = \tan^{-1}[(r_4 - r_2)/r_1] = \tan^{-1}(3/4) = 36.87°, \theta_4 = -90° \tag{3.75}$$

A. **The Newton-Raphson algorithm**

We now want to see if the Newton-Raphson method can provide the same result as we see in Eqs. (3.75) for the open posture of the mechanism at $\theta_2 = 90°$. The system of projected vector-loop Eqs. (3.54) can be rewritten as:

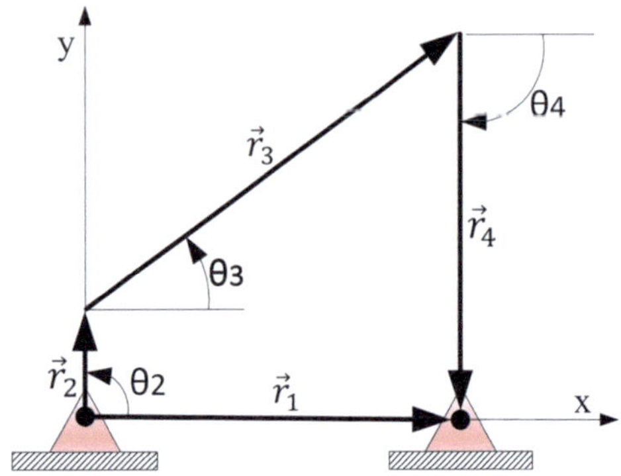

Fig. 3.33 The open posture of the four-bar linkage studied in Example 3.12 at $\theta_2 = 90°$

$$\begin{cases} r_2 \cos\theta_2 + r_3 \cos\theta_3 + r_4 \cos\theta_4 - r_1 = 0 \\ r_2 \sin\theta_2 + r_3 \sin\theta_3 + r_4 \sin\theta_4 = 0 \end{cases} \quad (3.76)$$

Based on which, functions f and g can be defined as follows:

$$\begin{cases} f(\theta_3, \theta_4) = r_2 \cos\theta_2 + r_3 \cos\theta_3 + r_4 \cos\theta_4 - r_1 \\ g(\theta_3, \theta_4) = r_2 \sin\theta_2 + r_3 \sin\theta_3 + r_4 \sin\theta_4 \end{cases} \quad (3.77)$$

Therefore, Eqs. (3.76) can be rewritten as:

$$\begin{cases} f(\theta_3, \theta_4) = 0 \\ g(\theta_3, \theta_4) = 0 \end{cases} \quad (3.78)$$

For the initial guess we use:

$$\theta_{3,0} = 40\,°, \theta_{4,0} = -85\,° \quad (3.79)$$

So,

$$\begin{Bmatrix} x_0 \\ y_0 \end{Bmatrix} = \begin{Bmatrix} \theta_{3,0} \\ \theta_{4,0} \end{Bmatrix} = \begin{Bmatrix} 40\,° \\ -85\,° \end{Bmatrix} \quad (3.80)$$

The Newton-Raphson algorithm starts with Eq. (3.65):

$$\vec{v}_1 = (x_1, y_1) = \begin{Bmatrix} x_1 \\ y_1 \end{Bmatrix} = \begin{Bmatrix} x_0 \\ y_0 \end{Bmatrix} + \begin{Bmatrix} \Delta x \\ \Delta y \end{Bmatrix} = \begin{Bmatrix} x_0 \\ y_0 \end{Bmatrix} - J^{-1}_{(x_0, y_0)} \times \begin{Bmatrix} f(x_0, y_0) \\ g(x_0, y_0) \end{Bmatrix} \quad (3.65)$$

In which the Jacobian matrix can be calculated using Eqs. (3.63) and (3.77) as follows:

$$J = \begin{bmatrix} \frac{\partial f}{\partial \theta_3} & \frac{\partial f}{\partial \theta_4} \\ \frac{\partial g}{\partial \theta_3} & \frac{\partial g}{\partial \theta_4} \end{bmatrix} = \begin{bmatrix} -r_3 \sin\theta_3 & -r_4 \sin\theta_4 \\ r_3 \cos\theta_3 & r_4 \cos\theta_4 \end{bmatrix} \tag{3.81}$$

Using the given values for the lengths of the links and the initial guess Eq. (3.80), one obtains:

$$J(40^\circ, -85^\circ) = \begin{bmatrix} -5\sin 40^\circ & -4\sin(-85^\circ) \\ 5\cos 40^\circ & 4\cos(-85^\circ) \end{bmatrix} = \begin{bmatrix} -3.214 & 3.985 \\ 3.830 & 0.349 \end{bmatrix} \tag{3.82}$$

Also the residual values calculated at the initial guess point are:

$$\begin{Bmatrix} f(x_0, y_0) \\ g(x_0, y_0) \end{Bmatrix} = \begin{Bmatrix} f(40^\circ, -85^\circ) \\ g(40^\circ, -85^\circ) \end{Bmatrix} = \begin{Bmatrix} 5\cos 40^\circ + 4\cos(-85^\circ) - 4 \\ 5\sin 40^\circ + 4\sin(-85^\circ) + 1\sin 90^\circ \end{Bmatrix} = \begin{Bmatrix} 0.179 \\ 0.229 \end{Bmatrix} \tag{3.83}$$

In the first Newton-Raphson iteration the estimations given by initial guess values are improved by using Eq. (3.65). Substituting the known values in Eq. (3.65) results in:

$$\begin{Bmatrix} x_1 \\ y_1 \end{Bmatrix} = \begin{Bmatrix} \theta_{3,1} \\ \theta_{4,1} \end{Bmatrix} = \begin{Bmatrix} x_0 \\ y_0 \end{Bmatrix} - \begin{bmatrix} -3.214 & 3.985 \\ 3.830 & 0.349 \end{bmatrix}^{-1} \times \begin{Bmatrix} 0.179 \\ 0.229 \end{Bmatrix} \text{(rad)}$$

It should be noted that the stated improvement terms in the above equation are in radians. By substituting the initial guess (3.80) and converting the last term to degrees one obtains:

$$\begin{Bmatrix} \theta_{3,1} \\ \theta_{4,1} \end{Bmatrix} = \begin{Bmatrix} 40^\circ \\ -85^\circ \end{Bmatrix} - \frac{1}{-3.214 \times 0.349 - 3.985 \times 3.830} \begin{bmatrix} 0.349 & -3.985 \\ -3.830 & -3.214 \end{bmatrix} \times \begin{Bmatrix} 0.179 \\ 0.229 \end{Bmatrix} \times \frac{180^\circ}{\pi}$$

or,

$$\begin{Bmatrix} \theta_{3,1} \\ \theta_{4,1} \end{Bmatrix} = \begin{Bmatrix} 40^\circ \\ -85^\circ \end{Bmatrix} + 3.497 \begin{Bmatrix} -0.850 \\ -1.418 \end{Bmatrix} = \begin{Bmatrix} 37.03^\circ \\ -89.96^\circ \end{Bmatrix} \tag{3.84}$$

Even though the estimation Eq. (3.84) was obtained after only one iteration, it is in good agreement with the exact solution for the open posture solution Eq. (3.75) which shows the efficiency of the Newton-Raphson algorithm.

B. **The closed-form solution**

First, the open posture

The open and the crossed postures of this mechanism can also be obtained using the closed-form solution discussed in Sect. 3.3.1. Recalling Fig. 3.4 and using Eq. (3.6), i.e.,

$$s^2 = r_1^2 + r_2^2 - 2r_1r_2\cos\theta_2 \tag{3.6}$$

results in:

$$s^2 = 4^2 + 1^2 - 8\cos 90°;\ \text{so}: s = 4.123\ \text{cm}$$

Then, using Eq. (3.7),

$$r_2^2 = s^2 + r_1^2 - 2sr_1\cos\beta \tag{3.7}$$

we get:

$$4.123^2 + 4^2 - 1^2 = 2 \times 4.123 \times 4\cos\beta;\ \text{so}: \beta = 14.037°$$

Also, using,

$$r_4^2 = s^2 + r_3^2 - 2sr_3\cos\psi \tag{3.8}$$

$$r_3^2 = s^2 + r_4^2 - 2sr_4\cos\lambda \tag{3.9}$$

result in:

$$4.123^2 + 5^2 - 4^2 = 2 \times 4.123 \times 5\cos\psi$$

$$4.123^2 + 4^2 - 5^2 = 2 \times 4.123 \times 4\cos\lambda$$

So,
$\psi = 50.906°$ and $\lambda = 75.965°$.
Since $0 \le \theta_2 \le \pi$, for open posture we use:

$$\theta_3 = \psi - \beta, \theta_4 = \pi - \beta - \lambda \tag{3.10}$$

So,

$$\begin{aligned} \theta_3 &= 50.906° - 14.037° = 36.87° \\ \theta_4 &= \pi - \beta - \lambda = 180° - 14.037° - 75.965° = 90.0° \end{aligned} \tag{3.85}$$

Notice that these are the exact results for the open posture case that we got before in Eq. (3.75)!

$$\theta_3 = \tan^{-1}(3/4) = 36.87°, \theta_4 = -90° \tag{3.75}$$

The opposite signs of θ_4 in Eqs. (3.85) and (3.75) refer to the fact that angle θ_4 shown in Fig. 3.4 is defined in the opposite direction of the θ_4 shown in Fig. 3.33.

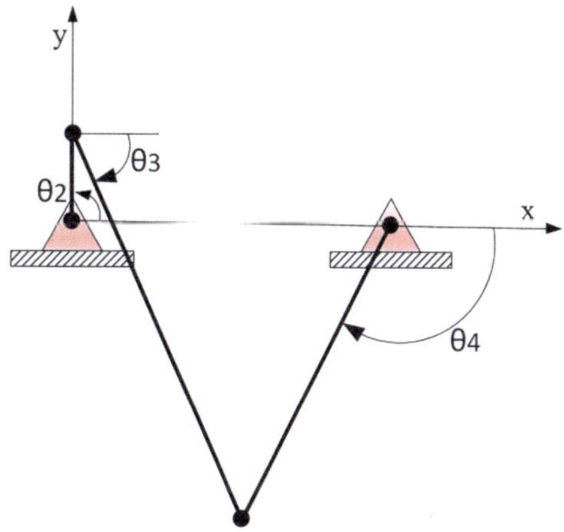

Fig. 3.34 The crossed posture of the four-bar linkage studied in Example 3.12 at $\theta_2 = 90°$

Now, the crossed posture
We recall,

$$\theta_3 = -\psi - \beta, \theta_4 = \pi - \beta + \lambda \tag{3.11}$$

Using the calculated ψ, β, and λ in Eqs. (3.11) we get:

$$\theta_3 = -50.906\,° - 14.037\,° = -64.943\,°$$

$$\theta_4 = 180\,° - 14.037\,° + 75.965\,° = 241.928\,° = -118.072\,°$$

The crossed posture of the mechanism at $\theta_2 = 90°$ is illustrated in Fig. 3.34.

3.5 Vector-Loop Equations for More Complex Mechanisms

Figure 3.35 illustrates a four-bar mechanism that also includes a pin A that is fixed on part 4 and can slide inside the slot in part 3. In this way, a one-DOF revolute joint between parts 3 and 4, that we usually have in a four-bar linkage, has been replaced by a two-DOF pin-in-the-slot joint.

We can draw the following vector loop for this mechanism:

The vector-loop equation corresponding to Fig. 3.36 can be written as follows:

$$\vec{r}_2 + \vec{r}_3 + \vec{r}_4 + \vec{r}_5 = \vec{r}_1 \tag{3.86}$$

Angular positions of vectors used in Eq. (3.86) are measured with respect to the positive direction of the x axis. The x axis is defined by the two fixed hinges O and C and its positive direction is from O to C. The y axis is obtained by 90° CCW

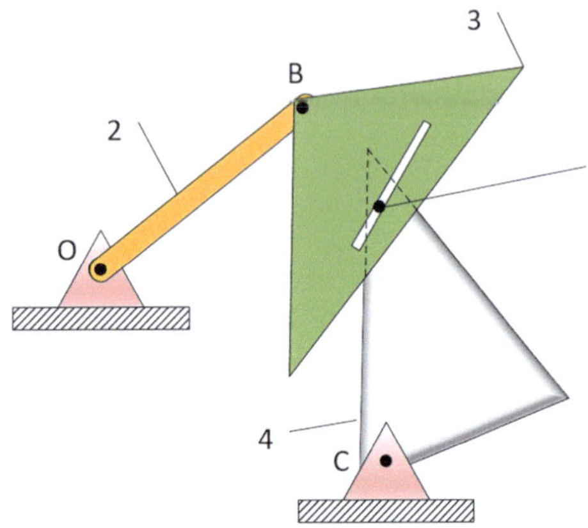

Fig. 3.35 A four-bar mechanism that includes a pin in the-slot joint

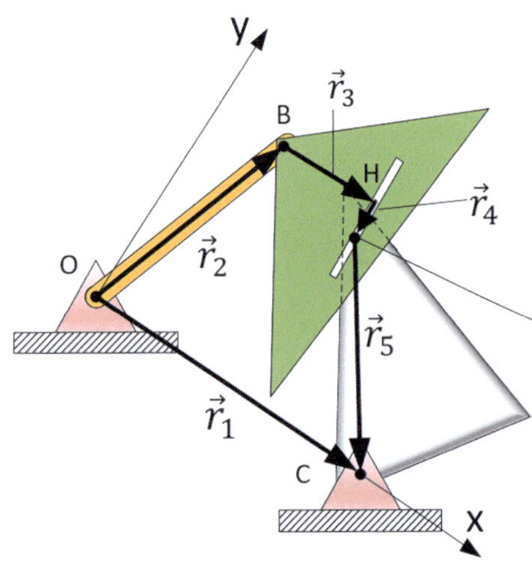

Fig. 3.36 Vector loop for the mechanism shown in Fig. 3.35

rotation of the x axis. All subscripts refer to the vector numbers and they might not be necessarily the same as the corresponding part numbers, as we see in Fig. 3.36.

In Eq. (3.86) $\vec{r}_2$ is along member 2, its length is equal to the known length of member 2 (i.e., OB) and its angular position is the same as that of member 2. Vector $\vec{r}_3$ is drawn on member 3 from pin B and it is perpendicular to the existing slot in this member. Since $\vec{r}_3$ is fixed on member 3, its angular position is the same as the angular position of member 3. The length of $\vec{r}_3$ is fixed and equal to BH (which is determined by the location and orientation of the slot). Vector $\vec{r}_4$ is along the slot so its angular position is 90° less than that of $\vec{r}_3$ (or member 3). The length of $\vec{r}_4$ is unknown. Vector $\vec{r}_5$ connects pin A to hinge C about which member 4 rotates. Since pin A is fixed on member 4, the length of $\vec{r}_5$ is known; however, its angular position is unknown and it can be defined as the angular position of member 4.

Projecting the vector-loop equation (3.86) on the x axis results in:

$$r_2 \cos\theta_2 + r_3 \cos\theta_3 + r_4 \cos\theta_4 + r_5 \cos\theta_5 = r_1 \cos\theta_1 \tag{3.87}$$

Since angles are measured from the x axis, clearly $\theta_1 = 0$ and Eq. (3.87) reduces to:

$$r_2 \cos\theta_2 + r_3 \cos\theta_3 + r_4 \cos\theta_4 + r_5 \cos\theta_5 = r_1 \tag{3.88}$$

Similarly, projecting the vector-loop equation (3.86) on the y axis results in:

$$r_2 \sin\theta_2 + r_3 \sin\theta_3 + r_4 \sin\theta_4 + r_5 \sin\theta_5 = 0 \tag{3.89}$$

Finally, the orientations of $\vec{r}_3$ and $\vec{r}_4$ are related to each other by the following constraint equation:

$$\theta_3 - \theta_4 = 90\,° \tag{3.90}$$

Therefore, we have three Eqs. (3.88), (3.89), and (3.90) for five variables θ_2, θ_3, θ_4, θ_5, and r_4. We can choose $5 - 3 = 2$ variables as inputs and then solve the system of the three equations for the remaining three unknown variables (or outputs). One may then conclude that the mechanism shown in Fig. 3.36 has two DOF.

The obtained number of DOF can be verified by using the Kutzbach's Eq. (2.10). In this case, we have four members, three revolute joints at O, B, and C and one two-DOF joint (the pin-in-the-slot joint) at A. Substitution of this data in Eq. (2.10) results in:

$$DOF = 3(n - 1) - 2j_1 - j_2 = 3(4 - 1) - 2 \times 3 - 1 = 2 \tag{3.91}$$

Equation (3.91) is compatible with the previously-mentioned vector-loop analysis.

3.6 Newton-Raphson Method for Solving More Complex Problems

A more complex mechanism is the six-member shaper mechanism discussed before in Sect. 2.5 and shown in Fig. 2.47. It is a quick-return mechanism that was modeled using SolidWorks Motion software as part of a capstone project of one of my students at USM, see Stevenson [7]. The mechanism has one crank, two sliders, a guide rod that directs slider 2, and a piston rod that pushes slider 1 to the right (the advance direction) or pulls it to the left (the return direction). Considering the

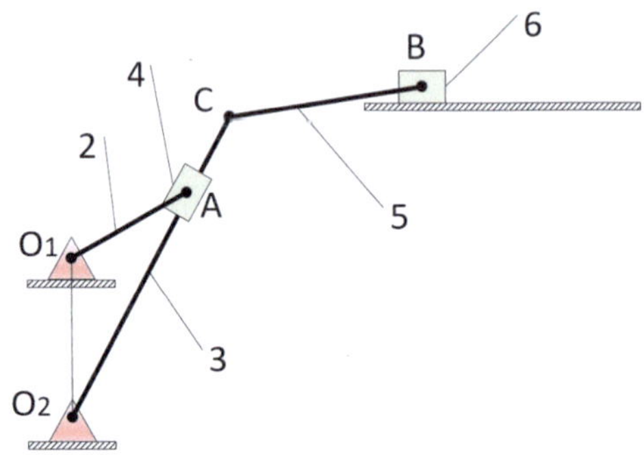

Fig. 3.37 The skeleton diagram of the six-member quick-return mechanism of Fig. 2.47

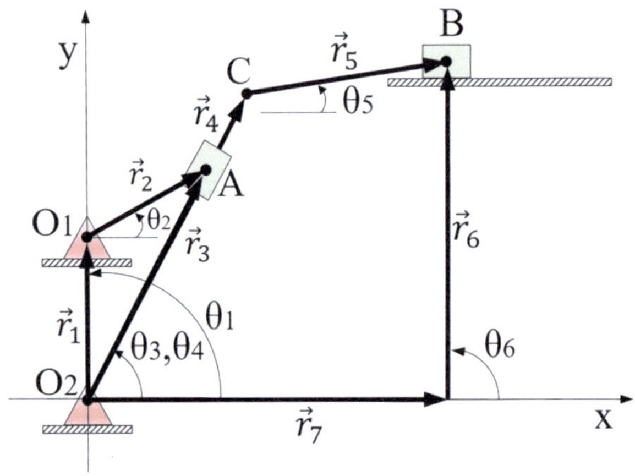

Fig. 3.38 The vector-loop diagram of the mechanism shown in Fig. 3.37

direction of the advance motion of slider 1, the quick-return characteristic of the mechanism is experienced if the crank rotates CW.

A skeleton diagram of this mechanism is shown in Fig. 3.37. In what follows, we create a vector loop for this mechanism, obtain vector-loop equations, and then solve the position problem using the Newton-Raphson method.

First, let us calculate the number of degrees of freedom of this mechanism using the Kutzbach's Eq. (2.10):

$$DOF = 3(n-1) - 2j_1 - j_2 = 3(6-1) - 2\times 7 - 0 = 1 \tag{3.92}$$

Figure 3.38 illustrates the vector-loop diagram of the mechanism shown in Fig. 3.37. In this diagram θ_2 is the mechanism's only DOF and it should be noted that vector $\vec{r}_4$ is the same as $\overrightarrow{O_2C}$. Furthermore, all angles are measured from the positive direction of the x axis. The corresponding vector-loop equations are as follows

$$\begin{cases} \vec{r}_1 + \vec{r}_2 = \vec{r}_3 \\ \vec{r}_4 + \vec{r}_5 = \vec{r}_6 + \vec{r}_7 \end{cases} \tag{3.93}$$

By projecting Eqs. (3.93) on the x and y axes one obtains:

$$\begin{cases} r_1 \sin\theta_1 + r_2 \sin\theta_2 = r_3 \sin\theta_3 \\ r_1 \cos\theta_1 + r_2 \cos\theta_2 = r_3 \cos\theta_3 \end{cases} \tag{3.94}$$

And,

$$\begin{cases} r_4 \sin\theta_4 + r_5 \sin\theta_5 = r_6 \sin\theta_6 \\ r_4 \cos\theta_4 + r_5 \cos\theta_5 = r_7 \cos\theta_7 \end{cases} \tag{3.95}$$

Equations (3.94) and (3.95) have 14 variables and parameters: r_1 to r_7 and θ_1 to θ_7. We assume that the lengths of all members are known. So we know r_1, r_2, r_4, r_5, and r_6. We notice that r_3 that determines the location of point A is not fixed. The same is true for r_7 that indicates the location of point B. Furthermore we know that,

$$\theta_1 = 90^\circ, \theta_6 = 90^\circ, \theta_3 = \theta_4, \theta_7 = 0 \tag{3.96}$$

So, we have five known link lengths r_1, r_2, r_4, r_5, and r_6, four angle constraints given in Eqs. (3.96), and four projected vector-loop Eqs. (3.94) and (3.95). Considering the total number of the quantities as 14 (7 r's and 7 θ's), the number of DOF of this mechanism would be:

$$DOF = 14 - 5 - 4 - 4 = 1 \tag{3.97}$$

This is the same as the result we got before in Eq. (3.92). Angle θ_2 that is the crank angle is taken as the input to the mechanism and it is its only DOF.

The four projected vector-loop Eqs. (3.94) and (3.95) should now be solved for the remaining four unknowns, i.e., r_3, θ_3, θ_5, and r_7. So, the vector of unknowns is:

$$\vec{v} = \begin{Bmatrix} r_3 \\ \theta_3 \\ \theta_5 \\ r_7 \end{Bmatrix} \tag{3.98}$$

In preparation for using the Newton-Raphson equation, we rewrite Eqs. (3.94) and (3.95) as follows:

$$\begin{cases} r_1 + r_2 \sin\theta_2 - r_3 \sin\theta_3 = 0 \\ r_2 \cos\theta_2 - r_3 \cos\theta_3 = 0 \\ r_4 \sin\theta_3 + r_5 \sin\theta_5 - r_6 = 0 \\ r_4 \cos\theta_3 + r_5 \cos\theta_5 - r_7 = 0 \end{cases} \tag{3.99}$$

Based on which, functions f_1, f_2, f_3, and f_4 can be defined as follows:

$$\begin{cases} f_1(r_3, \theta_3, \theta_5, r_7) = r_1 + r_2 \sin\theta_2 - r_3 \sin\theta_3 \\ f_2(r_3, \theta_3, \theta_5, r_7) = r_2 \cos\theta_2 - r_3 \cos\theta_3 \\ f_3(r_3, \theta_3, \theta_5, r_7) = r_4 \sin\theta_3 + r_5 \sin\theta_5 - r_6 \\ f_4(r_3, \theta_3, \theta_5, r_7) = r_4 \cos\theta_3 + r_5 \cos\theta_5 - r_7 \end{cases} \tag{3.100}$$

Therefore, Eqs. (3.99) can be rewritten as:

$$\begin{cases} f_1(r_3, \theta_3, \theta_5, r_7) = 0 \\ f_2(r_3, \theta_3, \theta_5, r_7) = 0 \\ f_3(r_3, \theta_3, \theta_5, r_7) = 0 \\ f_4(r_3, \theta_3, \theta_5, r_7) = 0 \end{cases} \tag{3.101}$$

Using an expanded 4 by 4 matrix version of Eq. (3.63) and Eqs. (3.100), the Jacobian matrix in this problem can be calculated as follows:

$$J = \begin{bmatrix} -\sin\theta_3 & -r_3\cos\theta_3 & 0 & 0 \\ -\cos\theta_3 & r_3\sin\theta_3 & 0 & 0 \\ 0 & r_4\cos\theta_3 & r_5\cos\theta_5 & 0 \\ 0 & -r_4\sin\theta_3 & -r_5\sin\theta_5 & -1 \end{bmatrix} \tag{3.102}$$

where the first row is associated with f_1, the second row is associated with f_2, and so on. The first column includes derivatives with respect to r_3, the second column includes derivatives with respect to θ_3, and similarly for θ_5 and r_7 [the order follows the order in Eq. (3.98)].

Example 3.13 We would like to analyze the posture of the mechanism shown in Fig. 3.37 at an input angle of $\theta_2 = 60°$ and for the following data:

$$r_1 = 3 \text{ in}, r_2 = 2 \text{ in}, r_4 = 7 \text{ in}, r_5 = 4.5 \text{ in}, \text{and } r_6 = 6 \text{ in} \tag{3.103}$$

As for the initial guesses we use:

$$\vec{v}_0 = \begin{Bmatrix} r_{3,0} \\ \theta_{3,0} \\ \theta_{5,0} \\ r_{7,0} \end{Bmatrix} = \begin{Bmatrix} 3.5 \\ 60° \\ 0 \\ 8 \end{Bmatrix} \tag{3.104}$$

A program similar to the one listed in Example 3.11 was used to solve this problem and the iteration results are given in Fig. 3.39. It can be seen that the algorithm quickly converges to the solution. The results of the fourth iteration are listed in Eq. (3.105):

$$\vec{v}^* = \begin{Bmatrix} r_{3,4} \\ \theta_{3,4} \\ \theta_{5,4} \\ r_{7,4} \end{Bmatrix} = \begin{Bmatrix} 4.8366 \\ 78.0675° \\ -10.8717° \\ 5.8665 \end{Bmatrix} \tag{3.105}$$

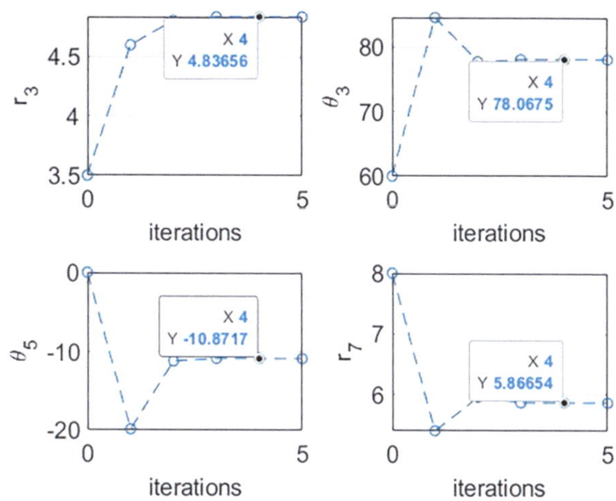

Fig. 3.39 Convergence of solution estimates of Eq. (3.99) using Eq. (3.104) as the initial guess

It can be easily verified that the obtained solution satisfies the projected vector-loop Eqs. (3.99). More examples of application of the Newton-Raphson method for solving complex mechanisms can be found in Stanišić [6].

3.7 Rolling Contact Angular Displacement

In this section we study rolling contact between friction disks. In Chaps. 6 and 7, we will see that the same formulations are also valid for pitch circles of gears in mesh. Parts of this section are based on the remarks made in Davis and Ghorashi [2, 3] that are based on a capstone project of one of my students at USM, see Davis [1].

3.7.1 External Contact and Fixed Axes

Consider two friction disks with fixed axes of rotation that are in rolling contact, as shown in Fig. 3.40. The radii of the friction disks 2 and 3 are R_2 and R_3, respectively. Figure 3.41 shows the positions of the disks after they start their rotations from the position at which point A_2 on disk 2 is in contact with point A_3 on disk 3. There is no sliding or rubbing of the disks on each other and their motions on each other are pure rolling. The mechanism has one DOF and the counter clockwise motion $\Delta\theta_2$ of disk 2 results in a clockwise motion $\Delta\theta_3$ of disk 3. Since the motion of the disks on each other is pure rolling, $\overset{\frown}{A_2C}$ on disk 2 should be equal to $\overset{\frown}{A_3C}$ on disk 3:

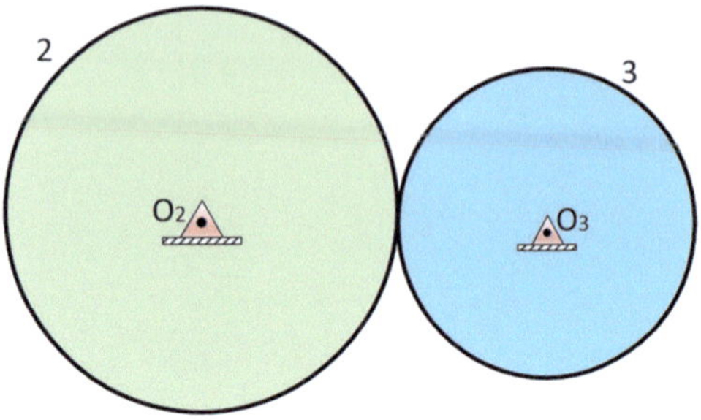

Fig. 3.40 Two friction disks with fixed axes of rotation and in external rolling contact

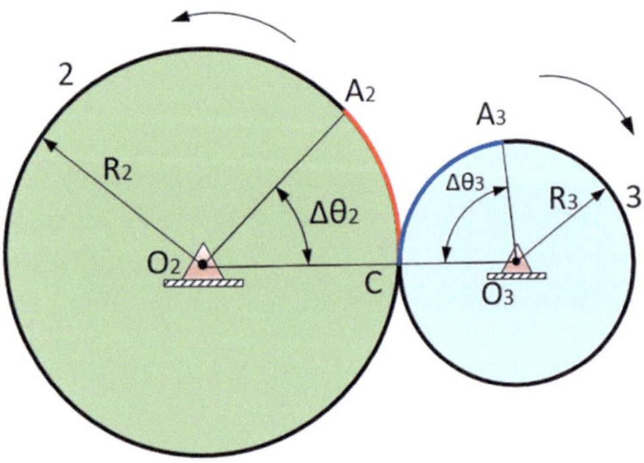

Fig. 3.41 The two friction disks shown in Fig. 3.40, after some pure rolling motion

$$\widehat{A_2C} = \widehat{A_3C} \tag{3.106}$$

If $\Delta\theta_2$ and $\Delta\theta_3$ are measured in radians, one can write the following equations:

$$\begin{aligned} \widehat{A_2C} &= R_2|\Delta\theta_2| \\ \widehat{A_3C} &= R_3|\Delta\theta_3| \end{aligned} \tag{3.107}$$

Therefore,

$$R_2|\Delta\theta_2| = R_3|\Delta\theta_3| \tag{3.108}$$

For external contact of friction disks or gears, which is what we see in Fig. 3.41, the CCW rotation of disk 2 is always accompanied with the CW rotation of disk 3. Considering the CCW rotation to be positive, for external contact, Eq. (3.108) can be rewritten as:

$$R_2\Delta\theta_2 = -R_3\Delta\theta_3 \tag{3.109}$$

3.7.2 External Contact and a Moving Axis

In Figs. 3.40 and 3.41, the centers of friction disks or gears, i.e., points O_2 and O_3 are fixed. In planetary gears (to be studied in Chap. 7), it frequently happens that the axis of a gear is not fixed and it can move on a circular path. In the mechanism shown in Fig. 3.42, the center point O_2 is still fixed, but hinge O_3 can rotate on a circular orbit around it. In this case, disk 2 would be the sun around which planet 3 rotates. Since O_3 is no longer fixed, in order to ensure that its circular motion about O_2 happens, the two center points should be connected by an arm 4. If we hold arm 4 fixed, the mechanism reduces to the one shown in Fig. 3.41 and it still has one DOF. So, if the arm is free to rotate, as is the case in Fig. 3.42, the mechanism will have two DOF and therefore needs two inputs to operate.

We concluded that if in Fig. 3.42 we hold arm 4 fixed, the mechanism reduces to the one shown in Fig. 3.41. This observation is similar to viewing the motion of the mechanism in Fig. 3.42 with respect to the arm. In this case, the arm is not fixed and the motion of the mechanism is analyzed with respect to the arm. Since motion is analyzed with respect to the arm, what an arm-attached observer notices is that the arm is stationary which is nothing but what we have in Fig. 3.41. Therefore, for the mechanism shown in Fig. 3.42, Eq. (3.109) is still valid, but *with respect to the arm*. In other words, by writing Eq. (3.109) with respect to the arm *we are making the arm stationary* so we can still use Eq. (3.109), that was developed for the stationary arm, for the case in Fig. 3.42 in which the arm is not stationary (*if the arm is stationary, fine, if not, make it stationary!*). So, for the mechanism in Fig. 3.42 we can write:

$$R_2 \Delta\theta_{2/4} = -R_3 \Delta\theta_{3/4} \tag{3.110}$$

In which the symbol: "/4" means: with respect to arm 4. The negative sign in Eq. (3.110) is used for external contact and implies opposite directions of rotation of

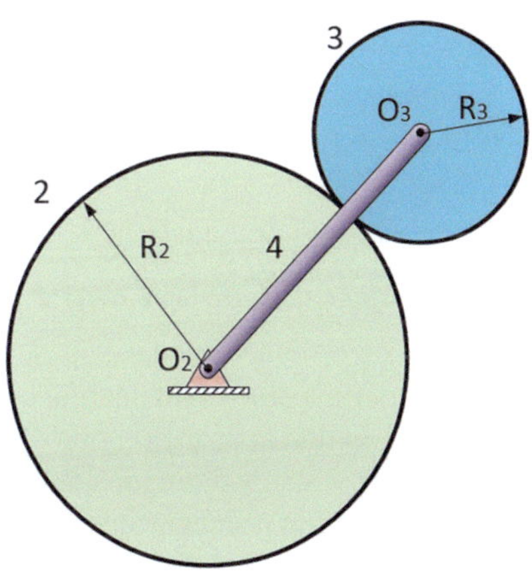

Fig. 3.42 Two friction disks in external rolling contact with only disk 2 having a fixed axis of rotation

the friction disks (or gears) with respect to the arm. Eq. (3.110) can be expanded as follows:

$$R_2(\Delta\theta_2 - \Delta\theta_4) = -R_3(\Delta\theta_3 - \Delta\theta_4) \tag{3.111}$$

It can be seen that if arm 4 is fixed, $\Delta\theta_4 = 0$, and Eq. (3.111) reduces to Eq. (3.109), as expected.

$$R_2\Delta\theta_2 = -R_3\Delta\theta_3 \tag{3.109}$$

Example 3.14 In the mechanism shown in Fig. 3.42 we hold disk 2 stationary. Find the angular displacement ratio $\Delta\theta_3/\Delta\theta_4$ if $R_2 = 2$ in and $R_3 = 4$ in.

Since disk 2 is fixed, $\Delta\theta_2 = 0$. Therefore, the rolling contact Eq. (3.111) reduces to

$$R_2(0 - \Delta\theta_4) = -R_3(\Delta\theta_3 - \Delta\theta_4)$$

Or:

$$(R_2 + R_3)\Delta\theta_4 = R_3\Delta\theta_3$$

And finally we get the following result:

$$\frac{\Delta\theta_3}{\Delta\theta_4} = \frac{R_2 + R_3}{R_3} \tag{3.112}$$

Using the given numerical values in Eq. (3.112) we obtain the angular displacement ratio $\Delta\theta_3/\Delta\theta_4$ as follows:

$$\frac{\Delta\theta_3}{\Delta\theta_4} = \frac{2 + 4}{4} = 1.5$$

Therefore, in such a mechanism disk 3 rotates in the same direction as arm 4, but 50% faster. A simple physical model can be easily used to verify this result.

3.7.3 *Internal Contact and Fixed Axes*

Now consider the case of an internal contact of friction disks with fixed axes shown in Fig. 3.43. In this case, the friction disk or gear 3 has a fix revolute joint at O_3. This disk is rolling without sliding inside the internal friction ring or ring gear 2 and is in contact with it at an instantaneous contact point C. The friction ring has a fix revolute joint at O_2. The radius of disk 3 is R_3 and the inner radius of the ring disk 2 is R_2. The disks start their rotations at a position in which point A_2 inside ring disk 2 is in contact with point A_3 on the friction disk 3. There is no sliding or rubbing of the disks on each other and their motions on each other are pure rolling. The mechanism has one DOF and the counter clockwise motion $\Delta\theta_3$ of disk 3 results in a counter clockwise motion $\Delta\theta_2$ of disk 2. Since the motion of the disks on each other is pure rolling, $\overset{\frown}{A_2C}$ on disk 2 should be equal to $\overset{\frown}{A_3C}$ on disk 3:

$$\overset{\frown}{A_2C} = \overset{\frown}{A_3C} \tag{3.113}$$

If $\Delta\theta_2$ and $\Delta\theta_3$ are measured in radians, one can write the following equations:

$$\overset{\frown}{A_2C} = R_2|\Delta\theta_2|; \overset{\frown}{A_3C} = R_3|\Delta\theta_3| \tag{3.114}$$

Therefore,

$$R_2|\Delta\theta_2| = R_3|\Delta\theta_3| \tag{3.115}$$

In Fig. 3.43, i.e., for the internal contact of friction disks or gears with fix axes, the CCW motion of disk 3 is always accompanied with the CCW motion of internal disk 2. Considering the CCW motion to be positive, for internal contact Eq. (3.115) can be rewritten as follows:

$$R_2\Delta\theta_2 = R_3\Delta\theta_3 \tag{3.116}$$

3.7.4 Internal Contact and a Moving Axis

In the previous case, the centers of friction disks or gears, i.e., points O_2 and O_3 were fixed. In planetary gears it frequently happens that the axis of one gear is not fixed and

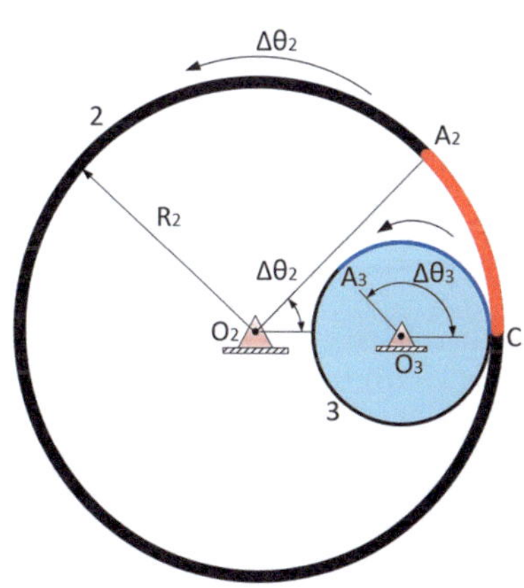

Fig. 3.43 Two fixed axes friction disks in internal contact after some pure rolling motion

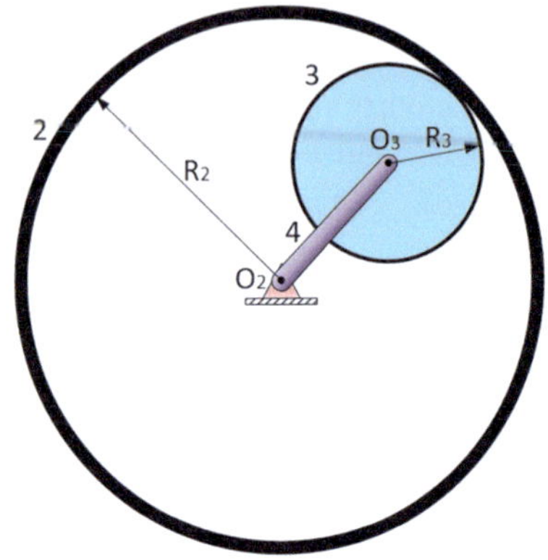

Fig. 3.44 Two friction disks in internal rolling contact with only disk 2 having a fixed axis of rotation

can move on a circular orbit (see Chap. 7). Let us assume that O_2 is still fixed, but O_3 can rotate on a circular orbit around O_2, as we see in Fig. 3.44. In this case, disk 2 would be the ring inside which planet 3 rotates. Since hinge point O_3 is no longer fixed, in order to ensure that its circular motion about O_2 happens, the two center points should be connected by an arm 4. If we hold arm 4 fixed, the mechanism reduces to that of Fig. 3.43 that still has one DOF. So, if the arm is free to rotate, the mechanism will have two DOF.

As mentioned before, if we hold arm 4 fixed, the mechanism reduces to the one we discussed in Sect. 3.7.3. This observation is also similar to viewing with respect to the arm. In the present case, the arm is not fixed and the motion with respect to arm 4 is studied. Since this motion is analyzed with respect to the arm, what an arm-attached observer notices is that the arm is stationary, that is nothing but what we observed in Sect. 3.7.3. Therefore, Eq. (3.116) is still valid, but with respect to the arm and we can write the following:

$$R_2\Delta\theta_{2/4} = R_3\Delta\theta_{3/4} \tag{3.117}$$

In which the symbol "/4" again means relative to (or with respect to) arm 4. The positive sign is used for internal contact and implies the same directions of rotation of the friction disks with respect to the arm. Equation (3.117) can be expanded as follows:

$$R_2(\Delta\theta_2 - \Delta\theta_4) = R_3(\Delta\theta_3 - \Delta\theta_4) \tag{3.118}$$

It can be seen that if arm 4 is fixed, $\Delta\theta_4 = 0$, and Eq. (3.118) reduces to Eq. (3.116), as expected.

$$R_2\Delta\theta_2 = R_3\Delta\theta_3 \tag{3.116}$$

Example 3.15 In the friction disk problem with the arm shown in Fig. 3.44, we fix ring disk 2. Calculate the ratio of angular displacements $\Delta\theta_3/\Delta\theta_4$ if $R_2 = 6$ in and $R_3 = 2$ in.

Since disk 2 is fixed, $\Delta\theta_2 = 0$ and Eq. (3.118) reduces to:

$$R_2(0 - \Delta\theta_4) = R_3(\Delta\theta_3 - \Delta\theta_4) \tag{3.119}$$

Or:

$$(-R_2 + R_3)\Delta\theta_4 = R_3\Delta\theta_3$$

And finally we get the following result:

$$\frac{\Delta\theta_3}{\Delta\theta_4} = \frac{-R_2 + R_3}{R_3} \tag{3.120}$$

Using the given numerical values we get:

$$\frac{\Delta\theta_3}{\Delta\theta_4} = \frac{-6+2}{2} = -2$$

Therefore, in such a mechanism disk 3 rotates in the opposite direction of arm 4 and 100% faster. Again, a simple physical model can be used to verify this result.

3.7.5 *Rack and Pinion*

Now, we consider the rack and pinion mechanism shown in Fig. 3.45. This mechanism has one DOF and it converts the translational motion of rack *AB* moving in linear bearings (equivalent to a gear with infinite pitch radius; see Chap. 6) to the rotational motion of pinion 2 and vice versa. If the instantaneous contact point of the rack and pinion is shown by *C*, and there is no slipping between the rack and the pinion during motion, a positive rotation of the pinion that is considered to be CCW generates the following changes in the lengths of segments *AC* and *CB* on the rack:

$$\Delta AC = -R_2\Delta\theta_2; \Delta CB = R_2\Delta\theta_2 \tag{3.121}$$

It should be clear that a CCW rotation of the pinion reduces length *AC* but increases length *CB* by the same amount (that is equal to the length of the arc $R_2\Delta\theta_2$ on the pinion, if no slipping happens between parts 2 and 3). If the pinion rotates CW, $\Delta\theta_2$ would be negative and the above equations predict an increase in the *AC* length and a decrease in the *CB* length, as expected.

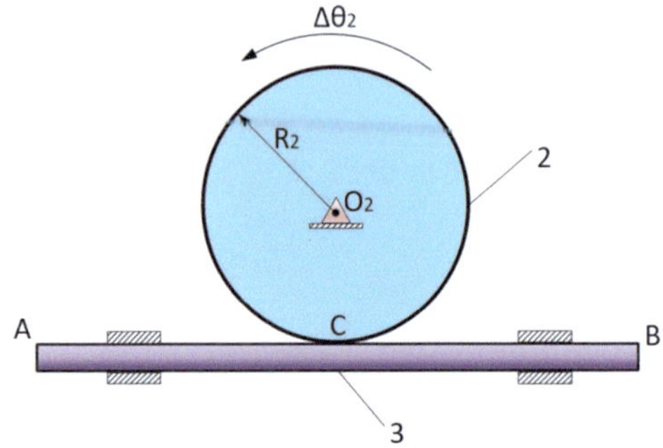

Fig. 3.45 Rack and pinion mechanism

Fig. 3.46 Rack and pinion in a can crusher machine (USM term project)

Rack and pinion is used, e.g., in a can crusher, seen in Fig. 3.46. In this case, the input torque T_{in} is applied by the user, and the resisting load F_{out} is the force imposed by the can while it is being crushed.

References

1. Davis, M. (2015). *Design and manufacture of a planetary gearset cutaway demonstration setup*. Capstone Project Report, Department of Engineering, University of Southern Maine.
2. Davis, M. & Ghorashi, M. (2016). A hybrid hands-on and simulation method for teaching the operation of automatic transmission systems. *ASEE northeastern section conference, Kingston, RI*. Awarded second place in the best paper award competition.
3. Davis, M., & Ghorashi, M. (2018). A holistic approach for deciphering the operation of planetary gear train systems. *ASEE's Journal of Engineering Technology, 35*, 16–25.
4. Ghorashi, M. (1994). Limit analysis of circular plates subjected to arbitrary rotational symmetric loadings. *International Journal of Mechanical Sciences, 36*(2), 87–94.

5. Ghorashi, M. (2016). *Statics and rotational dynamics of composite beams*. Springer International Publishing AG.
6. Stanišić, M. L. (2015). *Mechanisms and machines: Kinematics, dynamics, and synthesis*. Cengage Learning.
7. Stevenson, B. (2018). *Design and manufacture of several mechanisms to demonstrate the concept of mobility*. Capstone Project Report, Department of Engineering, University of Southern Maine.

Chapter 4
Velocity Analysis

4.1 Introduction

Velocity is the first time derivative of the position function. Unlike position analysis, velocity analysis is a linear problem and therefore it does not need application of numerical methods the way position analysis does. In this chapter, a variety of methods for performing velocity analysis of mechanisms are introduced. They include the algebraic, kinematic coefficients, relative motion, and instantaneous centers methods. Locked positions and limit positions of a mechanism are also obtained as by-products of velocity analysis. Several illustrative problems have been solved that provide examples for use of each of these methods.

4.2 Algebraic Method for Velocity Analysis

The algebraic method for velocity analysis is based on the introduction of a position function for the point of interest and then taking a derivative of the mentioned function with respect to time. As an example, consider the slider-crank mechanism shown in Fig. 4.1.

The corresponding vector-loop diagram for this mechanism is shown in Fig. 4.2. Rotation of crank 2 in this figure is the input motion to the mechanism. The crank rotates at an angular velocity ω_2 which can be expressed as follows:

$$\omega_2 = \frac{d\theta_2}{dt} \tag{4.1}$$

where $d\theta_2$ is the differential angular displacement of link 2.

Our aim is to calculate the velocity of the slider v_B at any crank angle θ_2. We recall Eq. (3.28) for the position of the slider x_B in terms of θ_2.

M. Ghorashi, *Introduction to Machine Dynamics*, Mechanical Engineering Series,
https://doi.org/10.1007/978-3-031-94736-0_4

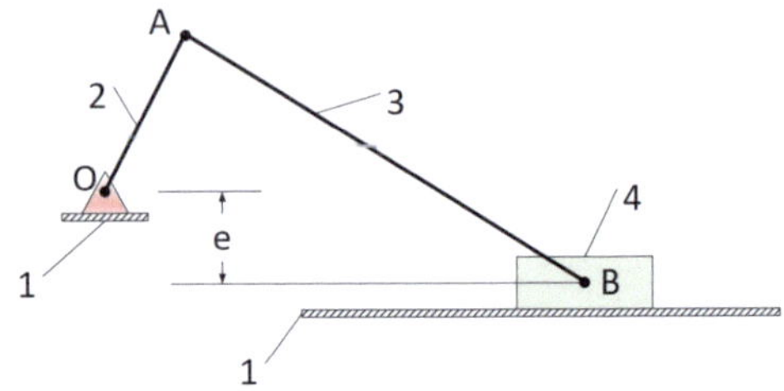

Fig. 4.1 Eccentric slider-crank mechanism

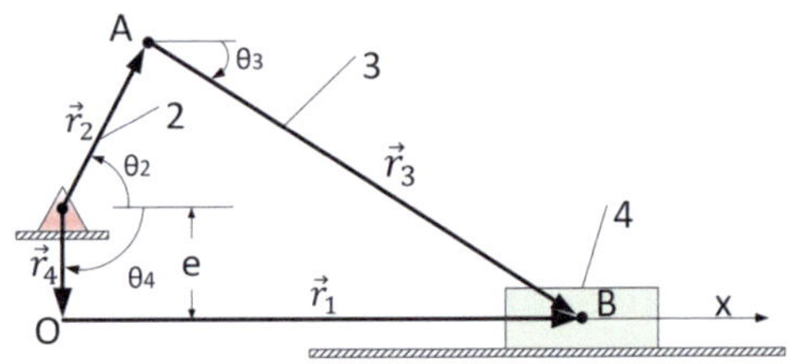

Fig. 4.2 Vector-loop diagram of the mechanism shown in Fig. 4.1

$$x_B = r_2 \cos\theta_2 \pm \sqrt{r_3^2 - (e + r_2 \sin\theta_2)^2} \tag{3.28}$$

The plus sign is used when the slider is on the right-hand side of the revolute joint O, and the minus sign is used when it is on its left-hand side. For no eccentricity ($e = 0$) and for $\theta_2 = 0$, Eq. (3.28) results in the following expected solutions:

$$x_B = r_2 \pm r_3 \tag{4.2}$$

Now, using Eq. (3.28) with the positive sign we would like to take a derivative of this equation with respect to time. To this end, we use the chain rule:

$$\frac{dx_B}{dt} = \frac{dx_B}{d\theta_2}\frac{d\theta_2}{dt} \tag{4.3}$$

In Eq. (4.3), the term

$$x'_B = \frac{dx_B}{d\theta_2} = \frac{d(\text{Output})}{d(\text{Input})} \tag{4.4}$$

represents how the position of the slider (the output) changes with changes of the crank angle (the input). This term in Eq. (4.4) is called the first-order kinematic coefficient of x_B. It is independent of time and the angular velocity of the crank. We will talk more about the kinematic coefficients later in Sect. 4.4.

Now to calculate v_B, we take the derivative of Eq. (3.28) with respect to time. For its first term, and using the chain rule shown in Eq. (4.3), we get:

$$\frac{d(r_2\cos\theta_2)}{dt}=\frac{d(r_2\cos\theta_2)}{d\theta_2}\frac{d\theta_2}{dt}=-r_2\dot{\theta}_2\sin\theta_2 \tag{4.5}$$

Showing the angular velocity $\dot{\theta}_2$ of the crank by ω_2, Eq. (4.5) can be rewritten as:

$$\frac{d(r_2\cos\theta_2)}{dt}=-r_2\omega_2\sin\theta_2 \tag{4.6}$$

For calculating the derivative of the second term in Eq. (3.28) we again use the chain rule. To this end we name the term under the square root, u:

$$u=r_3^2-(e+r_2\sin\theta_2)^2 \tag{4.7}$$

So the time derivative of the second term in Eq. (3.28) is:

$$\begin{aligned}\frac{d\left[\sqrt{r_3^2-(e+r_2\sin\theta_2)^2}\right]}{dt}&=\frac{d(\sqrt{u})}{du}\frac{du}{d\theta_2}\frac{d\theta_2}{dt}=\frac{1}{2\sqrt{u}}\\&\times[-2(e+r_2\sin\theta_2)r_2\cos\theta_2]\dot{\theta}_2\end{aligned} \tag{4.8}$$

Adding the two terms in Eqs. (4.6) and (4.8) and using Eq. (4.7) result in:

$$v_B=\frac{dx_B}{dt}=-r_2\omega_2\sin\theta_2-\frac{(e+r_2\sin\theta_2)\,r_2\omega_2\cos\theta_2}{\sqrt{r_3^2-(e+r_2\sin\theta_2)^2}} \tag{4.9}$$

Equation (4.9) gives the velocity of the slider v_B at any crank angle θ_2 during motion, if the positive sign in the position Eq. (3.28) is used. If the negative sign in Eq. (3.28) is used, we get a similar equation in which the second term in Eq. (4.9) would be positive. Therefore, the complete velocity function v_B can be written as follows:

$$v_B=\frac{dx_B}{dt}=-r_2\omega_2\sin\theta_2\mp\frac{(e+r_2\sin\theta_2)\,r_2\omega_2\cos\theta_2}{\sqrt{r_3^2-(e+r_2\sin\theta_2)^2}} \tag{4.10}$$

In which the negative sign is used if x_B in Eq. (3.28) has the positive sign and vice versa. If there is no eccentricity, i.e., $e=0$ and $\lambda=r_2/r_3$ (i.e., the crank to the connecting rod length ratio) then Eq. (4.10) for the velocity of the piston and point B can be rewritten as:

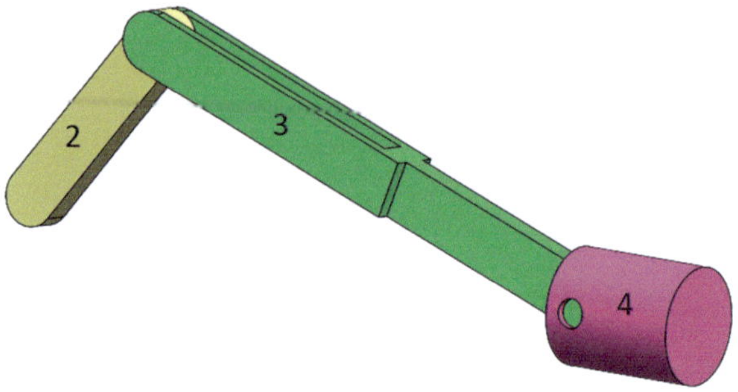

Fig. 4.3 The slider-crank mechanism modeled in SolidWorks, Lee [2]. © 2015 SDC Publications (reproduced with permission)

$$v_B = -r_2\omega_2\left[\sin\theta_2 \pm \frac{\lambda\sin 2\theta_2}{2\sqrt{1-(\lambda\sin\theta_2)^2}}\right];\lambda = r_2/r_3 \tag{4.11}$$

In which the sign to be used is the same as the sign that is used for x_B in Eq. (3.28). Also, for constant ω_2 we have: $\theta_2 = \omega_2 t$.

In the velocity Eq. (4.11), it can be seen that the velocity of the slider (or piston) at the two limit positions: $\theta_2 = 0, \pi$ is zero. These are named the top dead center and the bottom dead center, for the same reason. Furthermore, since Eq. (4.11) for v_B includes periodic functions $\sin\theta_2$ and $\sin 2\theta_2$, it is clear that the velocity of the slider changes periodically with θ_2.

Example 4.1 Consider a slider-crank mechanism with no eccentricity ($e = 0$) and with $r_2 = 75$ mm, and $r_3 = 200$ mm. The crank rotates at -2000 rpm (CW) and starts its motion at $\theta_2 = 0$ and $\theta_3 = 0$ (i.e., at the top dead center).

A. Plot the velocity of the slider represented by point B in Fig. 4.1 in one cycle as a function of time by MATLAB and using Eq. (4.11). Compare this result with the one obtained from SolidWorks simulation.
B. Perform manual calculation at $t = 0.0281$ s and compare the result with that of part A.

This problem has been modeled using SolidWorks in Lee [2] and a slightly modified model is shown in Fig. 4.3.

Part A: This problem was solved using SolidWorks Motion and the velocity of the slider obtained from SolidWorks, together with the output of a MATLAB script that uses Eq. (4.11) are given in Fig. 4.4. We see that the results obtained by the two methods are in perfect agreement with each other.

Part B: Manual calculation at $t = 0.0281$ s, considering the CW motion of the crank is given below:

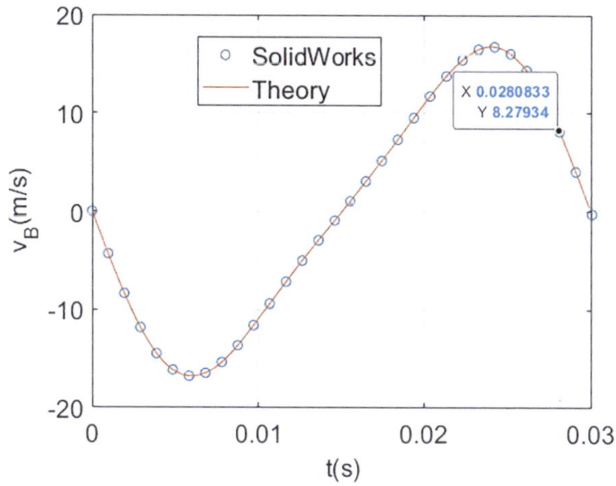

Fig. 4.4 Velocity of the slider in Example 4.1 obtained from SolidWorks together with the output of a MATLAB script that uses Eq. (4.11)

$$\omega_2 = -2000\ \text{rpm} = -2000\ \text{rev/min} \times \frac{2\pi\ \text{rad/rev}}{60\ \text{s/min}} = -209.44\ \text{rad/s}$$

$$\theta_2 = \omega_2 t = -209.44\ \text{rad/s} \times 0.02808\ \text{s} = -5.8811\ \text{rad} = -336.96^\circ$$

Also considering Eq. (4.11) we have:

$$\lambda = \frac{r_2}{r_3} = \frac{75}{200} = 0.375$$

Due to the stated zero initial values of θ_2 and θ_3, in Eq. (4.11) the positive sign should be used:

$$v_B = -r_2\omega_2 \left[\sin\theta_2 + \frac{\lambda \sin 2\theta_2}{2\sqrt{1-(\lambda\sin\theta_2)^2}} \right]$$

Substituting the numerical values of parameters in this equation results in: $v_B = 8.292$ m/s that compares well with the data shown in Fig. 4.4.

It should be pointed out that in Fig. 4.4, the velocity of the slider is zero at three locations. At $t = 0$, the motion starts when the slider is at the top dead center. At $t = 0.015$ s the slider reaches its bottom dead center which corresponds to:

$$\theta_2 = \omega_2 t = -209.44\ \text{rad/s} \times 0.015\ \text{s} = -3.1416\ \text{rad} = -180.00^\circ$$

And finally, at $t = 0.03$ s the slider returns to the top dead center which corresponds to: $\theta_2 = 360°$.

While the position problem discussed in Chap. 3 is nonlinear and requires Taylor series expansion and Newton-Raphson method for linearization, the velocity problem discussed in Chap. 4 is linear. In order to show this matter, in the next section we solve another slider-crank velocity problem.

4.3 Velocity Analysis by the Vector-Loop Method

In Chap. 3, we used the vector-loop method for position analysis of mechanisms. In this chapter, we extend its use to velocity analysis. Let us illustrate the use of the vector-loop method in velocity analysis by an example.

Example 4.2 Consider the slider-crank mechanism shown in Fig. 4.5 in which eccentricity is zero. If ω_2 is the input angular velocity, prove that solving for v_B and ω_3 is a linear problem.

Figure 4.6 illustrates the corresponding vector-loop diagram for Fig. 4.5.

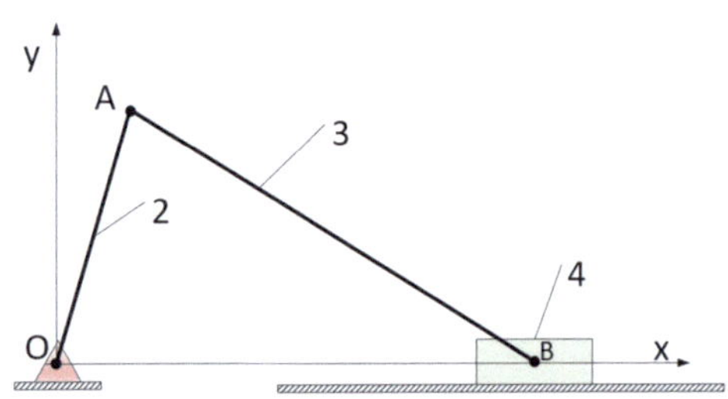

Fig. 4.5 Slider-crank mechanism with no eccentricity; $e = 0$

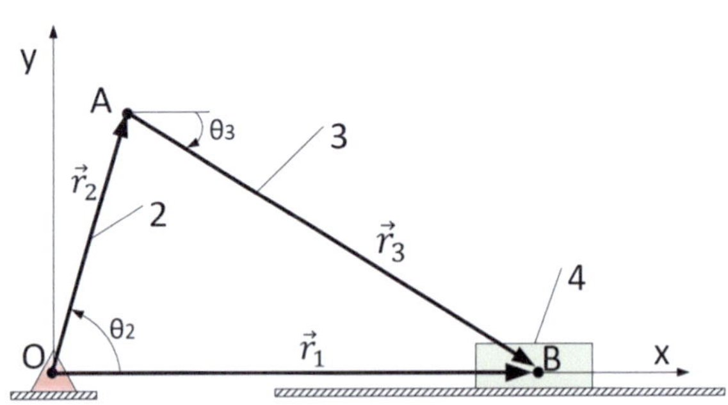

Fig. 4.6 Vector-loop diagram of the slider-crank mechanism with no eccentricity shown in Fig. 4.5

The corresponding vector-loop equation is:

$$\vec{r}_2 + \vec{r}_3 = \vec{r}_1 \tag{4.12}$$

Projecting Eq. (4.12) on the direction of $\vec{r}_1$ results in,

$$r_2 \cos\theta_2 + r_3 \cos\theta_3 = r_1 = x_B \tag{4.13}$$

Projecting Eq. (4.12) on the direction perpendicular to $\vec{r}_1$ gives,

$$r_2 \sin\theta_2 + r_3 \sin\theta_3 = 0 \tag{4.14}$$

Equations (4.13) and (4.14) are two nonlinear position equations for r_1 and θ_3 that should be solved (using Chap. 3 methods) at any value of the input variable θ_2. Once the position analysis is complete, to find the unknown velocity $v_B = \dot{r}_1 = \dot{x}_B$ of the slider and the unknown angular velocity ω_3 of the connecting rod, we take time derivatives of Eqs. (4.13) and (4.14) using the chain rule to get:

$$\begin{cases} -r_2\omega_2 \sin\theta_2 - r_3\omega_3 \sin\theta_3 = v_B \\ r_2\omega_2 \cos\theta_2 + r_3\omega_3 \cos\theta_3 = 0 \end{cases} \tag{4.15}$$

Moving the unknown terms of Eqs. (4.15) to the LHS and the known terms to the RHS we get:

$$\begin{cases} r_3\omega_3 \sin\theta_3 + v_B = -r_2\omega_2 \sin\theta_2 \\ -r_3\omega_3 \cos\theta_3 + 0 \times v_B = r_2\omega_2 \cos\theta_2 \end{cases} \tag{4.16}$$

Equations (4.16) form a linear system of equations in terms of the unknown velocities v_B and ω_3. So, while the position problem (4.13) and (4.14) is nonlinear, the velocity problem is linear and it can be solved easier. We will see in Chap. 5 that the same is correct for acceleration and the acceleration equations are also linear. The order of calculation is very important. The velocity equations given above have θ_3 that is obtained in the position analysis. So the velocity problem cannot be solved before the position problem. We will see that the same relation is true between the velocity problem and the acceleration problem: the acceleration problem cannot be solved before the velocity problem. So here is the order of calculations: position, velocity, and finally acceleration.

Because of the linear nature of Eqs. (4.16), they can be rewritten in the following matrix form:

$$\begin{bmatrix} r_3 \sin\theta_3 & 1 \\ -r_3 \cos\theta_3 & 0 \end{bmatrix} \begin{Bmatrix} \omega_3 \\ v_B \end{Bmatrix} = \begin{Bmatrix} -r_2\omega_2 \sin\theta_2 \\ r_2\omega_2 \cos\theta_2 \end{Bmatrix} \tag{4.17}$$

Since all of the elements in the coefficient matrix and the RHS vector are known, Eq. (4.17) can be solved for the vector of unknowns that includes ω_3 and v_B.

Finally, if the position Eqs. (4.13) and (4.14) are rewritten as:

$$\begin{cases} -r_3\cos\theta_3 + r_1 = r_2\cos\theta_2 \\ -r_3\sin\theta_3 + 0\times r_1 = r_2\sin\theta_2 \end{cases} \tag{4.18}$$

It can be easily verified that the coefficient matrix developed in the matrix equation (4.17) for velocity is nothing but the Jacobian matrix of the corresponding position equations!

Example 4.3 In the slider-crank mechanism shown in Fig. 4.5, the crank rotates at $\omega_2 = -100$ rpm (CW), and it is known that $r_2 = 1$ in and $r_3 = 5$ in. Calculate ω_3 (the angular velocity of link 3) and v_B (the linear velocity of the slider) at an instant when $\theta_2 = 40°$.

So, the inputs are θ_2 and ω_2 and we are looking for ω_3 and v_B. First we solve the nonlinear position problem (4.13) and (4.14) and calculate $r_1$and θ_3. Using Eq. (4.14) and the given data we get:

$$r_2\sin\theta_2 + r_3\sin\theta_3 = 0 \rightarrow \sin\theta_3 = -0.1286$$

That provides two answers:

$$\theta_3 = -7.386\,°, \theta_3 = 187.386\,° \tag{4.19}$$

The $\theta_3 = -7.386°$ *answer*
We substitute $\theta_3 = -7.386°$ into Eq. (4.13) to get:

$$r_2\cos\theta_2 + r_3\cos\theta_3 = r_1 \rightarrow r_1 = 5.725 \text{ in}$$

Converting the unit of the input angular velocity to rad/s we get:

$$\omega_2 = -100 \text{ rpm} = -100 \text{ rev/min} \times \frac{2\pi \frac{\text{rad}}{\text{rev}}}{60\frac{\text{s}}{\text{min}}} = -10.472 \text{ rad/s}$$

Substituting these results into the linear velocity Eq. (4.17) gives:

$$\begin{bmatrix} r_3\sin\theta_3 & 1 \\ -r_3\cos\theta_3 & 0 \end{bmatrix} \begin{Bmatrix} \omega_3 \\ v_B \end{Bmatrix} = \begin{Bmatrix} -r_2\omega_2\sin\theta_2 \\ r_2\omega_2\cos\theta_2 \end{Bmatrix}$$
$$\rightarrow \begin{bmatrix} 5\sin(-7.386\,°) & 1 \\ -5\cos(-7.386\,°) & 0 \end{bmatrix} \begin{Bmatrix} \omega_3 \\ v_B \end{Bmatrix} = \begin{Bmatrix} -(-10.472)\sin 40\,° \\ -10.472\cos 40\,° \end{Bmatrix} \tag{4.20}$$

Focusing on the second row of Eq. (4.20) we get:

$$\omega_3 = 1.618 \text{ rad/s}$$

Since $\omega_3 > 0$, the direction of rotation of link 3 at this instant (when $\theta_2 = 40°$ and $\theta_3 = -7.386°$) is CCW. Now, we use the first row of Eq. (4.20) to get:

$$v_B = 7.771 \text{ in/s}$$

Since $v_B > 0$, at this instant the slider is moving in the positive x direction. This answer corresponds to the posture shown in Fig. 4.7.

The $\theta_3 = 187.386°$ *answer*

We substitute $\theta_3 = 187.386°$ into Eq. (4.13) to get:

$$r_2 \cos\theta_2 + r_3 \cos\theta_3 = r_1 \rightarrow r_1 = -4.192 \text{ in}$$

Substituting these results into the linear velocity Eq. (4.17) gives:

$$\begin{bmatrix} 5\sin(187.386°) & 1 \\ -5\cos(187.386°) & 0 \end{bmatrix} \begin{Bmatrix} \omega_3 \\ v_B \end{Bmatrix} = \begin{Bmatrix} -(-10.472)\sin 40° \\ -10.472\cos 40° \end{Bmatrix} \tag{4.21}$$

Considering the second row of Eq. (4.21) we get:

$$\omega_3 = -1.618 \text{ rad/s}$$

Since $\omega_3 < 0$, the direction of rotation of link 3 at this instant (when $\theta_2 = 40°$ and $\theta_3 = 187.386°$) is CW. Now, we use the first row of Eq. (4.21) to get:

$$v_B = 5.691 \text{ in/s}$$

Since $v_B > 0$, at this instant the slider is moving in the positive x direction. This solution corresponds to the posture shown in Fig. 4.8.

The problem has two solutions and the question as to which one would actually happen depends on the way the slider-crank mechanism has been initially assembled.

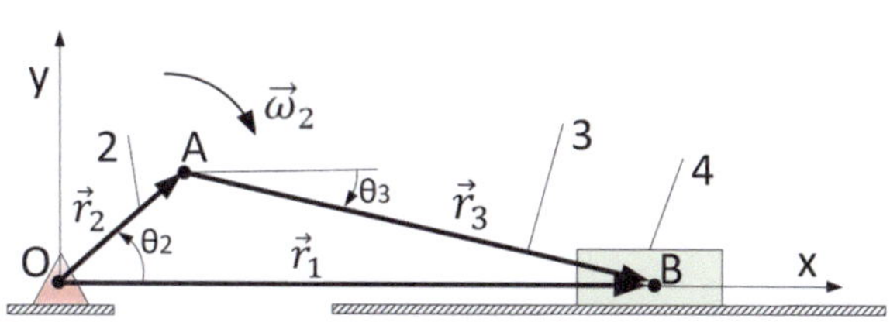

Fig. 4.7 The $\theta_3 = -7.386°$ solution of Example 4.3

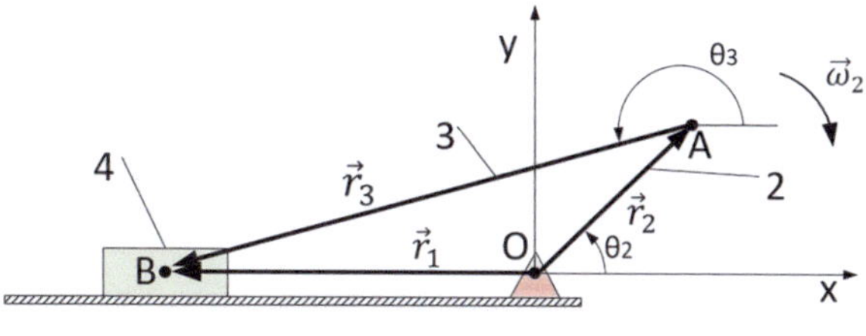

Fig. 4.8 The $\theta_3 = 187.386°$ solution of Example 4.3

Solution by the Cramer's Rule
In general, for solving the following system of linear equations:

$$\begin{bmatrix} a & b \\ c & d \end{bmatrix}\begin{Bmatrix} x \\ y \end{Bmatrix} = \begin{Bmatrix} A \\ B \end{Bmatrix} \tag{4.22}$$

we can use the Cramer's rule that results in the following solutions:

$$x = \frac{\begin{vmatrix} A & b \\ B & d \end{vmatrix}}{\begin{vmatrix} a & b \\ c & d \end{vmatrix}}; y = \frac{\begin{vmatrix} a & A \\ c & B \end{vmatrix}}{\begin{vmatrix} a & b \\ c & d \end{vmatrix}} \tag{4.23}$$

In which the determinant of the coefficient matrix is defined as follows:

$$\begin{vmatrix} a & b \\ c & d \end{vmatrix} = ad - bc \tag{4.24}$$

Now, let us go back to the velocity equations developed before that were linear and, because of that, could be expressed in the following matrix form:

$$\begin{bmatrix} r_3 \sin\theta_3 & 1 \\ -r_3\cos\theta_3 & 0 \end{bmatrix}\begin{Bmatrix} \omega_3 \\ v_B \end{Bmatrix} = \begin{Bmatrix} -r_2\omega_2\sin\theta_2 \\ r_2\omega_2\cos\theta_2 \end{Bmatrix} \tag{4.17}$$

After completing the solution of the position problem (4.13) and (4.14), all of the elements in the coefficients matrix and the RHS vector of the velocity problem (4.17) are known. Therefore, the velocity Eq. (4.17) can be solved for the vector of unknowns that includes ω_3 and v_B. We verified before that the coefficients matrix in Eq. (4.17) for velocity is nothing but the Jacobian matrix of the nonlinear position Eqs. (4.18):

$$J = \begin{bmatrix} r_3 \sin\theta_3 & 1 \\ -r_3\cos\theta_3 & 0 \end{bmatrix} \tag{4.25}$$

The determinant of this Jacobian matrix is:

$$|J| = r_3\cos\theta_3 \tag{4.26}$$

The solutions of Eq. (4.17), using Cramer's rule, can therefore be written as,

$$\omega_3 = \frac{\begin{vmatrix} -r_2\omega_2\sin\theta_2 & 1 \\ r_2\omega_2\cos\theta_2 & 0 \end{vmatrix}}{|J|}; v_B = \frac{\begin{vmatrix} r_3\sin\theta_3 & -r_2\omega_2\sin\theta_2 \\ -r_3\cos\theta_3 & r_2\omega_2\cos\theta_2 \end{vmatrix}}{|J|} \tag{4.27}$$

That result in,

$$\omega_3 = -\frac{r_2\omega_2 \cos\theta_2}{r_3 \cos\theta_3}; v_B = \frac{r_2\omega_2 \sin(\theta_3 - \theta_2)}{\cos\theta_3} \tag{4.28}$$

4.4 Velocity Analysis Using First-Order Kinematic Coefficients

Recall the slider-crank mechanism and the corresponding vector-loop diagram shown in Figs. 4.1 and 4.2, respectively. Our aim is to calculate the velocity of the slider v_B at any crank angle θ_2. We recall Eq. (3.28) for the position of the slider, x_B in terms of θ_2.

$$x_B = r_2 \cos\theta_2 \pm \sqrt{r_3^2 - (e + r_2 \sin\theta_2)^2} \tag{3.28}$$

The plus sign is used when the slider is on the right-hand side of the revolute joint O, and the minus sign is applied otherwise. In order to calculate v_B, we use the chain rule and obtain:

$$v_B = \dot{x}_B = \frac{dx_B}{dt} = \frac{dx_B}{d\theta_2}\frac{d\theta_2}{dt} = x'_B\omega_2 \tag{4.29}$$

In Eq. (4.29), the term,

$$x'_B = \frac{dx_B}{d\theta_2} = \frac{d(\text{Output})}{d(\text{Input})} \tag{4.4}$$

represents how the position of the slider (the output) changes with changes of the crank angle (the input). This term, i.e., x'_B, is called the first-order kinematic coefficient of x_B. While the velocity of the slider depends on the input angular velocity ω_2 that is given to the mechanism, x'_B is independent of time and ω_2. Therefore, the first-order kinematic coefficient is an inherent property of the geometry of the mechanism.

Equation (4.29) expresses the derivative of x_B with respect to time (i.e., the velocity of the slider or v_B) in terms of the first-order kinematic coefficient of x_B, i.e., x'_B, and the input angular velocity ω_2.

$$v_B = \dot{x}_B = x'_B\omega_2 \tag{4.30}$$

Similar to Eq. (4.29), for the output angle θ_3, the angular velocity ω_3 can be expressed as,

$$\omega_3 = \dot{\theta}_3 = \frac{d\theta_3}{dt} = \frac{d\theta_3}{d\theta_2}\frac{d\theta_2}{dt} = \theta'_3\omega_2 \tag{4.31}$$

in which,

$$\theta'_3 = \frac{d\theta_3}{d\theta_2} = \frac{d(\text{Output})}{d(\text{Input})} \tag{4.32}$$

The derivative θ'_3 is the first-order kinematic coefficient of θ_3 and it is equal to the change in angle θ_3 for unit positive change in the input angle θ_2. This quantity is only a function of the geometry of the mechanism and it is independent of the input angular velocity and time.

Similar to Eq. (4.30) we have:

$$\omega_3 = \theta'_3\omega_2 \tag{4.33}$$

In general, the first-order kinematic coefficient of an output variable is defined as follows:

$$\begin{aligned}\text{Output velocity} = {} & \text{First order kinematic coefficient of output} \\ & \times \text{Input velocity}\end{aligned} \tag{4.34}$$

In other words, for any output variable: ( ) in any mechanism we can write the following,

$$\dot{(\ )} = (\ )' \times \text{input velocity} \tag{4.35}$$

Equivalently, the first-order kinematic coefficient of a variable can be thought of as the ratio of the output velocity over the input velocity,

$$\text{First order kinematic coefficient of an output variable} = \frac{\text{Output velocity}}{\text{Input velocity}} \tag{4.36}$$

Using Eqs. (4.30) and (4.33), the following first-order kinematic coefficients can be written for a slider-crank mechanism that follow the general definition given in Eq. (4.36):

$$x'_B = \frac{v_B}{\omega_2};\ \theta'_3 = \frac{\omega_3}{\omega_2} \tag{4.37}$$

Example 4.4 Consider the slider-crank mechanism with no eccentricity ($e = 0$) that we discussed in Example 4.1 and is illustrated in Fig. 4.9. We know that $r_2 = 75$ mm, and $r_3 = 200$ mm ($\lambda = 0.375$). The crank rotates at -209.44 rad/s (CW) and it starts

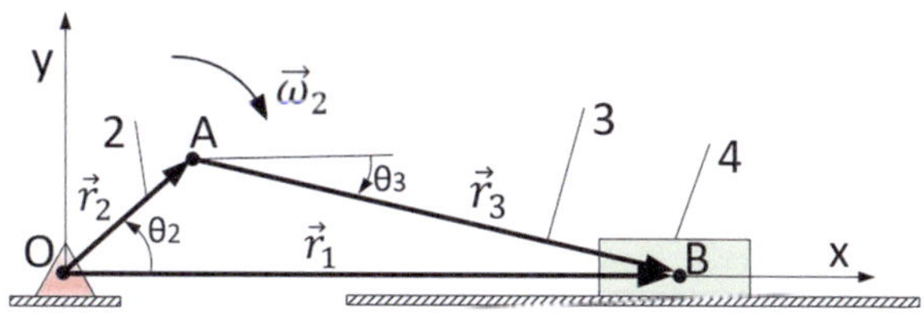

Fig. 4.9 Vector-loop diagram of the slider-crank mechanism

its motion at $\theta_2 = 0$ and $\theta_3 = 0$. We would like to calculate the velocity of the slider when $\theta_2 = 45°$ using the following methods:

A. Algebraic method; Eq. (4.11)
B. The vector loop and the matrix solution; Eq. (4.28)
C. First-order kinematic coefficients; Eqs. (4.37)

A. **Algebraic method**; Eq. (4.11): Due to the stated zero initial values of θ_2 and θ_3, the positive sign in Eq. (4.11) should be used.

$$v_B = -r_2\omega_2\left[\sin\theta_2 \pm \frac{\lambda\sin 2\theta_2}{2\sqrt{1-(\lambda\sin\theta_2)^2}}\right];\lambda = r_2/r_3 \tag{4.11}$$

Substituting the given values in Eq. (4.11) (notice $\omega_2 = -\ 209.44$ rad/s) we get: $v_B = 14.16$ m/s. Since the calculated value is positive, we conclude that when $\theta_2 = 45°$, the slider is moving in the positive x direction that is in harmony with Fig. 4.9 and the direction of motion of the crank.

B. **The vector loop and the matrix solution;** Eq. (4.28): Using the vector-loop equations and the matrix method we got:

$$\omega_3 = -\frac{r_2\omega_2\cos\theta_2}{r_3\cos\theta_3}, v_B = \frac{r_2\omega_2\sin(\theta_3-\theta_2)}{\cos\theta_3} \tag{4.28}$$

To calculate v_B using this method, we need to obtain θ_3 as the first step. Recalling Eq. (4.14)

$$r_2\sin\theta_2 + r_3\sin\theta_3 = 0 \tag{4.14}$$

We get:

$$\sin\theta_3 = -0.265$$

Therefore: $\theta_3 = -\ 15.367°$ or $\theta_3 = -\ 164.633°$. Between these two, the solution that is in harmony with the given initial condition of the motion of the mechanism

($\theta_2 = 0$ and $\theta_3 = 0$) is: $\theta_3 = -15.367°$. (It should be pointed out that the $\theta_3 = -164.633°$ answer is in harmony with the $\theta_2 = 0$ and $\theta_3 = 180°$ initial conditions.) Using $\theta_3 = -15.367°$ in Eq. (4.28) we get, $v_B = 14.16$ m/s. This result is in harmony with what we got using the first method in part A.

C. **First-order kinematic coefficients**; Eqs. (4.37): Recall the vector-loop equation:

$$\vec{r}_2 + \vec{r}_3 = \vec{r}_1 \tag{4.12}$$

Its projected components on the $\vec{r}_1$ direction and the direction perpendicular to $\vec{r}_1$ are:

$$r_2 \cos\theta_2 + r_3 \cos\theta_3 = r_1 \tag{4.13}$$

$$r_2 \sin\theta_2 + r_3 \sin\theta_3 = 0 \tag{4.14}$$

Using Eq. (4.14) we already got: $\theta_3 = -15.367°$. Similarly, Eq. (4.13) results in:

$$r_1 = x_B = 0.246\text{m}$$

In order to calculate the first-order kinematic coefficients x'_B and θ'_3, we take derivatives of Eqs. (4.13) and (4.14) with respect to the input angle θ_2. First we move the output terms to the left and the input terms to the right. So we get:

$$\begin{cases} -r_3 \cos\theta_3 + r_1 = r_2 \cos\theta_2 \\ -r_3 \sin\theta_3 + 0 \times r_1 = r_2 \sin\theta_2 \end{cases} \tag{4.18}$$

Recalling, Eqs. (4.4) and (4.32), we can take derivatives of Eqs. (4.18) with respect to the input angle θ_2 using the chain rule to get:

$$\begin{cases} r_3\theta'_3 \sin\theta_3 + x'_B = -r_2 \sin\theta_2 \\ -r_3\theta'_3 \cos\theta_3 + 0 \times x'_B = r_2 \cos\theta_2 \end{cases}$$

Which can be written in the following matrix form:

$$\begin{bmatrix} r_3\sin\theta_3 & 1 \\ -r_3 \cos\theta_3 & 0 \end{bmatrix} \begin{Bmatrix} \theta'_3 \\ x'_B \end{Bmatrix} = \begin{Bmatrix} -r_2 \sin\theta_2 \\ r_2 \cos\theta_2 \end{Bmatrix} \tag{4.38}$$

The obtained coefficient matrix is the Jacobian matrix of the mechanism we have seen before.

$$J = \begin{bmatrix} r_3\sin\theta_3 & 1 \\ -r_3 \cos\theta_3 & 0 \end{bmatrix} \tag{4.25}$$

Using Cramer's rule, i.e., Eq. (4.23), and recalling (4.26), the solutions of Eq. (4.38) can be written as follows:

$$\theta_3' = \frac{\begin{vmatrix} -r_2 \sin\theta_2 & 1 \\ r_2 \cos\theta_2 & 0 \end{vmatrix}}{|J|}, x_B' = \frac{\begin{vmatrix} r_3 \sin\theta_3 & -r_2 \sin\theta_2 \\ -r_3 \cos\theta_3 & r_2 \cos\theta_2 \end{vmatrix}}{|J|} \tag{4.39}$$

or,

$$\theta_3' = -\frac{r_2 \cos\theta_2}{r_3 \cos\theta_3}, x_B' = \frac{r_2 \sin(\theta_3 - \theta_2)}{\cos\theta_3} \tag{4.40}$$

Substituting the provided data into the kinematic coefficients (4.40) gives:

$$\theta_3' = -\frac{r_2 \cos\theta_2}{r_3 \cos\theta_3} = -0.275 \text{ rad/rad}; x_B' = \frac{r_2 \sin(\theta_3 - \theta_2)}{\cos\theta_3} = -67.62 \text{ mm/rad} \tag{4.41}$$

The reason why these values are negative is because a small increase of θ_2 (i.e., moving the crank CCW) when it is close to 45° causes the slider to go to the left and link 3 to rotate CW (i.e., it decreases x_B and θ_3). Now, we substitute Eqs. (4.40) into Eqs. (4.33) and (4.30) to get ω_3 and v_B, respectively,

$$\omega_3 = -\frac{r_2\omega_2 \cos\theta_2}{r_3 \cos\theta_3}; v_B = \frac{r_2\omega_2 \sin(\theta_3 - \theta_2)}{\cos\theta_3} \tag{4.42}$$

Equations (4.42) are identical to Eqs. (4.28) we used in Part B; so the final answer for the slider velocity using the method of kinematic coefficients is the same as well: $v_B = 14.16$ m/s.

Similarly, one could use Eqs. (4.30) and (4.41) to get v_B at $\theta_2 = 45°$:

$$v_B = x_B'\omega_2 = -\frac{67.62\text{mm}}{\text{rad}} \times \left(-209.44\frac{\text{rad}}{\text{s}}\right) = 14.16\text{m/s}$$

Finally, one could use Eqs. (4.33) and (4.41) to get ω_3 at $\theta_2 = 45°$:

$$\omega_3 = \theta_3'\omega_2 = -\frac{0.275\text{rad}}{\text{rad}} \times \left(-209.44\frac{\text{rad}}{\text{s}}\right) = 57.60 \text{ rad/s}$$

The positive value of ω_3 makes sense as it means at $\theta_2 = 45°$ link 3 is moving in the CCW direction which can be verified by examining Fig. 4.9.

4.5 Calculating Dead (Locked) Positions and Limit Positions

Recall the case of a slider-crank mechanism with no eccentricity shown in Fig. 4.6.

Using the vector-loop diagram shown in Fig. 4.6, the velocity equations have been developed before and because of the linear nature of the equations, they were written in matrix form as follows:

$$\begin{bmatrix} r_3 \sin\theta_3 & 1 \\ -r_3\cos\theta_3 & 0 \end{bmatrix} \begin{Bmatrix} \omega_3 \\ v_B \end{Bmatrix} = \begin{Bmatrix} -r_2\omega_2\sin\theta_2 \\ r_2\omega_2\cos\theta_2 \end{Bmatrix} \tag{4.17}$$

Solutions of Eq. (4.17) were derived before as:

$$\omega_3 = -\frac{r_2\omega_2\cos\theta_2}{r_3\cos\theta_3},\ v_B = \frac{r_2\omega_2\sin(\theta_3-\theta_2)}{\cos\theta_3} \tag{4.28}$$

in which we used $|J|$ in the denominators when we used the Cramer's rule, see Eqs. (4.23) and (4.27) where,

$$J = \begin{bmatrix} r_3\sin\theta_3 & 1 \\ -r_3\cos\theta_3 & 0 \end{bmatrix}; \tag{4.25}$$

$$|J| = r_3\cos\theta_3 \tag{4.26}$$

Dead Positions It can be seen that if the determinant of the Jacobian matrix is zero, i.e., $|J| = 0$, the output variables ω_3 and v_B increase without bound which is physically impossible. The only way $|J| = 0$ is physically possible is when the numerators of these fractions are also zero. But this is possible only if the input angular velocity is zero, i.e., $\omega_2 = 0$. So, we conclude that if $|J| = 0$, the input should also be zero. In other words, if $|J| = 0$, no motion is possible (as no input movement can be applied). For this reason such a position is called a dead or locked position. In this case, using Eq. (4.26) we find the dead positions as:

$$\theta_3 = 90°, 270° \tag{4.43}$$

Figure 4.10 illustrates a dead position of a slider-crank mechanism at $\theta_3 = 270°$.

It should be clear that an input torque applied on crank 2 at $\theta_3 = 90°, 270°$ cannot move the mechanism. Since link 3 is a two-force member (see Chap. 10) at these positions there will be no force component applied from link 3 on slider 4 that can move it horizontally! Therefore, indeed at these θ_3 values, the mechanism has dead or locked positions.

Limit Positions A slider-crank mechanism also has limit positions at which a part in the mechanism momentarily stops and then its motion direction reverses. The top dead center and the bottom dead center in a slider-crank mechanism are examples of

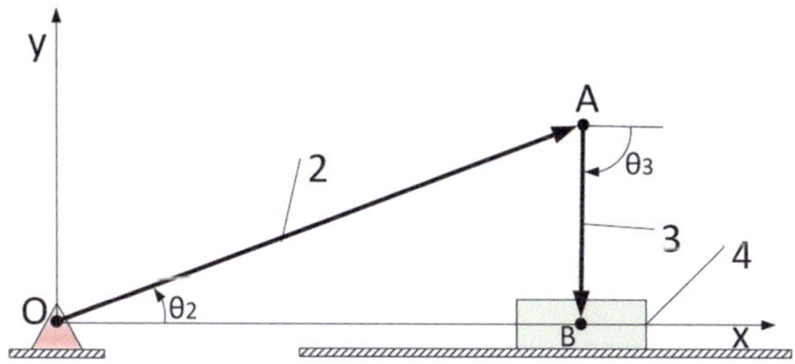

Fig. 4.10 A dead position of a slider-crank mechanism at $\theta_3 = 270°$

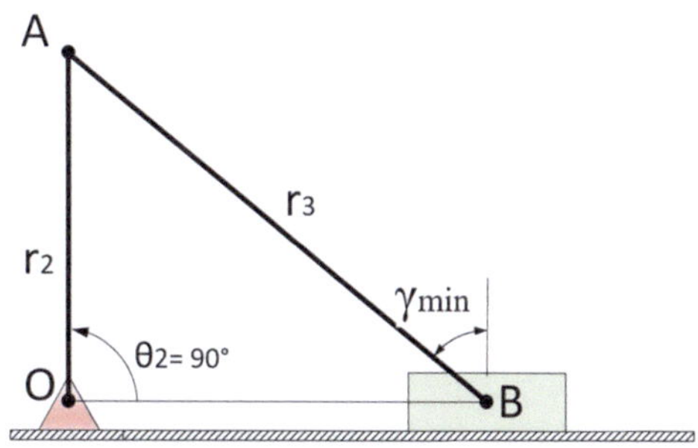

Fig. 4.11 The minimum transmission angle position and a limit position of link 3 at $\theta_2 = 90°$, for the no eccentricity case

limit positions (see Figs. 3.12 to 3.15). For limit position to happen, a part should momentarily have zero velocity. Recalling Eqs. (4.37):

$$x'_B = \frac{v_B}{\omega_2}; \theta'_3 = \frac{\omega_3}{\omega_2} \tag{4.37}$$

So, a temporary stop at a limit position for the slider or link 3 happens if we have, respectively,

$$\dot{x}_B = 0 \text{ or } \omega_3 = 0 \tag{4.44}$$

According to Eq. (4.37), if either of these velocities is zero, its corresponding first-order kinematic coefficient x'_B or θ'_3 will also be zero. So, for the mechanism to have limit positions, the corresponding first-order kinematic coefficients or, equivalently, the respective output velocity terms should be zero. Recalling the output velocities given in Eqs. (4.28), the limit positions that correspond to the reversal of the direction of motion of link 3 happen at:

$$\omega_3 = 0 \rightarrow r_2\omega_2 \cos\theta_2 = 0 \rightarrow \theta_2 = 90°, 270° \tag{4.45}$$

Therefore, the reversal of the direction of motion of link 3 happens at $\theta_2 = 90°$, 270°, see Figs. 4.11 and 4.12. Such positions also correspond to when the transmission angle of the mechanism is maximum or minimum (see the discussion about Figs. 3.18 and 3.19).

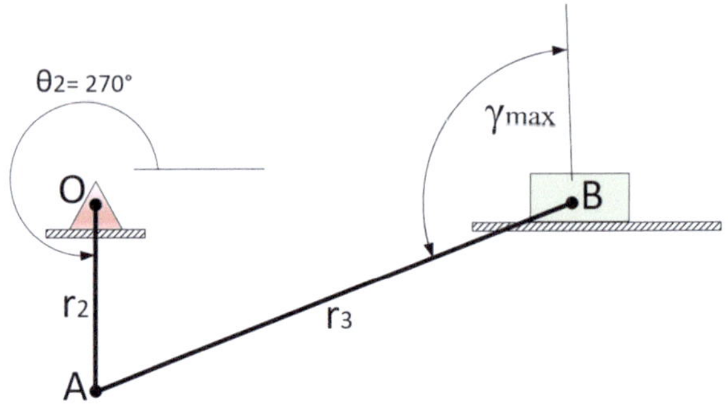

Fig. 4.12 The maximum transmission angle position and a limit position of link 3 at $\theta_2 = 270°$, for the no eccentricity case

Similarly, the limit positions of the motion of the slider can be calculated. Those limit positions happen at $\dot{x}_B = 0$ or $x'_B = 0$. Therefore, considering Eqs. (4.28), the limit positions of the slider happen if we have:

$$r_2\omega_2 \sin(\theta_3 - \theta_2) = 0 \rightarrow \sin(\theta_3 - \theta_2) = 0 \tag{4.46}$$

or equivalently when:

$$\theta_3 = \theta_2, \theta_3 - \theta_2 = 180\,° \tag{4.47}$$

Figures 3.12 to 3.15 illustrate such postures and they illustrate the top dead center and the bottom dead center.

Changing the Input In the previous analysis of a slider-crank mechanism θ_2 was the input and x_B the output. One can now change the input and use x_B as the new input. (This is what happens in an internal combustion engine where the combustion force moves the piston and therefore x_B is the input to the machine.). In such a case, θ_2 and θ_3 are the outputs and they are functions of x_B. Recalling the projected vector-loop equations:

$$r_2 \cos\theta_2 + r_3 \cos\theta_3 = r_1 = x_B \tag{4.13}$$

$$r_2 \sin\theta_2 + r_3 \sin\theta_3 = 0 \tag{4.14}$$

one can take derivatives with respect to x_B to get the following results:

$$\begin{cases} -r_2\theta_2' \sin\theta_2 - r_3\theta_3' \sin\theta_3 = 1 \\ r_2\theta_2' \cos\theta_2 + r_3\theta_3' \cos\theta_3 = 0 \end{cases} \tag{4.48}$$

In which the new kinematic coefficients θ'_2 and θ'_3 are defined with respect to the new input, x_B:

$$\theta_2' = \frac{d\theta_2}{dx_B}; \theta_3' = \frac{d\theta_3}{dx_B} \tag{4.49}$$

Rewriting Eq. (4.48) in matrix form results in:

$$\begin{bmatrix} -r_2 \sin\theta_2 & -r_3 \sin\theta_3 \\ r_2 \cos\theta_2 & r_3 \cos\theta_3 \end{bmatrix} \begin{Bmatrix} \theta_2' \\ \theta_3' \end{Bmatrix} = \begin{Bmatrix} 1 \\ 0 \end{Bmatrix} \tag{4.50}$$

It can be easily verified that the coefficient matrix in Eq. (4.50) is the Jacobian matrix of the projected vector-loop Eqs. (4.13) and (4.14), if θ_2 and θ_3 are chosen as the output variables.

$$J = \begin{bmatrix} -r_2 \sin\theta_2 & -r_3 \sin\theta_3 \\ r_2 \cos\theta_2 & r_3 \cos\theta_3 \end{bmatrix} \tag{4.51}$$

In Eq. (4.51) the first column includes derivatives of Eqs. (4.13) and (4.14) with respect to θ_2 and the second column includes derivatives of the same equations with respect to θ_3. Its determinant is:

$$|J| = -r_2 r_3 \sin(\theta_2 - \theta_3) \tag{4.52}$$

Equation (4.50) can be solved for the unknown kinematic coefficients θ_2' and θ_3' by using the Cramer's rule [see Eq. (4.23)] to result in:

$$\theta_2' = \frac{\begin{vmatrix} 1 & -r_3 \sin\theta_3 \\ 0 & r_3 \cos\theta_3 \end{vmatrix}}{|J|}; \theta_3' = \frac{\begin{vmatrix} -r_2 \sin\theta_2 & 1 \\ r_2 \cos\theta_2 & 0 \end{vmatrix}}{|J|}$$

Multiplication of the obtained kinematic coefficients θ_2' and θ_3' by the input velocity v_B result in the following output angular velocities ω_2 and ω_3:

$$\omega_2 = \theta_2' v_B = -\frac{r_3 \cos\theta_3 v_B}{-r_2 r_3 \sin(\theta_2 - \theta_3)}; \omega_3 = \theta_3' v_B = \frac{r_2 \cos\theta_2 v_B}{r_2 r_3 \sin(\theta_2 - \theta_3)} \tag{4.53}$$

Dead Positions Again it can be seen that if the determinant of the Jacobian matrix, i.e., Eq. (4.52) is zero, the output variables ω_2 and ω_3 increase without bound which is physically impossible. The only way $|J| = 0$ is physically possible is when the numerators of Eqs. (4.53) are also zero. But this is possible only if the input velocity of the slider is zero, i.e., $v_B = 0$. So, we conclude that if $|J| = 0$, the input should also be zero. In other words, if $|J| = 0$, no motion is possible (as no input movement can be applied). For this reason, such a position is called a dead or locked position. In this case, the dead position happens if:

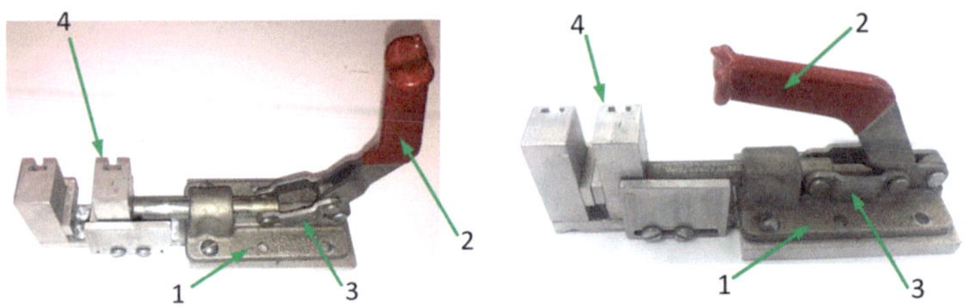

Fig. 4.13 The slider-crank toggle clamp mechanism; left: not locked, right: locked 1: body, 2. lever, 3. connecting link, 4. slider or clamping pad

$$r_2 r_3 \sin(\theta_2 - \theta_3) = 0 \rightarrow \sin(\theta_2 - \theta_3) = 0 \tag{4.54}$$

or when,

$$\theta_3 = \theta_2; \theta_3 - \theta_2 = 180\,^\circ \tag{4.47}$$

Figures 3.12 to 3.15 illustrate such postures and they are referred to as top dead center and bottom dead center. These postures are also frequently used in mechanical locks (see Fig. 4.13) as they literally do not allow any motions. In Fig. 4.13, rotation of the lever downward brings the connecting link 3 in alignment with the hinged segment of lever 2 and effectively satisfies condition (4.47) for the dead or locked position. Interestingly, it is clear that if any axial force is applied to the slider (input) since the members are collinear, and no moment arm about the revolute joints exist, the mechanism cannot move and hence it is locked. These locked positions of the mechanism with x_B as the input are therefore the same as the limit positions of x_B when θ_2 was the input.

Limit Positions The limit positions of θ_2 when x_B is the input happen when $\omega_2 = 0, \theta_2' = 0$. Referring to Eqs. (4.53), the limit positions of θ_2 happen when:

$$\cos\theta_3 = 0 \rightarrow \theta_3 = 90\,^\circ, 270\,^\circ \tag{4.43}$$

This posture is the same as the one shown in Fig. 4.10. Therefore, this limit position of θ_2 with x_B as the input is the same as the locked position of the mechanism when θ_2 was the input.

Finally, the limit positions of θ_3 when x_B is the input happen when $\omega_3 = 0, \theta_3' = 0$. Referring to Eqs. (4.53), the limit positions of θ_3 happen when:

$$\cos\theta_2 = 0 \rightarrow \theta_2 = 90\,^\circ, 270\,^\circ \tag{4.45}$$

This posture is the same as the one shown in Figs. 4.11 and 4.12. Therefore, this limit position of θ_3 with x_B as the input is the same as the limit position of θ_3 when θ_2 was the input. This analysis is summarized in Table 4.1.

Table 4.1 Classification of limit and dead positions in a slider-crank mechanism depending on the chosen input and output

		Dead Positions	Output Variable and Limit	
			x_B	θ_3
Input Variable	θ_2	$\theta_3 = 90°, 270°$	$\theta_3 = \theta_2$ $\theta_3 - \theta_2 = 180°$	$\theta_2 = 90°, 270°$
Input Variable	x_B	$\theta_3 = \theta_2$ $\theta_3 - \theta_2 = 180$	$\theta_3 = 90°, 270°$	$\theta_2 = 90°, 270°$
			θ_2	θ_3
			Output Variable and Limit	

Fig. 4.14 The pin-in-the-slot mechanism and its vector-loop diagram

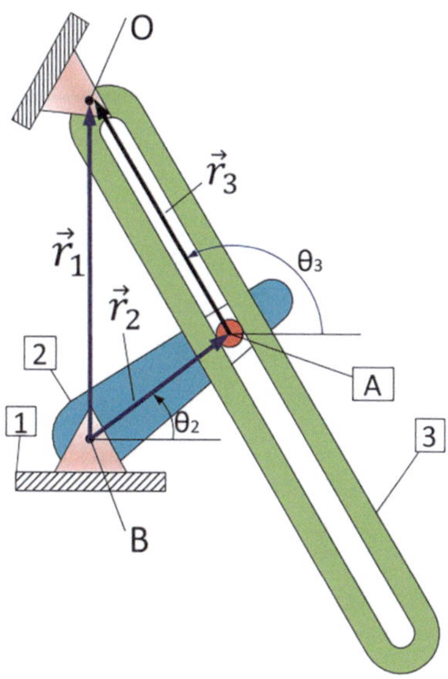

Example 4.5 Consider the mechanism and the vector-loop diagram shown in Fig. 4.14 in which pin A is rigidly attached to member 2 and can slide inside the slot of member 3. Calculate the limit and the dead positions of the mechanism in the following cases: A. Assuming θ_2 is the input B. Assuming θ_3 is the input.

The vector-loop equation corresponding to the vector-loop diagram in Fig. 4.14 is:

$$\vec{r}_2 + \vec{r}_3 = \vec{r}_1 \tag{4.55}$$

The corresponding projected vector-loop equations in the $\vec{r}_1$ direction and the direction perpendicular to $\vec{r}_1$ are as follows:

$$\begin{cases} r_2 \sin\theta_2 + r_3 \sin\theta_3 = r_1 \\ r_2 \cos\theta_2 + r_3 \cos\theta_3 = 0 \end{cases} \tag{4.56}$$

A. **Assuming θ_2 is the input**

Rearranging Eqs. (4.56) in such a way that the unknown terms including r_3 and θ_3 are moved to the left-hand side and the known terms are moved to the right-hand side we get:

$$\begin{cases} r_3 \sin\theta_3 = r_1 - r_2 \sin\theta_2 \\ r_3 \cos\theta_3 = -r_2 \cos\theta_2 \end{cases} \tag{4.57}$$

Equations (4.57) can be solved for r_3 and θ_3 and it concludes solution of the position problem. In order to calculate the first-order kinematic coefficients, we take derivatives of Eqs. (4.57) with respect to the input angle θ_2 and obtain:

$$\begin{cases} r_3\theta_3' \cos\theta_3 + r_3' \sin\theta_3 = -r_2 \cos\theta_2 \\ -r_3\theta_3' \sin\theta_3 + r_3' \cos\theta_3 = r_2 \sin\theta_2 \end{cases} \tag{4.58}$$

where r_3' and θ_3' are, respectively, the first-order kinematic coefficients of r_3 and θ_3 with respect to the input angle θ_2. Equations (4.58) are two equations for two unknowns: r_3' and θ_3'. These equations can be written in matrix form as follows:

$$\begin{bmatrix} r_3\cos\theta_3 & \sin\theta_3 \\ -r_3 \sin\theta_3 & \cos\theta_3 \end{bmatrix} \begin{Bmatrix} \theta_3' \\ r_3' \end{Bmatrix} = \begin{Bmatrix} -r_2 \cos\theta_2 \\ r_2 \sin\theta_2 \end{Bmatrix} \tag{4.59}$$

where,

$$J = \begin{bmatrix} r_3\cos\theta_3 & \sin\theta_3 \\ -r_3 \sin\theta_3 & \cos\theta_3 \end{bmatrix} \tag{4.60}$$

is the Jacobian matrix in which the first column includes derivatives of the position Eqs. (4.57) with respect to θ_3 and the second column includes derivatives of the position equations with respect to r_3. To find the dead or locked positions, we need to find postures at which the determinant of the Jacobian matrix becomes zero:

$$|J| = r_3\left(\cos^2\theta_3 + \sin^2\theta_3\right) = r_3 \tag{4.61}$$

Equation (4.61) shows that the determinant of the Jacobian matrix is never zero; so the mechanism has no dead or locked position and if θ_2 is the input angle the motion never stops. In order to find the limit positions of the mechanism we solve Eq. (4.59) using Cramer's rule and calculate the kinematic coefficients r_3' and θ_3'. For θ_3' we have:

$$\theta_3' = \frac{\begin{vmatrix} -r_2\cos\theta_2 & \sin\theta_3 \\ r_2\sin\theta_2 & \cos\theta_3 \end{vmatrix}}{|J|} = \frac{-r_2\cos(\theta_2 - \theta_3)}{r_3} \tag{4.62}$$

The limit positions of link 3 happen when $\theta_3' = 0$; using Eq. (4.62) we get:

$$\theta_2 - \theta_3 = 90\,°, 270\,° \tag{4.63}$$

So, the limit positions of the orientation of link 3 when rotation of link 2 is the input, happen when these two links are perpendicular to each other. This matter was illustrated before in Figs. 2.44 and 2.45 for calculating the time ratio for this quick-return mechanism (see Sect. 2.5).

Now, in order to find the limit positions of r_3, we calculate the kinematic coefficient r_3' by solving Eq. (4.59) using Cramer's rule:

$$r_3' = \frac{\begin{vmatrix} r_3 \cos\theta_3 & -r_2 \cos\theta_2 \\ -r_3 \sin\theta_3 & r_2 \sin\theta_2 \end{vmatrix}}{|J|} = \frac{r_2 r_3 \sin(\theta_2 - \theta_3)}{r_3} \tag{4.64}$$

Limit positions of r_3 happen when $r_3' = 0$. Therefore we get:

$$\theta_2 - \theta_3 = 0, 180\,° \tag{4.65}$$

In other words, the limit positions of r_3 (i.e., the limit positions of pin A in Fig. 4.14 with respect to O) when rotation of link 2 is the input, happen when links 2 and 3 are collinear (i.e., $\vec{r}_2$ is vertically up or down). This result is completely expected!

B. **Assuming θ_3 is the input**

In the same problem, we now assume that θ_3 is the input. The projected vector-loop equations have not changed:

$$\begin{cases} r_2 \sin\theta_2 + r_3 \sin\theta_3 = r_1 \\ r_2 \cos\theta_2 + r_3 \cos\theta_3 = 0 \end{cases} \tag{4.56}$$

But Eqs. (4.56) should now be solved for θ_2 and r_3. In order to calculate the first-order kinematic coefficients, we take derivatives of Eqs. (4.56) with respect to the input angle θ_3:

$$\begin{cases} r_2\theta_2' \cos\theta_2 + r_3' \sin\theta_3 = -r_3 \cos\theta_3 \\ -r_2\theta_2' \sin\theta_2 + r_3' \cos\theta_3 = r_3 \sin\theta_3 \end{cases} \tag{4.66}$$

where θ_2' and r_3' are the first-order kinematic coefficients of θ_2 and r_3 with respect to the input variable, θ_3. Eqs. (4.66) are two equations for two unknowns: θ_2' and r_3'. These equations can be written in the following matrix form:

$$\begin{bmatrix} r_2 \cos\theta_2 & \sin\theta_3 \\ -r_2 \sin\theta_2 & \cos\theta_3 \end{bmatrix} \begin{Bmatrix} \theta_2' \\ r_3' \end{Bmatrix} = \begin{Bmatrix} -r_3 \cos\theta_3 \\ r_3 \sin\theta_3 \end{Bmatrix} \tag{4.67}$$

where

$$J = \begin{bmatrix} r_2 \cos\theta_2 & \sin\theta_3 \\ -r_2 \sin\theta_2 & \cos\theta_3 \end{bmatrix} \tag{4.68}$$

is the Jacobian matrix in which the first column includes derivatives of the position equations (4.56) with respect to θ_2 and the second column includes derivatives of the position equations with respect to r_3. To find the dead or locked positions, we need to find the positions at which the determinant of the Jacobian matrix, i.e.,

$$|J| = r_2 \cos(\theta_2 - \theta_3) \tag{4.69}$$

becomes zero. The result is,

$$\theta_2 - \theta_3 = 90°, 270° \tag{4.70}$$

These dead positions of the mechanism when θ_3 is the input, happen when the two links 2 and 3 are perpendicular to each other. This is the same as the limit position (4.63) for the orientation of link 3 when θ_2 was the input. So, we get the same condition, but with a different meaning.

Finally, in order to find the limit positions, we use Eq. (4.67) and calculate the first-order kinematic coefficients r_3' and θ_2', using Cramer's rule. First we get:

$$\theta_2' = \frac{\begin{vmatrix} -r_3 \cos\theta_3 & \sin\theta_3 \\ r_3 \sin\theta_3 & \cos\theta_3 \end{vmatrix}}{|J|} = -\frac{r_3}{r_2 \cos(\theta_2 - \theta_3)} \tag{4.71}$$

That has no root and no limit position. However, for the values given in Eq. (4.70), the denominator of Eq. (4.71) is zero. So, practically, Eq. (4.70) *is* a limiting state of motion. Now, for calculating the limit positions of r_3, we use Eq. (4.67) and obtain r_3':

$$r_3' = \frac{\begin{vmatrix} r_2 \cos\theta_2 & -r_3 \cos\theta_3 \\ -r_2 \sin\theta_2 & r_3 \sin\theta_3 \end{vmatrix}}{|J|} = \frac{r_2 r_3 \sin(\theta_3 - \theta_2)}{r_2 \cos(\theta_2 - \theta_3)} \tag{4.72}$$

To get the limit positions of r_3 we set $r_3' = 0$ and get:

$$\theta_2 - \theta_3 = 0, 180° \tag{4.73}$$

Therefore, the limit positions of r_3 (i.e., the limit positions of pin A in Fig. 4.14 with respect to O) when orientation of link 3 is the input, happen when links 2 and 3 are collinear. This is a completely expected result and is the same condition as we found for the limit position of r_3 when θ_2 was the input variable. This analysis is summarized in Table 4.2.

Table 4.2 Classification of the limit and the dead positions in a pin-in-the-slot mechanism depending on the chosen input and output

		Dead Positions	Output Variable and Limit	
			θ_3	r_3
Input Variable	θ_2	None Existed	$\theta_2 - \theta_3 = 90°, 270°$	$\theta_2 - \theta_3 = 0, 180°$
	θ_3	$\theta_2 - \theta_3 = 90°, 270°$	None Existed	$\theta_3 - \theta_2 = 0, 180°$
			θ_2	r_3
			Output Variable and Limit	

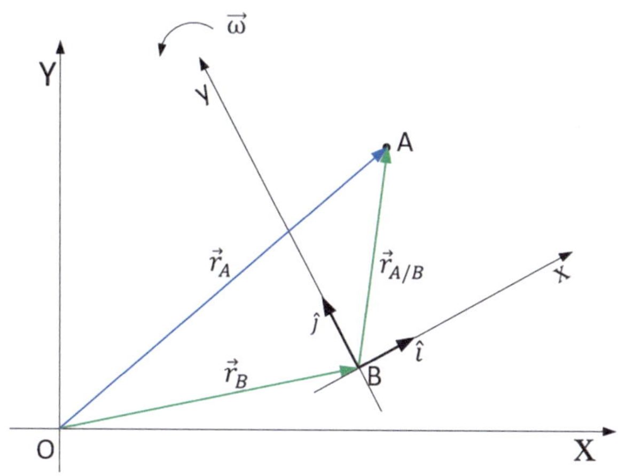

Fig. 4.15 The stationary observer O and a translating and rotating observer B are analyzing the velocity of point A

4.6 Velocity Analysis Using the Relative Motion Method

The subject of motion analysis using a moving observer is usually covered in detail in textbooks on engineering dynamics like Meriam and Kraige [3]. Here, a brief review of the relative motion method is presented. Figure 4.15 illustrates point A whose motion is analyzed by a moving observer B and a stationary observer O. The translating and rotating observer B moves at velocity $\vec{v}_B$ and its attached coordinate system x, y, z rotates at an angular velocity $\vec{\omega}$. A stationary observer O uses the X, Y, Z coordinate system and can observe both A and B and record their absolute velocities.

Using Fig. 4.15 we can write:

$$\vec{r}_A = \vec{r}_B + \vec{r}_{A/B} \tag{4.74}$$

or,

$$\vec{r}_A = \vec{r}_B + \vec{r} \tag{4.75}$$

Focusing on the motion in the x, y plane, we have:

$$\vec{r} = \vec{r}_{A/B} = x\hat{i} + y\hat{j} \tag{4.76}$$

where x and y are the coordinates of point A with respect to the moving coordinate system at B. Therefore,

$$\vec{r}_A = \vec{r}_B + x\hat{i} + y\hat{j} \tag{4.77}$$

In order to calculate $\vec{v}_A$, one may take a derivative of Eq. (4.77) with respect to time. Since $\vec{r}_A$ and $\vec{r}_B$ are the absolute position vectors of A and B with respect to the stationary observer at O, their time derivatives are absolute velocities of A and B, respectively, i.e.: $\vec{v}_A$ and $\vec{v}_B$. Furthermore, the moving coordinate system at B is rotating; so the time derivatives of its unit vectors are not zero and we have:

$$\vec{v}_A = \vec{v}_B + \dot{x}\hat{i} + x\frac{d\hat{i}}{dt} + \dot{y}\hat{j} + y\frac{d\hat{j}}{dt} \tag{4.78}$$

Considering Fig. 4.16, it is well known that the velocity of a point that performs circular motion at angular velocity $\vec{\omega}$ is:

$$\vec{v}_A = \frac{d\vec{r}_A}{dt} = \vec{\omega} \times \vec{r}_A \tag{4.79}$$

From this simple example, one concludes that the time derivative of a vector with constant magnitude that rotates at rate $\vec{\omega}$ can simply be obtained by the cross product of the angular velocity of rotation of that vector and the vector itself. One may calculate the time derivatives of the unit vectors $\hat{i}$ and $\hat{j}$ in Eq. (4.78) in a similar way.

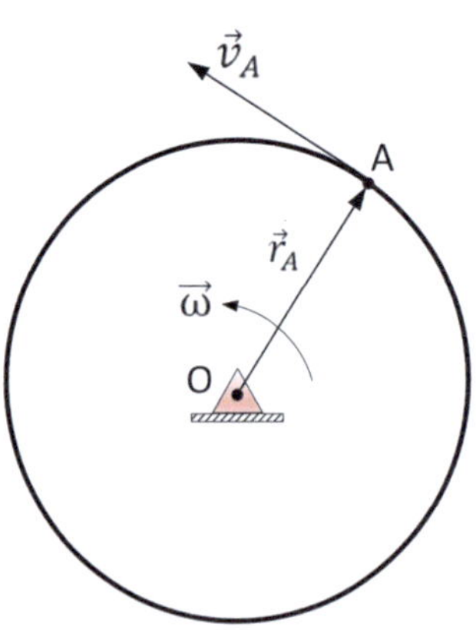

Fig. 4.16 Motion of a point A on a circular trajectory

Since these vectors have constant length and they rotate at angular velocity $\vec{\omega}$, their time derivatives are:

$$\frac{d\hat{i}}{dt} = \vec{\omega} \times \hat{i}; \quad \frac{d\hat{j}}{dt} = \vec{\omega} \times \hat{j} \tag{4.80}$$

Using Eqs. (4.80), one can rewrite Eq. (4.78) as follows:

$$\vec{v}_A = \vec{v}_B + \dot{x}\hat{i} + x\vec{\omega} \times \hat{i} + \dot{y}\hat{j} + y\vec{\omega} \times \hat{j}$$

or,

$$\vec{v}_A = \vec{v}_B + \dot{x}\hat{i} + \dot{y}\hat{j} + \vec{\omega} \times \left(x\hat{i} + y\hat{j}\right) \tag{4.81}$$

But, $\dot{x}\hat{i} + \dot{y}\hat{j}$ is the velocity of point A relative to the observer at B and it is shown by $\vec{v}_{rel}$:

$$\vec{v}_{rel} = \dot{x}\hat{i} + \dot{y}\hat{j} \tag{4.82}$$

Substituting Eqs. (4.76) and (4.82) into Eq. (4.81) we get:

$$\vec{v}_A = \vec{v}_B + \vec{v}_{rel} + \vec{\omega} \times \vec{r} \tag{4.83}$$

This is the complete relative velocity equation with three terms on the right-hand side. Among these terms the only one that is truly observed by B is $\vec{v}_{rel}$. The remaining terms are added to correct the misjudgment of the observer due to its translational and rotational motions.

There are three notable special cases that should be considered when using Eq. (4.83):

Case 1. Points A and B belong to the same rigid body and the moving coordinate system is attached to that rigid body. In this case, B observes that A is stationary so we have,

$$\vec{v}_{rel} = 0 \tag{4.84}$$

And Eq. (4.83) reduces to:

$$\vec{v}_A = \vec{v}_B + \vec{\omega} \times \vec{r} \tag{4.85}$$

Case 2. The observer B performs translational motion. In this case, $\omega = 0$ and Eq. (4.83) reduces to:

$$\vec{v}_A = \vec{v}_B + \vec{v}_{rel} \tag{4.86}$$

Case 3. Points A and B belong to two different objects, but they are momentarily collocated. In this case r = 0, and Eq. (4.83) reduces to:

$$\vec{v}_A = \vec{v}_B + \vec{v}_{rel} \tag{4.87}$$

Example 4.6 Figure 4.17 illustrates a slider-crank mechanism at an instant when crank link 2 is perpendicular to the direction of motion of the slider. It is also known that $\theta_3 = -30°$, link 2 is 5 in long, and its angular velocity is -10 rad/s (CW). Calculate the velocity of the slider and the angular velocity of link 3 at this instant.

Since B and C belong to the same rigid body, using special case 1 we can write Eq. (4.85) for points B and C.

$$\vec{v}_C = \vec{v}_B + \vec{\omega} \times \vec{r} \tag{4.88}$$

In Eq. (4.88), the moving coordinate system is located at B and it is attached to link BC. So,

$$\vec{\omega} = \vec{\omega}_{BC}; \ \vec{r} = \overrightarrow{BC} \tag{4.89}$$

Referring to Eq. (4.88), since at the illustrated moment both $\vec{v}_B$ and $\vec{v}_C$ are horizontal, but apparently $\vec{\omega} \times \vec{r}$ is not parallel to them, the only way this equation can be satisfied is that we have:

$$\vec{\omega}_{BC} = 0 \tag{4.90}$$

Hence, one may conclude that in the particular posture illustrated in Fig. 4.17 the connecting rod 3 has zero angular velocity (even though it may have non-zero angular acceleration, see Chap. 5) and from Eqs. (4.88) and (4.79) we get:

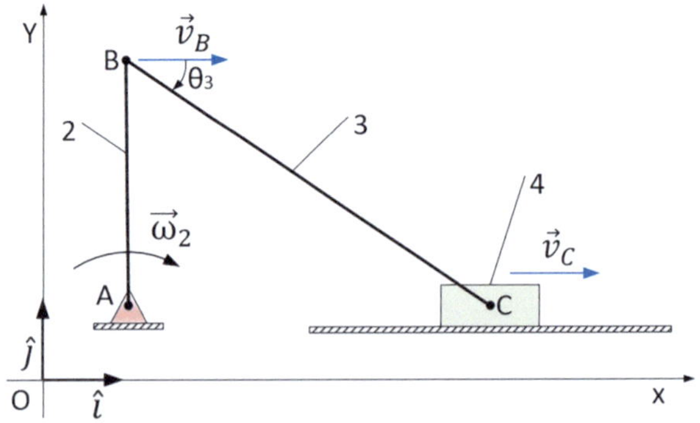

Fig. 4.17 A slider-crank mechanism when the crank is perpendicular to the direction of motion of the slider

$$\vec{v}_C = \vec{v}_B = \vec{\omega}_2 \times \overrightarrow{AB} \tag{4.91}$$

The same conclusion is obtained using the concept of instantaneous center that is discussed in Sect. 4.7. Substituting the given values for the parameters results in:

$$\vec{v}_C = \vec{v}_B = -10\hat{k} \times 5\hat{j} = 50\hat{i} \text{ in/s} \tag{4.92}$$

The obtained direction of motion of the slider is in harmony with what is expected from exploring the mechanism geometry in Fig. 4.17.

4.7 Velocity Analysis Using Instantaneous Centers

The concept of instantaneous center provides a very simple and conceptually understandable way for calculating velocity at points of interest. It is at the core of many other concepts and laws in kinematics of mechanisms like the fundamental law of gearing (see Chap. 6).

4.7.1 *Instantaneous Centers and Their Locations*

Instantaneous center (abbreviated as I.C.) of two objects i and j at an instant is a point that has the same velocity (zero or non-zero) on the two objects or on their imaginary extensions. If the objects are i and j, the corresponding I.C. is shown by I_{ij}. Furthermore, there is no distinction between I_{ij} and I_{ji}. So, I_{ij} is a point that, as far as velocity is concerned, can be regarded as belonging to objects i and j. In particular, if one of the objects is the frame or ground (#1), that is stationary, the I_{1j} point would be called the instantaneous center of zero velocity. This is a point about which object j is instantaneously rotating. Point I_{ij} can also be regarded as where object i is rotating about with respect to an observer that is attached to object j (and vice versa). Finding a point that has the same velocity in two different objects has many applications including in finding the pitch point for two gears in mesh (see Chap. 6).

Rolling without Slipping Figure 4.18 illustrates a disk that is rolling without slipping on fix ground and in planar motion. The existence of the ground constraint would allow point C (that is the instantaneous point of contact) to have velocity only in the parallel to the ground direction. However, the velocity of the instantaneous contact point C has to be zero. Otherwise, this velocity (being parallel to the ground direction) would result in sliding with respect to the ground. This result contradicts the original assumption made about the no slipping condition. Therefore, considering non-zero velocity for point C is incorrect and the contact point C is momentarily stationary. We can summarize what we know about point C: this point is

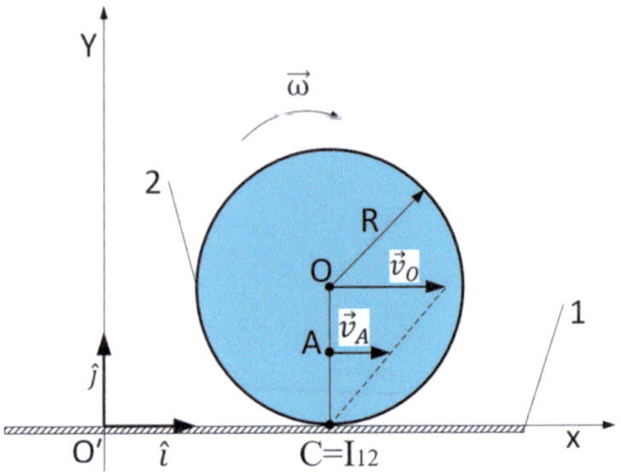

Fig. 4.18 Locating instantaneous center of zero velocity of a disk in the rolling without slipping motion

momentarily shared between objects 1 and 2, has zero velocity, and is the instantaneous center of zero velocity, I_{12}, of the disk. Since C is a momentarily fix point on the disk, the disk is momentarily rotating about it and using Eq. (4.79) the velocity of the center point of the disk can be written as:

$$\vec{v}_o = \vec{\omega} \times \overrightarrow{CO} \tag{4.93}$$

Using the right-hand rule for cross products we get:

$$\vec{v}_o = R\omega\hat{i} \tag{4.94}$$

Similarly, the velocity of a point A located at midpoint of the $\overrightarrow{CO}$ radius can be expressed as follows:

$$\vec{v}_A = \vec{\omega} \times \overrightarrow{CA} = 0.5R\omega\hat{i} \tag{4.95}$$

Therefore, as shown in Fig. 4.18, the velocities of points on the CO radius vary linearly with their distances from the point of contact.

Using the same arguments in Fig. 4.19, the velocities of points B and D could be expressed as follows:

$$\vec{v}_B = \vec{\omega} \times 2\overrightarrow{CO} \rightarrow \vec{v}_B = 2R\omega\hat{i} \tag{4.96}$$

$$\vec{v}_D = \vec{\omega} \times \overrightarrow{CD} \rightarrow \vec{v}_D = \sqrt{2}R\omega\left(\hat{i}\cos 45^\circ - \hat{j}\sin 45^\circ\right) \rightarrow \vec{v}_D = R\omega\left(\hat{i} - \hat{j}\right) \tag{4.97}$$

Example 4.7 Consider Fig. 4.19 and assume that the 4-cm-diameter disk is rolling without slipping. If the velocity at the center point O is 2 m/s, calculate velocities at points C, A, B, and D.

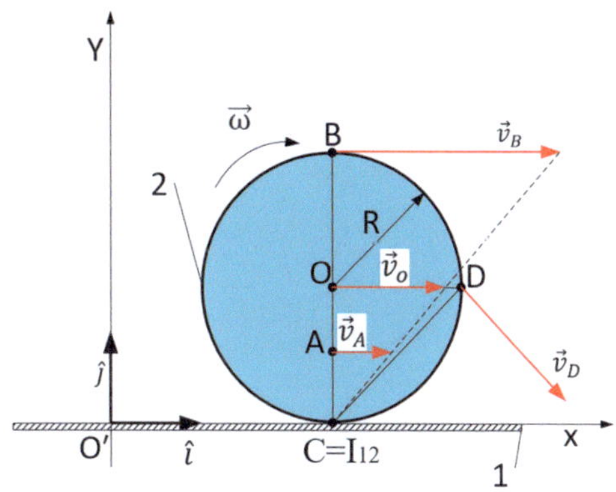

Fig. 4.19 Velocity calculation of points on a disk in the rolling without slipping motion

Since rolling is happening without slipping, point *C* has zero velocity and it is the instantaneous center of rotation of the disk.

Point *A* is half way between points *O* and *C*. Therefore, using Eq. (4.95) we have,

$$\vec{v}_A = 0.5\,\vec{v}_O = 1\hat{i}\ \text{m/s}$$

Similarly, we know that *BC=2OC*. So, using Eq. (4.96) we get,

$$\vec{v}_B = 2\,\vec{v}_O = 4\hat{i}\ \text{m/s}$$

And for point *D* we know $DC = \sqrt{2}OC$ and therefore,

$$v_D = \sqrt{2}v_O = 2\sqrt{2}\ \text{m/s}$$

Or, in vector form, using Eq. (4.97) we get:

$$\vec{v}_D = 2\left(\hat{i} - \hat{j}\right)\ \text{m/s}$$

Finally, to get the angular velocity of rotation of the disk, one can use Eq. (4.93):

$$\vec{v}_o = \vec{\omega} \times \overrightarrow{CO} \tag{4.93}$$

Therefore,

$$2\hat{i}\ \text{m/s} = \omega\hat{k} \times 0.02\hat{j} = -0.02\omega\hat{i}\text{m/s}$$
$$\omega = -100\ \text{rad/s (CW)}$$

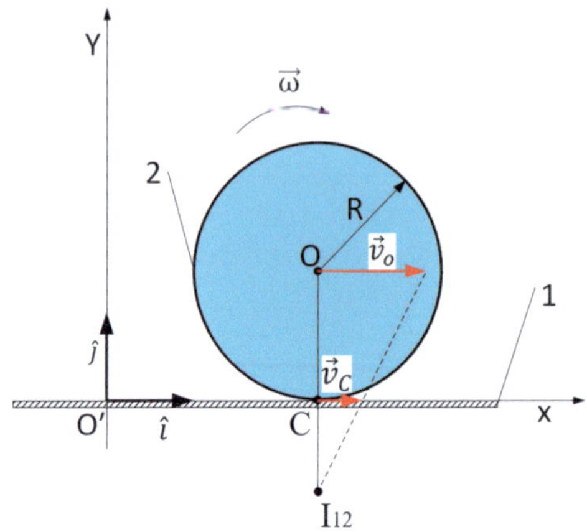

Fig. 4.20 Locating the instantaneous center of zero velocity of a disk in rolling with slipping motion when $\vec{v}_o$ and $\vec{v}_C$ are in the same direction

Rolling with Slipping Figure 4.20 illustrates a disk that is rolling and slipping. In this case, the instantaneous center of zero velocity will no longer be located at the contact point C (because point C has non-zero velocity and cannot be considered as point I_{12} that, by definition, has zero velocity). If the velocity of the contact point and the velocity of the center of the disk are known, the location of the instantaneous center of zero velocity I_{12} can be found by using the fact that velocity varies linearly with the distance from I_{12}. Therefore, for the velocity of an arbitrary point P on the disk we have:

$$\vec{v}_P = \vec{\omega} \times \overrightarrow{I_{12}P} \tag{4.98}$$

Consequently, by knowing $\vec{v}_o$ and $\vec{v}_C$, the location of I_{12} is found by using the linearity of velocity distribution from I_{12}, see Fig. 4.20. Having found I_{12}, the velocity of any point P on the disk can now be calculated using Eq. (4.98). In this case, I_{12} lies outside of the disk. If $\vec{v}_o$ and $\vec{v}_C$ are in opposite directions, as is seen in Fig. 4.21, I_{12} would fall inside the disk:

Example 4.8 In Fig. 4.20, considering $\vec{v}_o = 5\hat{i}$ m/s, $\vec{v}_C = 2\hat{i}$ m/s, and $R = 10$ cm, calculate the angular velocity of the disk and the location of I_{12}.

Since velocity changes linearly from the IC,

$$\omega = \frac{v_o - v_C}{R} = \frac{5-2}{0.1} = 30 \text{ rad/s (CW)}$$

And in vector form, we have $\vec{\omega} = -30\hat{k}$ rad/s. Now, using Eq. (4.98) for the point of contact C we get:

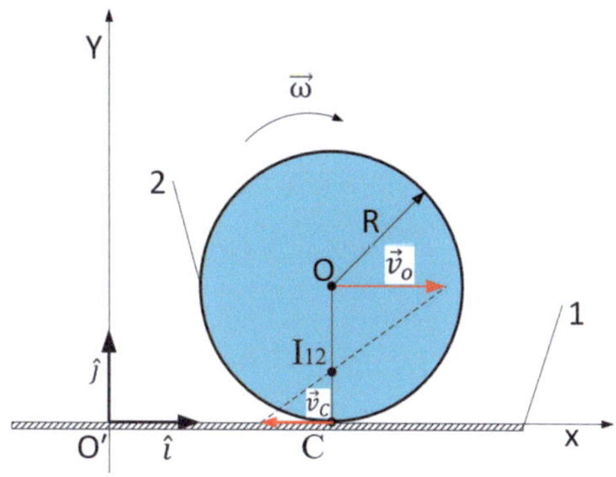

Fig. 4.21 Locating the instantaneous center of zero velocity of a disk in rolling with slipping motion when $\vec{v}_o$ and $\vec{v}_C$ are in opposite directions

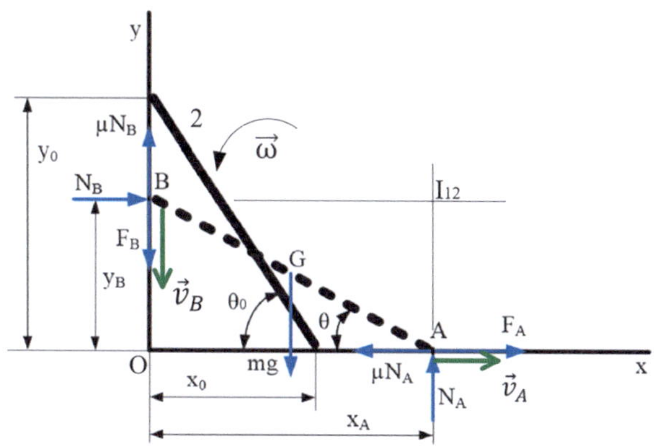

Fig. 4.22 Position of the instantaneous center in the case of a falling ladder. (Originally from Fig. 1 in Ghorashi and Ghorashi [1]). © 2020 Springer Nature (reproduced with permission)

$$\vec{v}_C = \vec{\omega} \times \overrightarrow{I_{12}C} \tag{4.99}$$

Substituting the known values of parameters in Eq. (4.99) we get:

$$2\hat{\imath}\ \text{m/s} = -30\hat{k} \times \overrightarrow{I_{12}C}$$

Therefore,

$$\overrightarrow{I_{12}C} = -6.667\hat{\jmath}\ \text{cm}$$

This result shows that I_{12} is 6.667 cm below the contact point (see Fig. 4.20).

Example 4.9 Figure 4.22 illustrates a ladder that is falling while it is maintaining its contact with the wall (the y axis) and the floor (the x axis), see Ghorashi and Ghorashi [1].

A. Locate the instantaneous center I_{12}.
B. If the velocity of point A is known, calculate the velocity of point B.

A. **Locate the instantaneous center I_{12}.**

Equation (4.98) is still valid for the velocity of an arbitrary point P on the ladder:

$$\vec{v}_P = \vec{\omega} \times \overrightarrow{I_{12}P} \tag{4.98}$$

Therefore, for the velocity of point A we can write,

$$\vec{v}_A = \vec{\omega} \times \overrightarrow{I_{12}A} \tag{4.100}$$

Since in Eq. (4.100) there is a cross product, we conclude that the position vector $\overrightarrow{I_{12}A}$ connecting I_{12} to A should be perpendicular to the velocity of point A, i.e., $\vec{v}_A$. Therefore, I_{12} should be on a line perpendicular to the horizontal x direction; that passes through the instantaneous position of the contact point A. A similar argument for contact point B results in:

$$\vec{v}_B = \vec{\omega} \times \overrightarrow{I_{12}B} \tag{4.101}$$

So, I_{12} is also located on a line perpendicular to the vertical y direction that passes through the instantaneous position of contact point B. As a result, I_{12} is at the intersection of the mentioned two lines as shown in Fig. 4.22.

B. **If the velocity of point A is known, calculate the velocity of point B.**

By dividing the scalar forms of Eqs. (4.100) and (4.101) one obtains:

$$\frac{v_A}{v_B} = \frac{I_{12}A}{I_{12}B} \tag{4.102}$$

Equation (4.102) simply states that velocity of a point on the ladder varies linearly from the instantaneous position of I_{12}. This result is compatible with what we saw in Examples 4.7 and 4.8 for a rolling or a rolling and sliding disk. At any instantaneous location of the ladder, Eq. (4.102) can provide the v_B value. Furthermore, the instantaneous value of the angular velocity ω can be calculated using Eq. (4.100).

Sometimes there is a misconception that point O should be the instantaneous center of the ladder; after all, it *is* the center of the coordinate system! However, $\overrightarrow{OA}$ is parallel to $\vec{v}_A$ and they cannot form a vector product like what we see in Eq. (4.100). Therefore, O is not the instantaneous center of zero velocity of the ladder.

Example 4.10 In the falling ladder problem shown in Fig. 4.22, consider $v_A = 10$ in/s and $\theta = 30°$. If the ladder is 3 ft. long, calculate v_B and the angular velocity ω of the ladder at this instant.

Recalling Eq. (4.102) and Fig. 4.22 we can write:

$$\frac{v_A}{v_B} = \frac{I_{12}A}{I_{12}B} \rightarrow \frac{v_A}{v_B} = \tan\theta \tag{4.103}$$

Substituting the given values we get:

$$v_B = 17.32 \text{ in/s}$$

To get the angular velocity of the ladder, we use Eq. (4.100):

$$\vec{v}_A = \vec{\omega} \times \overrightarrow{I_{12}A} \tag{4.100}$$

where

$$I_{12}A = L\sin\theta \rightarrow I_{12}A = 3 \text{ ft} \times 12 \text{ in/ft} \times \sin 30^\circ = 18 \text{ in}$$

Substituting the calculated $I_{12}A$ value into Eq. (4.100) gives,

$$v_A = 10 \text{ in/s} = 18\omega \rightarrow \omega = 0.556 \text{ rad/s}$$

Number of Instantaneous Centers

Mechanisms with more than two members have more than one instantaneous center. In general, if a mechanism has n members, each member can have an instantaneous center with any other member except for itself. Therefore, any member in that mechanism can be related to $n - 1$ other members and the total number of possibilities would be $n(n - 1)$. However, there is no distinction between I_{ij} and I_{ji}; so if we eliminate double counting, the total number of instantaneous centers of an n-member mechanism is:

$$\text{Number of IC's of an } n \text{ member mechanism} = \frac{n(n-1)}{2} \tag{4.104}$$

A four-bar linkage and a slider-crank mechanism, for example, each has four members and therefore, each has six instantaneous centers. Locating instantaneous centers correctly and quickly requires implementation of a few rules. One of them is the fact that for each three objects i, j, and k their corresponding instantaneous centers I_{ij} I_{jk}, and I_{ki} lie on a straight line. This is a result of the Aronhold-Kennedy theorem of three centers to be discussed next.

4.7.2 Aronhold-Kennedy Theorem of Three Centers

According to the Aronhold-Kennedy theorem of three centers, for any three objects i, j, and k (whether they are connected to one another or not) their corresponding instantaneous centers I_{ij}, I_{jk}, and I_{ki} lie on a straight line (i.e., they are collinear). The theorem can be proven by contradiction. Consider the three objects i, j, and k shown in Fig. 4.23. Now assume an observer is attached to object k. Clearly this observer would see object k as stationary and fixed, as if k is the ground object 1. We further assume that, as shown in Fig. 4.23, the three instantaneous centers I_{ij}, I_{jk}, and I_{ki} do not lie on a straight line (i.e., angle γ is not zero) and show that this assumption will result in a contradiction. Since relative to the chosen observer at k, points I_{jk}, and I_{ki} both are stationary (as they move with k) these points act like fixed hinges. That said, the velocity of point I_{ij} (shared between i and j) if we first consider it to belong to object i, should be perpendicular to the line connecting I_{ij} to the center of rotation with respect to our observer that is I_{ki}. So this velocity, shown by $\vec{v}_{ij,i}$ is perpendicular to $\overrightarrow{I_{ki}I_{ij}}$. With a similar argument one can prove that the velocity of point I_{ij}, if we now consider it to belong to j, and show this velocity by $\vec{v}_{ij,j}$, should be perpendicular to line $\overrightarrow{I_{jk}I_{ij}}$. But according to the definition of an instantaneous center, the two velocities $\vec{v}_{ij,i}$ and $\vec{v}_{ij,j}$ should be equal. Looking at Fig. 4.23, they cannot be equal if angle γ is not zero. So, the original assumption that angle γ is not zero was incorrect and indeed, $\gamma = 0$. As a result, points I_{ik}, I_{kj} and I_{ij} are collinear; which is what we wanted to prove.

Example 4.11 Use the Aronhold-Kennedy theorem to locate all of the instantaneous centers of the four-bar linkage shown in Fig. 4.24.

By using Eq. (4.104) with $n = 4$, we know that a four-bar linkage has six instantaneous centers. Four of these ICs are simply the four hinges (or revolute joints) that connect the bars to each other (I_{23} and I_{34}) and to the fix base (I_{12} and I_{14}). The remaining two (i.e., I_{24} and I_{13}) can be obtained by using the Aronhold-Kennedy theorem of three centers.

Locating I_{24} We know that I_{23}, I_{34}, and I_{24} are collinear. So the unknown instantaneous center I_{24} has to be on the line that connects I_{23} to I_{34}. In other words, <u>we</u>

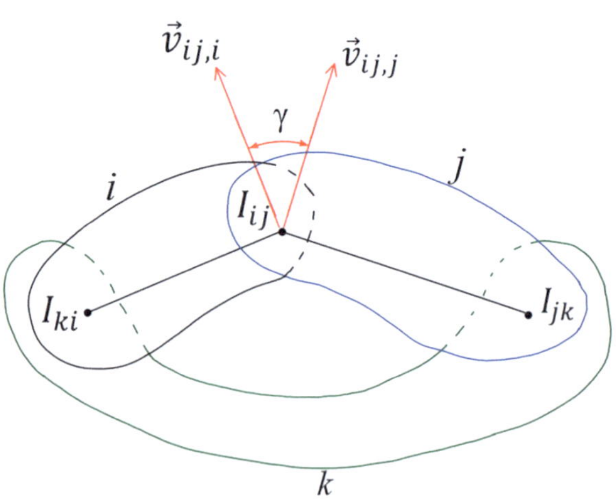

Fig. 4.23 Illustration of the Aronhold-Kennedy theorem of three centers (proof by contradiction)

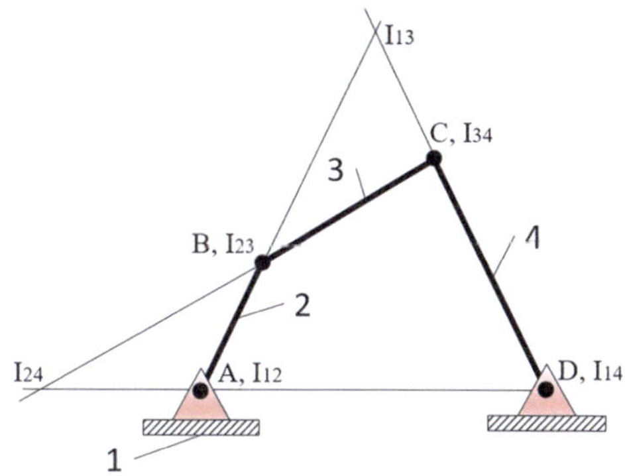

Fig. 4.24 A four-bar linkage and its six instantaneous centers

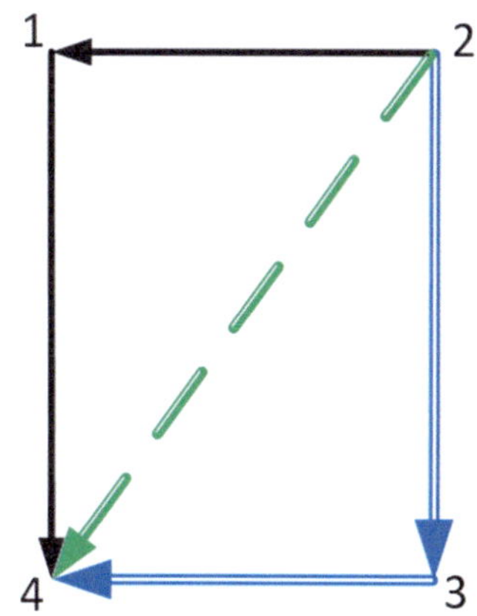

Fig. 4.25 Use of Aronhold-Kennedy theorem to locate I_{24} in Examples 4.11 and 4.12

connect I_{23} to I_{34} in order to find a locus for the third instantaneous center, I_{24}, whose index is obtained by dropping the common index, 3 between I_{23} and I_{34}. In a similar way, connecting I_{12} to I_{14} generates another line on which I_{24} (i.e., the instantaneous center without the common index 1) is located. Therefore, I_{24} is located at the intersection of the obtained two lines (see Fig. 4.24). The procedure for locating I_{24} is also illustrated in Fig. 4.25.

Locating I_{13} The same method can be used for locating the last unknown instantaneous center, i.e., I_{13}. The instantaneous center I_{13} is a point about which link 3 rotates at the shown instant. Since it is shared with the ground (i.e. link #1) this point is momentarily stationary and therefore it is an instantaneous center of zero velocity. We know that based on the Aronhold-Kennedy theorem I_{34}, I_{14}, and I_{13} are collinear. So the unknown IC, i.e., I_{13}, has to be on the line that connects I_{34} to I_{14}. In other words, we connect I_{34} to I_{14} and then drop the common index, i.e., 4, to find the locus of I_{13} that is the third instantaneous center on this line. With a similar argument, I_{13} should also be on the line that connects I_{23} and I_{12}. Therefore, I_{13} will be located at the intersection of the obtained two lines (see Fig. 4.24). Having

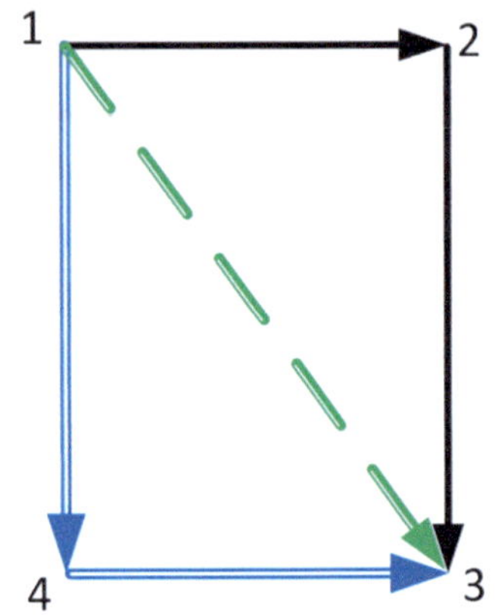

Fig. 4.26. Use of Aronhold-Kennedy theorem to locate I_{13} in Examples 4.11 and 4.12

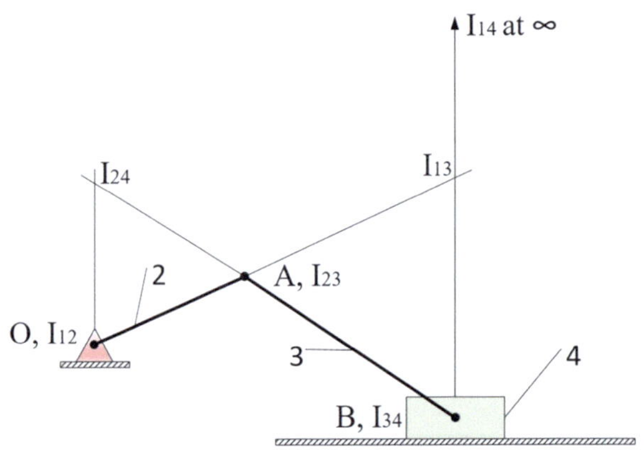

Fig. 4.27 A slider-crank mechanism and its six instantaneous centers

found that, now all six instantaneous centers of the four-bar linkage are known. The procedure for locating I_{13} is also illustrated in Fig. 4.26.

Example 4.12 Use the Aronhold-Kennedy theorem to locate all of the instantaneous centers of the slider-crank mechanism shown in Fig. 4.27.

By using Eq. (4.104) with $n = 4$ we know that a slider-crank mechanism has six instantaneous centers. Three of the ICs shown in Fig. 4.27 are simply the three hinges (or revolute joints) that connect the bars to each other (I_{23}) to the fix base (I_{12}) and to the slider (I_{34}). The fourth instantaneous center is I_{14} that is on the common normal of the slider and base 1. Since slider 4 is translating, I_{14} that is the point about which slider 4 is turning, is at infinity. The remaining two instantaneous centers (i.e., I_{24} and I_{13}) can be obtained by using the Aronhold-Kennedy theorem of three centers.

Locating I_{24} We know that based on the Aronhold-Kennedy theorem, I_{23}, I_{34}, and I_{24} are collinear. So the unknown instantaneous center I_{24} has to be on the line that connects I_{23} to I_{34}. In other words, we connect I_{23} to I_{34} and then drop the common

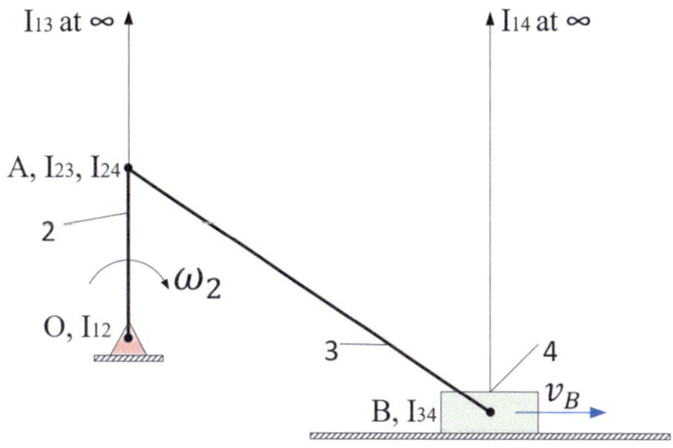

Fig. 4.28 Instantaneous centers when crank is perpendicular to the direction of motion of the slider

index, 3 to find the third instantaneous center (i.e., I_{24}) that is located on this line. Connecting I_{12} to I_{14} in a similar way generates a line on which I_{24}, i.e., the instantaneous center without the common index (i.e., 1) is located. To connect I_{12} to I_{14} (that is at infinity) a line from I_{12} parallel to the I_{14} direction line is drawn. Therefore, I_{24} is located at the intersection of the obtained two lines (see Fig. 4.27). The procedure for locating I_{24} is also illustrated in Fig. 4.25.

Locating I_{13} The same method can be used for finding the last unknown instantaneous center, i.e., I_{13}. We know that based on the Aronhold-Kennedy theorem, I_{34}, I_{14}, and I_{13} are collinear. So the unknown IC, i.e., I_{13} has to be on the line that connects I_{34} to I_{14}. In other words, we connect I_{34} to I_{14} (that is at infinity) and then drop the common index, i.e., 4 to find the locus of I_{13} that is the third IC on this line. With a similar argument, I_{13} should also be on the line that connects I_{23} and I_{12}. Therefore, I_{13} will be at the intersection of the obtained two lines (see Fig. 4.27). Having found that, now all six instantaneous centers of the slider-crank mechanism are known. The procedure for locating I_{13} is also illustrated in Fig. 4.26.

Example 4.13 Consider the slider-crank mechanism illustrated in Fig. 4.28 in which the crank angular velocity is ω_2 and at this instant crank 2 is perpendicular to the direction of motion of the slider. A. Calculate the angular velocity of the connecting rod 3. B. Determine the velocity of the slider v_B.

Comparing Figs. 4.27 and 4.28 we see that in Fig. 4.28 the two lines on which I_{13} should lie are parallel with each other and never intersect. So the instantaneous center I_{13} is at infinity. But I_{13} is a point that is shared with member 1 (that is stationary) so, it is a point in link 3 that is momentarily fixed and therefore, connecting rod 3 is rotating about it. (In general, using the same argument one concludes that I_{1k} is a point about which a member k momentarily rotates.) Now that I_{13} is momentarily at infinity, rotation of connecting rod 3 about it can be interpreted as translation! (Translation is equivalent to rotation about a point that is at infinite distance.) Therefore, at this instant the angular velocity of link 3 is zero and all of its points have the same velocity:

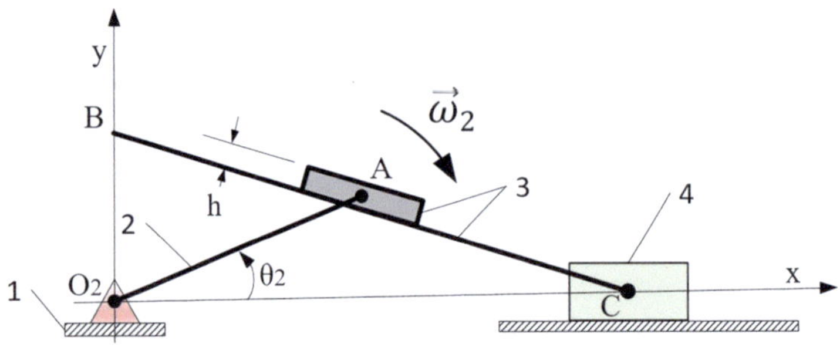

Fig. 4.29 A straight-line generator

$$\omega_3 = 0 \rightarrow v_A = v_B$$

However, we know that: $v_A = OA.\ \omega_2$ therefore,

$$v_B = OA.\omega_2$$

Example 4.14 Figure 4.29 illustrates a straight-line generator where $O_2A = AB = AC = 4$ cm and distance h is negligible. Crank 2 rotates at $\omega_2 = -30$ rad/s (CW) and causes point B to move on a straight vertical line.

Link BC is a straight rigid link and link 2 is hinged to BC at point A. At an instant when $\theta_2 = 15°$,

A. Locate all of the instantaneous centers of the mechanism
B. Based on the result of part A calculate the velocity of point B.
C. Using the algebraic method calculate the velocities of points A, B, and C.
D. Using the method of kinematic coefficients, calculate the velocity of point B.

A. **Locate all of the instantaneous centers of the mechanism**

This mechanism has four members and therefore six instantaneous centers. Using the foregoing discussions and the Aronhold-Kennedy theorem of three centers, we can locate all of the instantaneous centers as shown in Fig. 4.30.

B. **Using the result of part A calculate velocity of point *B*.**

Since point B belongs to link 3, and this link is instantaneously rotating about I_{13}, the velocity of point B can be expressed as follows:

$$\vec{v}_B = \vec{\omega}_3 \times \overrightarrow{I_{13}B} \tag{4.105}$$

Triangle O_2AC is an isosceles triangle ($O_2A = AC = 4$ cm) and during motion it remains to be an isosceles triangle, so we conclude that:

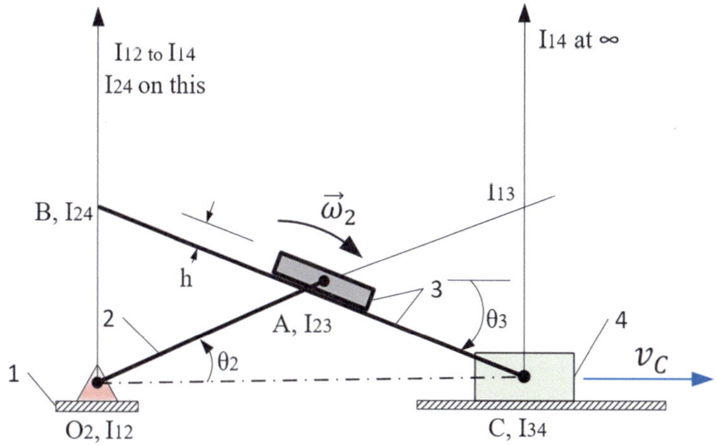

Fig. 4.30 The instantaneous centers of a straight-line generator

$$\theta_2 = -\theta_3 \tag{4.106}$$

And by taking a time derivative one obtains,

$$\omega_2 = -\omega_3 \tag{4.107}$$

So while links 2 and 3 have identical angular speeds, they rotate in opposite directions (see Fig. 4.30). Therefore, ω_3 is CCW and we have:

$$\omega_3 = 30 \text{ rad/s}$$

Substitution of ω_3 in Eq. (4.105) results in:

$$\vec{v}_B = 30\hat{k} \times 2 \times 4 \cos 15° \left(-\hat{i}\right) = -231.822\hat{j} \text{ cm/s}$$

So point B moves down on a straight line, as expected.

C. **Using the algebraic method, calculate the velocities of points *A*, *B*, and *C*.**

Point A is moving on a circular path about O_2. Therefore, its velocity is perpendicular to O_2A and its magnitude is:

$$v_A = \omega_2 \times O_2A = 30 \text{ rad/s} \times 4 \text{ cm} = 120 \text{ cm/s}$$

Considering triangle O_2AB for the position of point B we can write:

$$x_B = 0,\ y_B = 2O_2A \sin\theta_2 \tag{4.108}$$

Taking a derivative of Eq. (4.108) with respect to time gives:

$$v_B = 2O_2A\dot{\theta}_2 \cos\theta_2 \tag{4.109}$$

Substituting the given numerical values in Eq. (4.109) results in:

$$v_B = 2 \times 4\ \text{cm} \times (-30)\cos 15° \rightarrow v_B = -231.822\ \text{cm/s}$$

The negative sign indicates that point B moves in the negative y direction. Finally, considering triangle O_2AC for the position of point C we can write:

$$x_C = 2O_2A \cos\theta_2 \tag{4.110}$$

Taking a derivative of Eq. (4.110) with respect to time gives:

$$v_C = -2O_2A\dot{\theta}_2 \sin\theta_2 \tag{4.111}$$

Substituting the given numerical values in Eq. (4.111) results in:

$$v_C = -2 \times 4\ \text{cm} \times (-30)\sin 15° \rightarrow v_C = 62.117\ \text{cm/s}$$

The positive sign indicates that point C moves in the positive x direction.

D. **Using the method of kinematic coefficients, calculate the velocity of point *B*.**

In order to use the method of kinematic coefficients, again we consider triangle O_2AB and for the position of point B we write:

$$y_B = 2O_2A \sin\theta_2 \tag{4.112}$$

Taking a derivative with respect to the input angle θ_2 we get:

$$y'_B = 2O_2A \cos\theta_2 \tag{4.113}$$

Substituting the given numerical values in Eq. (4.113) gives:

$$y'_B = 2 \times 4\text{cm} \times \cos 15° \rightarrow y'_B = 7.7274\ \text{cm/rad}$$

Since y'_B is positive, one concludes that by increasing θ_2 the y_B value also increases. This observation is in harmony with the way we see the mechanism functions at $\theta_2 = 15°$ in Fig. 4.29. In order to calculate v_B, we use Eq. (4.35) for y_B,

$$v_B = \dot{y}_B = y'_B \times \omega_2 \tag{4.114}$$

Substituting the given numerical values in Eq. (4.114) results in:

$$v_B = 7.7274 \frac{\text{cm}}{\text{rad}} \times \left(-30 \ \frac{\text{rad}}{\text{s}} \right) = -231.822 \ \text{cm/s}$$

We see that the calculated v_B values using different methods are identical.

Example 4.15 In the transversal section of a car suspension system shown in Fig. 4.31 locate the roll center of the car. Roll canter is the instantaneous center of zero velocity of the vehicle body 4. So we are looking for I_{14}.

Figure 4.31 illustrates how sequential application of the Aronhold-Kennedy theorem can provide us with the location of I_{14} that is the roll center of the car. First, each one of the eight members is numbered. Then, assuming rolling contact where tires meet the road, I_{12} and I_{13} are located at tire contact points with the road. Next, the revolute joints are recognized as the trivial ICs. So the revolute joint shared between members 6 and 4 is I_{46} and so on. In this way, we locate I_{46}, I_{26}, I_{25}, I_{45}, I_{48}, I_{47}, I_{37}, and I_{38}. Now, in order to locate I_{24} we need to connect I_{46} to I_{26} and also connect I_{45} to I_{25}. According to Aronhold-Kennedy theorem of three centers, point I_{24} is at the intersection of these two lines. In a similar way, point I_{34} is located on the opposite side of the vehicle. Finally, to find the roll center I_{14}, we need to connect I_{24} to I_{12} and also connect I_{34} to I_{13}. According to Aronhold-Kennedy theorem of three centers, point I_{14} is at the intersection of these two lines.

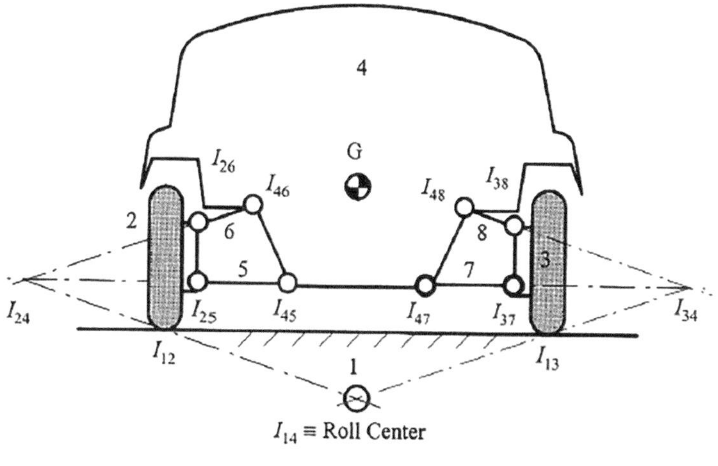

Fig. 4.31 Roll center geometry of an automotive independent suspension system. Page 354 of Waldron and Kinzel [5]. © 2004 John Wiley & Sons, Inc. (reproduced with permission)

There is a roll center for the front suspension and another one for the rear suspension. The line that connects the two roll centers is the roll axis. Roll centers and roll axis change position during motion. "The location of the roll center relative to the center of mass of the vehicle governs the effect of inertial forces caused by cornering on the system. Obviously, if the vertical distance between the roll center and the center of mass is large, the moment produced by the lateral acceleration will be large. A suspension geometry that brings the roll center progressively closer to the center of mass with increasing body roll might be attractive, …" Waldron and Kinzel [5], page 353.

4.7.3 Angular Velocity Ratio Theorem

Consider the case of a cam and follower mechanism shown in Fig. 4.32. Cam and follower are in contact at point *C*. We are looking for their angular velocity ratio. Cam and follower mechanisms are used in many position control applications. An example is their use for opening and closing gas and exhaust valves in internal combustion engines. In Fig. 4.32, three objects can be seen; the base or ground 1, the cam 2 (i.e., the driver, hinged at *A*), and the follower 3 (i.e., the driven, hinged at *B*).

One Possible Misconception There is a possible misconception that the angular velocity ratio should be inversely proportional to the hinge distances from the contact point, i.e.,

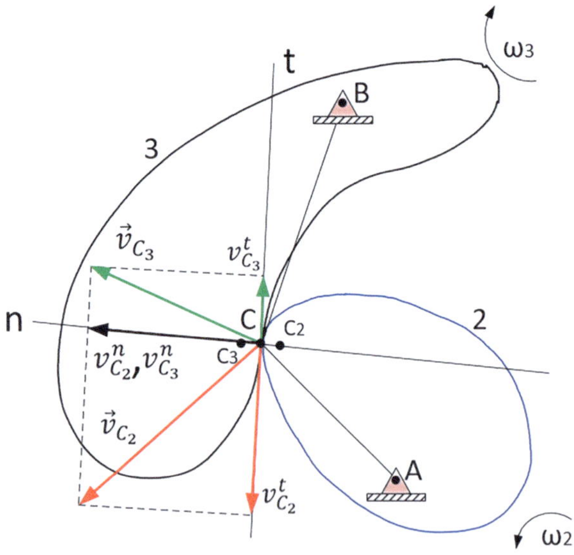

Fig. 4.32 Comparing velocities at points C_2 and C_3 that are infinitesimally distant, momentarily coincident, and belong to objects 2 and 3, respectively

$$\text{Possible misconception}: \frac{\omega_3}{\omega_2} = \frac{AC}{BC} \tag{4.115}$$

If C_2 is a point adjacent to C, but in object 2; and C_3 is a point adjacent to C, but in object 3, magnitudes of the velocities of points C_2 and C_3 can be written as:

$$v_{C_2} = AC_2\omega_2; v_{C_3} = BC_3\omega_3 \tag{4.116}$$

Since C_2 and C_3 are practically collocated, to get Eq. (4.115) the velocities of points C_2 and C_3 should be equal. But they clearly are not equal! Vector $\vec{v}_{C_2}$ is perpendicular to $\overrightarrow{AC}_2$ whereas $\vec{v}_{C_3}$ is perpendicular to $\overrightarrow{BC}_3$. Since in general $\overrightarrow{AC}_2$ and $\overrightarrow{BC}_3$ are not collinear, it is impossible that velocities $\vec{v}_{C_2}$ and $\vec{v}_{C_3}$ be parallel (that is a necessary condition for them to be equal). Therefore, Eq. (4.115) is not generally valid. The only exception to this rule is when the contact point lies on the center line (see Fig. 4.33) and the three points A, B, and C are collinear (they are the instantaneous centers I_{12}, I_{13}, and I_{23}, respectively). At such an instant, contact point C would coincide with I_{23} and the pitch point of the mechanism (to be introduced later in this section, see Figs. 4.34 and 4.35). This is the only instant at which the two objects do not slide on each other and are in pure rolling contact.

Contact at an Arbitrary Instant Consider Fig. 4.32 again in which contact of the objects happens at an arbitrary point. The two velocities $\vec{v}_{C_2}$ and $\vec{v}_{C_3}$ are not equal. They, however, have to have the same normal components, i.e.,

$$v^n_{C_2} = v^n_{C_3} \tag{4.117}$$

Otherwise, if we have:

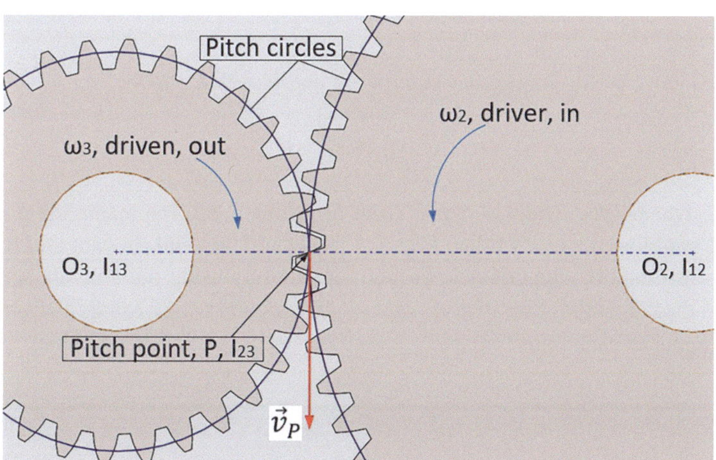

Fig. 4.33 Pure rolling condition in gears

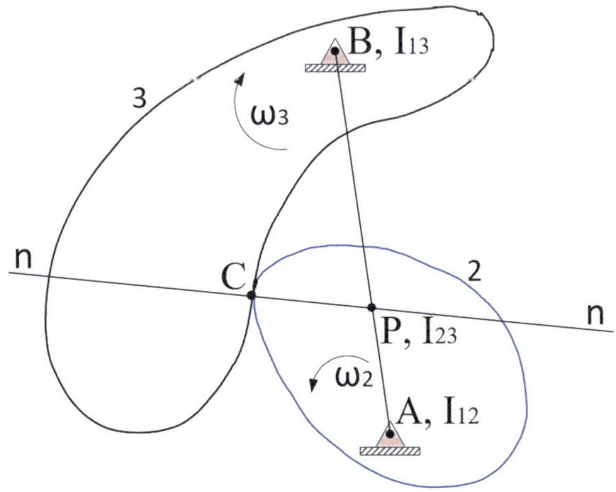

Fig. 4.34 Locating the instantaneous center, I_{23}, in a cam and follower mechanism, illustrating the angular velocity ratio theorem

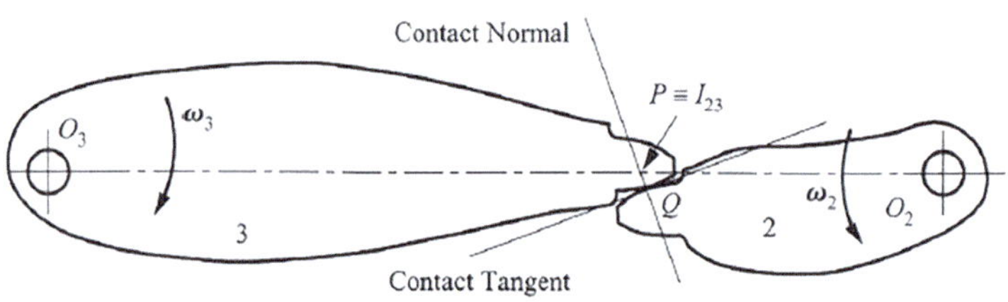

Fig. 4.35 Two gear teeth in contact and the pitch point that is the same as I_{23}. Page 461 of Waldron and Kinzel [5]. © 2004 John Wiley & Sons, Inc. (reproduced with permission)

$$v^n_{C_2} > v^n_{C_3} \tag{4.118}$$

then member 2 would bump into member 3; and else if we have:

$$v^n_{C_2} < v^n_{C_3} \tag{4.119}$$

then the two bodies would separate from each other. So, for maintaining contact without bumping into each other, Eq. (4.117) should be valid. In cam and follower mechanisms maintaining contact is usually assured by using pre-stressed springs.

As to the tangential velocities $v^t_{C_2}$ and $v^t_{C_3}$, we see in Fig. 4.32 that in general they are not equal to each other:

$$v^t_{C_2} \neq v^t_{C_3} \tag{4.120}$$

This difference between the tangential components is the cause of sliding and rubbing of a cam and follower on each other. Sliding can happen at high contact forces that are unavoidable when high levels of power are to be transmitted. The same phenomenon also happens in gear meshes and it will be analyzed in more detail in Chap. 6. As a result of sliding, wear on contacting surfaces can happen and to manage it, contact surfaces of cams, followers, and gears are usually hardened.

Correct Equation for the Angular Velocity Ratio Theorem Referring to Fig. 4.34, let us formally locate I_{23}. Since in this mechanism there are three objects, using Eq. (4.104) we know that the number of ICs is 3. Two of these instantaneous centers i.e. I_{12} and I_{13} can be easily found as the fixed hinges that 2 and 3 share with the ground. The third one is I_{23}. For objects in contact like 2 and 3, the corresponding instantaneous center is on their common normal. If the objects were in pure rolling contact (no sliding), as we saw in Fig. 4.18 for the case of a rolling disk, the instantaneous center would be at the contact point. In the present case of a cam and follower mechanism, it is impossible that the cam and follower can roll without sliding. Therefore, I_{23} will not be at the contact point, but it is located on the common normal of the contacting surfaces of the objects drawn at the contact point. Now, using the Aronhold-Kennedy theorem, I_{23} is also located on the line connecting I_{12} to I_{13} (drop the common index). So, the unknown instantaneous center, I_{23} is at the intersection of these two lines as shown in Fig. 4.34. That same point is shown in Fig. 4.35 for two gear teeth in contact and it is also called the pitch point.

Referring back to Fig. 4.34, since point P belongs to member 2, its velocity can be expressed as follows,

$$v_P = v_{I_{23}} = I_{12}I_{23}.\omega_2 \tag{4.121}$$

Similarly, since point P also belongs to member 3 we have,

$$v_P = v_{I_{23}} = I_{13}I_{23}.\omega_3 \tag{4.122}$$

Combining Eqs. (4.121) and (4.122) one obtains,

$$\frac{\omega_3}{\omega_2} = \frac{I_{12}I_{23}}{I_{13}I_{23}} \tag{4.123}$$

This result is the mathematical expression of the angular velocity ratio theorem applied to a cam and follower mechanism with members 2 and 3. In words, Eq. (4.123) states that the ratio of the angular velocities of the two objects is equal to the inverse of the corresponding ratio of the lengths of the segments that their common instantaneous center generates on the centerline AB in Fig. 4.34. It should be noted that Eq. (4.123) is also the basis of what is called the fundamental law of gearing (see Chap. 6).

If member 2 is the input and member 3 is the output, Eq. (4.123) can be written in a more detailed form as follows:

$$\frac{\omega_{out}}{\omega_{in}} = \frac{\omega_3}{\omega_2} = \frac{I_{12}I_{23}}{I_{13}I_{23}} = \frac{T_2}{T_3} = \frac{1}{M.A.} \tag{4.124}$$

In which *M.A.* refers to the Mechanical Advantage of the mechanism to be discussed in Chap. 11. Eq. (4.124) reveals that the angular velocity ratio of members 2 and 3 depends only on the location of their common IC, i.e., I_{23}, along the centerline *AB* in Fig. 4.34. In general, as the mechanism moves, point I_{23} also moves along the line of centers (i.e., *AB*, aka $I_{12}I_{13}$) and therefore the ratio given by Eq. (4.124) changes during motion.

Using the foregoing discussion we conclude that in gears the angular velocity ratio completely depends on the location of the pitch point, i.e., the common instantaneous center of the input and the output members. In gears, as we will see in Chap. 6, one major objective is to find proper gear-tooth profiles in order to keep the speed ratio constant (for which the location of the pitch point or I_{23} should remain fixed during motion). In Chap. 6 we will see how this objective is accomplished by using involute of a circle as the profile of gear teeth.

Contact at Pitch Point (I_{23}) If for the gear teeth shown in Fig. 4.35 contact happens to be at the pitch point *P* (the way it is in Fig. 4.33), a vector form of Eq. (4.123) can be written as follows:

$$\vec{v}_P = \vec{\omega}_2 \times \overrightarrow{O_2P} = \vec{\omega}_3 \times \overrightarrow{O_3P};\ \vec{v}_P = \vec{v}_{C_2} = \vec{v}_{C_3} \tag{4.125}$$

Where the second part of Eq. (4.125) refers to C_2 and C_3, introduced in Fig. 4.32. We already knew, by Eq. (4.117), that the normal components of velocities at C_2 and C_3 are *always* equal. Now, using Eq. (4.125) we conclude that *at this instant* the tangential components of the velocities at C_2 and C_3 are also equal:

$$\text{When contact happens at the pitch point}: v^t_{C_2} = v^t_{C_3} \tag{4.126}$$

So, we conclude that at an instant when contact happens to be at the pitch point, *P* (as shown in Fig. 4.33) there is no sliding and the two gear teeth are momentarily in pure rolling. Therefore, two gears in mesh are in rolling contact only at the pitch point. At this instant, the contact point *C* is also the instantaneous center of the two objects 2 and 3, i.e., I_{23}, as it is a point that has the same velocity when considered to be on either of them. This is similar to the case of the contact point in pure rolling of a disk that has zero velocity on either the disk or the ground, see Fig. 4.18. If the velocities of C_2 and C_3 at the contact point are not the same, while I_{23} will still be on the common normal drawn at the contact point, it will not be at the contact point itself. This is similar to the case of the contact point in rolling and sliding of a disk where the IC is located on the common normal at the contact point, not at the contact point itself, see Fig. 4.20.

Generalized Version of the Angular Velocity Ratio Theorem If the observer was no longer stationary and it was attached to an object k, the observer would feel that k is stationary and behave like our familiar fixed member, 1. Therefore, relative to that observer, an equation similar to Eq. (4.123) could be written with k replacing 1 and the angular velocities measured with respect to observer k rather than the stationary observer 1. So, for any three objects i,j, and k a generalized version of Eq. (4.123) relative to observer k would be,

$$\frac{\omega_{j/k}}{\omega_{i/k}} = \frac{I_{ki}I_{ij}}{I_{kj}I_{ij}} \tag{4.127}$$

Even though the foregoing derivation was only based on analyzing the motion of a cam and follower mechanism, Eq. (4.127) is indeed valid for any mechanism. This equation simply states that by determining the location of I_{ij} (general form of I_{23}) one can determine the speed ratio. In Eq. (4.127), the relative velocities could be expanded and therefore we can get the following result:

$$\frac{\omega_j - \omega_k}{\omega_i - \omega_k} = \frac{I_{ki}I_{ij}}{I_{kj}I_{ij}} \tag{4.128}$$

For $k = 1$ (i.e., the observer is attached to the stationary base) $\omega_k = 0$ and Eq. (4.128) reduces to the following:

$$\frac{\omega_j}{\omega_i} = \frac{I_{1i}I_{ij}}{I_{1j}I_{ij}} \tag{4.129}$$

Equation (4.129) is the general form of Eq. (4.123) for members i and j.

Example 4.16 Figure 4.36 illustrates a four-bar linkage at an instant when links 2 and 4 are both perpendicular to the fix link 1. We know that the input angular velocity is $\omega_{AB} = 2$ rad/s (CCW), $AB = 200$ mm, $BC = 400$ mm, and CD=300 mm.

A. Calculate the angular velocity of link CD.
B. Calculate the velocity of point I_{24}.

A. **Calculate the angular velocity of link *CD***

Using the angular velocity ratio theorem, i.e., Eq. (4.129), for members 2 and 4 one can write:

$$\frac{\omega_4}{\omega_2} = \frac{I_{12}I_{24}}{I_{14}I_{24}} \tag{4.130}$$

Now, noticing that triangles $I_{24}AB$ and $I_{24}DC$ are similar, one can write:

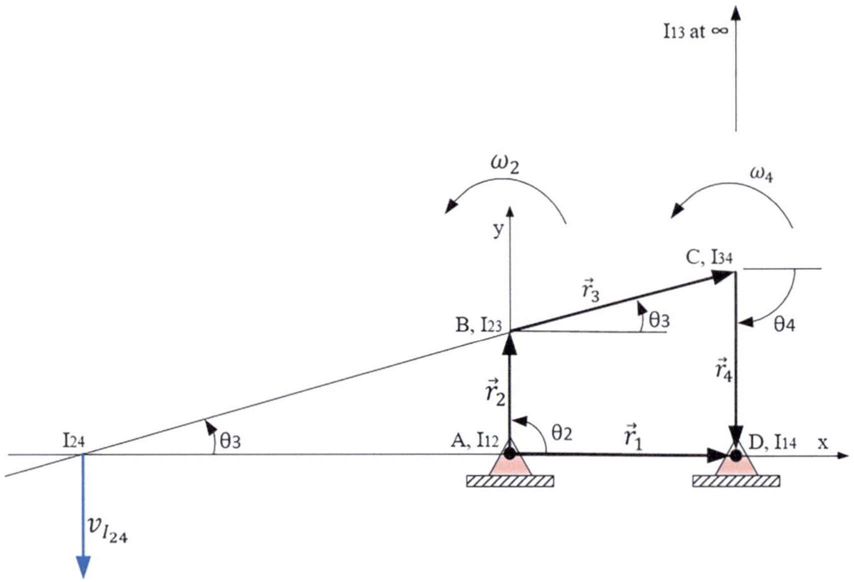

Fig. 4.36 Four-bar linkage with links 2 and 4 perpendicular to the fix link 1

$$\frac{I_{12}I_{24}}{I_{14}I_{24}} = \frac{AB}{CD} \tag{4.131}$$

Combining Eqs. (4.130), (4.131) and the provided data we get:

$$\frac{\omega_4}{\omega_2} = \frac{AB}{CD} \rightarrow \frac{\omega_4}{\omega_2} = \frac{200}{300} \tag{4.132}$$

Recalling $\omega_{AB} = \omega_2 = 2$ rad/s (CCW), we get: $\omega_4 = 1.333$ rad/s (CCW).

Another Way for Solving Part A

Noticing that at the instant shown in Fig. 4.36 links 2 and 4 are parallel, I_{13} is located at infinity. Therefore, link 3 is momentarily rotating about a point that is at infinity. This is another definition of translation. So we conclude that link 3 is momentarily translating and velocities of all points on it are momentarily identical. Therefore,

$$v_B = v_C \rightarrow \omega_2 r_2 = \omega_4 r_4 \tag{4.133}$$

From the provided data we know that:

$$\omega_2 r_2 = 2 \text{ rad/s} \times 200 \text{ mm} = 0.4 \text{ m/s}$$

Substitution of this result into Eq. (4.133) results in:

$\omega_4 = 1.333$ rad/s (CCW)
This is the same result as the one obtained using the first method.

B. **Calculate the velocity of point I_{24}**

To get the velocity of point I_{24} that belongs to both links 2 and 4, since link 2 rotates about I_{12}, one can write:

$$v_{I_{24}} = \omega_2 . I_{12} I_{24} \tag{4.134}$$

From the geometry of the mechanism in Fig. 4.36 we know that:

$$\sin \theta_3 = \frac{CD - AB}{BC} = \frac{300 - 200}{400}$$

Or,

$$\theta_3 = 14.478\,^\circ$$

Therefore, in triangle $I_{24}AB$ we get:

$$I_{24} I_{12} = \frac{AB}{\tan \theta_3} = \frac{200}{\tan 14.478\,^\circ} = 774.57 \text{ mm}$$

Substituting the calculated value of $I_{24}I_{12}$ into Eq. (4.134) results in:

$$v_{I_{24}} = \omega_2 . I_{24} I_{12} = 2 \text{ rad/s} \times 774.57 \text{ mm} = 1.549 \text{ m/s}$$

The direction of the velocity of point I_{24} is shown in Fig. 4.36 and it has been obtained based on the direction of rotation of link 2.

Another Way for Solving Part B
Velocity of point I_{24} can also be obtained by noticing that I_{24} belongs to link 4. Since link 4 rotates about I_{14}, one can write:

$$v_{I_{24}} = \omega_4 I_{24} I_{14} \tag{4.135}$$

But,

$$I_{24} I_{14} = I_{24} I_{12} + I_{12} I_{14} = I_{24} A + \text{BC} \times \cos \theta_3 = 774.57 + 400 \times \cos(14.478\,^\circ)$$

So,

$$I_{24} I_{14} = 1161.867 \text{ mm}$$

Substituting the calculated values of $I_{24}I_{12}$ and ω_4 into Eq. (4.135) results in:

$$v_{I_{24}} = \omega_4 I_{24} I_{14} = 1.333 \text{ rad/s} \times 1161.867 \text{ mm} = 1.549 \text{ m/s}$$

This result is the same as the one obtained before and confirms that I_{24} has the same velocity no matter whether it is considered to belong to link 2 or link 4.

4.7.4 Freudenstein's Theorem

Recall a four-bar linkage shown in Fig. 4.36 and Eq. (4.130) that expressed the angular velocity ratio theorem:

$$\frac{\omega_4}{\omega_2} = \frac{I_{12}I_{24}}{I_{14}I_{24}} \tag{4.130}$$

We are looking for a condition that, if satisfied, the angular velocity ratio given in Eq. (4.130) is a maximum or a minimum. Since usually the crank (i.e., link 2) speed is constant, we are looking for a condition at which the output (i.e., the follower) angular velocity is a maximum or a minimum. To find this condition, we first rewrite Eq. (4.130) as follows:

$$\frac{\omega_4}{\omega_2} = \frac{I_{24}I_{12}}{I_{24}I_{12} + I_{12}I_{14}} \tag{4.136}$$

The $I_{12}I_{14}$ part is a constant and it does not change during motion. So finding extremums of the function defined by Eq. (4.136) reduces to finding extremums of the following function:

$$\frac{\omega_4}{\omega_2} = \frac{x}{x + c} \tag{4.137}$$

where $x = I_{24}I_{12}$ changes during motion, but $c = I_{12}I_{14}$ does not. The first derivative of this function with respect to x is:

$$\frac{d\left(\frac{\omega_4}{\omega_2}\right)}{dx} = \frac{c}{(x + c)^2} > 0 \tag{4.138}$$

This expression is never zero and it is always positive. So the optimization problem can only have corner solutions. In other words, by increasing $x = I_{24}I_{12}$, the ratio $\frac{\omega_4}{\omega_2}$ also increases up to a point after which x can no longer increase (because of the size limitations of the mechanism). Therefore, answering the question, "What is the maximum value of $\frac{\omega_4}{\omega_2}$ during motion?" reduces to answering the question, "What is the maximum value of $x = I_{24}I_{12}$ during motion and at what condition it occurs?". The Freudenstein's theorem states that the maximum $x = I_{24}I_{12}$ value and

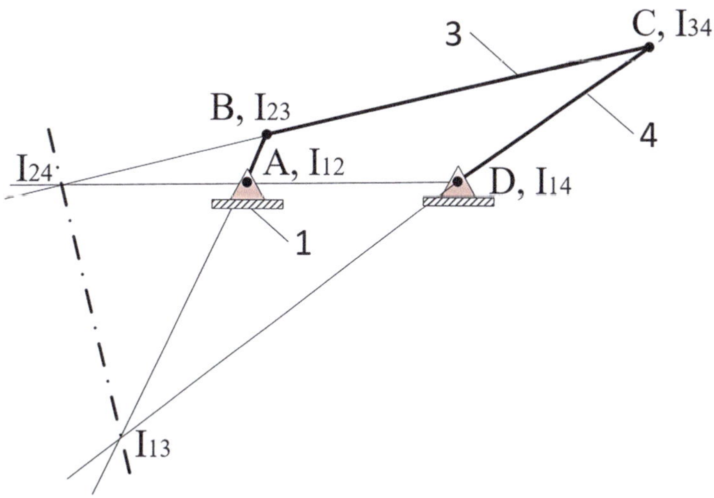

Fig. 4.37 Freudenstein's theorem condition for maximum $\frac{\omega_4}{\omega_2}$ in a four-bar linkage that happens when line $I_{24}I_{13}$ is perpendicular to the coupler link 3

therefore *the maximum* $\frac{\omega_4}{\omega_2}$ *value happen when the collineation axis,* i.e., *the line connecting the two non-trivial instantaneous centers* I_{24} *and* I_{13} *of the four-bar linkage is perpendicular to the coupler link 3, as seen in* Fig. 4.37. More discussion on this theorem and additional references can be found in Uicker et al. [4].

References

1. Ghorashi, A., & Ghorashi, M. (2020). Theoretical and computational analysis of the falling ladder problem. *SN (Springer Nature) Computer Science, 1*, 20. https://doi.org/10.1007/s42979-019-0019-7
2. Lee, H. H. (2015). *Engineering dynamics labs with SolidWorks motion 2015*. SDC Publications.
3. Meriam, J. L., & Kraige, L. G. (2007). *Engineering mechanics, dynamics* (6th ed.). Wiley.
4. Uicker, J. R., et al. (2016). *Theory of machines and mechanisms* (5th ed.). Oxford University Press.
5. Waldron, K. J., & Kinzel, G. L. (2004). *Kinematics, dynamics, and design of machinery* (2nd ed.). Wiley.

Chapter 5
Acceleration Analysis

5.1 Introduction

After learning about position and velocity analyses in Chaps. 3 and 4, respectively, now it's time to take a derivative of velocity with respect to time and analyze acceleration in mechanisms. Therefore, acceleration is the second time derivative of position. Unlike position analysis, velocity and acceleration analyses deal with linear problems and therefore they do not need application of numerical methods; the way position analysis does. In this chapter, the application of several methods for acceleration analysis is presented. They are the algebraic, vector loop, and relative motion methods. Several illustrative problems provide examples for use of each of these methods. One fundamental importance of the acceleration analysis is that it provides a link that connects kinematic analysis to kinetic analysis using Newton's second law. Once accelerations of mass center G of each member and the angular accelerations of each member are computed, calculating the corresponding inertia forces (i.e., $-m\vec{a}_G$) and inertia moments (i.e., $-I_G\vec{\alpha}$) can be easily done by just multiplying each acceleration term by its corresponding inertia and at the end change the sign of the obtained result. The inertia forces and the inertia moments can then be used in the dynamic analysis of the mechanism. Details of this matter are presented in Chaps. 12 and 13.

5.2 Algebraic Method for Acceleration Analysis

The algebraic method for acceleration analysis is based on the introduction of a position function for the point of interest and then taking a derivative of the mentioned function with respect to time to get the velocity function, and then taking a second derivative with respect to time to get the acceleration function.

M. Ghorashi, *Introduction to Machine Dynamics*, Mechanical Engineering Series,
https://doi.org/10.1007/978-3-031-94736-0_5

Let us use the algebraic method to calculate an approximate slider acceleration function for the slider-crank mechanism shown in Fig. 4.5. The velocity function of the slider was derived in Chap. 4 as follows:

$$v_B = -r_2\omega_2\left[\sin\theta_2 \pm \frac{\lambda \sin 2\theta_2}{2\sqrt{1-(\lambda\sin\theta_2)^2}}\right]; \lambda = r_2/r_3 \tag{4.11}$$

In which the crank-to-connecting-rod length ratio, λ, is usually less than 0.25. By applying this constraint, the square root term in Eq. (4.11) can be approximated as 1. Then by choosing the positive sign in Eq. (4.11), that is compatible with Fig. 4.5, the velocity Eq. (4.11) can be approximated as follows,

$$v_B \approx -r_2\omega_2(\sin\theta_2 + 0.5\lambda \sin 2\theta_2) \tag{5.1}$$

Taking a time derivative of Eq. (5.1) provides the following approximate acceleration function of the slider:

$$a_B \approx -r_2\omega_2^2(\cos\theta_2 + \lambda \cos 2\theta_2); \lambda = r_2/r_3 \tag{5.2}$$

In which for constant ω_2 we have: $\theta_2 = \omega_2 t$.

Example 5.1 Consider the slider-crank mechanism shown in Fig. 4.5 with no eccentricity ($e = 0$) and with r_2=75 mm and r_3=200 mm. The crank rotates at a constant angular velocity of -2000 rpm (CW) and starts its motion at $\theta_2 = 0$ and $\theta_3 = 0$ (i.e., at the top dead center). Plot the acceleration of the slider represented by point B in Fig. 4.5 in one cycle as a function of time by MATLAB and using Eq. (5.2). Compare this result with the one obtained from SolidWorks simulation.

This problem has been modeled using SolidWorks in Lee [1] and a slightly modified model is shown in Fig. 4.3.

In Example 4.1, the slider velocity for the provided data was analyzed. The corresponding solution for acceleration has been obtained by SolidWorks Motion software and it is shown in Fig. 5.1. This figure also includes the output of a MATLAB script that uses Eq. (5.2) for acceleration calculation. Even though the λ value in this case is a bit large, still the two results, are in good agreement with each other. Furthermore, the effect of the second harmonic term in Eq. (5.2) with twice the angular velocity could be observed. Finally, the shown data at $t = 0.027$ s corresponds to $\theta_2 = 324°$.

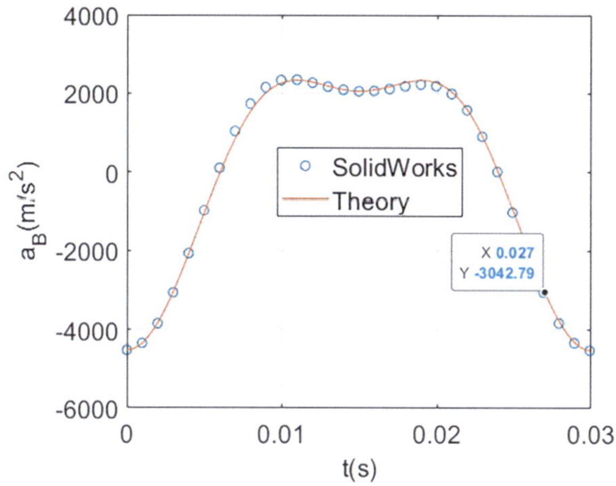

Fig. 5.1 Acceleration of the slider in Example 5.1 obtained from SolidWorks together with the output of a MATLAB script that uses Eq. (5.2)

5.3 Acceleration as a Vector

Recognizing that acceleration is a vector quantity will help us use it properly in Chaps. 12 and 13, where we would like to use it for analyzing forces and moments (that are vector quantities as well). Let us start with a simple example. Fig. 5.2 illustrates the motion of a disk that rotates with angular velocity $\vec{\omega}$ and angular acceleration $\vec{\alpha}$ about a hinge located at its center. We would like to analyze the motion of a point A on the circumference of the disk. We know that the velocity of point A located by the position vector $\vec{r}_A$ can be written as,

$$\vec{v}_A = \frac{d\vec{r}_A}{dt} = \vec{\omega} \times \vec{r}_A \tag{4.79}$$

Taking another derivative of Eq. (4.79) with respect to time will result in the acceleration of point A:

$$\vec{a}_A = \frac{d\vec{v}_A}{dt} = \frac{d\left(\vec{\omega} \times \vec{r}_A\right)}{dt} = \left(\frac{d\vec{\omega}}{dt} \times \vec{r}_A\right) + \vec{\omega} \times \frac{d\vec{r}_A}{dt} \tag{5.3}$$

Substituting:

$$\vec{\alpha} = \frac{d\vec{\omega}}{dt},\ \vec{v}_A = \frac{d\vec{r}_A}{dt} = \vec{\omega} \times \vec{r}_A$$

one obtains,

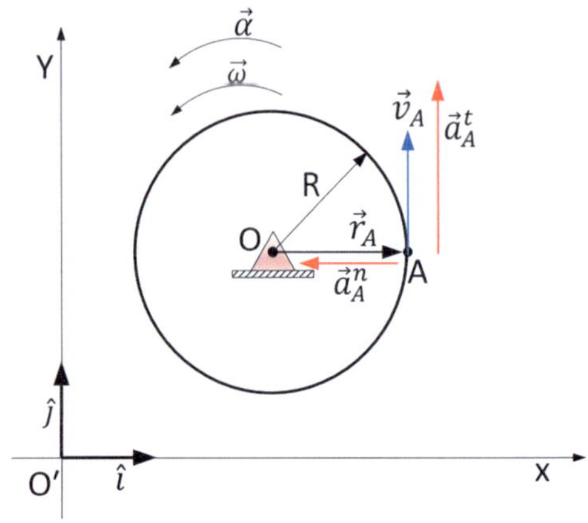

Fig. 5.2 Velocity and acceleration components of point *A* on the circumference of a rotating disk

$$\vec{a}_A = \left(\vec{\alpha} \times \vec{r}_A\right) + \vec{\omega} \times \left(\vec{\omega} \times \vec{r}_A\right) \tag{5.4}$$

The first term of Eq. (5.4), as we can see in Fig. 5.2, is tangent to the path of the particle *A* and for this reason it is called the tangential acceleration of this particle:

$$\vec{a}_A^{\,t} = \vec{\alpha} \times \vec{r}_A \tag{5.5}$$

This term is responsible for changing the magnitude of the velocity vector (i.e., changing speed). The second term is inward normal to the path of the particle *A* and that is why it is called the normal acceleration of this particle:

$$\vec{a}_A^{\,n} = \vec{\omega} \times \left(\vec{\omega} \times \vec{r}_A\right) = \vec{\omega} \times \vec{v}_A \tag{5.6}$$

This term is responsible for changing the direction of the velocity vector. As it was mentioned in Sect. 4.6, and by comparing Eqs. (4.79) and (5.6), we see that if $\vec{r}_A$ or $\vec{v}_A$ have constant magnitudes, their derivatives with respect to time are simply obtained by the cross product of the angular velocity of rotation of that vector, $\vec{\omega}$, and the vector itself.

Example 5.2 Referring to Fig. 5.2, calculate the velocity and the acceleration vectors of particle *A* if $\vec{r}_A = 2\hat{i}$ cm, $\omega = 10$ rad/s (CCW), and $\alpha = 2$ rad/s^2 (CCW).

For $\vec{v}_A$, by substituting the given numerical values of variables in Eq. (4.79) we get:

$$\vec{v}_A = \vec{\omega} \times \vec{r}_A = 10\hat{k} \times 2\hat{i} = 20\hat{j} \text{ cm/s}$$

And for the acceleration terms we use Eqs. (5.5) and (5.6):

$$\vec{a}^{\,t}_A = \vec{\alpha} \times \vec{r}_A = 2\hat{k} \times 2\hat{i} = 4\hat{j}\ \text{cm/s}^2$$

$$\vec{a}^{\,n}_A = \vec{\omega} \times \vec{v}_A = 10\hat{k} \times 20\hat{j} = -200\hat{i}\ \text{cm/s}^2$$

So the total acceleration of point A can be written as follows:

$$\vec{a}_A = \vec{a}^{\,t}_A + \vec{a}^{\,n}_A = 4\hat{j} - 200\hat{i}\ \text{cm/s}^2$$

Example 5.3 In the parallelogram four-bar linkage shown in Fig. 5.3, link 2 rotates at the constant angular velocity of -4 rad/s (CW). Calculate the velocity and the acceleration vectors of point C when $\theta_2 = 45\,°$. We know that: $AB = CD = 0.5$ m, and $BC = AD = 0.3$ m.

As we have seen before, since the geometry of the mechanism is a parallelogram and it will remain to be a parallelogram, link 3 will always remain parallel to the fix link, AD. Therefore, link 3 will have no rotational motion and it will be performing curvilinear translation (see Chap. 2). Consequently, the velocity and the acceleration vectors of point C are the same as the velocity and acceleration vectors of point B. Point B is moving on a circular path about A, so the problem reduces to what was described in Example 5.2 and we can write:

$$\vec{v}_B = \vec{v}_C = \vec{\omega}_2 \times \overrightarrow{AB} = -4\hat{k} \times 0.5\left(\cos 45\,°\,\hat{i} + \sin 45\,°\,\hat{j}\right) = \sqrt{2}\left(-\hat{j} + \hat{i}\right)\ \text{m/s}$$

Since the angular velocity of link 2 is constant, its angular acceleration is zero and the accelerations of points B or C will only have normal components. Using Eq. (5.6) we have:

$$\vec{a}_C = \vec{a}_B = \vec{a}^{\,n}_C = \vec{a}^{\,n}_B \rightarrow \vec{a}_C = \vec{\omega}_2 \times \vec{v}_C = -4\hat{k} \times \sqrt{2}\left(-\hat{j} + \hat{i}\right) = -4\sqrt{2}\left(\hat{i} + \hat{j}\right)\ \text{m/s}^2$$

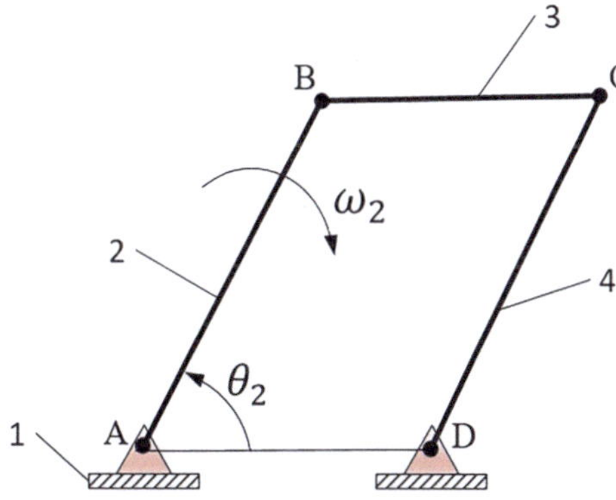

Fig. 5.3 Parallelogram four-bar linkage

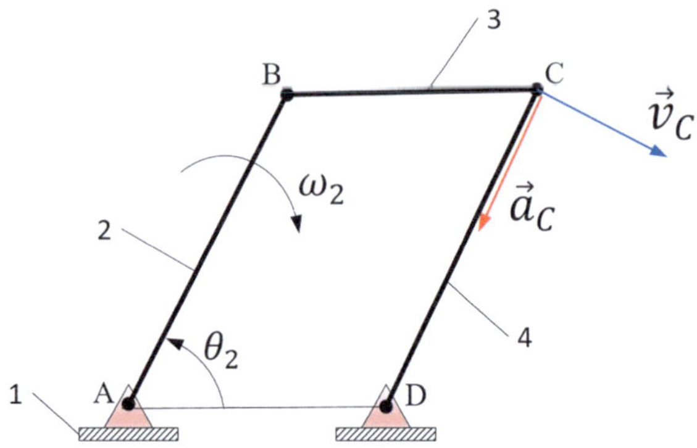

Fig. 5.4 Velocity and acceleration vectors at point *C* in Example 5.3

It can be easily verified that $\vec{a}_B$ and $\vec{a}_C$ are centripetal accelerations; i.e., they are vectors in the $\overrightarrow{BA}$ and $\overrightarrow{CD}$ directions, respectively; see Fig. 5.4.

5.4 Acceleration Analysis by the Vector-Loop Method

While in general a position problem is nonlinear and requires Taylor series expansion and Newton-Raphson method for linearization (see Chap. 3) the velocity and acceleration problems are linear. This matter was shown for velocity in Chap. 4. In what follows we solve the slider-crank acceleration problem and show how it reduces to the solution of a linear system of equations.

Example 5.4 Using a vector loop, perform acceleration analysis on the slider-crank mechanism with zero eccentricity shown in Fig. 4.5.

Figure 4.6 illustrates the vector-loop diagram for Fig. 4.5. And its corresponding vector-loop equation is

$$\vec{r}_2 + \vec{r}_3 = \vec{r}_1 \tag{4.12}$$

Projecting Eq. (4.12) on the direction of $\vec{r}_1$ and the one perpendicular to $\vec{r}_1$ gives,

$$r_2 \cos\theta_2 + r_3 \cos\theta_3 = r_1 = x_B \tag{4.13}$$

$$r_2 \sin\theta_2 + r_3 \sin\theta_3 = 0 \tag{4.14}$$

These are two nonlinear equations that for any value of the input variable θ_2 should be solved for two unknowns r_1 and θ_3. Once the position analysis is complete, and to find the unknown velocity $v_B = \dot{r}_1 = \dot{x}_B$ of the slider (or piston) and the unknown angular velocity ω_3 of the piston rod, we take time derivatives of Eqs. (4.13) and (4.14) to get:

$$\begin{cases} r_3\omega_3 \sin\theta_3 + v_B = -r_2\omega_2 \sin\theta_2 \\ -r_3\omega_3 \cos\theta_3 + 0\times v_B = r_2\omega_2 \cos\theta_2 \end{cases} \tag{4.16}$$

Because of the linear nature of Eqs. (4.16), they can be rewritten in the following matrix form:

$$\begin{bmatrix} r_3\sin\theta_3 & 1 \\ -r_3\cos\theta_3 & 0 \end{bmatrix} \begin{Bmatrix} \omega_3 \\ v_B \end{Bmatrix} = \begin{Bmatrix} -r_2\omega_2 \sin\theta_2 \\ r_2\omega_2 \cos\theta_2 \end{Bmatrix} \tag{4.17}$$

This is now a linear system of equations in terms of the unknown velocity v_B and the angular velocity ω_3. So, while the position problem is nonlinear, the velocity problem is linear and it can be solved easier. We will now see that the same is true for acceleration as well, and the acceleration equations will be shown to be linear. The order of calculation is very important. The velocity equations given above have θ_3 that is obtained from the position analysis. So the velocity problem cannot be solved before the position problem has already been solved. We are now going to see that the same relation is true between the velocity and the acceleration problems: the acceleration problem cannot be solved before the velocity one has already been solved. So here is the order of calculations: position, velocity, and then acceleration.

We now take another time derivative, this time from the velocity Eqs. (4.16) and use the chain rule to get:

$$\begin{cases} r_3\alpha_3 \sin\theta_3 + r_3\omega_3^2 \cos\theta_3 + a_B = -r_2\alpha_2 \sin\theta_2 - r_2\omega_2^2 \cos\theta_2 \\ -r_3\alpha_3 \cos\theta_3 + r_3\omega_3^2 \sin\theta_3 + 0\times a_B = r_2\alpha_2 \cos\theta_2 - r_2\omega_2^2 \sin\theta_2 \end{cases} \tag{5.7}$$

where θ_2, ω_2, and α_2 are the inputs given to the crank and are considered as known. At this stage, ω_3 and v_B are also known from solving the velocity problem, and finally θ_3 is known from solving the position problem. Therefore, Eqs. (5.7) have two unknowns, i.e., α_3 and a_B. Keeping all of the unknown terms in the LHS and moving all of the known terms to the RHS result in,

$$\begin{cases} r_3\alpha_3 \sin\theta_3 + a_B = -r_2\alpha_2 \sin\theta_2 - r_2\omega_2^2 \cos\theta_2 - r_3\omega_3^2 \cos\theta_3 \\ -r_3\alpha_3 \cos\theta_3 + 0\times a_B = r_2\alpha_2 \cos\theta_2 - r_2\omega_2^2 \sin\theta_2 - r_3\omega_3^2 \sin\theta_3 \end{cases} \tag{5.8}$$

Equations (5.8) can now be written in the following matrix form:

$$\begin{bmatrix} r_3\sin\theta_3 & 1 \\ -r_3\cos\theta_3 & 0 \end{bmatrix} \begin{Bmatrix} \alpha_3 \\ a_B \end{Bmatrix} = \begin{Bmatrix} -r_2\alpha_2 \sin\theta_2 - r_2\omega_2^2 \cos\theta_2 - r_3\omega_3^2 \cos\theta_3 \\ r_2\alpha_2 \cos\theta_2 - r_2\omega_2^2 \sin\theta_2 - r_3\omega_3^2 \sin\theta_3 \end{Bmatrix} \tag{5.9}$$

The acceleration Eq. (5.9) is linear for the two unknowns, α_3 and a_B and it can be easily verified that the coefficient matrix in Eq. (5.9) is nothing but the Jacobian matrix of the position and the velocity equations.

Example 5.5 In the slider-crank mechanism corresponding to Fig. 4.6 the crank rotates at constant rate $\omega_2 = -100$ rpm (CW), and it is known that $r_2 = 1$ in and $r_3 = 5$ in. Calculate the angular acceleration of link 3, α_3, and the linear acceleration of the slider, a_B at an instant when $\theta_2 = 40°$.

In Example 4.3, this problem was solved for the unknown velocity terms. Converting the unit of the input angular velocity to rad/s resulted in:

$$\omega_2 = -10.472 \text{ rad/s (CW)}$$

Position analysis revealed that at $\theta_2 = 40°$ we have two answers:

$$\theta_3 = -7.386°, \theta_3 = 187.386°$$

The $\theta_3 = -7.386°$ *answer resulted in:*

$$r_1 = 5.725\text{in}; \omega_3 = 1.618 \text{ rad/s}; v_B = 7.771\text{in/s}$$

Using the foregoing information in the second part of acceleration Eqs. (5.8), i.e.,

$$-r_3\alpha_3 \cos\theta_3 + 0 \times a_B = r_2\alpha_2 \cos\theta_2 - r_2\omega_2^2 \sin\theta_2 - r_3\omega_3^2 \sin\theta_3$$

gives,

$$-5\alpha_3 \cos(-7.386°) = 0 - (-10.472)^2 \sin 40° - 5 \times (1.618)^2 \sin(-7.386°)$$

So,

$$\alpha_3 = 13.877 \text{ rad/s}^2 \text{ (CCW)}$$

And now the first part of acceleration Eqs. (5.8), i.e.,

$$r_3\alpha_3 \sin\theta_3 + a_B = -r_2\alpha_2 \sin\theta_2 - r_2\omega_2^2 \cos\theta_2 - r_3\omega_3^2 \cos\theta_3$$

gives,

$$5(13.877)\sin(-7.386°) + a_B = 0 - (-10.472)^2 \cos 40° - 5 \times (1.618)^2 \cos(-7.386°)$$

So we get:

$$a_B = -88.068 \text{ in/s}^2$$

This answer corresponds to an assembly like the one that is shown in Fig. 4.7. *The* $\theta_3 = 187.386°$ *answer resulted in,*

$$r_1 = -4.192 \text{ in}; \omega_3 = -1.618 \text{ rad/s}; v_B = 5.691 \text{ in/s}$$

Using the foregoing information in the second part of acceleration Eqs. (5.8), i.e.,

$$-r_3\alpha_3 \cos\theta_3 + 0 \times a_B = r_2\alpha_2 \cos\theta_2 - r_2\omega_2^2 \sin\theta_2 - r_3\omega_3^2 \sin\theta_3$$

gives,

$$-5\alpha_3 \cos(187.386°) = 0 - (-10.472)^2 \sin 40° - 5 \times (-1.618)^2 \sin(187.386°)$$

So,

$$\alpha_3 = -13.877 \text{ rad}/s^2 \text{ (CW)}$$

And now the first part of acceleration Eqs. (5.8), i.e.,

$$r_3\alpha_3 \sin\theta_3 + a_B = -r_2\alpha_2 \sin\theta_2 - r_2\omega_2^2 \cos\theta_2 - r_3\omega_3^2 \cos\theta_3$$

gives,

$$5(-13.877) \sin(187.386°) + a_B = 0 - (-10.472)^2 \cos 40° - 5 \times (-1.618)^2 \cos(187.386°)$$

So,

$$a_B = -79.945 \text{ in}/s^2$$

This solution corresponds to an assembly like the one that is shown in Fig. 4.8. This problem has two solutions and the question as to which one would actually happen depends on the way the slider-crank mechanism has been initially assembled.

Solution by the Cramer's Rule

For the slider-crank acceleration Eq. (5.9) with Jacobian matrix:

$$J = \begin{bmatrix} r_3 \sin\theta_3 & 1 \\ -r_3 \cos\theta_3 & 0 \end{bmatrix} \tag{4.25}$$

and determinant:

$$|J| = r_3 \cos\theta_3 \tag{4.26}$$

using Cramer's rule (see Sect. 4.3) the answers can be written as,

$$\alpha_3 = \frac{-r_2\alpha_2\cos\theta_2 + r_2\omega_2^2\sin\theta_2 + r_3\omega_3^2\sin\theta_3}{r_3\cos\theta_3} \tag{5.10}$$

$$a_B = \frac{\begin{vmatrix} r_3\sin\theta_3 & -r_2\alpha_2\sin\theta_2 - r_2\omega_2^2\cos\theta_2 - r_3\omega_3^2\cos\theta_3 \\ -r_3\cos\theta_3 & r_2\alpha_2\cos\theta_2 - r_2\omega_2^2\sin\theta_2 - r_3\omega_3^2\sin\theta_3 \end{vmatrix}}{|J|} \tag{5.11}$$

5.5 Acceleration Analysis Using the Relative Motion Method

The subject of motion analysis using a moving observer is usually covered in detail in textbooks on engineering dynamics like Meriam and Kraige [2]. In Sect. 4.6 velocity analysis using the relative motion method was reviewed. In the present section, that discussion is extended to acceleration analysis. Figure 5.5 illustrates point A that its motion is analyzed by a moving observer B and a stationary observer O. The translating and rotating observer B moves at velocity $\vec{v}_B$ and acceleration $\vec{a}_B$ and its attached coordinate system x, y, z rotates at an angular velocity $\vec{\omega}$ and angular acceleration, $\vec{\alpha}$. A stationary observer O is attached to the X, Y, Z coordinate system and can observe both A and B and record their absolute velocities and absolute accelerations.

Recalling the relative velocity equation:

$$\vec{v}_A = \vec{v}_B + \dot{x}\hat{i} + \dot{y}\hat{j} + \vec{\omega} \times \left(x\hat{i} + y\hat{j}\right) \tag{4.81}$$

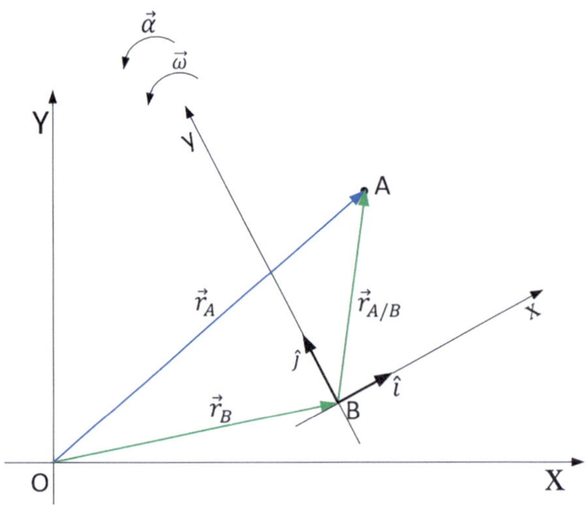

Fig. 5.5 Stationary observer O and a translating and rotating observer B analyzing acceleration of point A

we take another derivative with respect to time. Since $\vec{v}_A$ and $\vec{v}_B$ are absolute velocities, their time derivatives will be absolute accelerations and we get:

$$\vec{a}_A = \vec{a}_B + \ddot{x}\hat{i} + \ddot{y}\hat{j} + \dot{x}\frac{d\hat{i}}{dt} + \dot{y}\frac{d\hat{j}}{dt} + \frac{d\vec{\omega}}{dt} \times \left(x\hat{i} + y\hat{j}\right) + \vec{\omega} \times \left(\dot{x}\hat{i} + x\frac{d\hat{i}}{dt} + \dot{y}\hat{j} + y\frac{d\hat{j}}{dt}\right)$$

Now, recalling time derivatives of rotating unit vectors as,

$$\frac{d\hat{i}}{dt} = \vec{\omega} \times \hat{i};\ \frac{d\hat{j}}{dt} = \vec{\omega} \times \hat{j} \tag{4.80}$$

and noticing the definition of angular acceleration $\vec{\alpha}$,

$$\vec{\alpha} = \frac{d\vec{\omega}}{dt} \tag{5.12}$$

The $\vec{a}_A$ equation reduces to,

$$\begin{aligned}\vec{a}_A = \vec{a}_B + \ddot{x}\hat{i} + \ddot{y}\hat{j} + \dot{x}\vec{\omega} \times \hat{i} + \dot{y}\vec{\omega} \times \hat{j} + \vec{\alpha} \times \left(x\hat{i} + y\hat{j}\right)\\ + \vec{\omega} \times \left(\dot{x}\hat{i} + x\vec{\omega} \times \hat{i} + \dot{y}\hat{j} + y\vec{\omega} \times \hat{j}\right)\end{aligned}$$

But, $\widehat{\dot{x}i} + \widehat{\dot{y}j}$ is the velocity of point A relative to the observer B, shown by $\vec{v}_{rel.}$

$$\vec{v}_{rel} = \widehat{\dot{x}i} + \widehat{\dot{y}j} \tag{4.82}$$

Similarly, $\widehat{\ddot{x}i} + \widehat{\ddot{y}j}$ is the acceleration of point A relative to the observer B, shown by $\vec{a}_{rel}$,

$$\vec{a}_{rel} = \widehat{\ddot{x}i} + \widehat{\ddot{y}j} \tag{5.13}$$

And we have,

$$\vec{r} = \vec{r}_{A/B} = \widehat{xi} + \widehat{yj} \tag{4.76}$$

Therefore, we get the final form of the $\vec{a}_A$ equation in terms of observer B's motion and observations.

$$\vec{a}_A = \vec{a}_B + \vec{a}_{rel} + 2\vec{\omega} \times \vec{v}_{rel} + \vec{\alpha} \times \vec{r} + \vec{\omega} \times \left(\vec{\omega} \times \vec{r}\right) \tag{5.14}$$

This is the complete relative acceleration equation with five terms in the right-hand side. Among these acceleration terms the only acceleration that is truly observed by the observer B is $\vec{a}_{rel}$. The remaining terms are added to correct the misjudgment of the moving observer due to its translational and rotational velocities and accelerations. The term $2\vec{\omega} \times \vec{v}_{rel}$ is Coriolis acceleration and for discussions on its significance one can consult Meriam and Kraige [2] or Uicker et al. [3].

There are three notable special cases that should be considered when using Eq. (5.14):

Case 1. Points A and B belong to the same rigid body and the moving coordinate system is attached to that rigid body. In this case, B observes that A is stationary so we have,

$$\vec{v}_{rel} = 0; \ \vec{a}_{rel} = 0 \tag{5.15}$$

And Eq. (5.14) reduces to:

$$\vec{a}_A = \vec{a}_B + \vec{\alpha} \times \vec{r} + \vec{\omega} \times \left(\vec{\omega} \times \vec{r}\right) \tag{5.16}$$

Case 2. The observer B performs translational motion. In this case, $\omega = 0$, $\alpha = 0$ and Eq. (5.14) reduces to:

$$\vec{a}_A = \vec{a}_B + \vec{a}_{rel} \tag{5.17}$$

Case 3. Points A and B belong to two different objects, but they are momentarily collocated. In this case r = 0, and Eq. (5.14) reduces to:

$$\vec{a}_A = \vec{a}_B + \vec{a}_{rel} + 2\vec{\omega} \times \vec{v}_{rel} \tag{5.18}$$

Example 5.6 In Example 4.7, the velocity analysis of a disk that is rolling without slipping was presented. Referring to Fig. 5.6 and assuming that, as in Example 4.7, the diameter of the disk is 4 cm and the velocity at the center point O is 2m/s (which is constant) we would like to calculate accelerations at points C, B, and D.

Rolling is happening without slipping, so point C has zero velocity, but it doesn't mean this point has also zero acceleration! The velocity at C is only momentarily zero. It is not zero in a small time interval before or after point C became the disk contact point with the ground. Therefore, as we will formally prove, acceleration at C is not zero.

Considering the linear variation of velocity from point C in Example 4.7, it was shown that:

$$\vec{v}_B = 2\vec{v}_O = 4\hat{i}\,\text{m/s}; v_D = \sqrt{2}v_O = 2\sqrt{2}\,\text{m/s}; \vec{v}_D = 2\left(\hat{i} - \hat{j}\right)\text{m/s}; \omega = -100\,\text{rad/s (CW)}$$

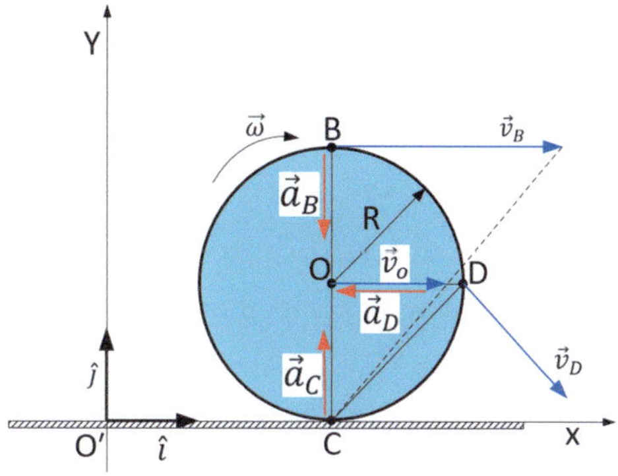

Fig. 5.6 Acceleration calculation of points on a disk in the rolling without slipping motion

Since rolling is happening with no slipping, the relation between the displacement of the center point O and the corresponding angular displacement of the disk, θ can be expressed as follows (see Sect. 2.2):

$$x_O = R\theta \tag{2.3}$$

Taking a derivative of Eq. (2.3) with respect to time gives:

$$v_O = R\omega \tag{5.19}$$

And another time derivative gives acceleration of point O:

$$a_O = R\alpha \tag{5.20}$$

Method 1 Since in this problem velocity of point O is constant, acceleration of this point is zero. Therefore, Eq. (5.20) requires that $\alpha = 0$. By attaching a moving coordinate system to the disk and center it at O, we can use Eq. (5.16) to write the following equation for acceleration of point C:

$$\vec{a}_C = \vec{a}_O + \vec{\alpha} \times \vec{r} + \vec{\omega} \times \left(\vec{\omega} \times \vec{r}\right) \tag{5.21}$$

In which $\vec{\omega}$ and $\vec{\alpha}$ are the angular velocity and the angular acceleration of the moving coordinate system (i.e. the disk), respectively. Furthermore, $\vec{r}$ is the position vector of the point of interest, C measured from the origin of the moving coordinate system, O, so $\vec{r} = \overrightarrow{OC}$. Substituting $a_O = 0$ and $\alpha = 0$ into Eq. (5.21) gives,

$$\vec{a}_C = 0 + 0 + \vec{\omega} \times \left(\vec{\omega} \times \overrightarrow{OC}\right) = \vec{\omega} \times \left(-\vec{v}_O\right) \tag{5.22}$$

We already know that,

$$\vec{v}_O = \vec{\omega} \times \overrightarrow{CO} = 2\hat{i}\ \mathrm{m/s}$$

Therefore, Eq. (5.22) reduces to,

$$\vec{a}_C = \vec{\omega} \times \left(-\vec{v}_O\right) = -100\hat{k}\ \mathrm{rad/s} \times \left(-2\hat{i}\ \mathrm{m/s}\right) = 200\hat{j}\ \mathrm{m/s^2}$$

Accelerations at points *B* and *D* in Fig. 5.6 can be calculated in a similar way.

Method 2 The problem could also be solved by noticing that through the eyes of an observer located at *O*, but now just translating, the motion of the disk is nothing but rotation at constant angular velocity about a fix point! Therefore, accelerations at points *B*, *C*, and *D* are all numerically equal to $R\omega^2 = 0.02 \times 100^2 = 200\ \mathrm{m/s^2}$ and, as we see in Fig. 5.6, they all point toward the center of the disk.

$$\vec{a}_C = 200\hat{j}\ \mathrm{m/s^2};\ \vec{a}_B = -200\hat{j}\ \mathrm{m/s^2};\ \vec{a}_D = -200\hat{i}\ \mathrm{m/s^2}$$

We see that this result for acceleration at *C* is identical to what we got using the first method. From this example and Example 4.7 one can conclude that in the case of a rolling disk without slipping and moving with constant speed, velocity varies linearly from the contact point *C*, and at any point it is perpendicular to the line connecting *C* to the point of interest; however, normal acceleration varies linearly from the center point *O* and it acts toward point *O*!

Example 5.7 Figure 5.7 illustrates a slider-crank mechanism at an instant when crank link 2 is perpendicular to the direction of motion of the slider. It is also known that $\theta_3 = -30°$, link 2 is 5 in long, and its angular velocity is -10 rad/s (CW) which is constant (so, $\vec{\alpha}_2 = 0$). The aim is to calculate the acceleration of points *B* and *C* and the angular acceleration of link 3 at this instant.

In Example 4.6, this problem was solved for velocity and we found out that link 3 is momentarily in translation and therefore the velocities of points *B* and *C* are equal. A summary of what we got in Example 4.6 is listed below:

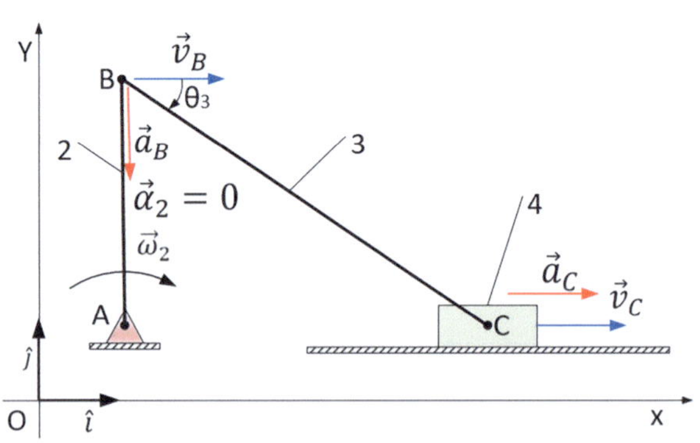

Fig. 5.7 A slider-crank mechanism when the crank is perpendicular to the direction of motion of the slider

$$\vec{\omega}_{BC} = 0 \tag{4.90}$$

$$\vec{v}_C = \vec{v}_B = 50\hat{i}\ \text{in/s} \tag{4.92}$$

Now the question is, "*Can we extend this result to acceleration and say since link 3 is momentarily in translation, we have the following?*"

$$\textit{Can we say?}\quad \vec{\alpha}_{BC} = 0;\ \vec{a}_C = \vec{a}_B \tag{5.23}$$

As we are going to see, the answer is: No!

Since $\vec{\alpha}_2 = 0$, we conclude that the tangential component of the acceleration of point B is zero:

$$\vec{a}^{\,t}_B = \vec{\alpha}_2 \times \vec{r}_B = 0$$

So, the only component of acceleration at point B is the normal one:

$$\vec{a}_B = \vec{a}^{\,n}_B = \vec{\omega}_2 \times \left(\vec{\omega}_2 \times \vec{r}_B\right) = \vec{\omega}_2 \times \vec{v}_B = -10\hat{k}\ \text{rad/s} \times 50\hat{i}\ \text{in/s} = -500\hat{j}\ \text{in/s}^2 \tag{5.24}$$

Now, to calculate $\vec{a}_C$ we use a moving coordinate system located at B and attached to link BC. Using Eq. (5.16), the acceleration of point C can be written in terms of acceleration of point B as follows:

$$\vec{a}_C = \vec{a}_B + \vec{\alpha} \times \vec{r} + \vec{\omega} \times \left(\vec{\omega} \times \vec{r}\right) \tag{5.25}$$

In which we have,

$$\vec{\alpha} = \vec{\alpha}_{BC};\ \vec{\omega} = \vec{\omega}_{BC};\ \vec{r} = \overrightarrow{BC}$$

Recalling Eq. (4.90), i.e., link BC is momentarily translating, and considering that the slider moves along the x direction, Eq. (5.25) for acceleration of point C reduces to:

$$\vec{a}_C = a_C\hat{i} = \vec{a}_B + \vec{\alpha}\,\vec{r} + 0 = -500\hat{j} + \alpha_{BC}\hat{k} \times \left(\frac{5}{\tan 30^\circ}\hat{i} - 5\hat{j}\right)$$

This is a vector equation with two unknowns, a_C and α_{BC}. Performing the cross product results in:

$$a_C\hat{i} = -500\hat{j} + \left(\frac{5\alpha_{BC}}{\tan 30^\circ}\hat{j} + 5\alpha_{BC}\hat{i}\right)$$

Equating the vector components on both sides of this vector equation gives us the following two scalar equations for the two unknowns, a_C and α_{BC}:

$$500 = \frac{5\alpha_{BC}}{\tan 30^\circ}; a_C = 5\alpha_{BC}$$

So,

$$\alpha_{BC} = 57.735 \text{ rad}/s^2(\text{CCW}); a_C = 288.675 \text{ in}/s^2 \rightarrow \vec{a}_C = 288.675\hat{i} \text{ in}/s^2 \quad (5.26)$$

By comparing Eqs. (5.23) and (5.26), we conclude that Eq. (5.23) is indeed incorrect and while the angular velocity of *BC* at this instant is momentarily zero, its angular acceleration is not zero.

Example 5.8 Figure 5.8 illustrates a four-bar linkage at an instant when links 2 and 4 are both perpendicular to the fix link 1. We know that the input angular velocity and angular acceleration are $\omega_{AB} = 2$ rad/s (CCW) and $\alpha_{AB} = -1$ rad/s^2 (CW), respectively. Also $AB = 200$ mm, $BC = 400$ mm, and CD=300 mm. We would like to calculate the angular accelerations of links 3 and 4 at the illustrated instant.

In Example 4.16, linear velocities and angular velocities of various elements in this problem were calculated. We also found out that link 3 is momentarily in translation and therefore the velocities of points *B* and *C* are equal, i.e.,

Fig. 5.8 Four-bar linkage with links 2 and 4 perpendicular to the fix link 1

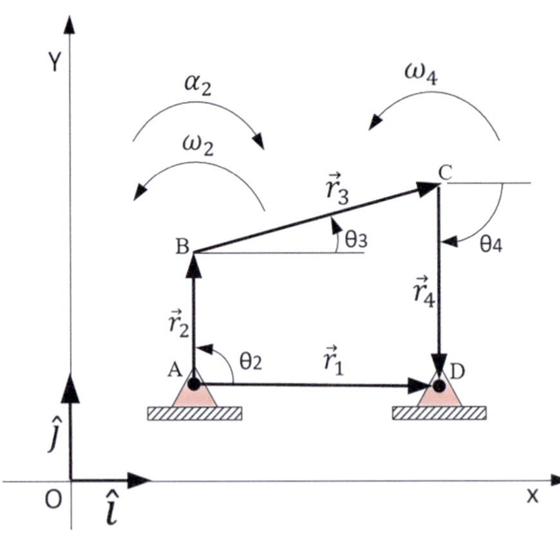

$$\vec{\omega}_{BC} = 0; \vec{v}_C = \vec{v}_B = -0.4\hat{i} \text{ m/s} \tag{5.27}$$

Furthermore, we got:

$$\theta_3 = 14.478\,°; \omega_4 = 1.333 \text{ rad/s (CCW)}$$

Now the question is, "*Can we extend this result to acceleration and say since link 3 is momentarily in translation, we have the following?*"

$$\textit{Can we say?} \quad \vec{\alpha}_{BC} = 0; \vec{a}_C = \vec{a}_B \tag{5.28}$$

As we are soon going to see, the answer is: No!

Since $\vec{\alpha}_2 = -1\,\hat{k} \text{ rad}/s^2$, using Eq. (5.5), the tangential component of acceleration of point B is:

$$\vec{a}^{\,t}_B = \vec{\alpha}_2 \times \vec{r}_B = \vec{\alpha}_2 \times \vec{r}_2 = -1\,\hat{k} \text{ rad}/s^2 \times 200\hat{j} \text{ mm} = 0.2\hat{i} \text{ m}/s^2$$

Equation (5.6) can now be used for calculation of the normal component of acceleration of point B:

$$\vec{a}^{\,n}_B = \vec{\omega}_2 \times \left(\vec{\omega}_2 \times \vec{r}_B\right) = \vec{\omega}_2 \times \vec{v}_B = 2\hat{k} \text{ rad/s} \times \left(-0.4\hat{i} \text{ m/s}\right) = -0.8\hat{j} \text{ m}/s^2$$

So, the total acceleration of point B is:

$$\vec{a}_B = \vec{a}^{\,t}_B + \vec{a}^{\,n}_B = 0.2\hat{i} - 0.8\hat{j} \text{ m}/s^2 \tag{5.29}$$

Now, to calculate $\vec{a}_C$ we use a moving coordinate system located at B and attached to link BC. Using Eq. (5.16), the acceleration of point C can be written in terms of acceleration of point B as follows:

$$\vec{a}_C = \vec{a}_B + \vec{\alpha} \times \vec{r} + \vec{\omega} \times \left(\vec{\omega} \times \vec{r}\right) \tag{5.30}$$

In which,

$$\vec{\alpha} = \vec{\alpha}_{BC}; \vec{\omega} = \vec{\omega}_{BC}; \vec{r} = \overrightarrow{BC}$$

Recalling that at this instant, $\vec{\omega}_{BC} = 0$, Eq. (5.30) reduces to the following one,

$$\vec{a}_C = \vec{a}_B + \vec{\alpha}_{BC} \times \overrightarrow{BC} = 0.2\hat{i} - 0.8\hat{j} + \alpha_{BC}\hat{k} \times 0.4\left[\cos(14.478^\circ)\hat{i} + \sin(14.478^\circ)\hat{j}\right] \tag{5.31}$$

Now, considering point C rotating about the fixed point D we can write,

$$\vec{a}_C = \vec{a}_C^{\,t} + \vec{a}_C^{\,n} \tag{5.32}$$

in which,

$$\vec{a}_C^{\,t} = \vec{\alpha}_4 \times \vec{r}_C = \vec{\alpha}_4 \times \left(-\vec{r}_4\right) = \alpha_4\,\hat{k}\ \text{rad}/s^2 \times 300\hat{j}\ \text{mm} = -0.3\alpha_4\hat{i}\ \text{m}/s^2$$

$$\vec{a}_C^{\,n} = \vec{\omega}_4 \times \left(\vec{\omega}_4 \times \vec{r}_C\right) = \vec{\omega}_4 \times \vec{v}_C = 1.333\hat{k}\ \text{rad/s} \times \left(-0.4\hat{i}\ \text{m/s}\right) = -0.533\hat{j}\ \text{m}/s^2$$

Substituting these results into Eq. (5.32) gives,

$$\vec{a}_C = \vec{a}_C^{\,t} + \vec{a}_C^{\,n} = -0.3\alpha_4\hat{i} - 0.533\hat{j}\ \text{m}/s^2 \tag{5.33}$$

And combining Eqs. (5.31) and (5.33) results in:

$$-0.3\alpha_4\hat{i} - 0.533\hat{j} = 0.2\hat{i} - 0.8\hat{j} + \alpha_{BC}\hat{k} \times 0.4\left[\cos(14.478^\circ)\hat{i} + \sin(14.478^\circ)\hat{j}\right]$$

This is a vector equation with two unknowns, α_4 and α_{BC}. Calculating the cross product results in:

$$-0.3\alpha_4\hat{i} - 0.533\hat{j} = 0.2\hat{i} - 0.8\hat{j} + 0.387\alpha_{BC}\hat{j} - 0.1\alpha_{BC}\hat{i}$$

This vector equation can be converted to the following two scalar equations for the two unknowns, α_4 and α_{BC}:

$$-0.3\alpha_4 = 0.2 - 0.1\alpha_{BC}$$
$$-0.533 = -0.8 + 0.387\alpha_{BC}$$

The second equation results in,

$$\alpha_{BC} = 0.690\ \text{rad}/s^2\ (\text{CCW}) \tag{5.34}$$

By substituting α_{BC} in the first equation we get:

$$\alpha_4 = -0.437\ \text{rad}/s^2\ (\text{CW})$$

Substituting α_4 into Eq. (5.33) gives:

$$\vec{a}_C = 0.131\hat{i} - 0.533\hat{j}\ \text{m}/s^2 \tag{5.35}$$

A review of Eqs. (5.29), (5.34), and (5.35) reveals that none of the expectations given in Eqs. (5.28) is met! In other words, while similar equations are valid for

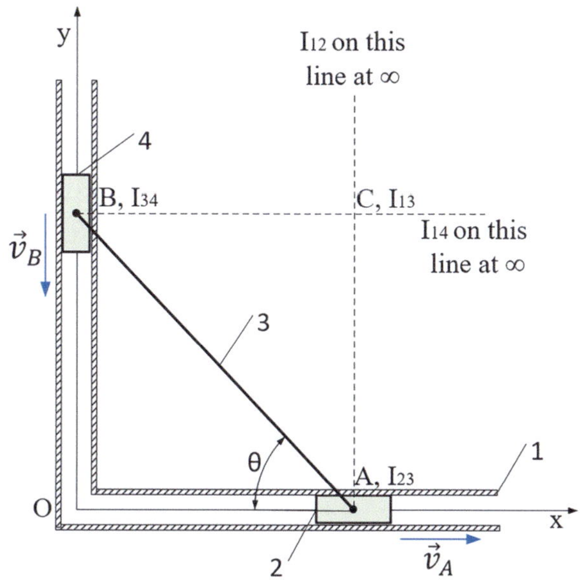

Fig. 5.9 Locating the instantaneous center I_{13}

velocity, as we see in Eqs. (5.27), those equations cannot be generalized to acceleration.

Example 5.9 that follows, is intended to present applications of various velocity and acceleration calculation methods.

Example 5.9 Figures 2.58 and 5.9 illustrate the trammel of Archimedes mechanism. In Fig. 5.9 revolute joints are used to connect sliders 2 and 4 to link *AB* (whose length is L). We know that $L = 10$ cm and slider 2 moves with a constant velocity $v_A = 2$ cm/s to the right. Calculate the following quantities at $\theta = 45°$:

A. The velocity of slider 4 using instantaneous centers, direct differentiation (the algebraic method), vector loop, and relative motion methods.
B. The acceleration of slider 4 using direct differentiation (algebraic), vector loop, and relative motion methods.

A. **The velocity of the slider**

Instantaneous Centers Method

The instantaneous center I_{13} (or point C in Fig. 5.9) can be located using the same reasoning as it was done for finding I_{12} in Example 4.9. At $\theta = 45°$ we have:

$$BC = CA = L\cos\theta = 10\cos 45° = 7.071 \text{ cm}$$

That is, points A and B are at the same distance from the instantaneous center I_{13}. Since both points A and B belong to link 3, and velocity of any point on this link varies linearly with its distance from I_{13}, the velocity of point B is equal to the velocity of point A. Here is a more formal proof:

$$v_A = CA.\omega_3 \rightarrow 2\ \text{cm/s} = (7.071\ \text{cm})\omega_3 \tag{5.36}$$

So,

$$\omega_3 = 0.283\ \text{rad/s (CCW)} \tag{5.37}$$

Similarly, for point B:

$$v_B = CB.\omega_3 \tag{5.38}$$

or,

$$v_B = (7.071\ \text{cm}) \times (0.283\ \text{rad/s}) = 2.00\ \text{cm/s}$$

This result for the magnitude of v_B is equal to the given speed of point A. We also notice that point B moves in the negative y direction.

Algebraic Method

An alternative way for solving this problem is by using the algebraic method. Considering Fig. 5.9, and using $x_A = x$ and $y_B = y$, one can write the Pythagorean theorem equation for triangle OAB as follows:

$$x^2 + y^2 = L^2 \tag{5.39}$$

Taking a derivative with respect to time gives:

$$2x\dot{x} + 2y\dot{y} = 0 \tag{5.40}$$

Using Eq. (5.40), $v_B = \dot{y}$ is calculated as follows:

$$v_B = \dot{y} = -\frac{x\dot{x}}{y} = -\frac{x}{y}v_A = -\frac{BC}{AC}v_A = -v_A = -2.00\ \text{cm/s} \tag{5.41}$$

We notice that for positive x and y, the sign of $\dot{y}$ is always negative. Therefore, we conclude that at this instant point B moves in the negative y direction with the same speed as point A. We also notice that the absolute value of $\dot{y}$ increases as y decreases and B approaches point O.

Vector-Loop Method

For applying the vector-loop method, we focus on using the vector-loop diagram shown in Fig. 5.10. The corresponding vector-loop equation can be written as follows:

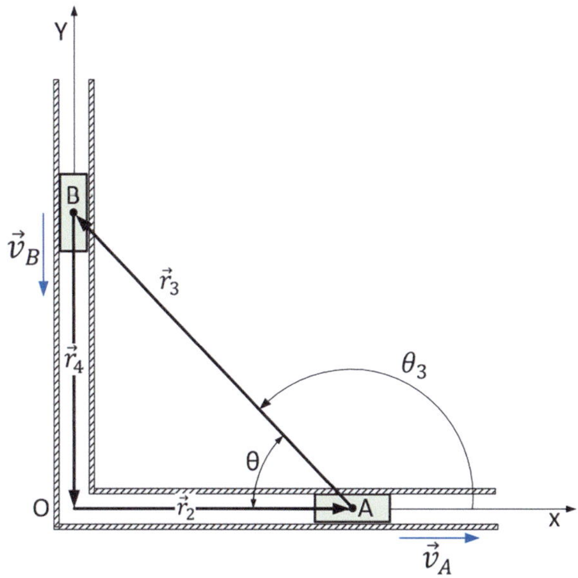

Fig. 5.10 The vector-loop diagram

$$\vec{r}_2 + \vec{r}_3 + \vec{r}_4 = 0 \tag{5.42}$$

Projecting Eq. (5.42) on the x and y directions and noticing that $\theta_2 = 0$ and $\theta_4 = -90°$, one obtains:

$$r_2 \cos 0 + r_3 \cos \theta_3 + r_4 \cos(-90°) = 0$$
$$r_2 \sin 0 + r_3 \sin \theta_3 + r_4 \sin(-90°) = 0$$

or,

$$\begin{aligned} r_2 + r_3 \cos \theta_3 &= 0 \\ r_3 \sin \theta_3 - r_4 &= 0 \end{aligned} \tag{5.43}$$

Let us first solve the position Eq. (5.43). Knowing that $r_3 = L = 10$ cm and $\theta_3 = 135°$, we conclude that:

$$r_2 = -r_3 \cos \theta_3 = -10 \cos 135° = 7.071 \text{ cm}$$

And:

$$r_4 = r_3 \sin \theta_3$$

So,

$$r_4 = 10 \sin 135° = 7.071 \text{ cm}$$

Therefore, one obtains the same results as before:

$$r_2 = r_4 = 7.071 \text{ cm}$$

To get the unknown velocities v_B and ω_3 we can take a time derivative of Eqs. (5.43) and obtain the following:

$$\begin{aligned} \dot{r}_2 - r_3\dot{\theta}_3 \sin\theta_3 = 0 \\ r_3\dot{\theta}_3 \cos\theta_3 - \dot{r}_4 = 0 \end{aligned} \quad (5.44)$$

Using the first one of Eqs. (5.44), the angular velocity of link AB is calculated as follows:

$$\omega_3 = \dot{\theta}_3 = \frac{\dot{r}_2}{r_3 \sin\theta_3} = \frac{2}{10 \sin 135°} = 0.283 \text{ rad/s (CCW)} \quad (5.45)$$

Substituting ω_3 in the second one of Eqs. (5.44) results in:

$$\dot{r}_4 = r_3\dot{\theta}_3 \cos\theta_3 = 10 \text{ cm} \times 0.283 \text{ rad/s} \times \cos 135° = -2.00 \text{ cm/s} \quad (5.46)$$

The positive sign in ω_3 is an indication of the CCW sense of rotation of link AB and the negative sign of $\dot{r}_4$ implies that slider B is moving in the negative y direction, as shown before.

Relative Motion Method

Finally, we can solve the same problem using the relative motion Eq. (4.83). Since points A and B belong to the same rigid body 3 (or bar AB) by attaching the moving coordinate system to that rigid body, and using A as the observer of motion of B, Eq. (4.83) reduces to:

$$\vec{v}_B = \vec{v}_A + \vec{\omega} \times \vec{r} \quad (5.47)$$

where $\vec{\omega} = \vec{\omega}_3$ and $\vec{r} = \overrightarrow{AB}$. Substituting the given data in Eq. (5.47) results in,

$$\vec{v}_B = 2\hat{i} + \omega_3\hat{k} \times 10\left(\cos 135°\hat{i} + \sin 135°\hat{j}\right)$$

or,

$$v_B\hat{j} = 2\hat{i} + 10\omega_3 \cos 135°\hat{j} - 10\omega_3 \sin 135°\hat{i}$$

which results in the following two algebraic equations:

$$2 - 10\omega_3 \sin 135^\circ = 0$$

$$v_B = 10\omega_3 \cos 135^\circ$$

The first one gives:

$$\omega_3 = 0.283 \text{ rad/s (CCW)}$$

And the second one results in:

$$v_B = -2 \text{ cm/s}$$

These are identical to the previous findings.

B. **The acceleration of the slider**

Algebraic Method

Considering Fig. 5.9, and using $x_A = x$ and $y_B = y$, we got:

$$x^2 + y^2 = L^2 \tag{5.39}$$

It is known that at $\theta = 45°$ we have,

$$x = y = L\cos\theta = 10\cos 45^\circ = 7.071 \text{ cm}$$

By taking one time derivative we got the following result:

$$2x\dot{x} + 2y\dot{y} = 0 \tag{5.40}$$

The final solution for velocity was given as,

$$v_B = \dot{y} = -\frac{x\dot{x}}{y} = -\frac{x}{y}v_A = -\frac{BC}{AC}v_A = -v_A = -2.00 \text{ cm/s} \tag{5.41}$$

Now, one may take another derivative of the velocity Eq. (5.40) with respect to time to get

$$(\dot{x})^2 + x\ddot{x} + (\dot{y})^2 + y\ddot{y} = 0 \tag{5.48}$$

Since the velocity of point A is constant, $\ddot{x} = 0$. Therefore Eq. (5.48) results in,

$$\ddot{y} = \frac{-\left[(\dot{x})^2 + (\dot{y})^2\right]}{y} \tag{5.49}$$

In Eq. (5.49) we notice that for positive y, the sign of $\ddot{y}$ is always negative. Furthermore, the absolute value of $\dot{y}$ increases as y decreases and B approaches point O. Substituting the known data into Eq. (5.49) gives,

$$\ddot{y} = \frac{-\left[(\dot{x})^2 + (\dot{y})^2\right]}{y} = \frac{-\left[(2)^2 + (2)^2\right]}{7.071} = -1.131\ \text{cm/s}^2$$

Interestingly, while the sliders have equal distances from the origin i.e., $x = y$, and the absolute values of the velocities of the sliders are also equal, $\dot{y} = -\dot{x}$, the same symmetry does not exist for accelerations. Indeed, we just proved that while $\ddot{x} = 0$, we have $\ddot{y} = -1.131\ \text{cm/s}^2$.

Vector-Loop Method

For applying the vector-loop method we focus on using the vector-loop diagram shown in Fig. 5.10. The corresponding vector-loop equation was written as follows:

$$\vec{r}_2 + \vec{r}_3 + \vec{r}_4 = 0 \tag{5.42}$$

The projection of the vector-loop equation on the x and y directions resulted in:

$$\begin{aligned} r_2 + r_3 \cos\theta_3 &= 0 \\ r_3 \sin\theta_3 - r_4 &= 0 \end{aligned} \tag{5.43}$$

Recalling, $r_3 = L = 10$ cm, $\theta_3 = 135°$ and $r_2 = r_4 = 7.071$ cm, taking a time derivative resulted in:

$$\begin{aligned} \dot{r}_2 - r_3\dot{\theta}_3 \sin\theta_3 &= 0 \\ r_3\dot{\theta}_3 \cos\theta_3 - \dot{r}_4 &= 0 \end{aligned} \tag{5.44}$$

With the following results:

$$\omega_3 = \dot{\theta}_3 = \frac{\dot{r}_2}{r_3 \sin\theta_3} = \frac{2}{10 \sin 135\,°} = 0.283\ \text{rad/s (CCW)} \tag{5.45}$$

$$\dot{r}_4 = r_3\dot{\theta}_3 \cos\theta_3 = 10\ \text{cm} \times 0.283\ \text{rad/scos}135\,° = -2.00\ \text{cm/s} \tag{5.46}$$

Now, let us take a time derivative of the velocity Eq. (5.44). Recall that $\dot{r}_2 =$ 2 cm/s (which is constant):

$$\begin{aligned} -r_3\ddot{\theta}_3 \sin\theta_3 - r_3\left(\dot{\theta}_3\right)^2 \cos\theta_3 &= 0 \\ r_3\ddot{\theta}_3 \cos\theta_3 - r_3\left(\dot{\theta}_3\right)^2 \sin\theta_3 - \ddot{r}_4 &= 0 \end{aligned} \tag{5.50}$$

The first one of Eqs. (5.50) gives,

$$-10\times\ddot{\theta}_3\sin 135^\circ - 10\times(0.283)^2\cos 135^\circ = 0$$

or,

$$\ddot{\theta}_3 = \alpha_3 = 0.080\ \mathrm{rad}/s^2\ (\mathrm{CCW})$$

Substituting α_3 into the second one of Eqs. (5.50) results in,

$$10\times 0.080\cos 135^\circ - 10\times(0.283)^2\sin 135^\circ - \ddot{r}_4 = 0$$

or,

$$\ddot{r}_4 = a_B = -1.132\ \mathrm{cm/s^2}$$

This is in harmony with the results obtained from using the algebraic method.

Relative Motion Method

Finally, we can solve the same problem using the relative motion Eq. (5.14). Since points A and B in Fig. 5.9 belong to the same rigid body 3 (or bar AB) by attaching the moving coordinate system to that rigid body, and using A as the observer of motion of B, Eq. (5.16) can be written as:

$$\vec{a}_B = \vec{a}_A + \vec{\alpha}\times\vec{r} + \vec{\omega}\times\left(\vec{\omega}\times\vec{r}\right) \tag{5.51}$$

or, more explicitly we have:

$$\vec{a}_B = \vec{a}_A + \vec{\alpha}_3\times\overrightarrow{AB} + \vec{\omega}_3\times\left(\vec{\omega}_3\times\overrightarrow{AB}\right) \tag{5.52}$$

Substituting the known values of variables into Eq. (5.52) gives,

$$\vec{a}_B = 0 + \alpha_3\hat{k}\times 10\left(\cos 135^\circ\hat{i} + \sin 135^\circ\hat{j}\right) + \omega_3\hat{k}$$
$$\times\left[\omega_3\hat{k}\times 10\left(\cos 135^\circ\hat{i} + \sin 135^\circ\hat{j}\right)\right]$$

Performing cross products and simplifying the result gives,

$$a_B\hat{j} = 10\alpha_3\cos 135^\circ\hat{j} - 10\alpha_3\sin 135^\circ\hat{i} - 10\omega_3^2\left(\cos 135^\circ\hat{i} + \sin 135^\circ\hat{j}\right)$$

The two associated algebraic equations are:

$$-10\alpha_3 \sin 135° - 10\omega_3^2 \cos 135° = 0$$

$$a_B = 10\alpha_3 \cos 135° - 10\omega_3^2 \sin 135°$$

Finally, using $\omega_3 = 0.283$ rad/s in these equations gives:

$$\alpha_3 = 0.080 \text{ rad}/s^2 \text{ (CCW)}$$

$$a_B = 10 \times 0.08 \cos 135° - 10 \times (0.283)^2 \sin 135° = -1.132 \text{ cm/s}^2$$

These results are identical to the corresponding ones obtained before using other methods.

References

1. Lee, H. H. (2015). *Engineering dynamics labs with SolidWorks motion 2015*. SDC Publications.
2. Meriam, J. L., & Kraige, L. G. (2007). *Engineering mechanics, dynamics* (6th ed.). Wiley.
3. Uicker, J. R., et al. (2016). Theory of machines and mechanisms (5th ed.). Oxford University Press.

Chapter 6
Spur Gears

6.1 Introduction

Power is usually generated at one location and consumed at another place. For power transmission in relatively long distances, we use belt and pulleys or chains and sprockets (see Chap. 11). Flat and V-belts are light, inexpensive, quiet, and tolerant of misalignments. They transmit power through friction and therefore, slipping can happen under shock, starting, and braking conditions. Timing belts, however, prevent slipping and in this way maintain the speed ratio as a constant. The effectiveness of belts drops at high speeds due to centrifugal forces.

For power transmission in shorter distances, we use gears. In gears no slipping can happen (i.e., they provide positive drive) but the teeth profile should be chosen in such a way that the speed ratio remains constant. Spur gears have teeth that are cut parallel to the axis of rotation of the shafts that they are mounted on. Therefore, spur gears are used for power transmission between parallel shafts. Rack is a straight gear and a rack and pinion mechanism can convert rotational motion of the pinion to the translational motion of the rack and vice versa. This mechanism was discussed in Sect. 3.7.5.

In this chapter, first the fundamental law of gearing is introduced. This law is an application of the angular-velocity-ratio theorem (discussed in Chap. 4) on gears. The relation between the fundamental law of gearing and the angular-velocity-ratio theorem provides a bridge between the instantaneous centers topic and the gears discussion. Then, gear terminology, related equations, and gear standards are introduced. The use of involute of a circle for maintaining constant speed ratio in two gears in mesh is discussed. Next, design methods for preventing interference and undercutting in meshed gears are presented. In this way, we make sure that in a gearset, motion is transmitted at constant speed ratio. Chapter 6 is concluded by quantifying contact ratio in a gear mesh and evaluating the effect of a change in the center distance of gears, on factors like pressure angle and backlash. As a historical note (brought to my attention by my son, Ali) the thesis of the very first engineering

M. Ghorashi, *Introduction to Machine Dynamics*, Mechanical Engineering Series,
https://doi.org/10.1007/978-3-031-94736-0_6

doctorate awarded in the U.S. was written *on the form of the teeth of wheels in spur gearing,* see Gibbs [1]. This chapter is closely related to that topic!

6.2 The Fundamental Law of Gearing

Gears are used to transmit power from one shaft to the other at constant speed ratio. This constant speed ratio property is very important because otherwise the driven gear would be continuously accelerated and decelerated and as a result would be subjected to cyclic stresses and noise. The constant speed ratio property in gears is achieved by using the involute of a circle profile for gear teeth.

Figures 6.1 and 6.2 illustrate a spur gear that includes a hub, a set screw, and a keyway which are used for installation of a shaft using a key. The shaft has a shoulder for securing the axial location of the gear. The set screw and the key help

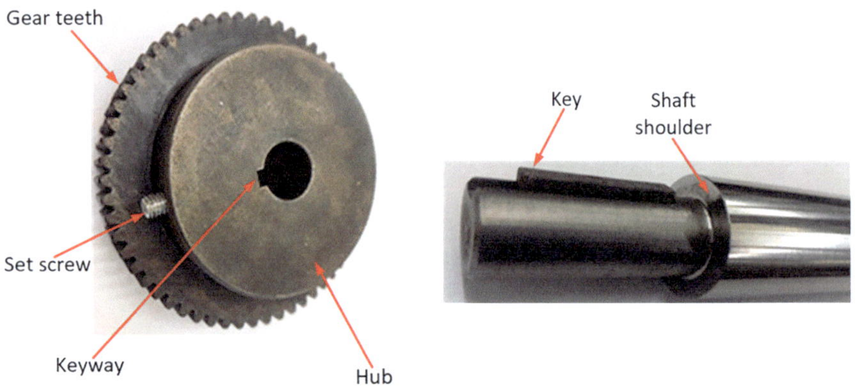

Fig. 6.1 A spur gear with a hub, a set screw, and a keyway (left); A matching shaft with a key and a shoulder (right)

Fig. 6.2 Partial assembly of the spur gear and the shaft shown in Fig. 6.1

in securing assembly parts together and provide means for power transmission between the shaft and the gear.

The fundamental law of gearing (that originates from the angular-velocity-ratio theorem) states that in the cam and follower mechanism shown in Fig. 4.34 (that is a generalized version of a gear train) we have the following equation for the speed ratio:

$$\frac{\omega_3}{\omega_2} = \frac{I_{12}I_{23}}{I_{13}I_{23}} \tag{4.123}$$

Analyzing this equation reveals that the speed ratio of members 2 and 3 only depends on the location of their common instantaneous center, I_{23}. In general, this ratio varies during motion. *So, for the speed ratio to remain constant, the common instantaneous center of the input and output (that is* I_{23} *here) should be fixed.* In gears, this point is called the pitch point. In a gear mesh, maintaining the speed ratio constant during motion is important as it ensures steady and non-jerky operation. So, for the speed ratio to remain constant during motion of a gear train and despite changes of the contact points, the gear teeth profiles should be designed in such a way that the resulting pitch point P (i.e., where the common normal line intersects the line of centers O_2O_3 in Fig. 4.34) remains unchanged (i.e., the intersection point remains the same while the contact points change during motion). In such a case, during motion the common normal line n-n will also be unchanged. Two gears in mesh that provide a constant speed ratio are said to be in conjugate action.

6.3 Gear Teeth Profile and Involute of a Circle

In order to ensure that the pitch point does not move during motion, the profile of gear teeth should be chosen to be an involute of a circle. This is the curve *AB* shown in Fig. 6.3 that is generated by the end of a string originally covering arc *DCA* that

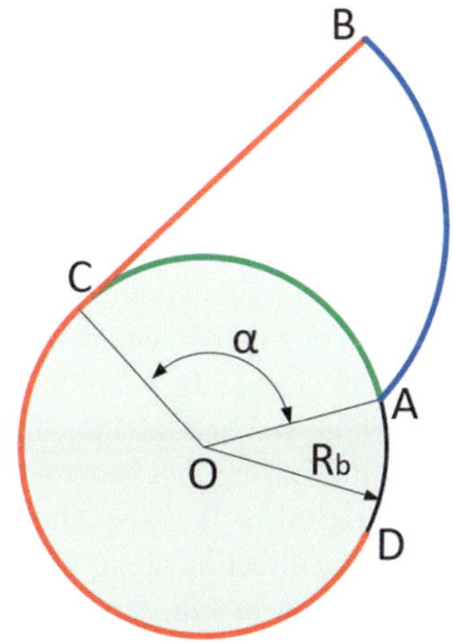

Fig. 6.3 Base circle and generation of the involute curve

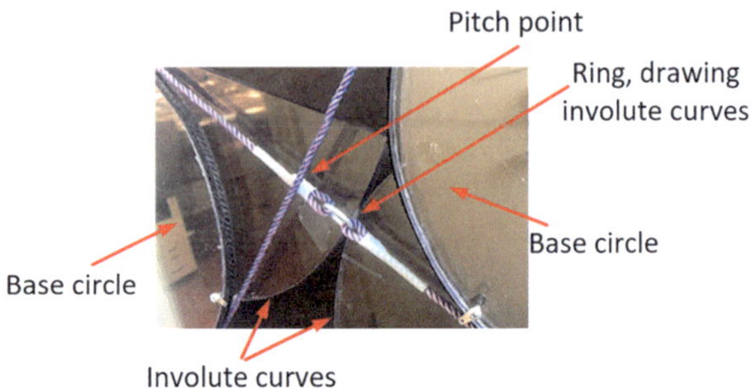

Fig. 6.4 A physical setup showing how the pitch point is fixed when gear profiles are involute of a circle (USM term project)

was wrapped around a disk (called the base circle); while it is unwrapped in a way that the string remains taut during unwrapping (as seen, BC is kept taut). During unwrapping, segment CA of the string is unwrapped and becomes the straight line segment CB, so:

$$CB = \widehat{CA} = R_b \alpha \tag{6.1}$$

In which R_b is the base circle radius.

In the physical model shown in Fig. 6.4, the illustrated ring simultaneously traces two involute curves—one on each of the shown base circles as they turn. The pitch point in this figure is at the point of intersection of the two strings (that are common tangents of the base circles).

6.4 Gear Terminology and AGMA Standards

Figures 6.5 and 6.6 illustrate basic gear and gear mesh nomenclature. In these figures, gear 2 is the driver and gear 3 is the driven gear, and as always, member 1 is the base or ground member. Gear 2 rotates in a CCW sense and causes the CW motion of gear 3. In Fig. 6.5, we see **base circles** in red with R_{b2} and R_{b3} radii. The common tangent of the base circles intersects the line of centers, O_2O_3 at the **pitch point**, P. The **Pitch circles** with R_2 and R_3 radii are imaginary circles that are centered at the center of each gear and pass through the pitch point. Two gears in mesh can be replaced by two friction disks with the same sizes as the two pitch circles that are tangent to each other at the pitch point. Such friction disks will result in the same speed ratio as the original meshed gears. Interestingly, while during motion the contact point(s) of gears change, the speed ratio is not dependent on the location of any momentary contact point, but on the location of the pitch point. (That does not change during motion.) Hence the speed ratio remains constant!

In Figs. 6.5 and 6.6 the common normal n-n on a pair of teeth in contact is along the virtual string that generates the involute profiles of these gear teeth. This line is also tangent to the base circles that create the two involute curves. Furthermore, this is the direction along which normal forces are transmitted between the two gear teeth shown in Fig. 6.6. That is why this line is also called the pressure line. In

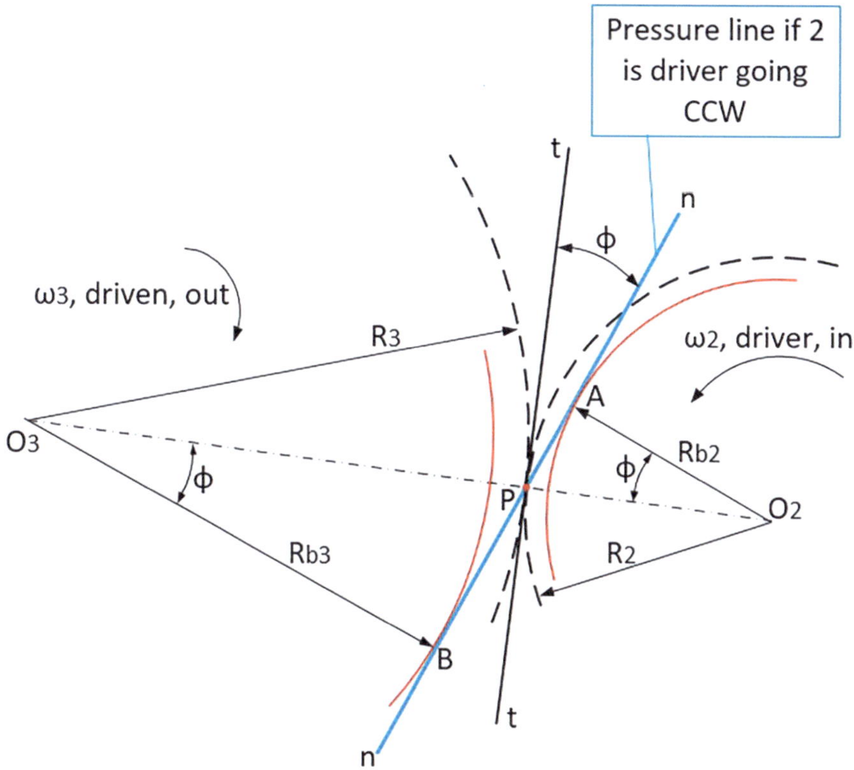

Fig. 6.5 Gear and gear mesh nomenclature

Fig. 6.5, the pressure line *n-n* is at an angle φ with respect to the common tangent of pitch circles, *t-t*. This angle is called the **pressure angle**, for obvious reasons. The line of centers, O_2O_3, the common tangent of pitch circles *t-t* and the pressure line *n-n* all intersect at the pitch point, *P*.

Figure 6.6 includes more gear and gear mesh nomenclature including: addendum, *a*, dedendum, *d*, **addendum circle** and **dedendum circle**. **Addendum** and **dedendum** are both measured from the pitch circle. Tooth space and tooth thickness are both measured on the pitch circle and their summation is called **circular pitch**, *p*. Circular pitch can also be defined as the distance from a point on the pitch circle on a tooth to the corresponding point on the adjacent tooth when measured on the pitch circle. If *R* is the radius of the pitch circle of a gear that has *N* teeth, we can write the following equation for the circular pitch:

$$p = \frac{2\pi R}{N} \tag{6.2}$$

The **diametral pitch**, *P*, of a gear is defined as the ratio of the **number of gear teeth**, *N*, over the diameter of the pitch circle, 2*R*.

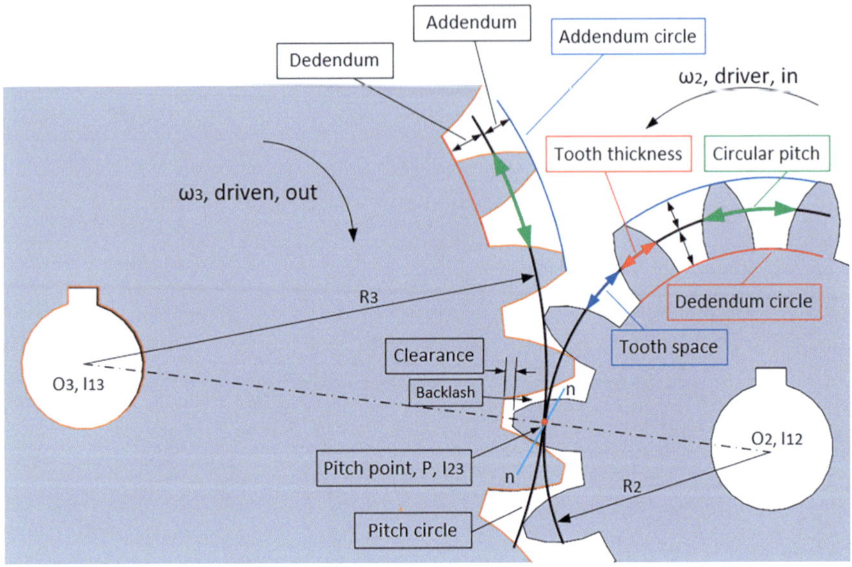

Fig. 6.6 More gear and gear mesh nomenclature

$$P = \frac{N}{2R} \tag{6.3}$$

Equations (6.2) and (6.3) result in:

$$Pp = \pi \tag{6.4}$$

Diametral pitch is used in the US Customary System (USCS) of units and its unit is teeth/inch. For the same gear size (i.e., pitch circle diameter) the more teeth a gear has, the higher the diametral pitch value will be.

Module, m, is the ratio of the diameter of the pitch circle, $2R$ over the number of gear teeth, N,

$$m = \frac{2R}{N} \tag{6.5}$$

Module is used in the SI system (French abbreviation for *Système International d'unités* or International System of units) and its unit is mm/tooth. For the same gear size (i.e., pitch circle diameter) the more teeth a gear has, the smaller the module value will be. The definitions of diametral pitch and module are inverse of each other. However, such a relation is of no practical use as the SI gears and the USCS gears are designed using different standards and there is no SI gear equivalent for a USCS gear and vice versa. Comparing Eqs. (6.2) and (6.5) we can conclude:

$$p = \pi m \tag{6.6}$$

Fig. 6.7 A 20-teeth spur gear with $P = 8$ teeth/in, 20° pressure angle, and 0.5 in shaft diameter; generated by SolidWorks

Fig. 6.8 A 30-teeth spur gear with $P = 8$ teeth/in, 20° pressure angle, and 0.5 in shaft diameter; generated by SolidWorks

For **proper mesh** of two gears, they should have the same N/R ratios (i.e., as gears become larger in size, they should have proportionally more teeth; like the ones shown in Figs. 6.7 and 6.8):

$$\frac{N_2}{R_2} = \frac{N_3}{R_3} \tag{6.7}$$

By comparing Eq. (6.7) with Eqs. (6.2), (6.3), and (6.5), we conclude that proper mesh of two gears happens if they have the same diametral pitch values, the same circular pitch values, and the same modules.

In Fig. 6.6, we also see **backlash** which is the gap between the teeth of two meshing gears. Tooth space should be a bit more than the tooth thickness so that gears can mesh properly. The difference is backlash, which is a common source of error especially in measurement devices. Such errors are especially observed when the direction of motion is reversed. In Fig. 6.6, gear 2 is the driver (input) and gear 3 is the driven (output) one. It should be clear that if the direction of motion of gear 2 is reversed, then there is a slight gap during which gear 2 can move, while gear 3 is stationary. This is the lost motion in the measurement due to backlash and it corresponds to a change in the angular position of the input gear that does not result in a corresponding motion of the output gear.

Recalling the angular-velocity-ratio theorem, i.e.,

$$\frac{\omega_3}{\omega_2} = \frac{I_{12}I_{23}}{I_{13}I_{23}} \tag{4.123}$$

And noticing that in the gear mesh shown in Fig. 6.6, point I_{23}, pitch point P, and the contact point of the pitch circles are identical, the **speed ratio** (i.e., the output angular velocity over the input angular velocity) can be rewritten in terms of the pitch circle radii as follows:

$$\frac{\omega_3}{\omega_2} = \frac{R_2}{R_3} \tag{6.8}$$

Equation (6.8) can also be obtained by replacing the gear mesh with pitch circle contact—as if the gears are now replaced with friction disks. In the latter case, if there is no slipping at the pitch point, the velocity of the pitch point on the two friction disks should be the same, that is:

$$R_2\omega_2 = R_3\omega_3 \tag{6.9}$$

Clearly Eqs. (6.8) and (6.9) are identical. It should be emphasized, however, that, in real gears slipping does happen (except when the contact point of the gears coincides with the pitch point, see Sect. 4.7.3 and Fig. 4.35). That is why to make gears more wear resistant they are hardened at the contact surface.

By combining Eqs. (6.7) and (6.8) and considering only the absolute values of angular velocities one obtains,

$$\frac{\omega_3}{\omega_2} = \frac{R_2}{R_3} = \frac{N_2}{N_3} \tag{6.10}$$

So, in calculating the speed ratios, the ratio of pitch circle radii can be replaced with the ratio of teeth numbers.

In Fig. 6.6, quantities like addendum, a, and dedendum, d, are illustrated and they are measured from the pitch circle. For each pressure angle, the addendum and the dedendum are expressed in terms of the diametral pitch, as recommended by the American Gear Manufacturer's Association (AGMA). AGMA sets national gearing standards on a vast set of topics that includes design, analysis, manufacturing, materials, quality, lubrication, maintenance, failure, heat treatment, and efficiency.

Table 6.1 includes the standard proportions of gear teeth. It can be seen that in each case, the addendum is a bit smaller than the dedendum so that there is enough room for the teeth to mesh properly. The difference between addendum and dedendum is called the **clearance**, c, shown in Fig. 6.6. The corresponding addendum and dedendum circles (with radii, $R + a$ and $R - d$) are also shown in that

Table 6.1 Summary of standard proportions of gear teeth as functions of diametral pitch and pressure angle

	Addendum, a	Dedendum, d	Whole depth, $a + d$	Clearance, $d - a$
Full-depth tooth, $\varphi = 14.5°$	$\frac{1}{P}$	$\frac{1.157}{P}$	$\frac{2.157}{P}$	$\frac{0.157}{P}$
Stub tooth, $\varphi = 20°$	$\frac{0.8}{P}$	$\frac{1}{P}$	$\frac{1.8}{P}$	$\frac{0.2}{P}$
Full-depth tooth, $\varphi = 20°, 22.5°, 25°$	$\frac{1}{P}$	$\frac{1.25}{P}$	$\frac{2.25}{P}$	$\frac{0.25}{P}$

Cleghorn and Dechev [2], Page 200; and Uicker et al. [3], Page 376

figure. The stub tooth is a bit shorter than the full-depth tooth. As a result, less bending moment (and stress) is applied on the stub tooth and there is less likelihood of interference (to be introduced later in this chapter).

Referring to gear 2 in Fig. 6.5 and considering the right triangle O_2AP, the relation between the base circle radius, the pitch circle radius, and the pressure angle can be obtained as follows:

$$R_{b2} = R_2 \cos \varphi \tag{6.11}$$

Using triangle O_3BP, a similar equation can be written for gear 3,

$$R_{b3} = R_3 \cos \varphi \tag{6.12}$$

Considering Eqs. (6.11) and (6.12), in general we can write:

$$R_b = R \cos \varphi \tag{6.13}$$

Figure 6.9 illustrates **contact forces** on spur gears in contact. It is observed that the pressure line *n-n* not only is the normal line on the contacting gear teeth involute profiles, but also it is the common tangent of the base circles. Furthermore, this line is the line of action along which gear contact force is transmitted. The smaller the pressure angle, the higher the tangential component of the transmitted force will be. (Or, for the same tangential component, less contact force would be developed.) This is the force that generates the transmitted torque. The contact force imposed by gear 2 on gear 3 is shown by F_{23}. Its tangential component is F_{23}^t and it is related to F_{23} by the pressure angle as follows:

$$F_{23}^t = F_{23} \cos \varphi \tag{6.14}$$

The corresponding **transmitted torque** from gear 2 on gear 3 would be,

$$T_3 = F_{23}^t R_3 = F_{23} R_3 \cos \varphi \tag{6.15}$$

And the output **power** is,

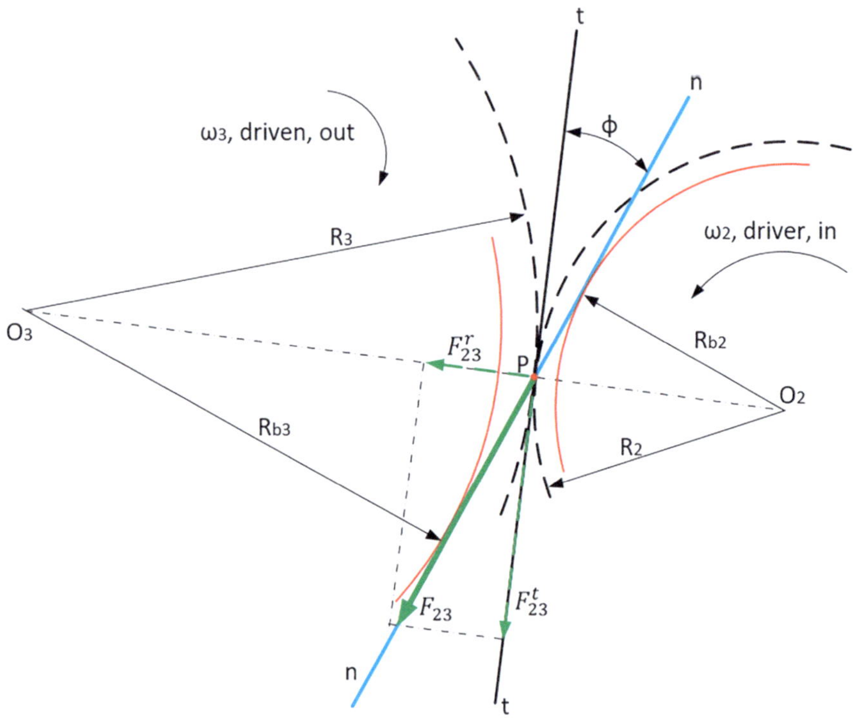

Fig. 6.9 Contact forces in gears in mesh

$$P_3 = T_3\omega_3 = F_{23}^t R_3\omega_3 = F_{23} R_3 \omega_3 \cos\varphi \tag{6.16}$$

Similarly, the input power is:

$$P_2 = T_2\omega_2 = F_{32}^t R_2\omega_2 = F_{32} R_2 \omega_2 \cos\varphi \tag{6.17}$$

where F_{32} is the contact force that gear 3 applies on gear 2 (which is equal and opposite to F_{23}; according to Newton's third law).

Furthermore, the radial component,

$$F_{23}^r = F_{23} \sin\varphi \tag{6.18}$$

in Fig. 6.9 does not transmit any power and should be supported by bearings. Therefore, the smaller it is, the better the design would be. Smaller values of radial component of the transmitted load are obtained at small pressure angles.

Considering Eqs. (6.14) to (6.16), in order to increase the transmitted force, torque, and power, small pressure angle values are preferred. That is why a pressure angle of 14.5° had been in common use for some time. However, small pressure angle values, as we will soon explore, come with a cost which is the possibility of

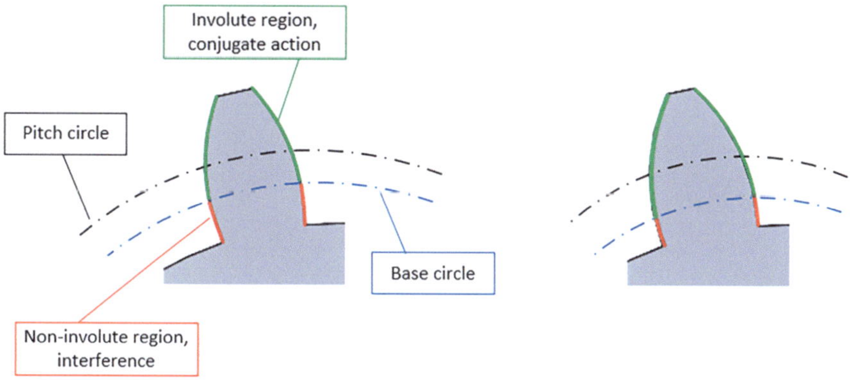

Fig. 6.10 Gear teeth profiles for pressure angles $\varphi = 14.5°$ (left) and $\varphi = 20°$ (right) generated by SolidWorks

interference (i.e., contact of gears in the non-involute region of the gear profiles). Today, most gears have a 20°, 22.5°, or 25° pressure angle.

Figure 6.10 illustrates that at $\varphi = 14.5°$ the gear is flatter and that is why the tangential force component is more than when $\varphi = 20°$. Also for $\varphi = 14.5°$, the region that is not involute of a circle (i.e., it is inside the base circle) is larger. That is why the likelihood of interference, when $\varphi = 14.5°$, is larger. This is one of the main reasons this pressure angle is now used less frequently.

Example 6.1 A spur gear has 24 teeth, a module of 2 mm, and a pressure angle of 20°. Calculate the pitch circle radius, the base circle radius, and the circular pitch.

Recalling, Eq. (6.5):

$$m = \frac{2R}{N} \tag{6.5}$$

the pitch circle radius is calculated as,

$$2 = \frac{2R}{24} \rightarrow R = 24 \text{ mm}$$

Now, for the base circle radius we can use Eq. (6.13),

$$R_b = R \cos \varphi \tag{6.13}$$

to get,

$$R_b = 24 \cos 20° = 22.55 \text{ mm}$$

Finally, for the circular pitch we can use Eq. (6.2),

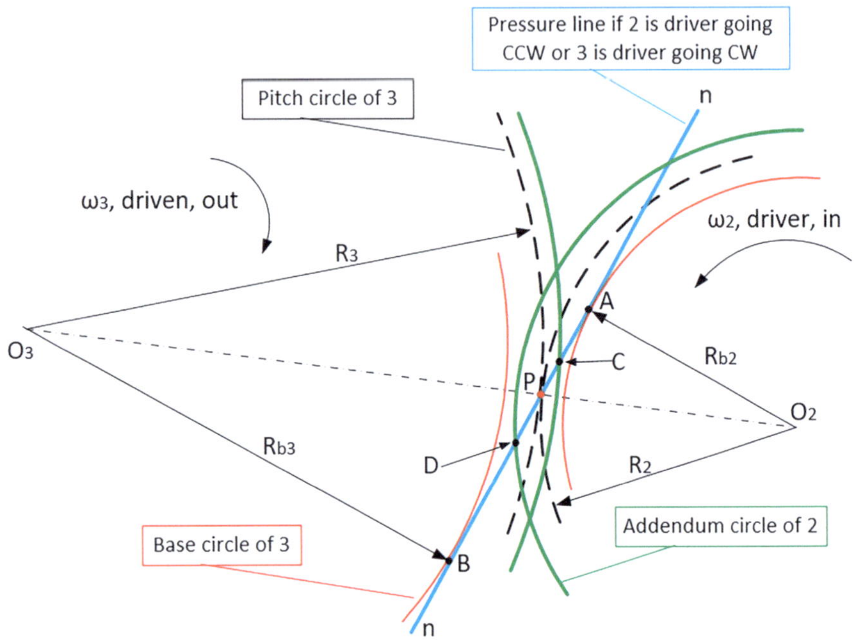

Fig. 6.11 Locating length of contact *CD* with respect to *AB* where conjugate action is possible

$$p = \frac{2\pi R}{N} \tag{6.2}$$

that results in:

$$p = \frac{2\pi \times 24}{24} = 6.28 \text{ mm/tooth}$$

Length of Contact and Its Relation to Interference

Figure 6.11 illustrates the base circles, the pitch circles, and the addendum circles of the two gears 2 and 3 centered at O_2 and O_3. In order to plot this figure, knowing the pitch circle radii, we add the two radii and locate the gear centers at a distance equal to the summation of the pitch circle radii. The pitch circles are also drawn. Then, knowing the pressure angle, we calculate the base circle radii using Eq. (6.13) and then draw the base circles in such a way that they are concentric with the coresponding pitch circles. The common tangent lines of the two base circles provide two possible pressure lines (only one is drawn). Depending on which gear is the driver and in what direction it rotates, we choose one of these lines as the correct pressure line for the case at hand (see Fig. 6.11). Segment *AB* on the obtained pressure line illustrates possible contact of the gear teeth that happens out of the base circles, and therefore, the corresponding contact surfaces are involute, the gears have conjugate action, no interference occurs, and the speed ratio of the gears is constant. The line of centers and the pressure line intersect at the pitch point, *P*.

After using the diametral pitch and Table 6.1 for calculating the addedndum and the dedendum values, one can draw the addendum and dedendum circles. Intersections of the addendum circles with the correct pressure line determine the points of start and end of contact for gear teeth, i.e., points *C* and *D*, respectively. So gear teeth contact starts at *C* and ends at *D*. If the *CD* line is completely within the *AB* line (as is the case in Fig. 6.11), all of the contact happens where gear teeth have involute profiles and therefore the gears have conjugate action. Otherwise, if contact extends beyond either of points of tangency with the base circles called the **interference points**, contact will happen on points inside the base circle (that are not part of the involute profile). Therefore, conjugate action is lost, we say interference has happened, and the speed ratio will not be constant during contact. In such a case, during manufacturing, the cutter teeth **undercut** the pinion teeth and weaken them, Waldron and Kinzel [4], page 479.

The larger the pressure angle is, the smaller the base circles would be (see Fig. 6.10) and the longer the *AB* line in Fig. 6.11 becomes. Therefore, the possibility of contact in the non-involute areas and having interference reduces. This is a major reason for using larger pressure angles these days compared to the traditional value of 14.5°. Of course, according to Eq. (6.14), this advantage comes at a cost which is the reduced value of the useful transmitted force, F^t_{23}.

One way to prevent interference is to use shorter teeth (stub teeth) with reduced addendum and dedendum values, see Table 6.1. In this way, the length of contact *CD* is reduced and it is more likely that line segment *CD* remains within line *AB*, in which case interference would not occur.

6.5 Interference Detection and Prevention Methods

So far, a qualitative discussion on the conditions under which interference happens has been presented. Now, the plan is to give a quantitative analysis on this matter. This analysis will also be used in calculating **contact ratio**, i.e., the average number of pairs of gear teeth that are in contact over time in a given gearset.

In the analysis of Fig. 6.11 we concluded that, for no interference, the *CD* line should be completely within the *AB* line. To write this condition as a mathematical inequality one cannot simply state it as follows:

$$AB > CD \tag{6.19}$$

Because such a condition does not guarantee that line segment *CD* is indeed within line *AB*—they might just have some overlap instead! Since each one of the points *C* and *D* on the n-n line has one degree of freedom, to make sure that line

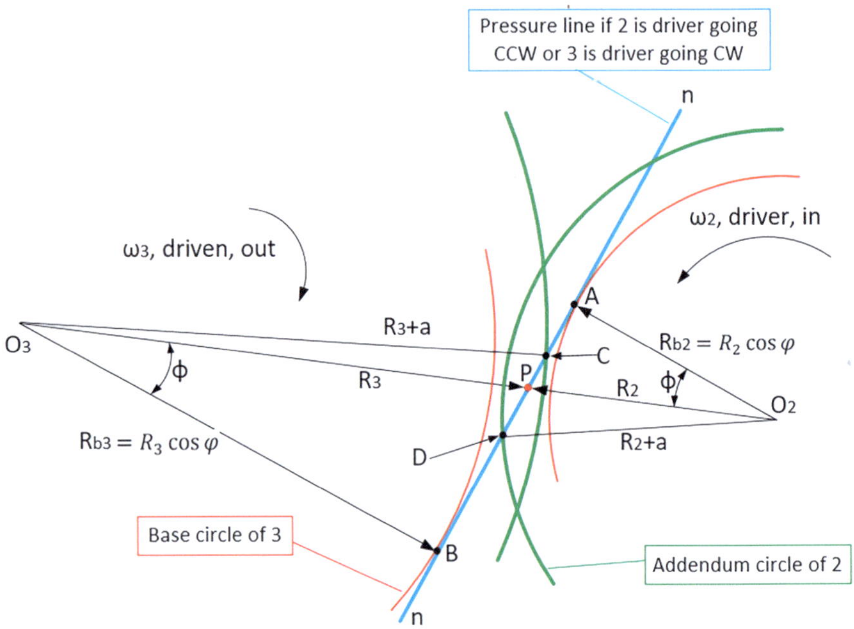

Fig. 6.12 Analyzing interference and contact ratio

segment CD is indeed within line AB, both of the following **no-interference conditions** should be satisfied:

$$AP > CP; BP > DP \tag{6.20}$$

Clearly, if conditions (6.20) are satisfied, condition (6.19) will also be valid. However, the converse is not correct: if (6.19) is satisfied one cannot deduce the validity of conditions (6.20).

Considering Fig. 6.12 and the right triangle O_2AP, for AP we can write,

$$AP = R_2 \sin \varphi \tag{6.21}$$

Similarly, for BP in triangle O_3BP we have,

$$BP = R_3 \sin \varphi \tag{6.22}$$

To get CP and DP in Fig. 6.12, that are used in (6.20), we notice that they can be written as,

$$CP = BC - BP; DP = DA - PA \tag{6.23}$$

The Pythagorean Theorem in right triangles O_3BC and O_2AD gives, respectively,

$$BC = \sqrt{(R_3 + a)^2 - (R_3 \cos\varphi)^2} \tag{6.24}$$

$$AD = \sqrt{(R_2 + a)^2 - (R_2 \cos\varphi)^2} \tag{6.25}$$

Combining Eqs. (6.21) to (6.25) we get:

$$CP = BC - BP = \sqrt{(R_3 + a)^2 - (R_3 \cos\varphi)^2} - R_3 \sin\varphi \tag{6.26}$$

$$DP = DA - PA = \sqrt{(R_2 + a)^2 - (R_2 \cos\varphi)^2} - R_2 \sin\varphi \tag{6.27}$$

So, the first one of the no-interference conditions (6.20) results in,

$$R_2 \sin\varphi > \sqrt{(R_3 + a)^2 - (R_3 \cos\varphi)^2} - R_3 \sin\varphi \tag{6.28}$$

and the second one of the no-interference conditions (6.20) gives,

$$R_3 \sin\varphi > \sqrt{(R_2 + a)^2 - (R_2 \cos\varphi)^2} - R_2 \sin\varphi \tag{6.29}$$

Condition (6.29) is similar to (6.28) and can be obtained from it by just switching 2 with 3 and 3 with 2.

Maximum Allowable Addendum

A review of conditions (6.28) and (6.29) reveals that by increasing the addendum, these conditions may at some point be violated (as expected). Now the question is, "*What is the range of addendum values that ensures no interference would happen?*" To answer this question, we solve conditions (6.28) and (6.29) for a's and obtain upper bounds for addendum to prevent interference:

$$a_2 < \sqrt{(R_2 + R_3)^2(\sin\varphi)^2 + (R_2 \cos\varphi)^2} - R_2 \tag{6.30}$$

$$a_3 < \sqrt{(R_2 + R_3)^2(\sin\varphi)^2 + (R_3 \cos\varphi)^2} - R_3 \tag{6.31}$$

Again, the second inequality, i.e., (6.31) can be obtained from the first one by just switching 2 with 3 and 3 with 2. The conclusion is: if the addendum values are less than those given on the RHS of (6.30) and (6.31), no interference happens.

Minimum Allowable Number of Gear Teeth

Now, consider a case in which the addendum values are obtained from Table 6.1 and they are no longer decision variables. *Let us calculate the minimum number of gear teeth that is needed to prevent interference*. We know that interference happens at contact points inside the base circle. The worst case for interference or undercutting is when a gear is meshed with a rack (as a rack has an infinitely large base circle).

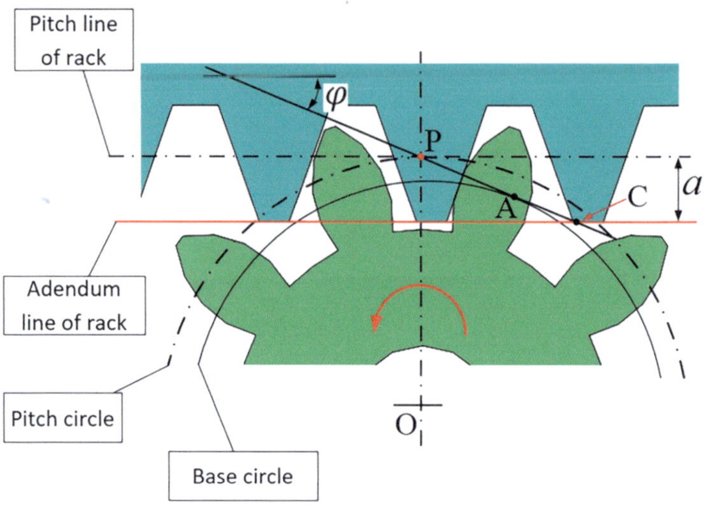

Fig. 6.13 Interference in a gear mesh with a rack

Therefore, if we design the gear to avoid interference with a rack, it will mesh with all other standard gears, Waldron and Kinzel [4], page 480.

In Fig. 6.13, a gear is driving a rack in the shown direction. We see that the addendum of the rack is larger than what it should be to prevent interference and consequently, point C is not between P and A [as it should be to avoid interference, see Fig. 6.12 and condition (6.20)]. Therefore, in this case we do have interference and the rack teeth touch the gear teeth at points that are inside its base circle.

To prevent interference, the maximum addendum length that corresponds to the minimum number of gear teeth (as we will soon see) happens when point A is coincident with point C, as we see in Fig. 6.14:

Considering the right triangle PHA we have;

$$a = AH = PA \sin \varphi \tag{6.32}$$

Similarly, in triangle OAP,

$$PA = OP \sin \varphi \tag{6.33}$$

Combining Eqs. (6.32) and (6.33) we get:

$$a = OP \sin^2 \varphi \tag{6.34}$$

Recalling $R = OP$ as the radius of the pitch circle, Eq. (6.34) can be rewritten as follows:

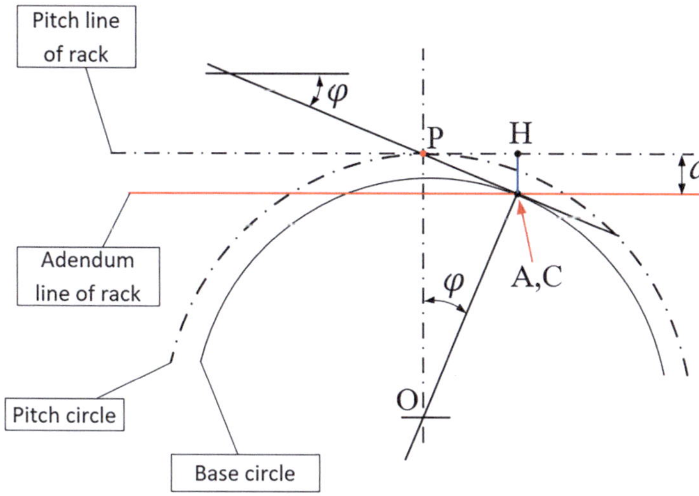

Fig. 6.14 Marginal interference in a gear mesh with a rack. (points *A* and *C* coincide)

$$a = R\sin^2\varphi \tag{6.35}$$

Now recall the standard addendum value given in Table 6.1 for full-depth tooth and pressure angles: $\varphi = 20°, 22.5°, 25°$:

$$a = \frac{1}{P} \tag{6.36}$$

And that the diametral pitch P is defined as:

$$P = \frac{N}{2R} \tag{6.3}$$

Combining Eqs. (6.36) and (6.3) gives,

$$a = \frac{2R}{N} \tag{6.37}$$

Equation (6.37) reveals that the maximum acceptable value of a, corresponds with the minimum acceptable value of N (the number of gear teeth). Combining Eqs. (6.37) and (6.35) results in,

$$N = \frac{2}{\sin^2\varphi} \tag{6.38}$$

Equation (6.38) provides the minimum acceptable number of gear teeth to avoid interference. For a continuously rotating gear, the value of N calculated from

Table 6.2 Variation of N_{min}, the minimum acceptable number of full-depth gear teeth to prevent interference as

φ	14.5°	20°	22.5°	25°
N_{min}	32	18	14	12

Eq. (6.38) should be an integer and the values of N calculated from this equation should be rounded up. We show this rounded up value by N_{min} and we know from Eq. (6.38) that N_{min} varies inversely with the pressure angle. Table 6.2 provides this relation for a few pressure angles.

Table 6.2 provides that by increasing the pressure angle we can build fewer gear teeth and still avoid interference. That is one of the reasons why using higher pressure angles is more popular.

6.6 Gear Mesh Contact Ratio

The contact ratio is the average number of pairs of gear teeth that are in contact at every instant during motion. If the contact ratio is equal to 1, right when one pair of teeth starts contact, the previous pair finishes it. If the contact ratio is 1.3, in 30% of time two pairs of teeth are in mesh and 70% of time only one pair of teeth are in contact. (In Fig. 6.6, we are observing an occasion in which two pairs of teeth are in contact.) The higher the contact ratio, the more gentle and gradual the transmission of power between the gear teeth will be. Using Fig. 6.12, contact ratio can be calculated as the ratio of the total contact length CD divided by the circular pitch of the base circle, p_b.

$$m_c = \frac{CD}{p_b} \tag{6.39}$$

From Fig. 6.12, we know that the total length of contact is:

$$CD = CP + PD \tag{6.40}$$

By substituting Eqs. (6.26) and (6.27) into Eq. (6.40), we obtain the following formula for the total length of contact:

$$CD = \sqrt{(R_3 + a)^2 - (R_3 \cos \varphi)^2} - R_3 \sin \varphi + \sqrt{(R_2 + a)^2 - (R_2 \cos \varphi)^2} - R_2 \sin \varphi \tag{6.41}$$

The circular pitch of the base circle (base pitch) is defined in a way similar to the definition of the circular pitch of the pitch circle, i.e., Eq. (6.2). Recalling Eq. (6.13) one obtains:

$$p_b = \frac{2\pi R_{b2}}{N_2} = \frac{2\pi R_{b3}}{N_3} = \frac{2\pi R_b}{N} = p\cos\varphi \tag{6.42}$$

Substituting Eqs. (6.41) and (6.42) into (6.39) results in the following equation for the **contact ratio**:

$$m_c = \frac{\sqrt{(R_3 + a)^2 - (R_3\cos\varphi)^2} - R_3\sin\varphi + \sqrt{(R_2 + a)^2 - (R_2\cos\varphi)^2} - R_2\sin\varphi}{(2\pi R_b/N)} \tag{6.43}$$

Equation (6.43) is valid only if there is no interference, i.e., in Fig. 6.12 segment *CD* is completely within line *AB*. Decreasing the pressure angle reduces the length of *AB* and increases the possibility of interference. Reducing addendum, as seen in Fig. 6.12, shortens *CD* and reduces the possibility of interference. However, reducing addendum decreases the contact ratio as well which is not a favorite outcome. That is why it is good to find the maximum possible addendum that can be used without interference, as it was done in Eqs. (6.30) and (6.31).

A contact ratio of 1.40 is generally accepted as the practical minimum value, Cleghorn and Dechev [2], page 217; and contact ratios less than 1.20 should be avoided, Waldron and Kinzel [4], page 470. As it can be seen in Eq. (6.43), one way for increasing the contact ratio is to increase the number of gear teeth, *N*. And for gears in external contact m_c should be between 1.2 and 1.6.

For a rack and pinion set the rack pitch radius is infinite and therefore Eq. (6.41) cannot be used. Instead, if pinion is #2, we should use the following equation for *CD*, Cleghorn and Dechev [2], page 220:

$$CD = \sqrt{(R_2 + a)^2 - (R_2\cos\varphi)^2} - R_2\sin\varphi + \frac{a}{\sin\varphi} \tag{6.44}$$

For internal contact, the contact ratio is higher than that of the external contact. If gear 3 is the internal (ring) gear, the length of contact is, Cleghorn and Dechev [2], page 221:

$$CD = -\sqrt{(R_3 - a)^2 - (R_3\cos\varphi)^2} + R_3\sin\varphi + \sqrt{(R_2 + a)^2 - (R_2\cos\varphi)^2} - R_2\sin\varphi \tag{6.45}$$

For a ring gear, the addendum circle is inside its pitch circle, and the teeth are concave rather than convex. This matter results in lower stresses between pinion and the ring gear since surfaces have more contact points. Ring gears are used in planetary gears and contribute to the compactness of the gear box. Because of the tooth shape, the strength of ring gear teeth is much greater than that of the external pinion. Like in external contacting gears, the line of action is tangent to both base circles and passes through the pitch point (where the two pitch circles touch), Waldron and Kinzel [4], page 474.

Example 6.2 A gear with 22 teeth is in external contact mesh with another gear with 42 teeth. Gears have full-depth teeth, the diametral pitch is 16 teeth/in, and the pressure angle is 20°

A. Check to see if there is any interference.
B. Calculate the contact ratio.
C. Calculate the maximum addendum that would prevent interference.

A. **Check to see if there is any interference.**

Considering the given parameters,

$$N_2 = 22, N_3 = 42, P = 16 \text{ teeth/in}, \varphi = 20\,^\circ$$

The pitch circle radii are obtained using Eq. (6.3):

$$P = \frac{N}{2R} \tag{6.3}$$

So,

$$R_2 = \frac{N_2}{2P} = \frac{22}{2 \times 16} = 0.6875 \text{ in}; R_3 = \frac{N_3}{2P} = \frac{42}{2 \times 16} = 1.3125 \text{ in}$$

Referring to Table 6.1 the addendum is:

$$a = \frac{1}{P} = \frac{1}{16} = 0.0625 \text{ in}$$

Now, referring to Fig. 6.12, one can check to see if the no-interference conditions in (6.20) are satisfied:

$$AP > CP; BP > DP \tag{6.20}$$

Substituting the given data in Eqs. (6.26) and (6.27) we get:

$$CP = \sqrt{(R_3 + a)^2 - (R_3 \cos \varphi)^2} - R_3 \sin \varphi = 0.1589 \text{ in}$$

$$DP = \sqrt{(R_2 + a)^2 - (R_2 \cos \varphi)^2} - R_2 \sin \varphi = 0.1458 \text{ in}$$

Furthermore, Eqs. (6.21) and (6.22) give:

$$AP = R_2 \sin\varphi = 0.2351 \text{ in}; BP = R_3 \sin\varphi = 0.4489 \text{ in}$$

It can be observed that the no-interference conditions (6.20) are satisfied.

B. **Calculate the contact ratio.**

We can use either the contact ratio Eq. (6.39) or Eq. (6.43) for the case of external contact. Using Eq. (6.39), i.e.,

$$m_c = \frac{CD}{p_b} \tag{6.39}$$

For the numerator we have,

$$CD = CP + PD = 0.1589 + 0.1458 = 0.3047 \text{ in}$$

The radius of the base circle of gear 2 can now be calculated using Eq. (6.11):

$$R_{b2} = R_2 \cos\varphi = 0.6875 \ \cos 20° = 0.6460 \text{ in}$$

So, the circular pitch of the base circle is calculated using Eq. (6.42),

$$p_b = \frac{2\pi R_{b2}}{N_2} = 0.1845 \text{ in}$$

Therefore, Eq. (6.39) for the contact ratio results in:

$$m_c = \frac{CD}{p_b} = \frac{0.3047}{0.1845} = 1.652$$

This result is equivalent to saying that on average one and 0.652 tooth pairs are in contact. That means that one pair of teeth are fully engaged and at almost 65% of time another pair is either engaging or disengaging.

C. **Calculate the maximum addendum that would prevent interference.**

The acceptable maximum addendum values that would prevent interference for each of the gears are obtained from the right-hand sides of (6.30) and (6.31):

$$\begin{aligned} a_{2,max} &= \sqrt{(R_2 + R_3)^2(\sin\varphi)^2 + (R_2\cos\varphi)^2} - R_2 \\ a_{3,max} &= \sqrt{(R_2 + R_3)^2(\sin\varphi)^2 + (R_3\cos\varphi)^2} - R_3 \end{aligned} \tag{6.46}$$

Substituting the numerical values of the parameters in Eqs. (6.46) results in,

$$a_{2,max} = 0.2534 \text{ in}; a_{3,max} = 0.0978 \text{ in}$$

Interestingly, in this case, the standard addendum value from Table 6.1 is as follows:

$$a = \frac{1}{P} = 0.0625 \text{ in}$$

which is less than both of the obtained acceptable upper bounds for no-interference. Therefore, using the AGMA standard value for addendum does not create interference and the provided result by the standard is compatible with our calculations for suitable upper bounds for addendum.

Example 6.3 Two spur gears cut full depth are in external contact as we see in Fig. 6.15. Their center-to-center distance is 15 in long, their pressure angle is 20°, and the contact force transmitted between the gears along the line of action (i.e., the common normal on contacting teeth profiles) is 10 lbf. Knowing their number of teeth as $N_2 = 20$, $N_3 = 40$, and the input angular velocity given to gear 2 as $\omega_2 = 100$ rpm CCW, calculate,

A. the pitch circle radii of the gears,
B. the diametral pitch of the gears,
C. the output torque on gear 3 (lbf.ft),
D. the transmitted power (lbf.ft./s and hp) assuming no loss.

A. **The pitch circle radii of the gears**

To get the pitch circle radii, we remember that the center-to-center distance is the summation of the two pitch circle radii, i.e.,

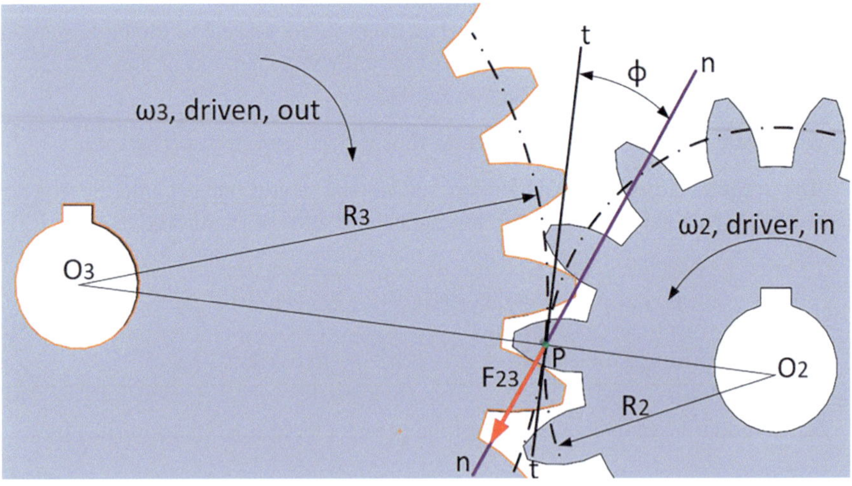

Fig. 6.15 Basic parameters of two gears in mesh

$$R_2 + R_3 = O_2O_3 \tag{6.47}$$

which in this case it is 15 in. Now, recall Eq. (6.10),

$$\frac{\omega_3}{\omega_2} = \frac{R_2}{R_3} = \frac{N_2}{N_3} \tag{6.10}$$

Substituting the provided data into Eq. (6.10) results in,

$$\frac{\omega_3}{100} = \frac{R_2}{R_3} = \frac{20}{40}$$

Therefore, $\omega_3 = 50$ rpm CW, and the direction of rotation of gear 3 is opposite that of the driver gear 2 because the gears have external contact. We also get,

$$R_2 = 0.5R_3 \tag{6.48}$$

By solving Eqs. (6.47) and (6.48), the radii of the pitch circles are obtained as follows:

$$R_2 = 5 \text{ in}; R_3 = 10 \text{ in}$$

B. **The diametral pitch of the gears**

Having calculated the radii of the pitch circles, one can now calculate the diametral pitch using Eq. (6.3):

$$P = \frac{N}{2R} \tag{6.3}$$

Therefore we get,

$$P = \frac{N_2}{2R_2} = \frac{N_3}{2R_3} = \frac{20}{2 \times 5} = 2 \text{ teeth/in}$$

C. **The output torque on gear 3 (lbf.ft)**

To get the output torque on gear 3 we use Eq. (6.15),

$$T_3 = F_{23}^t R_3 = F_{23} R_3 \cos \varphi \tag{6.15}$$

Substituting the available numerical values we get:

$$T_3 = 10 \text{ lbf} \times 10 \text{ in} \times \cos 20^\circ = 93.97 \text{ lbf.in} = 93.97 \text{ lbf.in} \times \frac{1 \text{ ft}}{12 \text{ in}} = 7.83 \text{ lbf.ft}$$

D. **The transmitted power (lbf.ft/s and hp) assuming no loss**

For calculating output power we first convert the angular velocity of gear 3 from rpm to rad/s:

$$\omega_3 = 50\ \text{rpm} = 50\ \frac{\text{rev}}{\text{min}} \times \frac{2\pi\ \text{rad}}{1\ \text{rev}} \times \frac{1\ \text{min}}{60\ \text{s}} = 5.24\ \text{rad/s}$$

$$P_3 = T_3\omega_3 = 7.83\ \text{lbf.ft} \times 5.24\ \text{rad/s} = 41.03\ \text{lbf.ft/s}$$

For converting this amount of power to horse power (hp) we recall:

$$1\ \text{hp} = 550\ \text{lbf.ft/s}$$

Therefore,

$$P_3 = 41.03\ \text{lbf.ft/s} \times \frac{1\ \text{hp}}{550\ \text{lbf.ft/s}} = 0.075\ \text{hp}$$

6.7 Impacts of Changing the Center Distance

Consider gears 2 and 3 illustrated in Fig. 6.15 that have base circle radii R_{b2} and R_{b3}, respectively. The gears are manufactured to have a pressure angle φ and their standard center distance is,

$$R_2 + R_3 = O_2O_3 \tag{6.47}$$

Using Eq. (6.13), the center distance Eq. (6.47) can be rewritten in terms of the base circle radii and the pressure angle,

$$O_2O_3 = R_2 + R_3 = \frac{R_{b2}}{\cos\varphi} + \frac{R_{b3}}{\cos\varphi} = \frac{1}{\cos\varphi}(R_{b2} + R_{b3}) \tag{6.49}$$

Now, if the center distance changes slightly by Δ, the speed ratio does not change (as it only depends on the ratio of the number of gear teeth and those numbers have not changed). The base circle radii do not change either as they are determined when the gears are manufactured. However, the effective pitch circle radii change to R'_2 and R'_3. Here a prime refers to the new value of a quantity after the center distance change has happened. These new effective pitch circle radii should still satisfy the speed ratio Eq. (6.10) so we get,

$$\frac{\omega_3}{\omega_2} = \frac{\omega'_3}{\omega'_2} = \frac{R_2}{R_3} = \frac{R'_2}{R'_3} = \frac{R_{b2}}{R_{b3}} = \frac{N_2}{N_3} \tag{6.50}$$

Furthermore, after the center distance change has happened, the effective pressure angle changes to φ'. Therefore, the center distance Eq. (6.49) changes to,

$$O_2O_3 + \Delta = R_2' + R_3' = \frac{1}{\cos\varphi'}(R_{b2} + R_{b3}) \tag{6.51}$$

Dividing Eq. (6.51) by Eq. (6.49) we get:

$$\frac{O_2O_3 + \Delta}{O_2O_3} = \frac{R_2' + R_3'}{R_2 + R_3} = \frac{\cos\varphi}{\cos\varphi'} \tag{6.52}$$

Using Eqs. (6.52) and (6.50), one can calculate the effective pitch circle radii R_2' and R_3', after the change in center distance has happened. Also using Eq. (6.52), and by just entering the change of the center distance, Δ, one can easily calculate the resulting effective pressure angle, φ'. If the center distance increases, i.e., $\Delta > 0$, one could see that $\cos\varphi' < \cos\varphi$ and since cosine is a decreasing function in the $[0,\pi]$ domain, we conclude that: $\varphi' > \varphi$. So, increasing the center distance of the two gears results in an increase of the effective pitch circle radii and the effective pressure angle. Furthermore, increasing the pressure angle decreases the tangential transmitted force F_{23}^t, decreases the contact ratio, and there will be less likelihood of interference, but with more backlash.

References

1. Gibbs, J. W. (1863). *On the form of the teeth of wheels in spur gearing*, Ph.D. thesis, Yale University.
2. Cleghorn, W. L., & Dechev, N. (2015). *Mechanics of machines* (2nd ed.). Oxford University Press.
3. Uicker, J. R., et al. (2016). *Theory of machines and mechanisms* (5th ed.). Oxford University Press.
4. Waldron, K. J., & Kinzel, G. L. (2004). *Kinematics, dynamics, and design of machinery* (2nd ed.). Wiley.

Chapter 7
Fixed-Axis and Planetary Gear Trains

7.1 Introduction

Chapters 7 is dedicated to the analysis of fixed-axis and planetary gear trains. They are used in many applications including clocks and watches, drills, automotive and helicopter power transmission, and power generation in wind turbines. In many cases, gear trains are used to allow us to choose how we would like to use our available power in a way that it suits the situation at hand. For example, an automotive transmission allows us to operate a car at low speed and high torque (at start) or at high speed and low torque (in a highway). In both fixed-axis and planetary gear trains, a major focus of the discussion is on methods that can be used for calculating the speed ratio (that is also related to mechanical advantage and torque ratio (see Chap. 11)). We will see that speed ratio can be calculated using the rolling contact equations discussed in Chap. 3.

7.2 Simple and Compound Fixed-Axis Gear Trains

In fixed-axis gear trains all of the axes of the gears are fixed and supported by bearings that are attached to the frame. This is contrary to planetary gear trains in which not all of the gear axes are fixed. Figures 7.1 and 7.2 illustrate **simple fixed-axes gear trains**. The word *simple* refers to the existence of only one gear on each shaft.

In **compound fixed-axis gear trains** shown in Figs. 7.3 and 7.4, the two gears that are mounted on the middle shaft form an integral part that is called a **compound gear**. This gear train has two stages (i.e., two sets of gear meshes) and if the small gear on the left is the driver, in each stage a certain amount of speed reduction would happen. Compound gears can achieve the necessary gear ratios in a restricted space by coupling large and small gears on the same axle. If instead of a compound gear in

M. Ghorashi, *Introduction to Machine Dynamics*, Mechanical Engineering Series,
https://doi.org/10.1007/978-3-031-94736-0_7

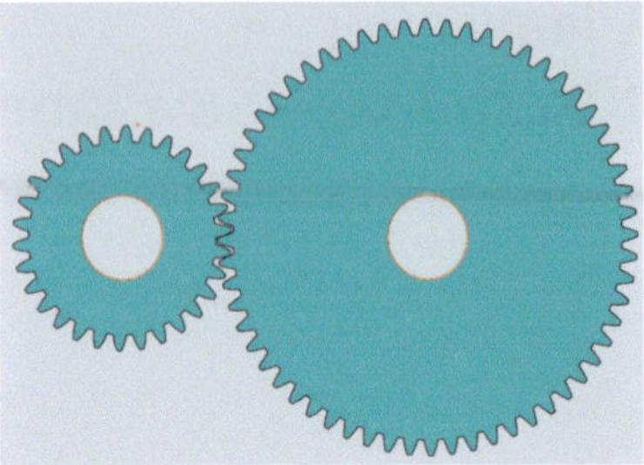

Fig. 7.1 A simple fixed-axis gear train modeled by SolidWorks

Fig. 7.2 Physical model of a simple fixed-axis gear train (USM term project)

Fig. 7.3 Physical model of a compound fixed-axis gear train with two stages of speed reduction and torque increase (USM term project)

the illustrated mechanisms, simple gears were used to provide a large speed reduction, the final (output) gear would have a very large diameter compared to the first (input) gear. Figure 7.4 illustrates the top view of a gear train similar to the one shown in Fig. 7.3. It also includes a simplified symbolic representation (aka the skeleton diagram) of the top view of this gear train.

Gear trains are also widely used in manufacturing mechanical clocks as can be seen in Fig. 7.5. This mechanical clock was modeled using SolidWorks Motion software and manufactured as part of two capstone projects of my students at USM; Palmer [6] and Scott [7].

Figure 7.6 illustrates a 20 ft-long clock that was made of 300 recycled intermeshing beer bottles. The clock was manufactured by Stanley Clockworks in Pennsylvania.

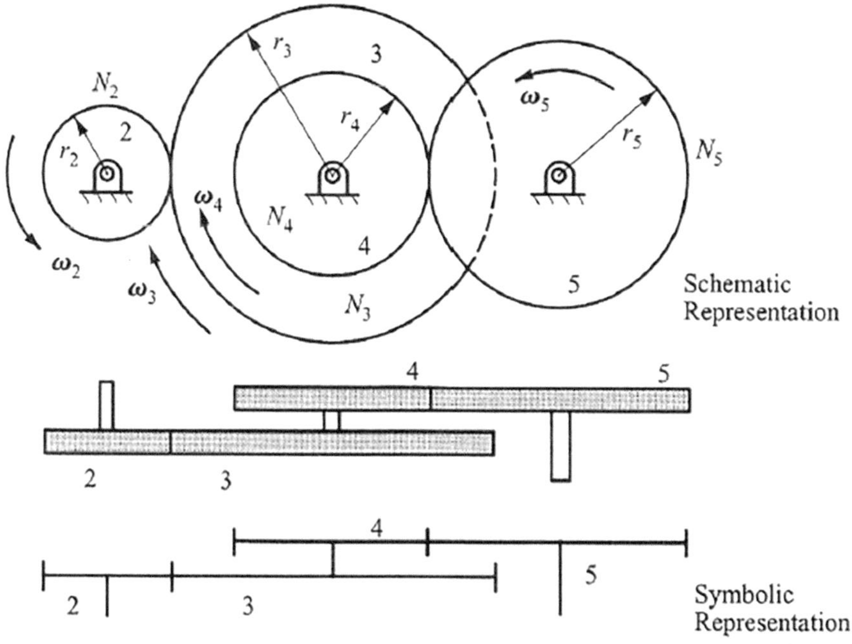

Fig. 7.4 Various views and methods of representation of a compound fixed-axis gear train with two stages of speed reduction and torque increase. Page 535 of Waldron and Kinzel [9]. © 2004 John Wiley & Sons, Inc. (reproduced with permission)

7.3 Analyzing Fixed-Axis Gear Trains

In a gear train, the **speed ratio** is defined as the ratio of the output gear (driven) angular velocity over the input gear (driver) angular velocity.

$$\text{speed ratio} = \frac{\omega_{out,driven}}{\omega_{in,driver}} \tag{7.1}$$

The **gear ratio** is defined as the ratio of the input gear (driver) angular velocity over the output gear (driven) angular velocity. In many applications, speed reducers are used for which the gear ratio is more than one.

$$\text{gear ratio} = \frac{\omega_{in,driver}}{\omega_{out,driven}} = \frac{1}{\text{speed ratio}} \tag{7.2}$$

Finally, the **torque ratio** is the ratio of the torque on the output shaft (driven) over the torque on the input shaft (driver). This ratio is also the mechanical advantage (M.A.) of the mechanism,

Fig. 7.5 A mechanical clock that includes several fixed-axis gears, Palmer [6] and Scott [7]

Fig. 7.6 Stanley Clockworks' beer bottle clock when it was exhibited at Philadelphia International Airport's Terminal A. Public domain picture, https://www.flickr.com/photos/chriswaits/5620435602/in/photostream/, accessed 1/4/2025

$$\text{torque ratio} = \text{M.A.} = \frac{T_{out,driven}}{T_{in,driver}} \tag{7.3}$$

If losses are negligible and the output power can be assumed to be equal to the input power, the torque ratio will be equal to the gear ratio,

$$\frac{T_{out,driven}}{T_{in,driver}} = \frac{\omega_{in,driver}}{\omega_{out,driven}} \tag{7.4}$$

We will see it is possible that the driver and the driven gears rotate in opposite directions. In such a case the speed ratio would be negative. This matter is correct for gear ratio and torque ratio as well; even though usually we just care about the absolute value of the those ratios and do not report the negative signs.

In most applications the aim is either speed change or torque change. Using the involute profile for gear teeth assures us that the speed or torque ratios remain constant during motion.

Referring to Fig. 6.15 and what we studied in Chap. 6, one can write speed ratio and gear ratio in terms of teeth numbers and pitch circle radii as well.

$$\text{speed ratio} = \frac{\omega_{out,driven}}{\omega_{in,driver}} = \frac{\omega_3}{\omega_2} = \frac{N_2}{N_3} = \frac{R_2}{R_3} \tag{6.10}$$

$$\text{gear ratio} = \frac{\omega_{in,driver}}{\omega_{out,driven}} = \frac{\omega_2}{\omega_3} = \frac{N_3}{N_2} = \frac{R_3}{R_2} \tag{7.5}$$

And again, *if losses are negligible*, the output power can be assumed to be equal to the input power and for the torque ratio one obtains,

$$\frac{T_{out,driven}}{T_{in,driver}} = \text{gear ratio} = \frac{\omega_{in,driver}}{\omega_{out,driven}} = \frac{\omega_2}{\omega_3} = \frac{N_3}{N_2} = \frac{R_3}{R_2} \tag{7.6}$$

As we can see in Fig. 6.15, the directions of rotation of the two gears that are in external contact are opposite to each other.

Example 7.1 The gears in the speed reducer shown in Fig. 6.15 have $N_2 = 20$ and $N_3 = 40$. Calculate the speed ratio, the gear ratio and, if losses are negligible, the torque ratio.

Using Eqs. (6.10), (7.5), and (7.6) we have,

$$\text{speed ratio} = \frac{\omega_3}{\omega_2} = \frac{N_2}{N_3} = \frac{20}{40} = 0.5;\ \text{gear ratio} = \frac{\omega_2}{\omega_3} = \frac{N_3}{N_2} = \frac{40}{20} = 2$$

$$\text{torque ratio} = \text{M.A.} = \frac{T_{out,driven}}{T_{in,driver}} = \frac{\omega_2}{\omega_3} = 2$$

Fig. 7.7 A helical gear with an internal bushing to form a journal bearing. No torque is transmitted to the shaft to be located at the bearing location

Fig. 7.8 Helical gears with keyways in order to transmit torque and power to their own shafts (not shown)

For transmitting torque or power to their shafts, gears should have a keyway or an internal spline. If at the contact surface of a gear with a shaft only a bushing exists (see Fig. 7.7), no torque or power transmission can be done and between them only radial or frictional forces exist. This matter does not, however, change the speed ratio or the gear ratio of the gear train that only depends on the number of gear teeth, as we have seen in Eqs. (6.10) and (7.5). Keys or splines are used in order to transmit power between a gear and its shaft (see Figs. 7.8 and 7.9).

Figure 7.10 illustrates a two-stage speed reduction compound gear train. As we saw in Figs. 7.3 and 7.4, the compound gear 34 is a single part which is rigidly connecting gears 3 and 4 together. The angular velocity ω_{34} is the angular velocity of this compound gear.

Considering gear 2 to be the input, the speed ratio of the first stage (that includes gears 2 and 3) is:

Fig. 7.9 Helical gear with internal spline in order to transmit torque and power to the shaft (not shown)

Fig. 7.10 Two-stage speed reduction (torque increase) if gear 2 is the input (USM term project)

$$\frac{\omega_{34}}{\omega_2} = \frac{N_2}{N_3} = \frac{R_2}{R_3} \tag{7.7}$$

The speed ratio of the second stage (that includes gears 4 and 5) is:

$$\frac{\omega_5}{\omega_{34}} = \frac{N_4}{N_5} = \frac{R_4}{R_5} \tag{7.8}$$

By multiplying Eqs. (7.7) and (7.8) we obtain the speed ratio of the whole gear train as:

$$\frac{\omega_5}{\omega_2} = \frac{N_2 N_4}{N_3 N_5} = \frac{R_2 R_4}{R_3 R_5} \tag{7.9}$$

where 5 is the output gear and 2 is the input gear. In stage 1, gear 2 is the input (driver) and gear 3 is the output and in stage 2, gear 4 is the input (driver) and gear 5 is the output. So Eq. (7.9) for the speed ratio can be rewritten and (for a simple multi-stage gear train) generalized as follows:

$$\frac{\omega_{out}}{\omega_{in}} = \frac{\prod N_{in}}{\prod N_{out}} = \frac{\prod R_{in}}{\prod R_{out}} \tag{7.10}$$

The capital *pi* symbol in Eq. (7.10) means product of the terms that follow this symbol. So the numerator of the second fraction includes the product of the number of gear teeth of all of the input gears (driver) in a gear mesh and the denominator includes the product of the number of gear teeth of all of the output (driven) gears in a gear mesh. Equivalently, the products of the radii of the pitch circles could be used instead of the number of gear teeth, as we see in Eq. (7.10). This result can also be interpreted as the product of speed ratios at each stage.

Example 7.2 In the gear train shown in Fig. 7.10 assume that we have: $N_2 = 20$, $N_3 = 40$, $N_4 = 20$, and $N_5 = 40$.

A. Calculate the speed ratio of the gear train if gear 2 is the input.
B. If $\omega_2 = 100$ rpm (CCW), calculate the output angular velocity, ω_5.

A. **Calculate the speed ratio of the gear train.**

Using either Eq. (7.9) or (7.10) we have:

$$\frac{\omega_5}{\omega_2} = \frac{N_2 N_4}{N_3 N_5} = \frac{20 \times 20}{40 \times 40} = 0.25$$

B. **If $\omega_2 = 100$ *rpm* (CCW), calculate the output angular velocity, ω_5.**

Having known $\omega_2 = 100$ rpm (CCW), and using the result of part A of this example we get:

$$\omega_5 = 0.25\omega_2 = 25 \text{ rpm (CCW)}$$

Since in external contact of gears each stage reverses the direction of motion, we know that after two stages the direction of motion of 5 is the same as the direction of motion of 2. It can also be easily found out that the compound gear 34 rotates at $\omega_{34} = 50$ rpm (CW).

Not all gears in a gear train affect the speed ratio. If a third gear is used between the input and the output gears, as we see it in Fig. 7.11, its contributions in the numerator and the denominator of the speed ratio fraction cancel each other out:

$$\frac{\omega_{out}}{\omega_{in}} = \frac{\omega_4}{\omega_2} = \frac{\prod N_{in}}{\prod N_{out}} = \frac{N_2 N_3}{N_3 N_4} = \frac{N_2}{N_4} = \frac{R_2}{R_4} \tag{7.11}$$

In this case, gear 3 has no effect on the speed ratio. However, this gear changes the direction of rotation in such a way that the input gear and the output gear both turn in the same CCW direction. Gear 3 is called an **idler gear**. Idler gears are also used where two gears cannot be directly meshed because of their distance. The idler would then be used as a **spacer**, as an alternative to using large input and output gears.

Fig. 7.11 Two-stage speed increase (torque reduction), if gear 2 is the input

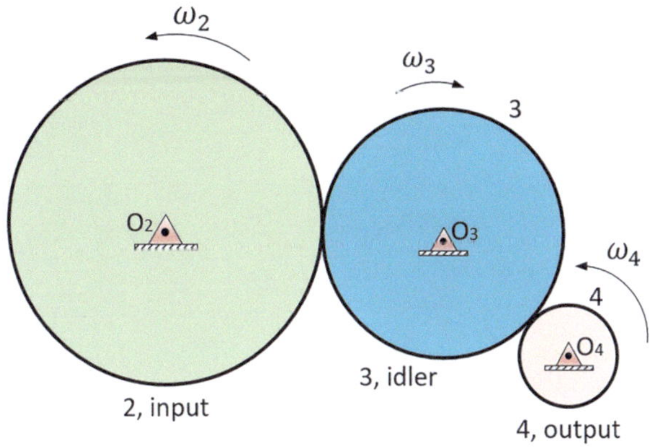

7.4 Simple and Compound Planetary Gear Trains

In fixed-axis gear trains, shafts of gears are held by bearings that are attached to the frame. This is a seriously limiting constraint. In planetary gear trains this constraint is lifted. The result is a gear train that is much more compact than the comparable fixed-axis gear train (especially when the input and output shafts are supposed to be collinear). Planetary gear trains provide a wide range of speed ratios and they are widely used in applications like automatic transmission of vehicles, drills, and wind turbines. In wind turbines, planetary gear trains provide a compact method for significantly increasing the speed of the slowly rotating blades (but at high torque) so that they can be used for efficient power generation.

A planetary gear train can have two inputs and provide one output. One of the two inputs may be zero, that is provided by a band brake. The non-zero input may be applied by a clutch. Figure 7.12 illustrates the basic elements of a Ford C4 transmission that includes the **sun gear**, the **planet gears**, and the **ring gear** as well as the **arm (carrier)** that holds the planet gears. These elements and their relative positions are illustrated in Fig. 7.12. Notice that three planet gears are used for balancing inertia forces during rotation.

Figure 7.13 illustrates a holistic approach to decipher the operation of planetary gear transmissions in which a Ford C4 transmission has been analyzed as an example. The research included converting the operation of this hydraulic transmission to a PLC-controlled equivalent that operated by pneumatic actuators (so that details of its operation could be seen). The project included theoretical calculations, SolidWorks simulations, design and manufacture, as well as experimentation. This research was introduced in Davis and Ghorashi [4, 5]. Both of these papers were based on a capstone project of one of my students at USM, Davis [3].

The gear train shown in Fig. 7.13 is an example of a **simple planetary gear train** in which only one planet gear is used per holding pin. Figure 7.14 shows an example of a **compound planetary gear train** in which compound planet gears are used instead. This arrangement allows the planet gears to mesh with different gears at the

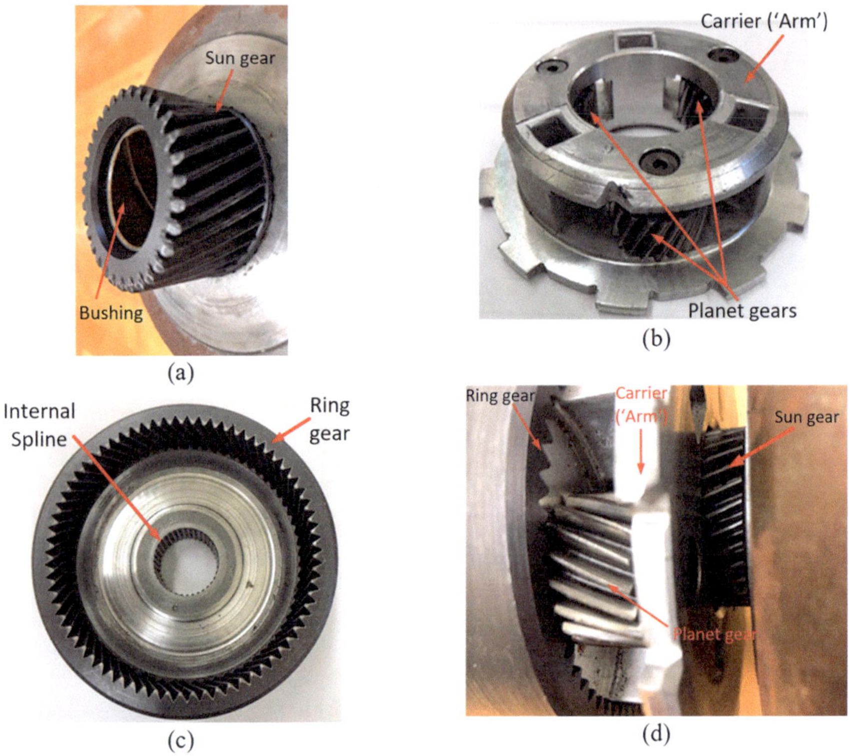

Fig. 7.12 Parts of a Ford C4 transmission: (**a**) sun gear; (**b**) planet gears and carrier ("arm"); (**c**) ring gear, and (**d**) their relative position

same time. Consequently, significantly higher gear ratios are obtained, compared to simple planetary gear trains. Use of compound planet gears is a way to combine multiple gear stages within a single planetary gear train. Figure 7.14 is an schematic of a Rohloff Speedhub 500/14 for bikes. It has 14 equally spaced gear ratios and its highest gear is 5.26 times its lowest gear. A major advantage of this gear train is that it has no external components that can make it vulnerable in dirty mountain roads.

7.5 Analyzing Planetary Gear Trains

In this section we analyze a few basic mechanisms with increasing order of difficulty in order to understand the basic methods of analyzing planetary gear trains. In each case, the rolling contact equations of Sect. 3.7 for external contact and internal contact are used for planetary gear trains. Unlike Chap. 6 (and the previous sections of this chapter) in which the focus of equations like (6.8) was on the absolute value

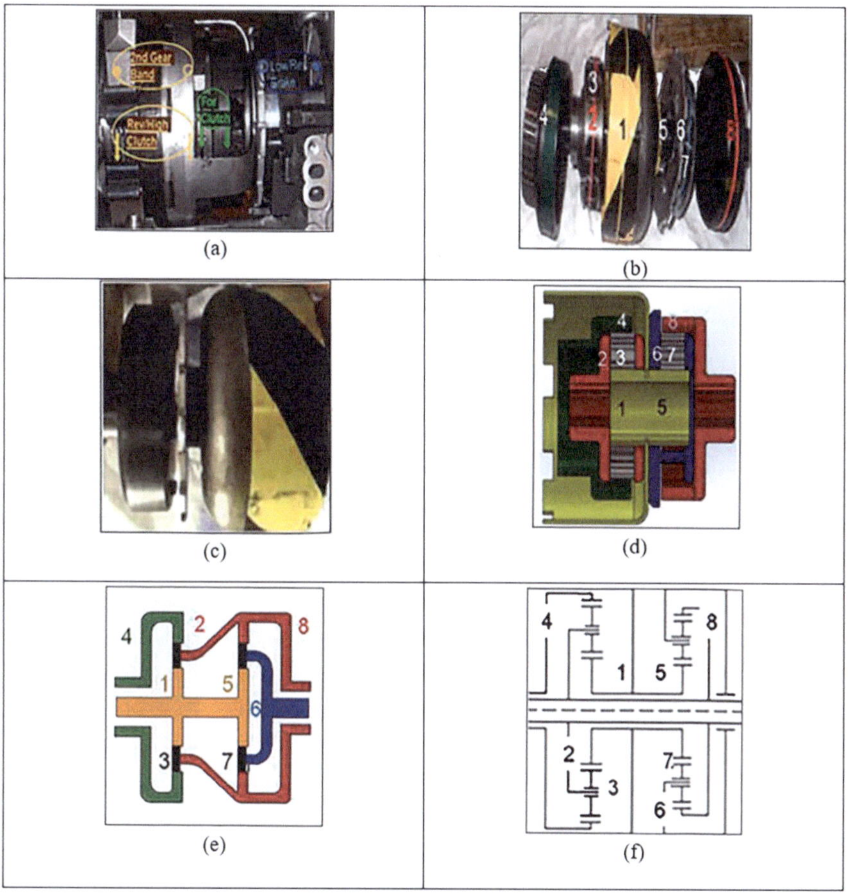

Fig. 7.13 (**a**) PLC-Controlled Ford C4 transmissions; (**b**) exploded view of the teaching aid; (**c**) assembled unpowered gearset; (**d**) SolidWorks model; (**e**) mechanism sketch; (**f**) skeleton diagram of the Simpson gear train, Davis and Ghorashi [5]. © 2018 American Society for Engineering Education (reproduced with permission)

of the speed ratio, in the present analysis we consider angular velocity as an algebraic quantity so that we can use its sign to determine any unknown direction of rotation.

7.5.1 External Contact and Fixed Axes

Consider two gears with fixed axes of rotation whose pitch circles are shown in Fig. 7.15. The radii of the pitch circles of gears 2 and 3 are R_2 and R_3, respectively. Recalling Eq. (3.109) we can write:

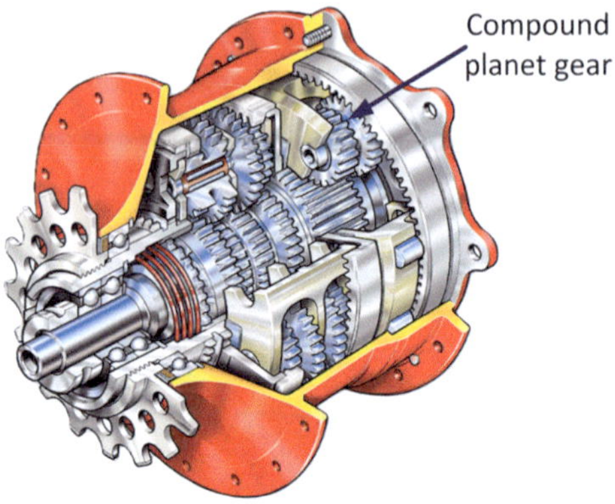

Fig. 7.14 Internal schematic of a Rohloff Speedhub 500/14 with its compound planet gears. Public domain picture, https://en.wikipedia.org/wiki/Rohloff_Speedhub#/media/File:Speed1c.png, accessed 1/5/2025 (note added)

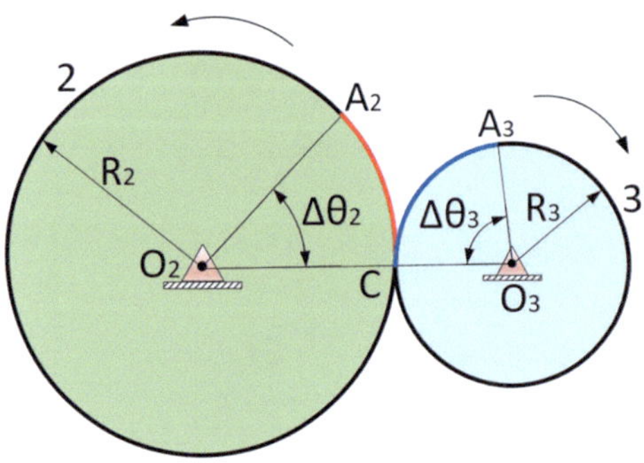

Fig. 7.15 Two fixed-axis gears in external contact and after performing the illustrated angular displacements

$$R_2.\Delta\theta_2 = -R_3.\Delta\theta_3 \tag{3.109}$$

The negative sign refers to the fact that in external contact the gears rotate in opposite directions. By dividing both sides by Δt, and converging it to zero, the corresponding equation for the angular velocities is obtained:

$$R_2\omega_2 = -R_3\omega_3 \tag{7.12}$$

7.5.2 *External Contact and a Moving Axis*

In planetary gears it frequently happens that the axis of a planet gear is not fixed and it can move on a circular orbit. In the mechanism shown in Figs. 7.16 and 7.17, the center point O_2 is still fixed, but O_3 can rotate on a circular orbit around it. In this case, gear 2 would be the sun gear around which planet gear 3 rotates. Since O_3 is no longer fixed, in order to ensure that its circular motion about O_2 happens, the two points should be connected by an arm 4. If we hold arm 4 fixed, the mechanism

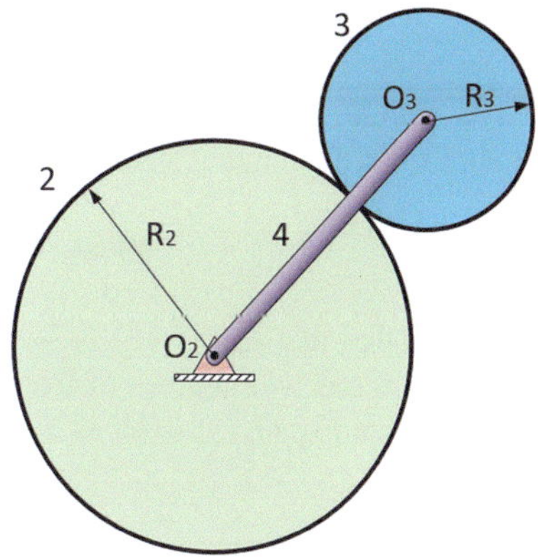

Fig. 7.16 Two gears in external contact with only gear 2 having a fixed axis of rotation

Fig. 7.17 Physical model of the mechanism shown in Fig. 7.16, Bartholomew [1]

reduces to the one studied in Sect. 7.5.1 and it still has one DOF. So, if the arm is free to rotate, as is the case in Fig. 7.16, the mechanism will have two DOF and therefore needs two inputs to operate. The mechanism shown in Fig. 7.17 was part of a capstone project of one of my students at USM; Bartholomew [1].

In Chap. 3, it was shown that in this case Eq. (3.109) was still valid, but with respect to the arm, and one could write,

$$R_2.\Delta\theta_{2/4} = -R_3.\Delta\theta_{3/4} \tag{3.110}$$

In which /4 means with respect to the arm. The negative sign in Eq. (3.110) is used for external contact and it implies opposite directions of rotation of gears 2 and 3 with respect to the arm. Equation (3.110) can be expanded as follows:

$$R_2(\Delta\theta_2 - \Delta\theta_4) = -R_3(\Delta\theta_3 - \Delta\theta_4) \tag{3.111}$$

By dividing both sides by Δt and converging Δt to zero, the corresponding equation for the angular velocities is obtained,

$$R_2(\omega_2 - \omega_4) = -R_3(\omega_3 - \omega_4) \tag{7.13}$$

We notice that Eq. (7.13) is one equation with three angular velocities, so the mechanism has two degrees of freedom. It can be seen that if the arm is fixed, i.e., $\omega_4 = 0$, then Eq. (7.13) reduces to,

$$R_2\omega_2 = -R_3\omega_3 \tag{7.12}$$

as expected. Also, using the relative velocity symbol, Eq. (7.13) can be rewritten as,

$$R_2\omega_{2/4} = -R_3\omega_{3/4} \tag{7.14}$$

Example 7.3 In the mechanism shown in Fig. 7.16, we fix the sun gear 2. Calculate the speed ratio $\frac{\omega_3}{\omega_4}$ if $R_2 = 2$ in and $R_3 = 4$ in.

Since gear 2 is fixed, $\omega_2 = 0$ and Eq. (7.13), i.e.,

$$R_2(\omega_2 - \omega_4) = -R_3(\omega_3 - \omega_4) \tag{7.13}$$

Reduces to,

$$(R_2 + R_3)\omega_4 = R_3\omega_3 \tag{7.15}$$

And finally we get,

$$\frac{\omega_3}{\omega_4} = \frac{R_2 + R_3}{R_3} > 1 \tag{7.16}$$

So, in this case, planet gear 3 rotates in the same direction as arm 4, but somewhat faster. Using the provided data, Eq. (7.16) gives,

$$\frac{\omega_3}{\omega_4} = \frac{2 + 4}{4} = 1.5$$

Therefore, for the given data the planet gear 3 rotates in the same direction as arm 4, but 50% faster. A simple physical model like the one shown in Fig. 7.17 can be used to verify this result!

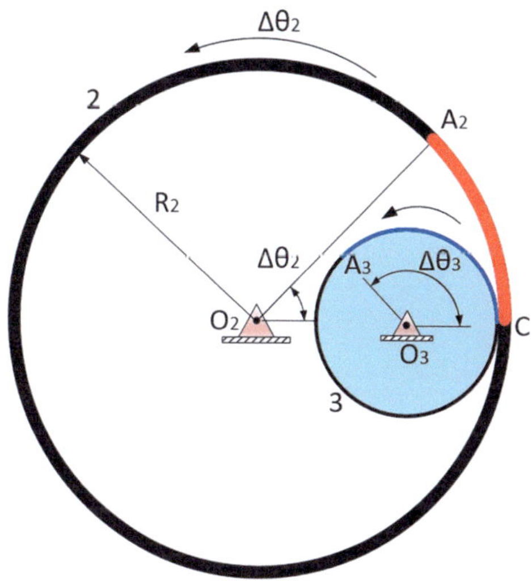

Fig. 7.18 Two fixed-axis gears in internal contact and after some angular displacement

7.5.3 Internal Contact and Fixed Axes

Now consider the case of an internal contact of gears with fixed axes shown in Fig. 7.18. In this case, gear 3 has a revolute joint at O_3. It is rotating inside the internal or ring gear 2 and the ring gear has a revolute joint at O_2. The pitch circle radius of gear 3 is R_3 and the radius of the pitch circle of the ring gear 2 is R_2. This problem has been discussed in Sect. 3.7.3 and it was concluded that for this case of internal contact the following equation is valid,

$$R_2 \Delta\theta_2 = R_3 \Delta\theta_3 \tag{3.116}$$

The positive sign in Eq. (3.116) indicates that in the case of internal contact, the gears rotate in the same direction. By dividing both sides of this equation by Δt and converging Δt to zero the corresponding equation for the angular velocities is obtained as follows:

$$R_2 \omega_2 = R_3 \omega_3 \tag{7.17}$$

7.5.4 Internal Contact and a Moving Axis

In Sect. 7.5.3, the centers of both gears, i.e., points O_2 and O_3 were fixed. In planetary gears, it frequently happens that the axis of one gear is not fixed and it can move on a circular orbit. Let us assume that O_2 is still fixed, but O_3 can rotate on a

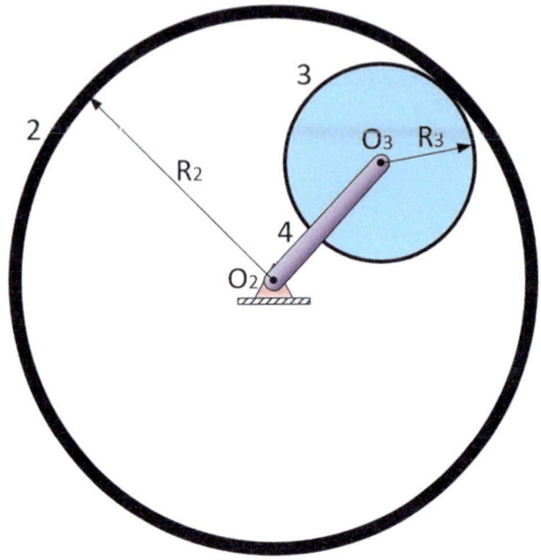

Fig. 7.19 Two gears in internal contact with only ring gear 2 having a fixed axis of rotation

circular orbit around O_2; as we see in Fig. 7.19. In this case, the hollow disk 2 would be the ring gear inside which planet gear 3 rotates. Since O_3 is no longer fixed, in order to ensure that its circular motion about O_2 happens, the two center points should be connected by an arm 4. If we hold arm 4 fixed, the mechanism reduces to that of Fig. 7.18 that still has one DOF. So, if the arm is free to rotate, the mechanism will have two DOF.

As mentioned before, if we hold arm 4 fixed, the mechanism reduces to the one discussed in Sect. 7.5.3. This is also similar to viewing the motion with respect to the arm when the arm is not fixed. Since this motion is analyzed with respect to the arm, what an arm-attached observer notices is that the arm is stationary, that is nothing but what was observed in the previous case. Therefore, Eq. (3.116) is still valid, but with respect to the arm and one can write the following:

$$R_2 \Delta\theta_{2/4} = R_3 \Delta\theta_{3/4} \tag{3.117}$$

In which /4 means relative to (or with respect to) the arm 4. The positive sign is used for internal contact and it implies same directions of rotation of the gears with respect to the arm. Equation (3.117) can be expanded as follows:

$$R_2(\Delta\theta_2 - \Delta\theta_4) = R_3(\Delta\theta_3 - \Delta\theta_4) \tag{3.118}$$

By dividing both sides of Eq. (3.118) by Δt and converging Δt to zero, the corresponding equation for the angular velocities is obtained,

$$R_2(\omega_2 - \omega_4) = R_3(\omega_3 - \omega_4) \tag{7.18}$$

We notice that this is one equation with three angular velocities so the mechanism has two degrees of freedom. It can be seen that if arm is fixed, $\omega_4 = 0$, and Eq. (7.18) reduces to,

$$R_2\omega_2 = R_3\omega_3 \tag{7.17}$$

as expected. Also, Eq. (7.18) can be rewritten using the relative velocity symbol as,

$$R_2\omega_{2/4} = R_3\omega_{3/4} \tag{7.19}$$

Example 7.4 In the internal gear problem with the arm shown in Fig. 7.19 ring gear 2 is fixed. Calculate the ratio $\frac{\omega_3}{\omega_4}$ if $R_2 = 6$ in and $R_3 = 2$ in.

Since ring gear 2 is fixed, $\omega_2 = 0$ and Eq. (7.18), i.e.,

$$R_2(\omega_2 - \omega_4) = R_3(\omega_3 - \omega_4) \tag{7.18}$$

reduces to,

$$(-R_2 + R_3)\omega_4 = R_3\omega_3 \tag{7.20}$$

and finally,

$$\frac{\omega_3}{\omega_4} = \frac{-R_2 + R_3}{R_3} < 0 \tag{7.21}$$

So, planet gear 3 rotates in the opposite direction of arm 4. Using the given data we get,

$$\frac{\omega_3}{\omega_4} = \frac{-6 + 2}{2} = -2$$

Therefore, in such a mechanism disk 3 rotates in the opposite direction of arm 4 and 100% faster. A simple physical model can be used to verify this result.

7.5.5 The General Case of Planetary Motion

Now the general case of planetary motion shown in Figs. 7.20 and 7.21 is considered. The physical model displayed in Fig. 7.21 is part of the capstone project of one of my students at USM; see Bartholomew [1]. This mechanism has all of the basic elements of a planetary gear train, i.e., ring gear 2, planet gear 3, arm 4, and sun gear 5. In this case, both external contact (between the sun and the planet gear) and the internal contact (between the planet and the ring gear) exist. It should be mentioned

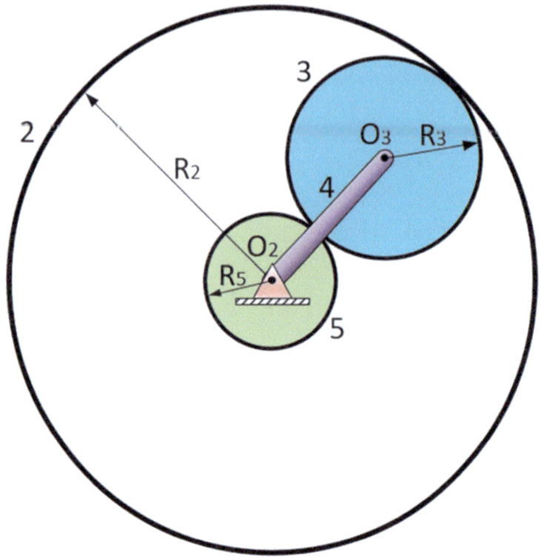

Fig. 7.20 The general case of a planetary gear train with ring gear 2, planet gear 3, arm 4, and sun gear 5

Fig. 7.21 Physical model of the general case of a planetary gear train shown in Fig. 7.20, Bartholomew [1]

that in the real life planetary gear trains there are more planet gears so that the inertia forces due to rotation are balanced. A model for such a gear train can be seen in Fig. 7.22. More discussion on balancing will be presented in Chap. 14.

As we saw in Figs. 7.4 and 7.13f, together with the view of a gear train in which gears are illustrated as circles, one can sketch another view of a gear train that includes more details about its axial dimensions. Such a skeleton diagram is also shown in Fig. 7.23. The side view of this mechanism has dash–dotted lines that show revolute joints where a gear revolves on a shaft. The hatched lines are used to show fixed bearings where the mechanism is attached to the fix base 1. The common axis of gears 2 and 5 is shown to be rigidly attached to arm 4.

For the mechanism shown in Fig. 7.20, clearly there is a geometrical constraint that should be satisfied:

Fig. 7.22 Physical model of the general case of a planetary gear train with balanced planet gears (USM term project)

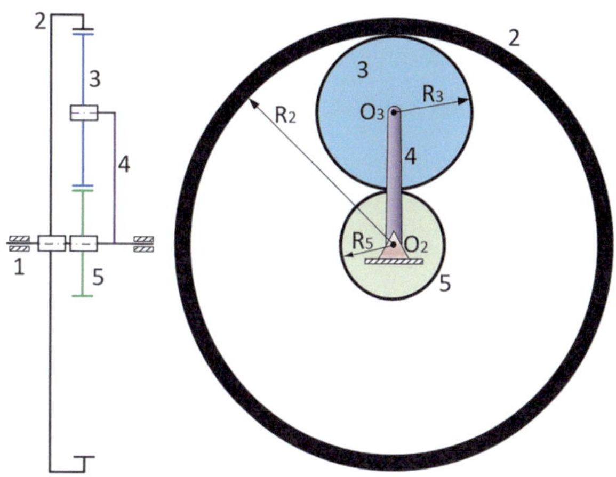

Fig. 7.23 The general case of a planetary gear train with ring gear 2, planet gear 3, arm 4, and sun gear 5, together with its side view skeleton diagram

$$R_5 + 2R_3 = R_2 \tag{7.22}$$

For the external contact between sun gear 5 and planet gear 3, one can write,

$$R_5\omega_{5/4} = -R_3\omega_{3/4} \tag{7.23}$$

Similarly, the internal contact between planet 3 and ring 2 can be expressed as follows:

$$R_3\omega_{3/4} = R_2\omega_{2/4} \tag{7.24}$$

Expanding Eqs. (7.23) and (7.24) results in,

$$R_5(\omega_5 - \omega_4) = -R_3(\omega_3 - \omega_4) \tag{7.25}$$

$$R_3(\omega_3 - \omega_4) = R_2(\omega_2 - \omega_4) \tag{7.26}$$

Equations (7.25) and (7.26) have four angular velocities, ω_2, ω_3, ω_4, and ω_5. Therefore, the mechanism has two degrees of freedom and requires two inputs. Depending on which angular velocities are given as input, the gear train will provide different responses, gear ratios, and mechanical advantages. This is the basic principle of operation of a power transmission system that operates using planetary gears. Combining Eqs. (7.25) and (7.26) we get:

$$R_5(\omega_5 - \omega_4) = -R_2(\omega_2 - \omega_4) \tag{7.27}$$

or,

$$\frac{\omega_2 - \omega_4}{\omega_5 - \omega_4} = -\frac{R_5}{R_2} \tag{7.28}$$

In what follows, a few choices for the two input variables are considered and for each choice the resulting mechanism response is calculated.

Case 1: Angular velocities ω_2 and ω_4 are given

For the mechanism shown in Fig. 7.20 assume that the ring gear is fixed (i.e., $\omega_2 = 0$) and the driving part of the gear train is the arm, so ω_4 is known. Considering the sun gear 5 as the output, one should solve Eq. (7.27), reminded below, for ω_5.

$$R_5(\omega_5 - \omega_4) = -R_2(\omega_2 - \omega_4) \tag{7.27}$$

Substituting $\omega_2 = 0$ into Eq. (7.27) gives,

$$R_5\omega_5 = (R_2 + R_5)\omega_4$$

Therefore, the speed ratio in this case could be expressed as,

$$\frac{\omega_{out}}{\omega_{in}} = \frac{\omega_5}{\omega_4} = \frac{R_2 + R_5}{R_5} > 1 \tag{7.29}$$

So the output (sun gear 5) rotates faster than the input (arm 4) and they move in the same direction. This kind of response is called, **overdrive**. Now using Eq. (7.26), i.e.,

$$R_3(\omega_3 - \omega_4) = R_2(\omega_2 - \omega_4) \tag{7.26}$$

and $\omega_2 = 0$, we get,

$$\frac{\omega_3}{\omega_4} = \frac{R_3 - R_2}{R_3} < 0 \tag{7.30}$$

So, the planet gear rotates in the opposite direction of the arm (driver).

Case 2: Angular velocities ω_2 and ω_5 are given

Now considering Fig. 7.20, assume that the sun gear is fixed (i.e., $\omega_5 = 0$) and the driving part of the gear train is the ring gear, so ω_2 is known. Considering arm 4 as the output, one can solve Eq. (7.27) for the output angular velocity, ω_4.

$$R_5(\omega_5 - \omega_4) = -R_2(\omega_2 - \omega_4) \tag{7.27}$$

or,

$$R_2\omega_2 = (R_2 + R_5)\omega_4$$

Therefore, the speed ratio in this case would be,

$$\frac{\omega_{out}}{\omega_{in}} = \frac{\omega_4}{\omega_2} = \frac{R_2}{R_2 + R_5} < 1 \tag{7.31}$$

So the output (arm 4) rotates slower than the input (ring 2) and they move in the same direction. This kind of response is called, **underdrive**. Now, using Eq. (7.25):

$$R_5(\omega_5 - \omega_4) = -R_3(\omega_3 - \omega_4) \tag{7.25}$$

and $\omega_5 = 0$, we get,

$$\frac{\omega_3}{\omega_4} = \frac{R_3 + R_5}{R_3} > 1 \tag{7.32}$$

So, the planet gear rotates in the same direction as the arm (driver), but at a faster pace. Finally, by multiplying Eqs. (7.31) and (7.32) we get:

$$\frac{\omega_3}{\omega_{in}} = \frac{\omega_3}{\omega_2} = \frac{R_2(R_3 + R_5)}{R_3(R_2 + R_5)} > 1 \tag{7.33}$$

Case 3: Angular velocities ω_4 and ω_5 are given

For the mechanism shown in Fig. 7.20, we now assume that the arm is fixed (i.e., $\omega_4 = 0$) and the driving part of the gear train is the sun gear, so ω_5 is known. Considering the ring gear 2 as the output, one can solve Eq. (7.27) for the output angular velocity ω_2.

$$R_5(\omega_5 - \omega_4) = -R_2(\omega_2 - \omega_4) \tag{7.27}$$

or,

$$R_5\omega_5 = -R_2\omega_2 \tag{7.34}$$

Therefore, the speed ratio in this case would be,

$$\frac{\omega_{out}}{\omega_{in}} = \frac{\omega_2}{\omega_5} = -\frac{R_5}{R_2} < 0 \tag{7.35}$$

So, the output (ring gear 2) rotates in the opposite direction of the input (sun gear 5) and at a slower rate. The response is therefore a **reverse** motion. Now, using Eq. (7.25):

$$R_5(\omega_5 - \omega_4) = -R_3(\omega_3 - \omega_4) \tag{7.25}$$

and $\omega_4 = 0$, we get,

$$\frac{\omega_3}{\omega_5} = -\frac{R_5}{R_3} < 0 \tag{7.36}$$

So, the planet gear rotates in the opposite direction of the sun gear (driver). Considering Eq. (7.35), we conclude that by using planetary gear trains we can get the reverse motion in the output without using a separate gear that is moved into action only for this purpose.

It is interesting to note that since arm 4 is fixed, this case is similar to a fixed-axis gear train with sun gear 5 as the input and ring gear 2 as the output. The corresponding speed ratio for this fixed-axis gear train with internal contact can be readily obtained by using the fixed-axis gear train Eq. (7.10):

$$\frac{\omega_{out}}{\omega_{in}} = \frac{\prod N_{in}}{\prod N_{out}} = \frac{\prod R_{in}}{\prod R_{out}} \tag{7.10}$$

and noticing that the planet gear 3 acts as an idler gear with no effect on the speed ratio. Therefore, again we get,

$$\frac{\omega_2}{\omega_5} = -\frac{R_5}{R_2} < 0 \tag{7.35}$$

The result,

$$\frac{\omega_3}{\omega_5} = -\frac{R_5}{R_3} < 0 \tag{7.36}$$

can also be easily obtained using the fixed axis Eq. (7.10).

This example demonstrated how by fixing or driving different members of a planetary gear train (in practice, using band brakes and clutches) one can obtain various speed ratios. The results of the foregoing cases 1 to 3 are summarized in the top part of Table 7.1:

Example 7.5 Figure 7.24 illustrates a planetary gear train for which we have $N_3 = N_5 = 10$, $N_2 = 30$, $P = 2.5$ teeth/in, $\varphi = 20°$, and $AB = 4$ in. If the input angular velocity given to the sun gear 5 is 60 rpm (CCW) and the ring gear is held fixed, calculate the angular velocity of arm 4 and planet gear 3.

The mechanism shown in Fig. 7.24 is structurally the same as the one seen in Fig. 7.20. However, we notice that in none of the three cases we just studied, we have

Table 7.1 Summary of the speed ratios obtained for the mechanism shown in Fig. 7.20

case	input	output	fixed	speed ratio	response type	planet gear motion
1	arm	sun	ring	$\frac{\omega_5}{\omega_4} = \frac{R_2 + R_5}{R_5} > 1$	overdrive	$\frac{\omega_3}{\omega_4} = \frac{R_3 - R_2}{R_3} < 0$
2	ring	arm	sun	$\frac{\omega_4}{\omega_2} = \frac{R_2}{R_2 + R_5} < 1$	underdrive	$\frac{\omega_3}{\omega_2} = \frac{R_2(R_3 + R_5)}{R_3(R_2 + R_5)} > 1$
3	sun	ring	arm	$\frac{\omega_2}{\omega_5} = -\frac{R_5}{R_2} < 0$	reverse	$\frac{\omega_3}{\omega_5} = -\frac{R_5}{R_3} < 0$
inv1	sun	arm	ring	$\frac{\omega_4}{\omega_5} = \frac{R_5}{R_2 + R_5} < 1$	underdrive	$\frac{\omega_3}{\omega_5} = \frac{R_5(R_3 - R_2)}{R_3(R_2 + R_5)} > 1$
inv2	arm	ring	sun	$\frac{\omega_2}{\omega_4} = \frac{R_2 + R_5}{R_2} > 1$	overdrive	$\frac{\omega_3}{\omega_4} = \frac{R_3 + R_5}{R_3} > 1$
inv3	ring	sun	arm	$\frac{\omega_5}{\omega_2} = -\frac{R_2}{R_5} < 0$	reverse	$\frac{\omega_3}{\omega_2} = \frac{R_2}{R_3} > 1$

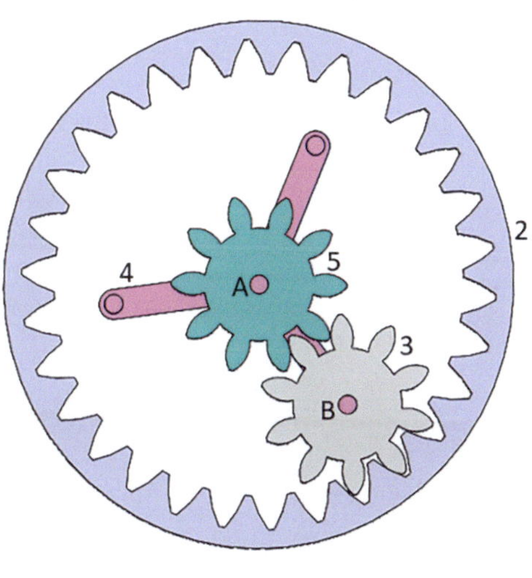

Fig. 7.24 Planetary gear train for Example 7.5

input sun gear and fix ring gear. Therefore, we need to solve this problem by going back to the rolling contact Eq. (7.27):

$$R_5(\omega_5 - \omega_4) = -R_2(\omega_2 - \omega_4) \tag{7.27}$$

Since the ring gear in this case is fixed, $\omega_2 = 0$, and we get:

$$\frac{\omega_{out}}{\omega_{in}} = \frac{\omega_4}{\omega_5} = \frac{R_5}{R_2 + R_5} < 1 \tag{7.37}$$

Interestingly (and expectedly) Eq. (7.37) is the inverse of Eq. (7.29) obtained for case 1:

$$\frac{\omega_{out}}{\omega_{in}} = \frac{\omega_5}{\omega_4} = \frac{R_2 + R_5}{R_5} > 1 \tag{7.29}$$

in which input and output were switched (sun gear 5 was output, arm 4 was input) and similar to this case the ring gear 2 was held fixed. So, we conclude that if input and output are switched, but no change is made to the fixed member, the speed ratio would be inverted.

Back to Eq. (7.37), and the provided data in this example, we get:

$$\frac{\omega_{out}}{\omega_{in}} = \frac{\omega_4}{60 \text{ rpm}} = \frac{R_5}{R_2 + R_5} = \frac{N_5}{N_2 + N_5} = \frac{10}{30 + 10}$$

Therefore,

$$\omega_4 = \frac{600}{40} = 15 \text{ rpm (CCW)}$$

Now, to get $\frac{\omega_3}{\omega_5}$ we substitute Eq. (7.37) into Eq. (7.25), repeated below:

$$R_5(\omega_5 - \omega_4) = -R_3(\omega_3 - \omega_4) \tag{7.25}$$

which gives the following equation for ω_3 in terms of ω_5:

$$\frac{\omega_3}{\omega_{in}} = \frac{\omega_3}{\omega_5} = \frac{R_5(R_3 - R_2)}{R_3(R_2 + R_5)} < 0 \tag{7.38}$$

Substituting the provided data in Eq. (7.38) gives:

$$\frac{\omega_3}{\omega_5} = \frac{R_5(R_3 - R_2)}{R_3(R_2 + R_5)} = \frac{N_5(N_3 - N_2)}{N_3(N_2 + N_5)} = \frac{10(10 - 30)}{10(30 + 10)} = \frac{-1}{2}$$

Recalling $\omega_5 = 60$ rpm (CCW), we get:

$$\omega_3 = -30 \text{ rpm (CW)}$$

It should be noted that Eq. (7.38) could also be obtained by dividing Eq. (7.30) for $\frac{\omega_3}{\omega_4}$ by Eq. (7.29) for $\frac{\omega_5}{\omega_4}$, in Case 1. Let us call this Case, inv1. Table 7.1 includes the obtained solutions for the previously mentioned Cases 1 to 3 and the Case inv1, just discussed. In a similar way, Cases inv2 and inv3 can be defined and their speed ratios calculated. Those cases have also been included in Table 7.1.

More complicated planetary gear trains use the four basic elements that have been discussed so far, i.e., sun gear, planet gear, ring gear, and arm (carrier) in a variety of ways. Figure 7.25 illustrates the skeleton diagram of the longitudinal view a Lèvai gear train type D.

Figure 7.26 illustrates a manufactured model of the gear train shown in Fig. 7.25. This model was simulated using SolidWorks Motion software and then manufactured as part of a capstone project of one of my students at USM, Verrill [8].

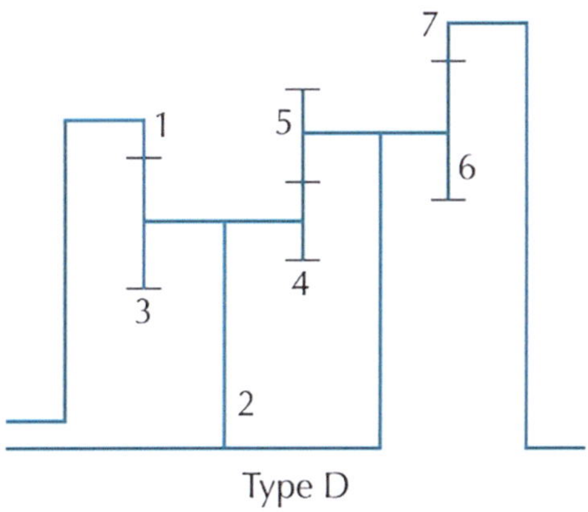

Fig. 7.25 Skeleton diagram of Lèvai type D gear train. Page 258 of Cleghorn and Dechev [2] © 2015 Oxford University Press (reproduced with permission)

Fig. 7.26 Lèvai type D gear train, Verrill [8]

References

1. Bartholomew, W. (2016). *Instructional learning aids for gear train analysis in machine dynamics*. Capstone Project Report, Department of Engineering, University of Southern Maine.
2. Cleghorn, W. L., & Dechev, N. (2015). *Mechanics of machines* (2nd ed.). Oxford University Press.
3. Davis, M. (2015). *Design and manufacture of a planetary gearset cutaway demonstration setup*. Capstone Project Report, Department of Engineering, University of Southern Maine.
4. Davis, M. & Ghorashi, M. (2016). A hybrid hands-on and simulation method for teaching the operation of automatic transmission systems. *ASEE northeastern section conference, Kingston, RI*. Awarded second place in the best paper award competition.
5. Davis, M., & Ghorashi, M. (2018). A holistic approach for deciphering the operation of planetary gear train systems. *ASEE's Journal of Engineering Technology, 35*, 16–25.
6. Palmer, M. (2013). *Design and manufacture of a mechanically driven clock*. Capstone Project Report, Department of Engineering, University of Southern Maine.
7. Scott, J. (2020). *Fine-tuning the design and manufacture of a mechanically driven clock*. Capstone Project Report, Department of Engineering, University of Southern Maine.
8. Verrill, A. (2018). *Modelling and manufacturing of a a Lèvai gear train type D*. Capstone Project Report, Department of Engineering, University of Southern Maine.
9. Waldron, K. J., & Kinzel, G. L. (2004). *Kinematics, dynamics, and design of machinery* (2nd ed.). Wiley.

Chapter 8
Helical, Worm, and Bevel Gears and Gear Trains

8.1 Introduction

In Chap. 8, as the last one of the three chapters dedicated to gears, more gear types and gear train types are introduced. They include helical, bevel, worm and worm gears. The geometry of gear teeth in these gears is more complicated than that of the spur gears and as a result more parameters need to be introduced in order to capture it.

Compared to spur gears, helical gears are less noisy. That is because in helical gears load transfer among gear teeth happens more gradually than in spur gears. Also helical gears have a higher load carrying capacity. In helical gear trains, the gear axes do not intersect; so they may be either parallel or skewed (i.e., they are not on the same plane). Bevel gears, worm and worm gears provide more options in terms of speed ratios and angles of the power transmission shafts with respect to each other. In bevel gear trains the shafts are intersecting, whereas in worm and worm gears they are skewed. An important application of bevel gears is their use in the differential. A differential allows wheels on the same axle in a car to spin at different speeds during a turn: the inner wheel turns slower and the outer wheel turns faster. Consequently, minimum tire slipping happens during a turn.

8.2 Helical Gears and Gear Trains

8.2.1 Geometry and Forces in Helical Gears

The teeth of a helical gear are similar to those of a spur gear, except they are cut at an angle called the **helix angle**, ψ (see Figs. 8.1 and 8.2). The helix angle of a helical gear is the angle between the gear teeth and the axis of the gear and it is similar to the **lead angle** of a screw (see Sect. 8.3). Helical gears can be **right-handed** or **left-**

M. Ghorashi, *Introduction to Machine Dynamics*, Mechanical Engineering Series,
https://doi.org/10.1007/978-3-031-94736-0_8

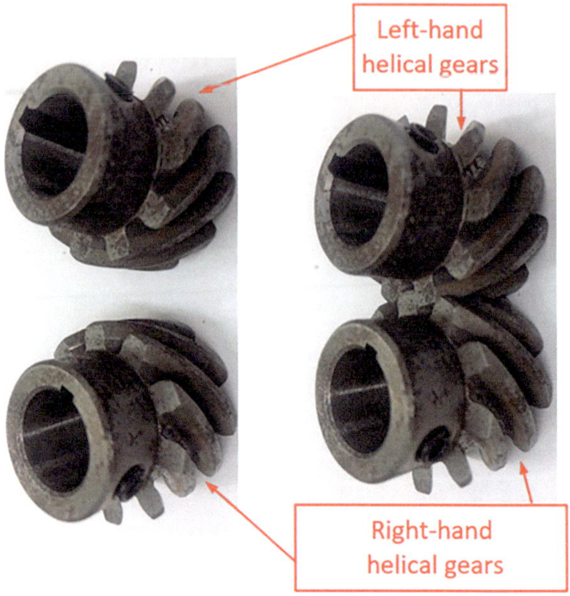

Fig. 8.1 Right-handed and left-handed helical gears

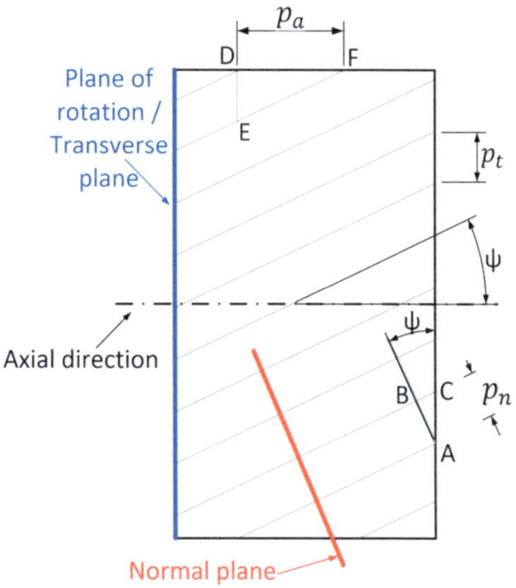

Fig. 8.2 The pitch cylinder of a left-handed helical gear, the transverse circular pitch, p_t, normal circular pitch, p_n, and axial pitch, p_a

handed (see Fig. 8.1). In Chap. 6, we saw that a pair of spur gears can be assembled only on parallel shafts. In helical gear trains the gear axes do not intersect so they may be either parallel or skewed (i.e., they are not on the same plane).

Figure 8.2 illustrates the pitch cylinder of a helical gear. It is the three-dimensional version of the pitch circle studied in Chap. 6. The pitch cylinder is an imaginary cylinder that provides a reference for defining tooth geometry. It is considered to roll without slipping on the pitch cylinder of a mating gear.

While a spur gear only has one type of circular pitch, shown by p, (see Chap. 6) depending on the direction of measurement, in a helical gear three pitch types are defined. These are **transverse circular pitch**, p_t, **normal circular pitch**, p_n, and the **axial pitch**, p_a, that are shown in Fig. 8.2. The transverse direction t refers to the tangential direction on the pitch cylinder in the plane of rotation (aka the transverse plane) of the gear. The normal direction n refers to the direction of cutting the gear teeth, and the axial direction a is along the axis of the gear. The axial pitch p_a is similar to the pitch of a screw and it is the distance between corresponding points on two adjacent gear teeth measured along the gear axis.

The equation of the transverse circular pitch, p_t, is similar to that of the circular pitch, p, in spur gears, i.e., Eq. (6.2):

$$p = \frac{2\pi R}{N} \tag{6.2}$$

So we have:

$$AC = p_t = \frac{2\pi R}{N} \tag{8.1}$$

in which R is the radius of the pitch cylinder. Also, considering the right triangle ABC in Fig. 8.2, it should be clear that the normal circular pitch can be expressed in terms of the transverse circular pitch and the helix angle as follows:

$$AB = p_n = p_t \cos\psi \tag{8.2}$$

Finally, in the right triangle DEF, the transverse circular pitch can be related to the axial pitch and the helix angle:

$$DE = p_t = p_a \tan\psi \tag{8.3}$$

Substituting Eq. (8.3) into Eq. (8.2) results in:

$$p_n = p_a \sin\psi \tag{8.4}$$

For the normal and the transverse diametral pitches (P_n and P_t, respectively) equations similar to Eq. (6.4):

$$Pp = \pi \tag{6.4}$$

can be written. Therefore we have:

$$P_t \cdot p_t = \pi \tag{8.5}$$

$$P_n \cdot p_n = \pi \tag{8.6}$$

Combining Eqs. (8.1) and (8.5) the **transverse diametral pitch** would be:

$$P_t = \frac{N}{2R} \tag{8.7}$$

And using Eqs. (8.1), (8.2), and (8.6), the **normal diametral pitch** can be expressed as follows:

$$P_n = \frac{N}{2R\cos\psi} \tag{8.8}$$

Comparing Eqs. (8.7) and (8.8) we get:

$$P_t = P_n \cos\psi \tag{8.9}$$

Figure 8.3 illustrates the direction of the contact force F_{23} applied by gear 2 on gear 3 and its three components, F_{23}^t, F_{23}^a, and F_{23}^r. As we see in this figure, helical gears have two pressure angles φ_t and φ_n. Each of these pressure angles refers to one of the main directions (t and n). In general, the transverse properties P_t and φ_t express the kinematics of helical gears, whereas the normal properties P_n and φ_n are used in stress analysis.

Using Fig. 8.3, we can obtain relations between the **transverse pressure angle** φ_t (measured on the plane of rotation), **normal pressure angle** φ_n (measured perpendicular to the tooth helix), and helix angle ψ. We see that transverse pressure angle φ_t and the helix angle ψ can be expressed as follows:

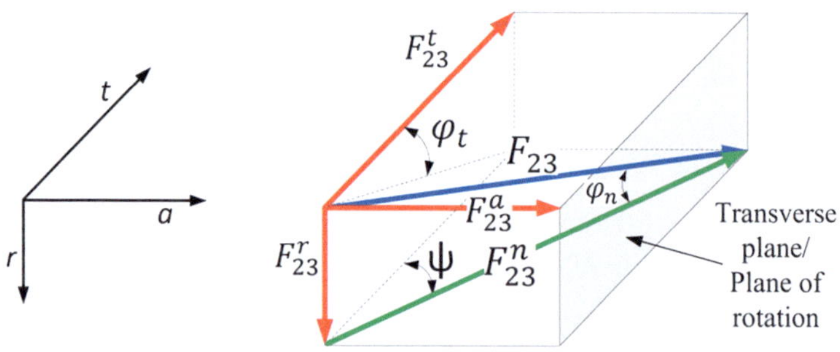

Fig. 8.3 Directions of the tangential, axial, and radial components of F_{23}, together with relations of the transverse pressure angle φ_t, normal pressure angle φ_n, and helix angle ψ

$$\tan \varphi_t = \frac{F_{23}^r}{F_{23}^t} \tag{8.10}$$

$$\tan \psi = \frac{F_{23}^a}{F_{23}^t} \tag{8.11}$$

These two angles, φ_t and ψ, are related to the normal pressure angle φ_n in the following way,

$$\tan \varphi_n = \frac{F_{23}^r}{F_{23}^n} = \frac{F_{23}^r}{F_{23}^t} \times \frac{F_{23}^t}{F_{23}^n} = \tan \varphi_t . \cos \psi \tag{8.12}$$

8.2.2 *Parallel-Axis Helical Gear Trains*

Figure 8.4 illustrates parallel-axis helical gears. For proper mesh, these gears should have:

1. the same helix angles, ψ;
2. opposite hands;
3. the same normal diametral pitch values (since they have the same helix angles, considering Eq. (8.9), they will also have the same transverse diametral pitch values).

In general, if the helix angles of the two gears are ψ_2 and ψ_3 and they have opposite hands, the **shaft intersection angle** Σ would be:

Fig. 8.4 Helix angle, ψ, in helical gears in mesh on parallel shafts in a power transmission gear train

$$\Sigma = \psi_2 - \psi_3 \tag{8.13}$$

For the stated case of same helix angles with opposite hands, one can see that Eq. (8.13) results in $\Sigma = 0$ which means the shafts are indeed parallel, as we see in Fig. 8.4.

If gear 2 is the driver and gear 3 is the driven, the **general speed ratio equation** is:

$$\frac{\omega_{driven,out}}{\omega_{driver,in}} = \frac{\omega_3}{\omega_2} = \frac{N_2}{N_3}$$

Recalling,

$$P_n = \frac{N}{2R\cos\psi} \tag{8.8}$$

And calculating the ratio of N's in terms of the ratio of R's using Eq. (8.8), the speed ratio can also be written as follows (we notice that the gears have the same normal diametral pitch values, P_n, so it cancels out.)

$$\frac{\omega_{driven,out}}{\omega_{driver,in}} = \frac{\omega_3}{\omega_2} = \frac{N_2}{N_3} = \frac{R_2\cos\psi_2}{R_3\cos\psi_3}$$

For parallel shafts, the helix angles ψ_2 and ψ_3 are equal, so this equation reduces to the following **speed ratio equation for parallel shafts**:

$$\frac{\omega_{driven,out}}{\omega_{driver,in}} = \frac{\omega_3}{\omega_2} = \frac{N_2}{N_3} = \frac{R_2}{R_3} \tag{8.14}$$

And for the **center distance** we have:

$$C = R_2 + R_3 \tag{8.15}$$

Now, substituting the transverse diametral pitch as given in Eq. (8.7),

$$P_t = \frac{N}{2R} \tag{8.7}$$

into Eq. (8.15) gives:

$$C = \frac{N_2}{2P_t} + \frac{N_3}{2P_t} = \frac{N_2 + N_3}{2P_t} \tag{8.16}$$

Finally, substituting Eq. (8.9),

$$P_t = P_n \cos \psi \tag{8.9}$$

into Eq. (8.16) we get:

$$C = \frac{N_2 + N_3}{2P_n \cos \psi} \tag{8.17}$$

In spur gears, as we see in Fig. 6.9, the transmission force between gears generates radial and tangential forces on the gears, but no axial force. In helical gears, as Fig. 8.5 illustrates, due to the helical shape of gear teeth some axial force is also generated. This force should be supported by thrust, ball, or tapered roller bearings that hold the shafts. Alternatively, one can use another helical gear with opposite hand on the same shaft so that its axial load offsets the axial load created by the other helical gear. The same philosophy is used in Herringbone gears that are compound gears each made of two segments with opposite hands. Therefore, the net axial force (thrust component) in a Herringbone gear is zero.

The direction of application of axial forces on a pair of helical gears depends on the hands of the gears, their rotation directions, as well as on which gear is the driver and which one is the driven. As we can see in Figs. 8.3 and 8.5, the total contact force applied by gear 2 on gear 3, named F_{23}, has three components: F_{23}^t, which is tangent to the pitch cylinder shown in Fig. 8.2, F_{23}^a, parallel to the gear axis, and F_{23}^r which is the radial component. Among these components, in Fig. 8.5, two components, i.e.,

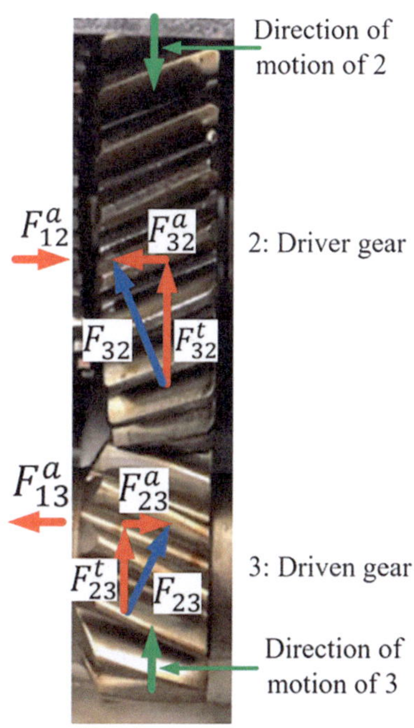

Fig. 8.5 Directions of tangential and axial forces on two helical gears in mesh

F^t_{23} and F^a_{23} are shown. It should be noted that forces F_{23} and F_{32} act at the contact point, but in this figure, they are rotated by 90°, otherwise F^t_{23} and F^t_{32} would be perpendicular to the plane of the figure and could not be seen! The axial force, F^a_{23}, applied by gear 2 on 3 is balanced by the axial force F^a_{13} applied by the bearings that support the shaft of gear 3, as seen in Fig. 8.5.

A similar analysis can be done for the reaction force F_{32} applied by gear 3 on 2 and its components F^t_{32}, F^a_{32}, and F^r_{32}. Again, the axial force F^a_{32} applied by gear 3 on 2 is balanced by the axial force F^a_{12} applied by the bearings that support the shaft of gear 2.

Example 8.1 Design two helical gears to be used on parallel shafts the way we see in Fig. 8.4. The shaft centers are to be about 4 in apart. The normal diametral pitch of the gears is $P_n = 8$ teeth/in, and the gears are supposed to have a normal pressure angle of $\varphi_n = 20°$. Consider a helix angle of $\psi = 25°$ and a speed ratio of 0.5. Calculate:

A. the transverse pressure angle φ_t,
B. the normal circular pitch, p_n,
C. the transverse circular pitch, p_t,
D. the transverse diametral pitch, P_t,
E. the proper gear tooth numbers, N_2 and N_3,
F. The radii of the pitch cylinders, R_2 and R_3,
G. The realized center distance C and
H. The realized speed ratio.

A. **The transverse pressure angle φ_t.**

Recalling,

$$\tan\,\varphi_n = \tan\varphi_t.\cos\psi \tag{8.12}$$

we get:

$$\tan\varphi_t = \frac{\tan\,\varphi_n}{\cos\psi} = \frac{\tan\,20°}{\cos 25°} = 0.4016 \longrightarrow \varphi_t = 21.88°$$

B. **The normal circular pitch, p_n**

Recalling,

$$P_n.p_n = \pi \tag{8.6}$$

we get:

$$p_n = \frac{\pi}{P_n} = \frac{\pi}{8} = 0.3927\ \text{in/tooth}$$

C. **The transverse circular pitch, p_t**

For the transverse circular pitch, p_t we have:

$$p_n = p_t \cos \psi \tag{8.2}$$

$$p_t = \frac{p_n}{\cos \psi} = \frac{0.3927}{\cos 25°} = 0.4333 \text{ in/tooth}$$

D. **The transverse diametral pitch, P_t**

Considering,

$$P_t . p_t = \pi \tag{8.5}$$

$$P_t = \frac{\pi}{p_t} = \frac{\pi}{0.4333} = 7.250 \text{ teeth/in}$$

E. **The proper gear tooth numbers**

Recalling,

$$\frac{\omega_{driven,out}}{\omega_{driver,in}} = \frac{\omega_3}{\omega_2} = \frac{N_2}{N_3} = \frac{R_2}{R_3} \tag{8.14}$$

and the given speed ratio we get,

$$\frac{\omega_3}{\omega_2} = \frac{R_2}{R_3} = 0.5 \longrightarrow \mathrm{R}_2 = 0.5\mathrm{R}_3$$

By using this result and the given center distance we have:

$$C = R_2 + R_3 = 4 \text{ in} \longrightarrow \mathrm{R}_3 = 2.667 \text{ in and } \mathrm{R}_2 = 1.333 \text{ in}$$

Now, recalling,

$$P_t = \frac{N}{2R} \tag{8.7}$$

we get,

$$N = 2RP_t \longrightarrow \mathrm{N}_2 = 2 \times 1.333 \text{ in} \times 7.250 = 19.33 \text{ teeth}$$

For continuous operation, we need to choose round numbers. So we choose: $N_2 = 19$ teeth. Similarly we have:

$$N_3 = 2 \times 2.667 \text{ in} \times 7.250 = 38.67 \text{ teeth}$$

and we choose: $N_3 = 39$ teeth.

F. **The radii of the pitch cylinders**

Using the chosen number of teeth we recalculate the radii of the pitch cylinders:

$$P_t = \frac{N}{2R} \tag{8.7}$$

So,

$$R = \frac{N}{2P_t} \rightarrow R_2' = \frac{19}{2 \times 7.250} = 1.310 \text{ in}$$

Similarly,

$$R_3' = \frac{39}{2 \times 7.250} = 2.690 \text{ in}$$

G. **The realized center distance**

The realized center distance of gears is:

$$C' = R_2' + R_3' = 4.0 \text{ in}$$

This center distance value is exactly equal to what we wanted it to be.

H. **The realized speed ratio**

Now that we know the teeth numbers, we can calculate the realized speed ratio as follows:

$$\frac{\omega_3}{\omega_2} = \frac{N_2}{N_3} = \frac{19}{39} = 0.4872$$

This result is close to the design objective of 0.5.

8.2.3 Crossed-Axis Helical Gear Trains

Helical gears can be installed for power transmission between non-parallel, non-intersecting (skewed) shafts (see Fig. 8.6). In crossed-axis helical gear trains, shaft axes are not parallel and same-hand gears are able to mesh. Figure 8.6 illustrates crossed-axis helical gear trains and the relations of the gear hands, directions of rotation, and the resulting thrust (axial force). For proper meshing, such **gears should have the same normal diametral pitch values**. However, the transverse diametral pitches and the helix angles can be different.

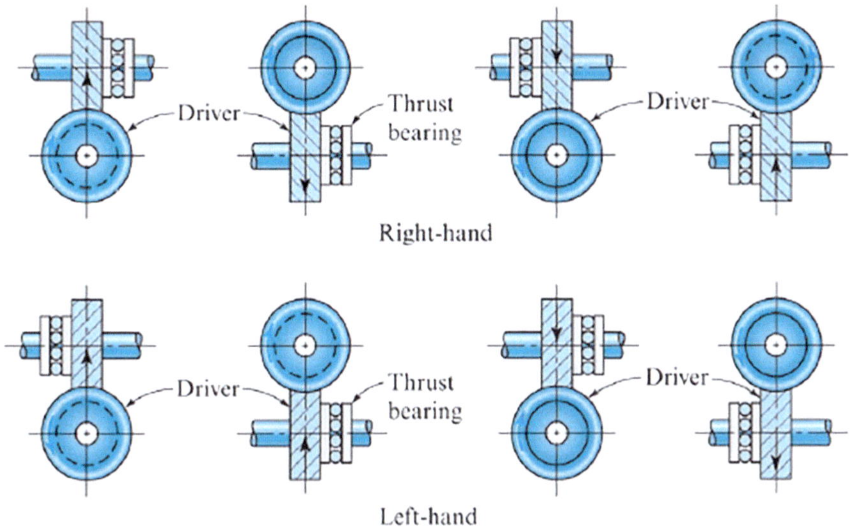

Fig. 8.6 Crossed-axis helical gear trains and the relations of the gear hands, directions of rotation, choice of driver, and the resulting thrust. Page 434 of Uicker et al. [1] © 2016 Oxford University Press (reproduced with permission)

If the angle between the two shafts is 90°, the shafts could both have right-hand gears with 45° helix angles. In general, if the helix angles of the two gears are ψ_2 and ψ_3 and they have **same hands**, their corresponding **shaft intersection angle** Σ will be,

$$\Sigma = \psi_2 + \psi_3 \tag{8.18}$$

For the stated case of 45° helix angles with the same hands, Eq. (8.18) results in $\Sigma = 90°$ which means the shafts are perpendicular to each other (but not intersecting). This is the common use of crossed-axis helical gear trains. Equation (8.18) is the same-hand version of Eq. (8.13) that is valid for gears having opposite hands (as was the case for the parallel shafts of Fig. 8.4).

If gear 2 is the driver and gear 3 is the driven, **the general speed ratio** equation is:

$$\frac{\omega_{driven,out}}{\omega_{driver,in}} = \frac{\omega_3}{\omega_2} = \frac{N_2}{N_3}$$

Recalling,

$$P_n = \frac{N}{2R\cos\psi} \tag{8.8}$$

And calculating the ratio of N's in terms of the ratio of R's using Eq. (8.8), the **speed ratio** can also be written as follows (we notice that the gears have the same normal diametral pitch, P_n, values so it cancels out).

$$\frac{\omega_{driven,out}}{\omega_{driver,in}} = \frac{\omega_3}{\omega_2} = \frac{N_2}{N_3} = \frac{R_2 \cos\psi_2}{R_3 \cos\psi_3} \tag{8.19}$$

It is emphasized that in crossed-axis helical gear trains the speed ratio is not equal to just $\frac{R_2}{R_3}$ as it was in the case of spur gears! However, if the helix angles ψ_2 and ψ_3 are equal (as they are for the 45° helix angles with the same hands and $\Sigma = 90°$), then Eq. (8.19) reduces to Eq. (8.14) that was obtained for the case of parallel shafts.

$$\frac{\omega_{driven,out}}{\omega_{driver,in}} = \frac{\omega_3}{\omega_2} = \frac{N_2}{N_3} = \frac{R_2}{R_3} \tag{8.14}$$

Also, for the **center distance** of the two crossed-axis helical gears we have:

$$C = R_2 + R_3 \tag{8.20}$$

Calculating R's using Eq. (8.8) and substituting the results into Eq. (8.20) gives:

$$C = \frac{N_2}{2P_n \cos\psi_2} + \frac{N_3}{2P_n \cos\psi_3} = \frac{1}{2P_n}\left(\frac{N_2}{\cos\psi_2} + \frac{N_3}{\cos\psi_3}\right) \tag{8.21}$$

If $\psi_2 = \psi_3 = \psi$, the center distance Eq. (8.21) reduces to the following that is the same as Eq. (8.17) obtained for parallel-axis helical gear trains:

$$C = \frac{N_2 + N_3}{2P_n \cos\psi} \tag{8.22}$$

Crossed-axis helical gears have limited contact surface and therefore they cannot be used for transmitting large power values (unlike parallel-axis helical gear trains). When transmitting large amounts of power between intersecting axes one can use bevel gears. For high gear ratios between non-intersecting (skewed) axes we can use worm and worm gears.

Correct **locations of thrust bearings** in crossed-axis helical gears are shown in Fig. 8.6. and they depend on whether a gear is driver or driven, what its hand is (right or left), and its direction of rotation.

Example 8.2 A left-handed helical gear with 15 teeth, normal diametral pitch of $P_n = 8$ teeth/in and a helix angle of $\psi_2 = 75°$ is to drive another gear. The design objective is to have a gear ratio of 4 and a shaft intersection angle of $\Sigma = 90°$. Calculate:

A. the helix angle of the mating gear and its hand,
B. the number of teeth on the mating gear,

C. the radii of the pitch cylinders, R_2 and R_3,
D. the realized center distance C, and,
E. the realized speed ratio.

A. **The helix angle of the mating gear and its hand.**

Recalling, Eq. (8.18) for shaft intersection angle when gears have the same hand,

$$\Sigma = \psi_2 + \psi_3 \tag{8.18}$$

we get:

$$90^\circ = 75^\circ + \psi_3$$

Therefore,

$$\psi_3 = 15^\circ$$

And the mating gear is left-handed as well.

B. **The number of teeth on the mating gear.**

As we learned in Sect. 7.3, gear ratio is the inverse of speed ratio. We know that the gear ratio is 4. Therefore, the speed ratio in this problem is 0.25. Now, recalling the following speed ratio equation,

$$\frac{\omega_{driven,out}}{\omega_{driver,in}} = \frac{\omega_3}{\omega_2} = \frac{N_2}{N_3} = \frac{R_2 \cos \psi_2}{R_3 \cos \psi_3} \tag{8.19}$$

we get,

$$0.25 = \frac{15}{N_3} \longrightarrow \mathrm{N}_3 = 60$$

C. **The radii of the pitch cylinders.**

Recalling,

$$P_n = \frac{N}{2R \cos \psi} \tag{8.8}$$

if we use it for the driving gear 2, we get:

$$8 = \frac{15}{2R_2 \cos 75^\circ} \longrightarrow \mathrm{R}_2 = 3.6222 \text{ in}$$

Similarly, for the driven gear 3 we have:

$$8 = \frac{60}{2R_3 \cos 15\,°} \rightarrow R_3 = 3.8823 \text{ in}$$

D. **The realized center distance.**

Having already calculated the radii of the pitch cylinders, we can now use Eq. (8.20) to get the shaft center distance:

$$C = R_2 + R_3 \tag{8.20}$$

Therefore,

$$C = R_2 + R_3 = 3.6222 + 3.8823 = 7.5045 \text{ in}$$

E. **The realized speed ratio.**

Recalling Eq. (8.19) for the speed ratio,

$$\frac{\omega_{driven,out}}{\omega_{driver,in}} = \frac{\omega_3}{\omega_2} = \frac{N_2}{N_3} = \frac{R_2 \cos \psi_2}{R_3 \cos \psi_3} \tag{8.19}$$

we already know that $\frac{N_2}{N_3} = 0.25$. Let us verify that the last segment of Eq. (8.19) provides the same result for speed ratio:

$$\frac{R_2 \cos \psi_2}{R_3 \cos \psi_3} = \frac{3.6222 \ \cos 75\,°}{3.8823 \cos 15\,°} = 0.2499$$

This result is compatible with our previous calculations and the stated gear ratio design requirement.

8.2.4 *Minimum Number of Gear Teeth to Prevent Interference*

Recall Eq. (6.35) from the discussion of spur gears:

$$a = R \sin^2 \varphi \tag{6.35}$$

In helical gears the same equation can be written, but on the transverse plane. Therefore we have,

$$a = R \sin^2 \varphi_t \tag{8.23}$$

Also on the normal plane (i.e., the gear cut direction) we have:

$$a = \frac{1}{P_n} \tag{8.24}$$

Equating Eqs. (8.23) and (8.24) one obtains:

$$\frac{1}{P_n} = R \sin^2 \varphi_t \tag{8.25}$$

Now, recalling Eq. (8.7) for the transverse diametral pitch,

$$P_t = \frac{N}{2R} \tag{8.7}$$

one can eliminate R between Eqs. (8.7) and (8.25) to get:

$$\frac{N}{2P_t} = \frac{1}{P_n \sin^2 \varphi_t}$$

Solving it for the number of gear teeth, N gives,

$$N = \frac{2P_t}{P_n \sin^2 \varphi_t} \tag{8.26}$$

Combining Eq. (8.26) and Eq. (8.9), i.e.,

$$P_t = P_n \cos \psi \tag{8.9}$$

results in,

$$N = \frac{2 \cos \psi}{\sin^2 \varphi_t} \tag{8.27}$$

For a continuously rotating gear the value of N calculated from Eq. (8.27) should be an integer, and to make sure interference does not happen, the values of N calculated from this equation should be rounded up. To calculate the minimum teeth number, N_{min}, that ensures no interference, we choose various helix angles ψ and standard values for the normal pressure angle φ_n (i.e., 14.5°, 20°, and 25°). For each case, using the chosen values of ψ and φ_n, the corresponding φ_t value is calculated using Eq. (8.12):

$$\tan \varphi_n = \tan \varphi_t . \cos \psi \tag{8.12}$$

And the minimum teeth number, N_{min}, that ensures no interference is obtained by using Eq. (8.27). The results of these calculations are given in Table 8.1. These

Table 8.1 Variation of N_{min} (the minimum acceptable number of full-depth gear teeth to prevent interference) with helix angle ψ and normal pressure angle, φ_n

Ψ		0°	20°	40°	60°	80°
φ_n	14.5°	32	27	15	5	1
	20°	18	15	9	3	1
	25°	12	10	6	3	1

Fig. 8.7 A worm and worm gear assembly ($N_2 = 5$, $N_3 = 75$)

results are applicable to individual helical gears regardless of the axis orientation of the meshing gear, Waldron and Kinzel [2], Page 503.

A review of Table 8.1 reveals that by increasing the normal pressure angle φ_n, we could build fewer gear teeth and still avoid interference! That is one of the reasons why we seldom use the 14.5° pressure angle anymore. Also by increasing the helix angle, the minimum number of gear teeth with no interference eventually reduces to one. This is the case of a single-threaded worm that will be discussed later in Sect. 8.3. Finally, comparison of Table 8.1 with Table 6.2 reveals that at zero helix angle (i.e., the case of spur gears) these tables provide identical N_{min} values, as expected.

8.3 Worms and Worm Gears

Figure 8.7 illustrates a **worm and worm gear mechanism**. Many of the geometrical characteristics of worm and worm gear mechanisms are a combination of those of crossed-axis helical gear trains on one hand and screws on the other hand. Usually, in a worm and worm gear mechanism the input shaft runs the worm. In Fig. 8.7, the helix angle between the worm threads (yes, threads like in screws, not teeth in gears)

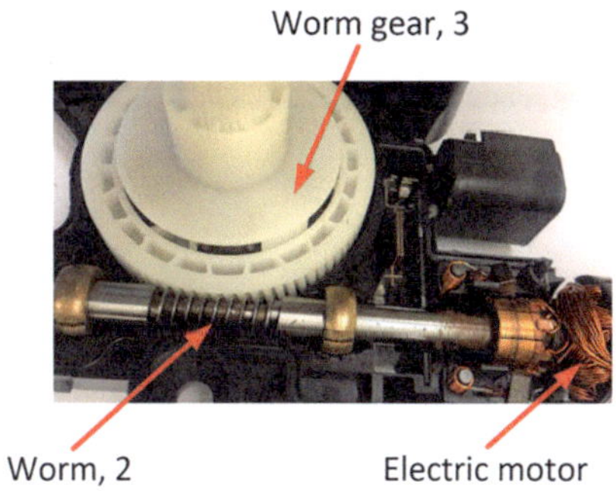

Fig. 8.8 Worm and worm gear mechanism used for raising or lowering a car window

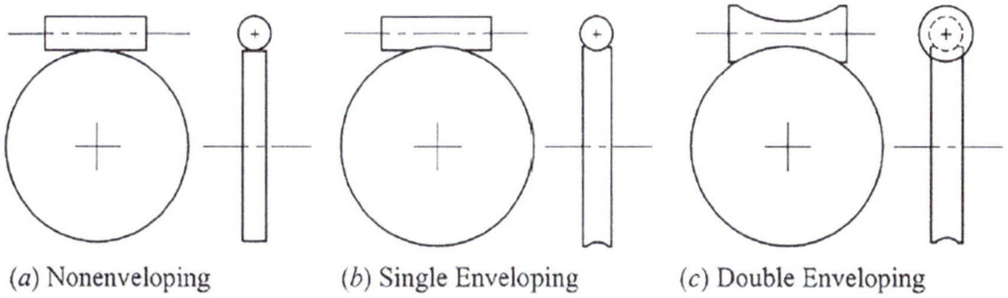

Fig. 8.9 Enveloping types in worm and worm gear mechanisms. Page 513 of Waldron and Kinzel [2] © 2004 John Wiley & Sons, Inc. (reproduced with permission)

and the axis of the worm is large. As a matter of fact, in worms the helix angle can be higher than 80°. The input shaft is perpendicular to the output shaft, that is run by the worm gear. In a worm and worm gear mechanism the two shafts do not intersect (they are skewed) and importantly, the mechanism normally cannot be back driven or run by switching the input and output elements (i.e., the worm gear cannot run the worm—it is **self-locking**). We will talk about this self-locking property at the end of this section.

Worm and worm gear mechanisms are used in applications where large speed reductions and torque increases are needed in a single stage (i.e., high gear ratio and torque ratio are the objectives). For example, Fig. 8.8 illustrates a worm and worm gear mechanism that is operated by a small low-torque high-speed electric motor in the input. The resulting increased torque in the output is suitable for raising or lowering a car window at reduced speed. The self-locking property of the worm and worm gear mechanism prevents that a person could force open a car window by pushing it down from outside of the car.

Figure 8.9 illustrates schematically the three types of worm and worm gear contact design. When the contact is nonenveloping (see Fig. 8.9a), the worm and worm gear have point contact. As a result, in this case only relatively small forces can be transmitted. In the single-enveloping design shown in Fig. 8.9b, the worm gear is cut to partially envelop the worm. The single-enveloping design can also be seen in Fig. 8.7. In the double-enveloping design shown in Fig. 8.9c, the worm is

Fig. 8.10 An octuple-threaded worm ($N_2 = 8$) with eight helix starts

also cut in the shape of an hour glass and meshed with a worm gear that also partially envelops the worm. The single- and double-enveloping worm and worm gear sets have line contact and as a result can transmit considerably more power than the nonenveloping worm gear set, Waldron and Kinzel [2], Page 513.

A **worm** is essentially a screw and its teeth are usually named as threads. Like other screws, a worm can be **single-threaded**, **double-threaded**, or **multiple-threaded**. The thread number is equal to the number of helix starts, as can be seen in Fig. 8.10 that shows an octuple-threaded worm ($N_2 = 8$) with eight helix starts.

A **worm gear** is essentially a helical gear, but as we saw in Fig. 8.9b and c, the worm gear can have a concave face so that it partially wraps around the worm. This feature increases the contact area between worm and worm gear. As a result, higher power can be transmitted.

Considering Fig. 8.7, one notices that worm and worm gear mechanism is similar to a crossed-axis helical gear train. Therefore, if worm and worm gear have the same hands, the shaft intersection angle is,

$$\Sigma = \psi_2 + \psi_3 \tag{8.18}$$

And considering that the shaft intersection angle in worm and gear mechanisms is 90°, we have,

$$\psi_2 + \psi_3 = 90\,^\circ \tag{8.28}$$

Usually the helix angle of worm is large and around 80 ° to 85°. Fig. 8.11 illustrates an octuple-threaded rod and a few of its characteristics, i.e., the axial pitch p_a, helix angle ψ_2, lead l, and lead angle λ. Considering Fig. 8.11, we have,

Fig. 8.11 An octuple-threaded rod ($N_2 = 8$)

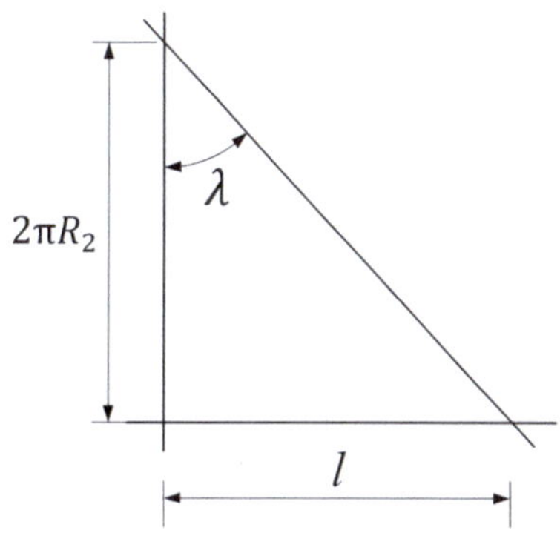

Fig. 8.12 Unwrapped view of one turn of a thread helix of Fig. 8.11

$$\psi_2 + \lambda = 90^\circ \tag{8.29}$$

Comparing Eq. (8.28) and Eq. (8.29) one can conclude that the **lead angle of the worm, λ, is equal to the helix angle of worm gear, ψ_3,**

$$\lambda = \psi_3 = 90^\circ - \psi_2 \tag{8.30}$$

The **lead of the worm**, l, is the axial movement of the worm if it rotates one turn in a receptive nut. Referring to Fig. 8.11, one concludes that the lead is equal to the thread number of the worm, N_2, multiplied by the axial pitch, p_a

$$l = N_2 p_a \tag{8.31}$$

Equation (8.31) is indeed identical to Eq. (2.6) discussed in Sect. 2.3.1. The relation between the lead l, and the **lead angle λ**, can be best observed in the unwrapped view of one turn of a thread helix of Fig. 8.11 as shown in Fig. 8.12; from which we have,

$$\tan \lambda = \frac{l}{2\pi R_2} \tag{8.32}$$

where R_2 is the pitch cylinder radius of the worm.

For proper assembly, and since worm and worm gear axes are perpendicular to each other, **the transverse circular pitch, p_t of the worm gear should be equal to the axial pitch of the worm, p_a**,

$$p_{t,3} = p_{a,2} \tag{8.33}$$

Substituting Eq. (8.1) for the definition of the transverse circular pitch, p_t, and Eq. (8.31) into Eq. (8.33) one obtains,

$$\frac{2\pi R_3}{N_3} = \frac{l}{N_2}$$

or,

$$\frac{N_2}{N_3} = \frac{l}{2\pi R_3} \tag{8.34}$$

This ratio is nothing but the speed ratio,

$$\frac{\omega_3}{\omega_2} = \frac{N_2}{N_3} \tag{8.35}$$

As an example, in the mechanism shown in Fig. 8.7, $N_2 = 5$ and $N_3 = 75$; therefore, the speed ratio is: 5/75 = 1/15. Now, comparing Eqs. (8.34), (8.35), and (8.19) we get,

$$\frac{\omega_3}{\omega_2} = \frac{N_2}{N_3} = \frac{R_2 \cos \psi_2}{R_3 \cos \psi_3} = \frac{l}{2\pi R_3} \tag{8.36}$$

It should be reminded that **N_2 is the thread number of the worm** and **N_3 is the number of gear teeth of the worm gear**.

Now, substituting Eq. (8.30) into Eq. (8.36) we obtain,

$$\frac{\omega_3}{\omega_2} = \frac{R_2 \sin \lambda}{R_3 \cos \lambda} = \frac{R_2}{R_3} \tan \lambda \tag{8.37}$$

Finally, using Eq. (8.32) and Eq. (8.37) one obtains:

$$\frac{\omega_3}{\omega_2} = \frac{l}{2\pi R_3} \tag{8.38}$$

Example 8.3 In a worm and worm gear application, the number of gear teeth of the worm gear is 50 and the worm is double-threaded. Calculate the speed ratio.

Using Eq. (8.35), one obtains,

$$\frac{\omega_3}{\omega_2} = \frac{N_2}{N_3} = \frac{2}{50} = \frac{1}{25}$$

Example 8.4 A worm and worm gear mechanism has a triple-thread worm and a 30-teeth worm gear. The worm has an axial pitch of 0.5 in and a pitch diameter of 1.25 in. Calculate:

A. the lead of the worm;
B. the lead angle of the worm;
C. the helix angle of the worm gear;
D. the pitch diameter of the worm gear.

A. **The lead of the worm.**

Recalling:

$$l = N_2 p_a \tag{8.31}$$

And using the provided data we get:

$$l = 3 \times 0.5 = 1.5 \text{ in}$$

B. **The lead angle of the worm.**

For the lead angle we have:

$$\tan\lambda = \frac{l}{2\pi R_2} \tag{8.32}$$

Therefore, in this case we have:

$$\tan\lambda = \frac{1.5}{1.25\pi} \longrightarrow \lambda = 20.905^\circ$$

C. **Helix angle of the worm gear.**

We know that the lead angle of the worm, λ, is equal to the helix angle of worm gear, ψ_3:

$$\lambda = \psi_3 = 90^\circ - \psi_2 \tag{8.30}$$

Therefore,

$$\psi_3 = 20.905^\circ$$

D. **The pitch diameter of the worm gear.**

Recalling,

$$\frac{\omega_3}{\omega_2} = \frac{N_2}{N_3} = \frac{R_2 \cos \psi_2}{R_3 \cos \psi_3} = \frac{l}{2\pi R_3} \tag{8.36}$$

We have:

$$\frac{3}{30} = \frac{1.5}{2\pi R_3}$$

Therefore, the pitch diameter of the worm gear is obtained as follows:

$$2R_3 = 4.775 \text{ in}$$

Being **self-locking** in an application like controlling the rudder of a ship, assures that rough seas cannot affect the rudder angle! For having the self-locking property, the coefficient of friction between the worm and the worm gear, μ, should be larger than cos φ_n tan λ, see page 813 of Budynas and Nisbett [3]. So for self-locking we should have:

$$\mu > \cos \phi_n \tan \lambda \tag{8.39}$$

It can be seen that for small lead angle values (i.e., large values of the helix angle of the worm) like what we see in Fig. 8.7, the self-locking condition (8.39) is likely to be satisfied. However, it is possible that a mechanism that was designed to be self-locking loses this property because of poor maintenance. A greasy worm and worm gear mechanism has a reduced coefficient of friction that may no longer satisfy condition (8.39). In this case, the mechanism would not be self-locking anymore.

8.4 Bevel Gears

We have seen that spur gears, helical gears, and worm and worm gears all are used for power transmission between shafts that do not intersect. **Straight-tooth bevel gears** (see Figs. 8.13 and 8.14) are the ones that can be used for power transmission between shafts whose axes do intersect. Bevel gears are usually installed at the free end of a shaft in a cantilever form (see Fig. 8.18). The pitch surfaces of bevel gears are cones and as we see in Fig. 8.14, the two gears' apexes are coincident.

Usually the **radius of a bevel gear** is given at its large end (see R_2 and R_3 in Fig. 8.14). Also the summation of the two **pitch cone angles** γ_2 and γ_3 is equal to the shaft intersection angle, Σ:

$$\Sigma = \gamma_2 + \gamma_3 \tag{8.40}$$

Fig. 8.13 A pair of straight-tooth bevel gears. Page 437 of Uicker et al. [1] © 2016 Oxford University Press (reproduced with permission)

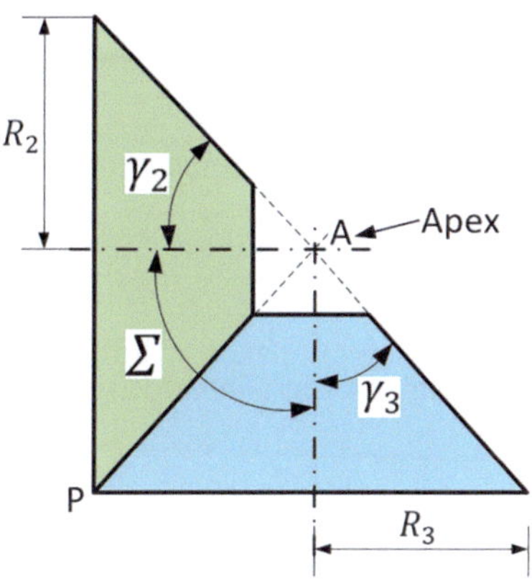

Fig. 8.14 Basic geometry of meshed straight-tooth bevel gears

In Fig. 8.14, if gear 2 is the driver and gear 3 is the driven one, the speed ratio can be written as follows,

$$\frac{\omega_{driven,out}}{\omega_{driver,in}} = \frac{\omega_3}{\omega_2} = \frac{N_2}{N_3} = \frac{R_2}{R_3} \tag{8.41}$$

The **AGMA standard** recommends 20° pressure angle for bevel gears with 14 or more teeth and 25° pressure angle for gears with 13 or fewer teeth to avoid interference and undercutting, see the ANSI/AGMA ISO 23509-B17 standard. If the number of gear teeth N_2 and N_3 (or just the speed ratio) and the shaft intersection angle, Σ are known, one can calculate their corresponding pitch cone angles γ_2 and γ_3. To this end, and by referring to Fig. 8.14, we can write sine equations for the right triangles built in each of the gears as follows:

$$\sin\gamma_2 = \frac{R_2}{AP};\ \sin\gamma_3 = \frac{R_3}{AP}$$

Dividing these two equations results in,

$$\frac{\sin\gamma_2}{\sin\gamma_3} = \frac{R_2}{R_3} \tag{8.42}$$

Substituting γ_3 from Eq. (8.40) into Eq. (8.42), one obtains,

$$\frac{\sin\gamma_2}{\sin(\Sigma - \gamma_2)} = \frac{R_2}{R_3} \tag{8.43}$$

or,

$$\frac{\sin\gamma_2}{\sin\Sigma\cos\gamma_2 - \sin\gamma_2\cos\Sigma} = \frac{R_2}{R_3}$$

Dividing the numerator and the denominator by $\cos\gamma_2$ gives,

$$\frac{\tan\gamma_2}{\sin\Sigma - \tan\gamma_2\cos\Sigma} = \frac{R_2}{R_3}$$

Using, the angular velocity Eq. (8.41) we can conclude:

$$\frac{\omega_{driven,out}}{\omega_{driver,in}} = \frac{\omega_3}{\omega_2} = \frac{N_2}{N_3} = \frac{R_2}{R_3} = \frac{\tan\gamma_2}{\sin\Sigma - \tan\gamma_2\cos\Sigma} \tag{8.44}$$

Equation (8.44) implicitly provides the pitch cone angle γ_2 in terms of the ratio of the gear teeth numbers N_2 and N_3 (i.e., the speed ratio) and the shaft intersection angle, Σ. After cross multiplication, Eq. (8.44) can be rewritten as follows:

$$(N_3 + N_2\cos\Sigma)\tan\gamma_2 = N_2\sin\Sigma$$

By dividing this equation by N_2, one could get an explicit equation for the pitch cone angle, γ_2 as follows:

$$\tan\gamma_2 = \frac{\sin\Sigma}{\frac{N_3}{N_2} + \cos\Sigma} \tag{8.45}$$

Similarly, for the pitch cone angle, γ_3 one obtains,

$$\tan \gamma_3 = \frac{\sin \Sigma}{\frac{N_2}{N_3} + \cos \Sigma} \tag{8.46}$$

If the shaft intersection angle, Σ, is 90°, i.e., the shafts are perpendicular to each other, then Eqs. (8.45) and (8.46) reduce to,

$$\tan \gamma_2 = \frac{N_2}{N_3}; \ \tan \gamma_3 = \frac{N_3}{N_2} \tag{8.47}$$

Example 8.5 Two straight-tooth bevel gears are to be assembled on intersecting shafts that are perpendicular to each other ($\Sigma = 90°$). The gear ratio is supposed to be 4 and the driver gear has $N_2 = 20$ teeth. Calculate the pitch angles of the two gears.

The speed ratio is inverse of gear ratio, so it is 0.25. Recalling the speed ratio Eq. (8.41):

$$\frac{\omega_{driven,out}}{\omega_{driver,in}} = \frac{\omega_3}{\omega_2} = \frac{N_2}{N_3} = \frac{R_2}{R_3} \tag{8.41}$$

we have:

$$\frac{N_2}{N_3} = 0.25 \rightarrow N_3 = 80 \text{ teeth}$$

Since the shaft intersection angle Σ is 90° we have,

$$\tan \gamma_2 = \frac{N_2}{N_3}; \ \tan \gamma_3 = \frac{N_3}{N_2} \tag{8.47}$$

$$\tan \gamma_2 = \frac{N_2}{N_3} \rightarrow \tan \gamma_2 = 0.25 \rightarrow \gamma_2 = 14.04\,°$$

Similarly,

$$\Sigma = \gamma_2 + \gamma_3 \tag{8.40}$$

Therefore,

$$90\,° = 14.04\,° + \gamma_3 \rightarrow \gamma_3 = 75.96\,°$$

Figure 8.15 illustrates a **flat face bevel gear** that is also called a **crown gear**; together with its bevel pinion. (Sometimes, in a gear mesh the smaller gear is called pinion and the bigger one is just called gear.) The crown gear has a pitch cone angle of 90° and is the bevel gear equivalent of a rack in spur gears.

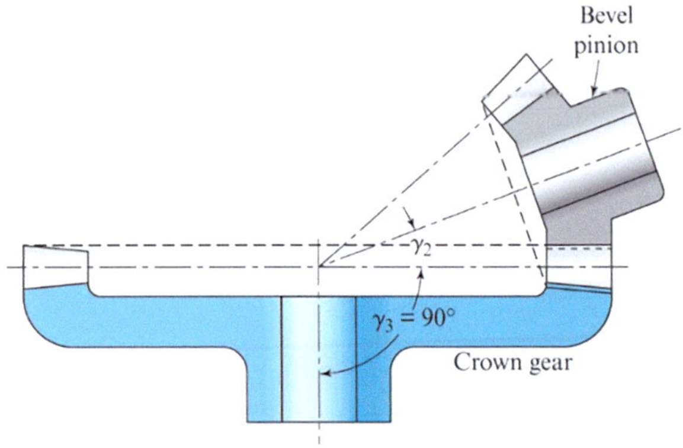

Fig. 8.15 A flat face bevel gear or a crown gear together with its bevel pinion. Page 443 of Uicker et al. [1] © 2016 Oxford University Press (reproduced with permission)

Fig. 8.16 Spiral bevel gears. Page 443 of Uicker et al. [1] © 2016 Oxford University Press (reproduced with permission)

Table 8.2 Comparing straight-tooth, flat, and quiet versions of cylindrical and conical gears

Geometry	Straight-tooth type	Flat version	Quieter version
Cylindrical	Spur	Rack	Helical
Conical	Straight-tooth bevel	Crown	Spiral

In Sect. 8.2, the advantages of using helical gears compared to spur gears in noise reduction at high speeds of operation were discussed. In a similar way, at high speeds, **spiral bevel gears** that are shown in Fig. 8.16 are advantageous to straight-tooth bevel gears. This matter as well as other comparisons between cylindrical and conical gears are summarized in Table 8.2.

Finally, **hypoid bevel gears** (shown in Fig. 8.17) are used in applications where the input and output shafts are perpendicular to each other, but they do not intersect. These gears are especially desirable in rear-wheel drive car differentials because hypoid gears lower the level of the drive shaft and lower the car body. Hypoid bevel gears are quieter than the spiral bevel gears.

Fig. 8.17 Hypoid bevel gears. Page 445 of Uicker et al. [1] © 2016 Oxford University Press (reproduced with permission)

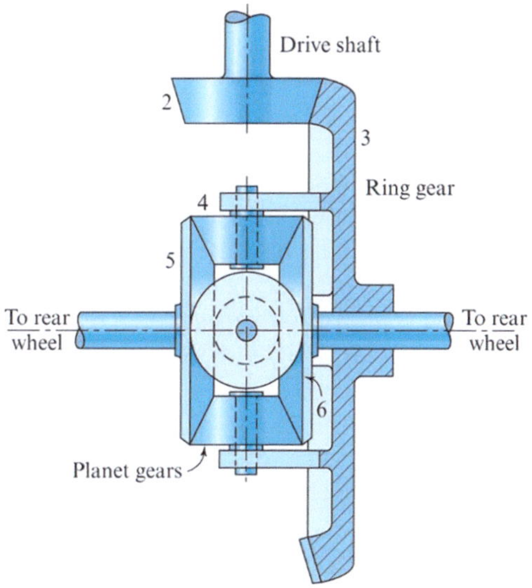

Fig. 8.18 Schematic drawing of a bevel gear automotive rear-axle differential. Page 451 of Uicker et al. [1] © 2016 Oxford University Press (reproduced with permission)

8.5 Differential

A differential, shown in Fig. 8.18, transmits an input power to two output shafts that can rotate at different speeds; that is why it is called: the differential. Figure 8.18 illustrates a bevel gear automotive differential for a rear-wheel drive (RWD) car. The drive shaft transmits the power generated in the engine, and then adjusted for torque and speed in the gearbox, to the differential. The drive shaft pinion as well as the ring gear are usually hypoid bevel gears. The power is then transmitted to the two rear shafts in such a way that in straight line motion the two rear wheels rotate at the same speed. However, on a curve, the speed of the outer wheel will be more than the speed of the inner wheel so that the wheels can turn without sliding.

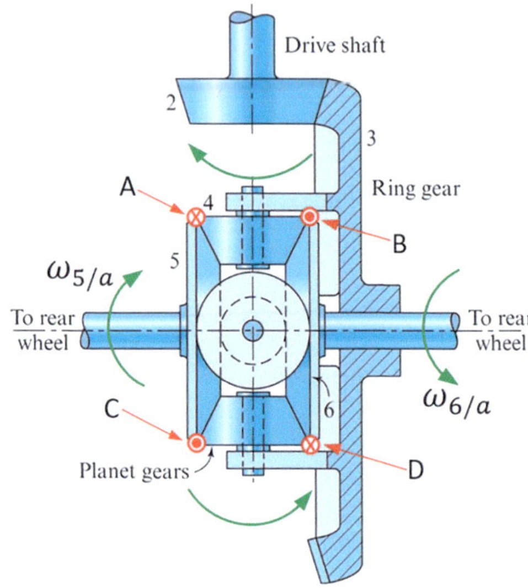

Fig. 8.19 Counter rotation of gears 5 and 6 when gear/arm 3 is held fixed. Page 451 of Uicker et al. [1] © 2016 Oxford University Press (reproduced with permission, labels added)

In Fig. 8.19, assume that we hold the drive shaft connected to pinion 2, and the arm or planet carrier (integral with ring gear 3) fixed and then try to turn gear 5 (that rotates with the left rear wheel) in the shown direction. Such a motion causes point A to go into the page and point C out of the page, then point B moves out of the page and point D goes into the page. As a result, we can see that gear 6 (that rotates the right rear wheel) would turn in the opposite direction of gear 5.

This phenomenon is named counter rotation and it originates from the speed ratio equation (stated in terms of the teeth numbers) between gears 5 and 6 when gear (arm) 3 is held fixed,

$$\frac{\omega_5}{\omega_6} = -\frac{N_5}{N_4}\frac{N_4}{N_6} \tag{8.48}$$

Therefore, if gear (arm) 3 is held fixed we get,

$$\frac{\omega_5}{\omega_6} = -1 \tag{8.49}$$

If gear (arm) 3 is not held fixed, the same equation will be valid, but with respect to the gear (arm) 3. So it can be rewritten as,

$$\frac{\omega_{5/a}}{\omega_{6/a}} = -1 \tag{8.50}$$

Expanding the relative velocity terms in Eq. (8.50) results in,

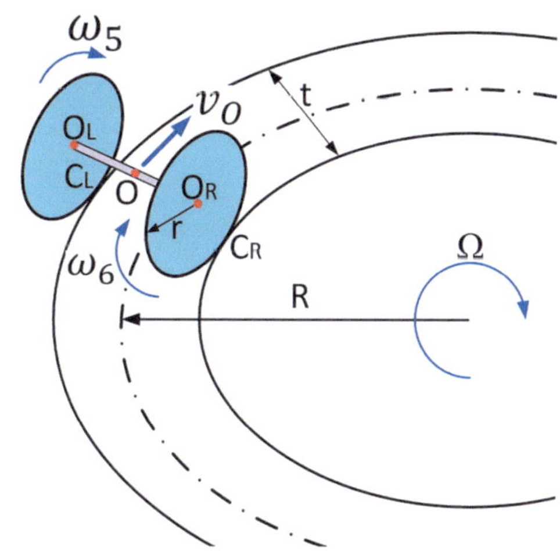

Fig. 8.20 Schematic view of the car rear axle, with the differential shown in Fig. 8.19, in a right turn

$$\frac{\omega_5 - \omega_a}{\omega_6 - \omega_a} = -1 \tag{8.51}$$

A cross multiplication gives:

$$\omega_a = \omega_3 = \frac{\omega_5 + \omega_6}{2} \tag{8.52}$$

This simple equation has an important meaning: The angular velocity of the gear (arm) 3 is always the average of the angular velocities of the two rear wheels. Therefore, when during a turn the inner wheel rotates slower, the outer one goes faster in such a way that Eq. (8.52) is still satisfied and minimal wheel sliding on the road happens. Consequently, on a straight road the two wheels both rotate at the same rate which is equal to the gear (arm) speed, ω_a. However, in a right turn (see Fig. 8.20) as ω_5 increases, according to Eq. (8.52), ω_6 decreases by the same amount.

Example 8.6 In an automotive differential shown in Fig. 8.19, $N_2 = 17$, $N_3 = 51$, $N_4 = 10$, and $N_5 = N_6 = 14$. The angular velocity of the drive shaft is 1500 rpm. Calculate the angular velocity of the left and right wheels in each of the following scenarios:

A. The vehicle is moving on a straight path.
B. The left wheel is jacked up and the right wheel rests on the road.

Considering, $\omega_2 = 1500 \text{ rpm} = 1500 \times \frac{2\pi}{60} \text{ rad/s} = 157.08 \text{ rad/s}$, the speed of rotation of the gear (arm) 3 can be calculated using the speed ratio equation,

$$\frac{\omega_{driven,out}}{\omega_{driver,in}} = \frac{\omega_3}{\omega_2} = \frac{N_2}{N_3} = \frac{R_2}{R_3} \tag{8.41}$$

So,

$$\omega_3 = 157.02 \times \frac{17}{51} = 52.34 \text{ rad/s}$$

A. **The vehicle is moving on a straight path.**

Recalling,

$$\omega_a = \omega_3 = \frac{\omega_5 + \omega_6}{2} \tag{8.52}$$

For this case in which the vehicle moves on a straight path,

$$\omega_5 = \omega_6$$

Therefore, using Eq. (8.52) we have,

$$\omega_5 = \omega_6 = \omega_3 = 52.34 \text{ rad/s}$$

This is also the angular velocity of the left and right wheels that are attached to gears 5 and 6.

B. **The left wheel is jacked up and the right wheel rests on the road.**

In this case, the right wheel is stationary and the left wheel is in rotation while it is jacked up. Therefore we have,

$$\omega_6 = 0$$

Using Eq. (8.52) and the obtained value of ω_3, i.e.,

$$\omega_3 = \frac{\omega_5 + \omega_6}{2}; \omega_3 = 52.34 \text{ rad/s}$$

one obtains,

$$\omega_5 = 104.68 \text{ rad/s}$$

Example 8.7 In an automotive differential shown in Fig. 8.19, $N_2 = 17$, $N_3 = 51$, $N_4 = 10$, and $N_5 = N_6 = 14$. The speed of the vehicle is $v_O = 25$ mph and it is making a 70-ft radius turn to the right, as we see in Fig. 8.20. The wheel track, t (distance between the center of the left tire to the center of the right tire measured along the same axle) is 65 in and the diameter of the tires is $2r = 15$ in. Assuming that the tires roll without slipping,

A. Calculate the angular velocities of the left and right wheels.
B. Calculate the angular velocity of the drive shaft.

A. **Calculate the angular velocities of the left and right wheels.**

First, the speed of the car is expressed by using the in/s unit,

$$v_O = 25 \text{ mph} = 25 \frac{\text{mi}}{\text{h}} \times \frac{5280 \text{ ft}}{1 \text{ mi}} \times \frac{12 \text{ in}}{1 \text{ ft}} \times \frac{1 \text{ h}}{3600 \text{ s}} = 440 \text{ in/s}$$

The angular velocity of rotation of the vehicle about the center of the path curve is shown by Ω and it can be calculated by using,

$$v_O = \Omega \text{R}$$

Therefore, Ω is,

$$\Omega = \frac{v_O}{R} = \frac{440 \text{ in/s}}{70 \text{ ft} \times 12 \text{ in/ft}} = 0.5238 \text{ rad/s}$$

The velocities of the center-tire points, O_L and O_R shown in Fig. 8.20, are now calculated as,

$$v_{O_L} = \Omega\left(\text{R} + \frac{\text{t}}{2}\right) = 0.5238\left(70 \text{ ft} \times 12 \frac{\text{in}}{\text{ft}} + \frac{65\text{in}}{2}\right) = 457.02 \text{ in/s}$$

$$v_{O_R} = \Omega\left(\text{R} - \frac{\text{t}}{2}\right) = 0.5238\left(70 \text{ ft} \times 12 \frac{\text{in}}{\text{ft}} - \frac{65\text{in}}{2}\right) = 422.97 \text{ in/s}$$

Rolling happens without slipping, so contact points C_L and C_R in Fig. 8.20 are instantaneous centers of zero velocity for the tires. Therefore, the following alternative equations for the velocities of the center-tire points O_L and O_R using angular velocities of the left and right wheels can be written,

$$v_{O_L} = \omega_5 r; v_{O_R} = \omega_6 r$$

The angular velocities of the left and right wheels are now calculated as follows:

$$\omega_5 = \frac{v_{O_L}}{r} = \frac{457.02 \text{ in/s}}{\frac{15}{2} \text{in}} = 60.94 \text{ rad/s}$$

$$\omega_6 = \frac{v_{O_R}}{r} = \frac{422.97 \text{ in/s}}{\frac{15}{2}\text{ in}} = 56.40 \text{ rad/s}$$

B. **Calculate the angular velocity of the drive shaft.**

Recalling,

$$\omega_a = \omega_3 = \frac{\omega_5 + \omega_6}{2} \tag{8.52}$$

the angular velocity of the ring gear is obtained as,

$$\omega_3 = \frac{60.94 + 56.40}{2} = 58.67 \text{ rad/s}$$

Finally, the angular velocity of the drive shaft ω_2 can be calculated by using:

$$\frac{\omega_{driven,out}}{\omega_{driver,in}} = \frac{\omega_3}{\omega_2} = \frac{N_2}{N_3} \tag{8.41}$$

So,

$$\omega_2 = 58.67 \times \frac{51}{17} = 176.01 \text{ rad/s}$$

References

1. Uicker, J. R., et al. (2016). *Theory of machines and mechanisms* (5th ed.). Oxford University Press.
2. Waldron, K. J., & Kinzel, G. L. (2004). *Kinematics, dynamics, and design of machinery* (2nd ed.). Wiley.
3. Budynas, R. G. & Nisbett, J. K. (2011). *Shigley's mechanical engineering design* (9th ed.). McGraw Hill.

Chapter 9
Function Generation and Mechanism Design

9.1 Introduction

Chapter 9 is focused on the design of mechanisms. The principal aim is to create mechanisms in such a way that they provide a desired input–output response. These mechanisms are called function generators. Computing the dimensions of such mechanisms (i.e., designing them) can be achieved by using the Freudenstein's equations. In this chapter, Freudenstein's equations are derived for both four-bar linkages and slider-crank mechanisms and they are used for mechanism design. Furthermore, cognate linkages that have identical coupler curves are introduced. Roberts-Chebyshev Theorem, discussed in this chapter, shows that for each four-bar linkage there are two more four-bar linkages (a total of three) that have the same coupler curves. Therefore, if in a particular application one of these solutions is not a proper choice due to constraints like space limitations, we can try any of the other two solutions that can provide an identical coupler curve.

9.2 Cognate Linkages and Roberts-Chebyshev Theorem

The middle link between the input link (crank) and the output link (follower) in Fig. 1.9 is called the coupler link. **Each point on the coupler link is called a coupler point**. During motion of a mechanism, each coupler point traces a curve called a **coupler curve**. It is clear that points on the crank and follower all trace circles or circular arcs, but on the coupler, coupler points can trace very different curves, as can be seen in the coupler curve shown in Fig. 9.1 (generated by SolidWorks). Cognate linkages are linkages that all generate the same coupler curve, while being different otherwise (see Fig. 9.1).

Roberts-Chebyshev theorem states that for every four-bar linkage there are two other four-bar linkages (collectively known as the **three cognate linkages**) that if

M. Ghorashi, *Introduction to Machine Dynamics*, Mechanical Engineering Series,
https://doi.org/10.1007/978-3-031-94736-0_9

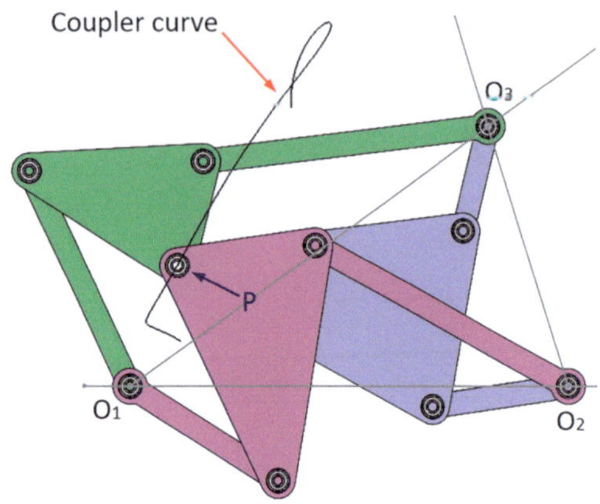

Fig. 9.1 The coupler curve at point P, shared by three cognate four-bar linkages drawn by SolidWorks, Hannon [2]

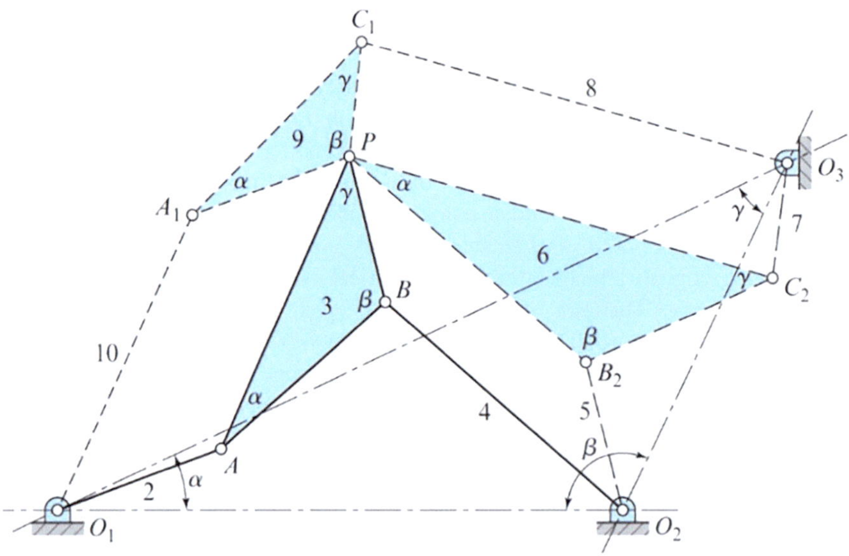

Fig. 9.2 Three cognate four-bar linkages, Page 489 of Uicker et al. [4]. © 2016 Oxford University Press (reproduced with permission)

shared a coupler point, during motion, they all generate the **same coupler curve** at the shared point, Koetsier and Ceccarelli [3]. This reference also includes a very interesting proof of this theorem.

To be more specific, consider the three four-bar linkages O_1ABO_2, $O_1A_1C_1O_3$, and $O_2B_2C_2O_3$, in Fig. 9.2 that have the illustrated triangular members as couplers. Notice that these triangles can be treated as rigid links, and also notice that all of the illustrated triangular members are similar to one another (they all have the same angles, α, β, and γ) and they are also similar to the triangle that is built by connecting the three hinge locations O_1,O_2, and O_3 to one another. Finally, the three couplers share the coupler point, P. With all this in mind, if we have any one of these three mechanisms, the other two could be drawn and built by tracing similar triangles, as shown in the Fig. 9.2.

Since P is shared by all of these three four-bar linkages, during motion identical coupler curves would be generated by point P no matter to which one of these four-bar linkages it is considered to belong. Therefore, if in an application using one of these linkages is not possible (due to, for example, space constraints), one may replace that linkage with any of the other two cognate linkages, if one only cares about the trajectory of motion of that particular point P.

Again, for this mechanism to work as intended, the three triangular coupler elements, as well as the triangle generated by the three fixed points O_1, O_2, and O_3, should all be similar triangles (i.e., have identical corresponding angles) as seen in Fig. 9.2.

The mechanism shown in Fig. 9.2 is comprised of ten links. As to the number of joints, each of the six revolute joints A, B, A_1, C_1, B_2, and C_2 connects two members. However, joint O_1 connects three members 1, 2, and 10, and therefore it is counted as $3 - 1 = 2$ revolute joints. The same is true for joints O_2, O_3, and P. Therefore, the total number of revolute joints (or j_1) is: $6 + 2 \times 4 = 14$. So, the Kutzbach's Eq. (2.10) results in,

$$m = 3(n - 1) - 2j_1 - j_2 = 3(10 - 1) - 2 \times 14 - 0 = -1$$

This result is clearly incorrect as this mechanism is another example of the parallel-members exception discussed in Chap. 2. The correct answer is 1, as each of the three mentioned four-bar linkages has one DOF. However, by choosing the posture of either of these linkages using its one DOF, the posture of the other two linkages will be fully known. So, the whole mechanism has only one DOF.

In Fig. 9.1, members of each of the three four-bar linkages are shown in a different color for clarity.

This mechanism was modeled using the SolidWorks Motion software and manufactured as part of a capstone project of one of my students at USM, Hannon [2]. The corresponding physical model of the same mechanism is shown in Fig. 9.3.

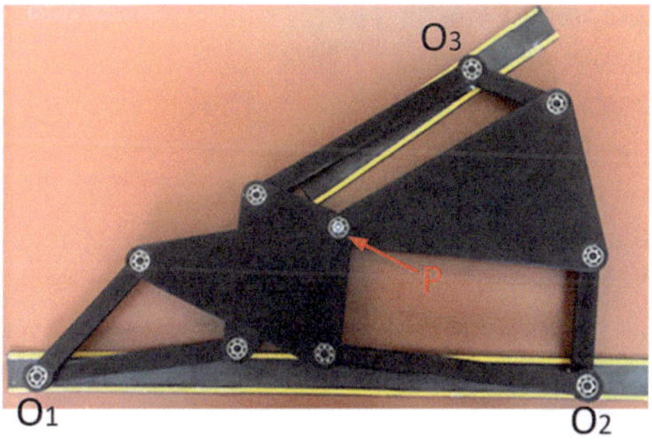

Fig. 9.3 Physical model of the three cognate linkages similar to the one shown in Fig. 9.2, Hannon [2]

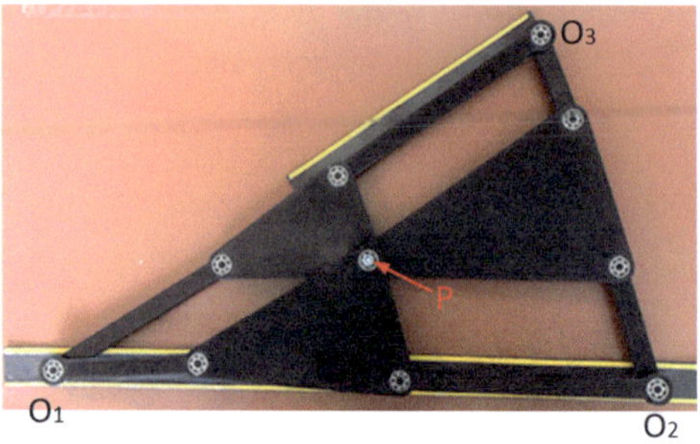

Fig. 9.4 Physical model of the Cayley diagram, Hannon [2]

Considering Fig. 9.3, joints O_2 and O_3 are now relaxed from their fixed connection points to the base and lines O_1O_2, O_2O_3, and O_3O_1 are made parallel with the sides of the coupler links as seen in Fig. 9.4. In this way, we get four similar triangles, three of which are the coupler triangles and the last one is determined by the new positions of points O_2 and O_3 together with the original position of O_1. The resulting configuration, shown in Fig. 9.4, is called the Cayley diagram. One can now move O_2 and O_3 back to their positions in Fig. 9.3 to regain the three cognate linkages that all generate the same coupler curve at the common point, P:

9.3 Function Generation by Four-Bar Linkages

So far in this book and in most cases it has been assumed that the dimensions of all members in mechanisms are already known and one was only supposed to **analyze** the response of a given mechanism to certain input motions. In this section, the aim is to **design** a mechanism and find its link dimensions in such a way that it responds, as demanded, to given inputs. The importance of design in engineering and its relation to analysis and science has been explained beautifully in Waldron and Kinzel [5], page 2: "The activity that distinguishes engineering from science is design. Science is the study of what is; engineering is the creation of what is to be." In design, "The basic procedure is to "guess" a set of dimensions and then use analysis to check the resulting performance. The dimensions are then adjusted to attempt to match closely the performance specifications, and the mechanism is analyzed again. The process is repeated until an acceptably close match to the specifications is achieved. Thus, a primary use of the analysis material is also in mechanism design."

In designing function generators, lengths of links are determined in such a way that the motions of two members are related by the requested function. Figure 9.5 illustrates a four-bar linkage function generator that has been designed, using the formulas to be discussed in this section, in such a way that the angular displacement of link 4 is half the angular displacement of link 2, and in this way effectively replacing a gear set. Figure 9.6 shows the realization of this aim. This mechanism was modeled using SolidWorks Motion software and manufactured as part of a capstone project of one of my students at USM, Badose [1].

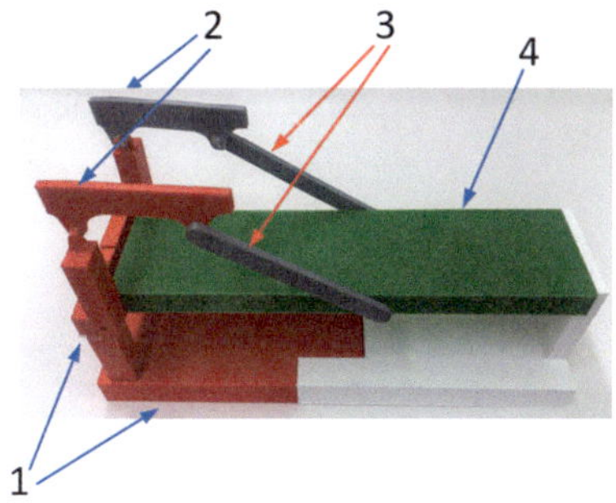

Fig. 9.5 Model of a bridge-lifting mechanism, Badose [1]

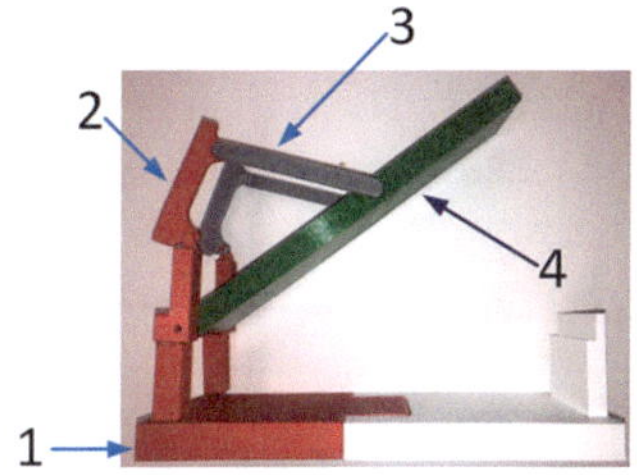

Fig. 9.6 Bridge-lifting mechanism in operation, Badose [1]

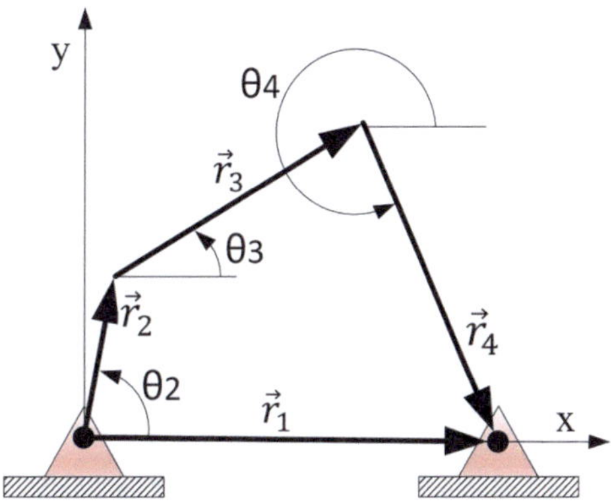

Fig. 9.7 A four-bar linkage and its corresponding vector-loop diagram

9.3.1 The Freudenstein's Equation for Four-Bar Linkages

Consider the general case of designing a four-bar linkage in which a given input value of the crank angle θ_2, is expected to result in a certain value of the follower angle, θ_4. Such a mechanism is called a **function generator** as it is a physical realization of the functional relationship between variables θ_2 and θ_4. The output variable θ_4 can be, e.g., the sine, the square, or a multiple of the input variable, θ_2.

For the vector loop of the four-bar linkage shown in Fig. 9.7, we would like to derive an equation that includes input angle θ_2, output angle θ_4, and the lengths of the links (that serve as design parameters). The aim is to calculate those lengths (and in this way, design the mechanism) in such a way that a desired functional

Table 9.1 Precision points for the design of a four-bar linkage.

Precision point input angle value	Precision point output angle value
$\theta_{2,\,1}$	$\theta_{4,\,1}$
$\theta_{2,\,2}$	$\theta_{4,\,2}$
$\theta_{2,\,3}$	$\theta_{4,\,3}$

relationship between the θ_2 and θ_4 values is precisely realized at a select few points (called precision points). Consider the following table for the precision points:

In $\theta_{i,\,j}$, the first index, *i*, is the variable number (2 or 4) and the second index, *j*, is the precision point number (1 to 3). It should be noted that the functional relationship is only approximately satisfied at points other than the precision points listed in Table 9.1. The reason is because designing the mechanism is achieved with only a limited number of available design degrees of freedom (or the number of parameters that we can choose to design the mechanism). In this case, the design degrees of freedom are the lengths of the four links so there are four design degrees of freedom. It will soon be clear why the number of precision points shown in Table 9.1 is only three, not four.

For the illustrated vector-loop diagram shown in Fig. 9.7, the corresponding vector-loop equation is:

$$\overrightarrow{AB} + \overrightarrow{BC} + \overrightarrow{CD} = \overrightarrow{AD} \tag{9.1}$$

in which, points *A*, *B*, *C*, and *D* are the hinge points from left to right. So, we have,

$$\vec{r}_2 + \vec{r}_3 + \vec{r}_4 = \vec{r}_1 \tag{9.2}$$

Projecting this vector equation in the direction of the fix base vector, $\vec{r}_1$ results in,

$$r_2 \cos\theta_2 + r_3 \cos\theta_3 + r_4 \cos\theta_4 = r_1 \tag{9.3}$$

Projecting Eq. (9.2) in the direction perpendicular to $\vec{r}_1$ gives,

$$r_2 \sin\theta_2 + r_3 \sin\theta_3 + r_4 \sin\theta_4 = 0 \tag{9.4}$$

Since the aim is to realize a functional relationship between θ_2 and θ_4, the final equation should only have θ_2 and θ_4 as variables. Therefore, one needs to eliminate θ_3 between Eqs. (9.3) and (9.4). The easiest way to accomplish this task is by moving the terms that include θ_3 to one side of those equations,

$$r_3 \cos\theta_3 = r_1 - r_2 \cos\theta_2 - r_4 \cos\theta_4 \tag{9.5}$$

$$r_3 \sin\theta_3 = -r_2 \sin\theta_2 - r_4 \sin\theta_4 \tag{9.6}$$

Then, by squaring and adding Eqs. (9.5) and (9.6) we get:

$$r_3^2(\sin^2\theta_3+\cos^2\theta_3)=(r_1-r_2\cos\theta_2-r_4\cos\theta_4)^2+(-r_2\sin\theta_2-r_4\sin\theta_4)^2$$

Using the following trigonometric identity that is valid for any angle α,

$$\sin^2\alpha+\cos^2\alpha=1$$

we see that θ_3 is eliminated and the resulting equation does not include θ_3. Expanding the RHS of the equation gives,

$$r_3^2=r_1^2+r_2^2+r_4^2+2r_2r_4(\cos\theta_2\cos\theta_4+\sin\theta_2\sin\theta_4)-2r_1r_2\cos\theta_2-2r_1r_4\cos\theta_4$$

In which, the same trigonometric identity has been used, this time for θ_2 and θ_4. Now, using another trigonometric identity i.e.:

$$\cos(\alpha-\beta)=\cos\alpha\cos\beta+\sin\alpha\sin\beta$$

we get:

$$r_3^2=r_1^2+r_2^2+r_4^2+2r_2r_4\cos(\theta_2-\theta_4)-2r_1r_2\cos\theta_2-2r_1r_4\cos\theta_4$$

or,

$$2r_1r_2\cos\theta_2+2r_1r_4\cos\theta_4+r_3^2-r_1^2-r_2^2-r_4^2=2r_2r_4\cos(\theta_2-\theta_4)$$

Dividing both sides by $2r_2r_4$ results in,

$$\frac{r_1}{r_4}\cos\theta_2+\frac{r_1}{r_2}\cos\theta_4+\frac{r_3^2-r_1^2-r_2^2-r_4^2}{2r_2r_4}=\cos(\theta_2-\theta_4) \tag{9.7}$$

or,

$$k_1\cos\theta_2+k_2\cos\theta_4+k_3=\cos(\theta_2-\theta_4) \tag{9.8}$$

This is the **Freudenstein's equation for designing a four-bar linkage** in which the non-dimensional parameters k_1, k_2, and k_3 are defined as follows:

$$k_1=\frac{r_1}{r_4},k_2=\frac{r_1}{r_2},k_3=\frac{r_3^2-r_1^2-r_2^2-r_4^2}{2r_2r_4} \tag{9.9}$$

While there are four unknown link lengths that should be calculated, the final Eq. (9.8), that relates input angle θ_2 to the output angle θ_4 has only three parameters. So the design problem has just three degrees of freedom. The reason is because one of the link lengths (say r_1) can be thought of as a scale factor. The choice of the scale factor r_1 only affects the size of the mechanism, not the functional relationship

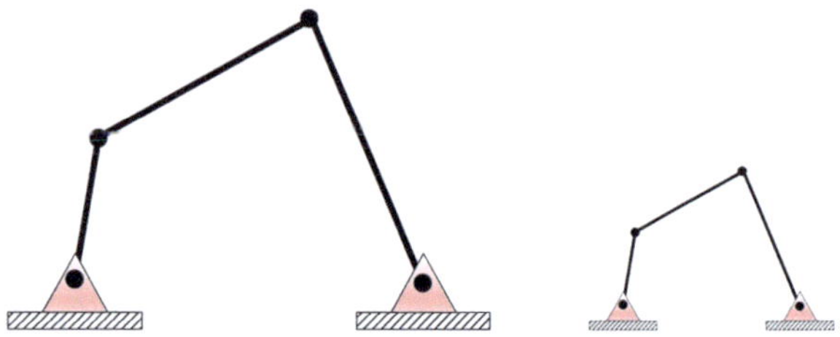

Fig. 9.8 Two four-bar linkages with proportional corresponding links

between θ_2 and θ_4! The response of these four-bar linkage mechanisms is **scale-invariant**. Figure 9.8 illustrates this matter. We see that the shown mechanisms in this figure are different from each other only by a scale factor. The similarity of geometries requires that these mechanisms have identical functional relationships between θ_2 and θ_4.

Now, we would like to use the Freudenstein's equation for designing a four-bar linkage. Since Eq. (9.8) has three unknowns (k_1, k_2, and k_3) one first needs to choose three precision points (shown parametrically in Table 9.1) at which the simulated relation between θ_2 and θ_4 is desired to be realized in a precise manner by the mechanism. Applying the precision points in the Freudenstein's equation, results in a system of three equations for three unknowns that should be solved for k_1, k_2, and k_3. Finally, by assuming a suitable value for the scale factor, r_1, and by using the k_1, k_2, and k_3 Eqs. (9.9), one can calculate the lengths of the three remaining links, i.e., r_2, r_3, and r_4. The designed mechanism accurately satisfies the desired functional relationship at the three precision points. The explained algorithm is now implemented.

By substituting the precision point values of Table 9.1 in the Freudenstein's Eq. (9.8), one obtains the following three linear equations for the three unknown parameters, k_1, k_2, and k_3:

$$\begin{cases} k_1 \cos\theta_{2,1} + k_2 \cos\theta_{4,1} + k_3 = \cos(\theta_{2,1} - \theta_{4,1}) \\ k_1 \cos\theta_{2,2} + k_2 \cos\theta_{4,2} + k_3 = \cos(\theta_{2,2} - \theta_{4,2}) \\ k_1 \cos\theta_{2,3} + k_2 \cos\theta_{4,3} + k_3 = \cos(\theta_{2,3} - \theta_{4,3}) \end{cases} \tag{9.10}$$

These equations can also be rewritten in the following matrix form:

$$\begin{bmatrix} \cos\theta_{2,1} & \cos\theta_{4,1} & 1 \\ \cos\theta_{2,2} & \cos\theta_{4,2} & 1 \\ \cos\theta_{2,3} & \cos\theta_{4,3} & 1 \end{bmatrix} \begin{Bmatrix} k_1 \\ k_2 \\ k_3 \end{Bmatrix} = \begin{Bmatrix} \cos(\theta_{2,1} - \theta_{4,1}) \\ \cos(\theta_{2,2} - \theta_{4,2}) \\ \cos(\theta_{2,3} - \theta_{4,3}) \end{Bmatrix} \tag{9.11}$$

After solving Eq. (9.11) and calculating the non-dimensional parameters k_1, k_2, and k_3, and then deciding about the scale factor, r_1, one can use Eqs. (9.9) for finding r_2, r_3, and r_4 and completing the design calculations. While this mechanism

Table 9.2 Precision points for design of a four-bar linkage

θ₂	θ₄
$\theta_{2,1} = 65°$	$\theta_{4,1} = 240°$
$\theta_{2,2} = 110°$	$\theta_{4,2} = 260°$
$\theta_{2,3} = 155°$	$\theta_{4,3} = 280°$

accurately satisfies the desired functional relationship at the precision points, it can only approximately satisfy this relationship at other points.

Example 9.1 Design a four-bar linkage (i.e., find the lengths of its links) that satisfies the precision points given in Table 9.2. Use $r_1 = 10$ in as the scale factor. Does the designed mechanism satisfy Grashof's law?

Substituting the coordinates of the precision points into Eq. (9.11), i.e,

$$\begin{bmatrix} \cos\theta_{2,1} & \cos\theta_{4,1} & 1 \\ \cos\theta_{2,2} & \cos\theta_{4,2} & 1 \\ \cos\theta_{2,3} & \cos\theta_{4,3} & 1 \end{bmatrix} \begin{Bmatrix} k_1 \\ k_2 \\ k_3 \end{Bmatrix} = \begin{Bmatrix} \cos(\theta_{2,1}-\theta_{4,1}) \\ \cos(\theta_{2,2}-\theta_{4,2}) \\ \cos(\theta_{2,3}-\theta_{4,3}) \end{Bmatrix} \tag{9.11}$$

gives,

$$\begin{bmatrix} 0.4226 & -0.5000 & 1 \\ -0.3420 & -0.1736 & 1 \\ -0.9063 & 0.1736 & 1 \end{bmatrix} \begin{Bmatrix} k_1 \\ k_2 \\ k_3 \end{Bmatrix} = \begin{Bmatrix} -0.9962 \\ -0.8660 \\ -0.5736 \end{Bmatrix}$$

Solving the matrix equation results in,

$$k_1 = 0.6171, k_2 = 1.8448, k_3 = -0.3346$$

By using these results, and assuming $r_1 = 10$ in, applying Eqs. (9.9), i.e.,

$$k_1 = \frac{r_1}{r_4}, k_2 = \frac{r_1}{r_2}, k_3 = \frac{r_3^2 - r_1^2 - r_2^2 - r_4^2}{2r_2r_4} \tag{9.9}$$

gives,

$$r_4 = 16.2042 \text{ in}, r_2 = 5.4207 \text{ in}, r_3 = 18.2531 \text{ in}$$

So all of the lengths of the links are obtained and the design is complete. It can be verified that the designed mechanism satisfies the Grashof's law Eq. (2.39) as:

$$r_2 + r_3 < r_1 + r_4$$

One can model the obtained solution in MATLAB or SolidWorks to verify the accuracy of the response and realizations of Table 9.2.

9.3.2 Ackerman Mechanism for Steering

In Sect. 8.5 it was mentioned that during a turn, in a rear-drive vehicle, the inner rear wheel should rotate slower that the outer rear wheel so that no slipping happens in the rear wheels during a turn. Now, referring to Fig. 9.9 let us focus on the front wheels. Assume that only the front wheels are steered and again the aim is that no slipping happens while the car makes a turn. Notice the definition of wheel track, t (distance between the center of the left tire to the center of the right tire measured along the same axle) and wheel base, w (distance between the centers of the front and rear tires). For no slipping in an R-radius turn, lines drawn perpendicular to the two front wheels and the two rear wheels should be concurrent (i.e., pass through the same point), as shown in Fig. 9.9. For this to happen, the steering angles of the front left wheel, φ and that of the front right wheel, ψ should be slightly different. Both of these angles are measured from the straight-ahead position in which case the turn radius is infinity ($R = \infty$). So the steering mechanism should satisfy the following condition,

$$@R = \infty \quad : \varphi = 0 \text{ and } \psi = 0 \tag{9.12}$$

The plan is to derive a functional relationship between φ and ψ that should be valid for any arbitrary turn radius, R, so that it ensures that the no slipping condition is realized. Considering Fig. 9.9, one can write the following equations for tangents of φ and ψ,

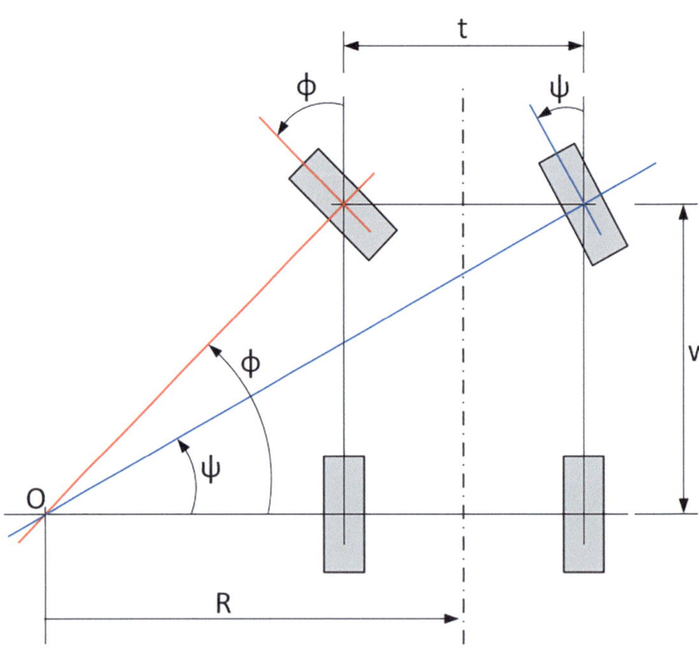

Fig. 9.9 Top view of a forward-steering vehicle during an R-radius left-hand turn

$$\tan \varphi = \frac{w}{R - t/2}; \ \tan \psi = \frac{w}{R + t/2} \tag{9.13}$$

In Eq. (9.13), t and w are known constant parameters of the vehicle. In order to find a functional relationship between φ and ψ we need to eliminate the turn radius, R between the two equations in (9.13). To this end, one can rewrite them as follows:

$$R - \frac{t}{2} = \frac{w}{\tan \varphi}; R + \frac{t}{2} = \frac{w}{\tan \psi} \tag{9.14}$$

Subtracting the two equations in (9.14) eliminates R and we get,

$$\frac{w}{\tan \psi} - \frac{w}{\tan \varphi} - t = 0 \tag{9.15}$$

Equation (9.15) is called the **Ackerman Effect** and it shows how, in a car with wheel track t and wheel base w the left and right front wheels should turn so that during steering no tire slipping happens.

Now that we know the functional relationship between the steering angles at the two front wheels, i.e., Eq. (9.15), we need to design a simplified steering mechanism that can assure this relationship is maintained during a turn and therefore the no-slipping condition of the tires is realized. Let us use the trapezoidal $ABCD$ four-bar linkage shown in Fig. 9.10 for this purpose. Links AB and CD are the left- and right-steering arms that turn (together with their corresponding wheels) about their respective kingpin axes, A and D.

The steer arm angles θ_2 and θ_4 are related to the wheel turn angles φ and ψ by the following equations:

$$\theta_2 = \theta_{2i} + \varphi; \theta_4 = \theta_{4i} + \psi \tag{9.16}$$

The initial values θ_{2i} and θ_{4i} correspond to Fig. 9.11 in which the vehicle is moving straight ahead. At this baseline position we know that $\varphi = 0$ and $\psi = 0$. Next, we choose the initial values θ_{2i} and θ_{4i} in such a way that the initial shape of

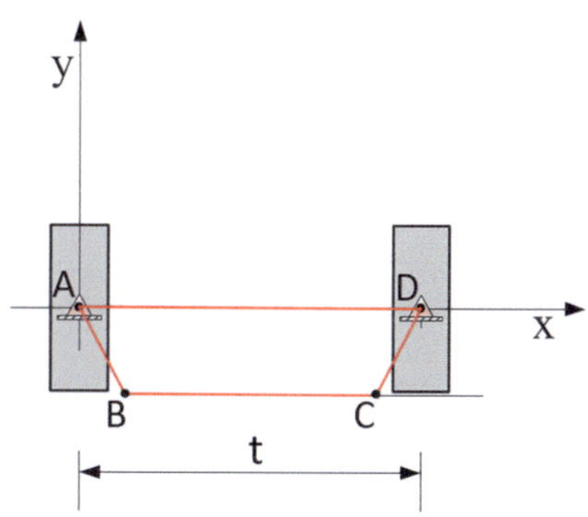

Fig. 9.10 Steering mechanism during the straight-ahead motion

the mechanism is an isosceles trapezoid. Then, we choose two more sets of values for φ and ψ in such a way that the response of the mechanism is symmetrical when turning right compared to turning left. Having decided about three sets of φ and ψ values, one can use Eq. (9.16) to calculate the three corresponding sets of θ_2 and θ_4 values. Finally, Freudenstein's Eq. (9.8) can be used in order to calculate the lengths of the members of the steering mechanism, and in this way design it. The following numerical example clarifies the details.

Example 9.2 Design a four-bar steering mechanism for a vehicle with 4 ft wheel track and 6 ft wheel base.

First, we need to make sure that the mechanism satisfies condition (9.12) for moving straight ahead,

$$@R = \infty \quad : \varphi = 0 \text{ and } \psi = 0 \tag{9.12}$$

To this end, and by referring to Fig. 9.11, we choose the initial values θ_{2i} and θ_{4i} in such a way that the initial shape of the mechanism is an isosceles trapezoid,

$$\theta_{2i} = 300\,^\circ; \theta_{4i} = 60\,^\circ \tag{9.17}$$

For the remaining two precision points, one can consider two turns: one to the left with $R = 25$ ft and another one to the right with $R = -25$ ft. Therefore, using, Eqs. (9.14):

$$R - \frac{t}{2} = \frac{w}{\tan\varphi}; R + \frac{t}{2} = \frac{w}{\tan\psi} \tag{9.14}$$

one can calculate proper φ and ψ values in such a way that no slipping happens. Using $R = 25$ ft, $t = 4$ ft, and $w = 6$ ft, Eq. (9.14) result in:

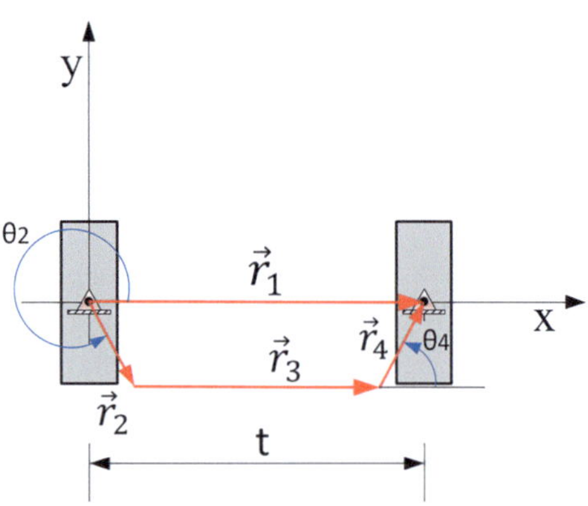

Fig. 9.11 Steering mechanism vector loop in the straight-ahead motion; (initial angles θ_{2i} and θ_{4i} correspond to $\varphi = 0$ and $\psi = 0$ in Fig. 9.9)

Table 9.3 The three precision points in terms of φ and ψ for designing the steering mechanism of Example 9.2

Precision Point #	R (ft)	φ	ψ
1	∞	0	0
2	25	14.62°	12.53°
3	−25	−12.53°	−14.62°

Table 9.4 The three precision points in terms of θ_2 and θ_4 for designing the steering mechanism of Example 9.2

Precision Point #	R (ft)	θ_2	θ_4
1	∞	300°	60°
2	25	314.62°	72.53°
3	−25	287.47°	45.38°

$$25-\frac{4}{2}=\frac{6}{\tan\varphi}\rightarrow\varphi=14.62\,^\circ$$

$$25+\frac{4}{2}=\frac{6}{\tan\psi}\rightarrow\psi=12.53\,^\circ$$

Now, the right turn is considered with $R = -25$ ft. Again Eqs. (9.14) give:

$$-25-\frac{4}{2}=\frac{6}{\tan\varphi}\rightarrow\varphi=-12.53\,^\circ$$

$$-25+\frac{4}{2}=\frac{6}{\tan\psi}\rightarrow\psi=-14.62\,^\circ$$

Table 9.3 includes all of the calculated precision points in terms of φ and ψ.

By using Eqs. (9.16) and (9.17) the obtained φ and ψ precision points in Table 9.3 can be translated into equivalent θ_2 and θ_4 values. Table 9.4 includes all of the calculated precision points in terms of θ_2 and θ_4.

Notice that in this problem the scale factor is, $r_1 = t$, where, $t = 4$ ft. Now, the Freudenstein's Eq. (9.11) is used to calculate the dimensions of the links of the steering mechanism. The following MATLAB script automates the necessary calculations that were done manually before.

```
Thet21=300; Thet22=314.62; Thet23=287.47;
Thet41=60; Thet42=72.53; Thet43=45.38;
A=[cosd(Thet21) cosd(Thet41) 1;...
 cosd(Thet22) cosd(Thet42) 1;...
 cosd(Thet23) cosd(Thet43) 1]
B=[cosd(Thet21-Thet41);cosd(Thet22-Thet42);cosd(Thet23-Thet43)]
K=inv(A)*B
r1=4
r2=r1/K(2,1)
r4=r1/K(1,1)
r3=sqrt(2*r2*r4*K(1, 3)+r1^2+r2^2+r4^2)
```

The obtained output is given in Eq. (9.18) and it provides the necessary dimensions of the links of the steering mechanism:

$$r_2 = 0.3269 \text{ ft}, r_3 = 3.6731 \text{ ft}, r_4 = 0.3269 \text{ ft} \tag{9.18}$$

Finally, as it was expected, in the straight-ahead position the obtained solution is an isosceles trapezoid that satisfies the initial conditions given in Eqs. (9.17).

9.4 The Freudenstein's Equation for Slider-Crank Mechanisms

The aim here is to design a slider-crank mechanism that can provide a desired crank angle versus slider motion response. Consider the vector-loop diagram of a slider-crank mechanism shown in Fig. 9.12 and its corresponding vector-loop equation:

$$\vec{r}_2 + \vec{r}_3 = \vec{r}_1 + \vec{r}_4 \tag{9.19}$$

Case A: The $\vec{r}_4$ vector is pointing down.

Projecting Eq. (9.19) on the direction of $\vec{r}_1$ results in,

$$r_2 \cos\theta_2 + r_3 \cos\theta_3 = r_1 \tag{9.20}$$

And projecting Eq. (9.19) on the direction perpendicular to $\vec{r}_1$ gives,

$$r_2 \sin\theta_2 + r_3 \sin\theta_3 = -r_4 \tag{9.21}$$

In order to find a functional relationship between the input variable, i.e., the crank angle θ_2, and the position of the slider, r_1, one needs to eliminate θ_3 between Eqs. (9.20) and (9.21). First, both equations are solved for the θ_3 components,

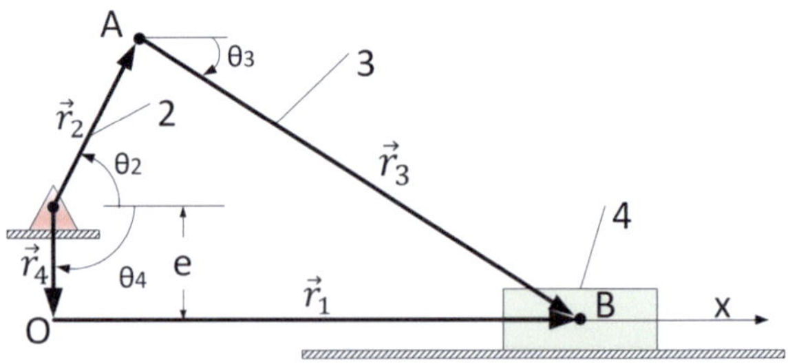

Fig. 9.12 A four-bar linkage and its corresponding vector loop ($\vec{r}_4$ is pointing down)

$$\begin{aligned} r_3 \cos\theta_3 &= r_1 - r_2 \cos\theta_2 \\ r_3 \sin\theta_3 &= -r_4 - r_2 \sin\theta_2 \end{aligned} \tag{9.22}$$

Now, squaring and adding these equations and using the following trigonometric identity,

$$\sin^2\alpha + \cos^2\alpha = 1$$

result in:

$$r_3^2 = r_1^2 + r_2^2 + r_4^2 - 2r_1r_2\cos\theta_2 + 2r_2r_4\sin\theta_2$$

or,

$$2r_1r_2\cos\theta_2 - 2r_2r_4\sin\theta_2 - r_2^2 + r_3^2 - r_4^2 = r_1^2 \tag{9.23}$$

This is the functional relationship between the input variable θ_2 and the position of the slider, r_1. This relation can be expressed more clearly if the coefficients are lumped as parameters K_i as follows:

$$K_1r_1\cos\theta_2 + K_2\sin\theta_2 + K_3 = r_1^2 \tag{9.24}$$

This is the Freudenstein's equation for designing a slider-crank mechanism in which parameters K_1, K_2, and K_3 are defined as follows and they are dimensional [unlike in the case of a four-bar linkage, see Eq. (9.9)]:

$$K_1 = 2r_2, K_2 = -2r_2r_4, K_3 = -r_2^2 + r_3^2 - r_4^2 \tag{9.25}$$

Case B: The $\vec{r}_4$ vector is pointing up.

This case is shown in Fig. 9.13. As we see, the vector-loop equation (9.19) and the projected Eq. (9.20) remain unchanged. However, Eq. (9.21), i.e.,

$$r_2\sin\theta_2 + r_3\sin\theta_3 = -r_4 \tag{9.21}$$

is modified as:

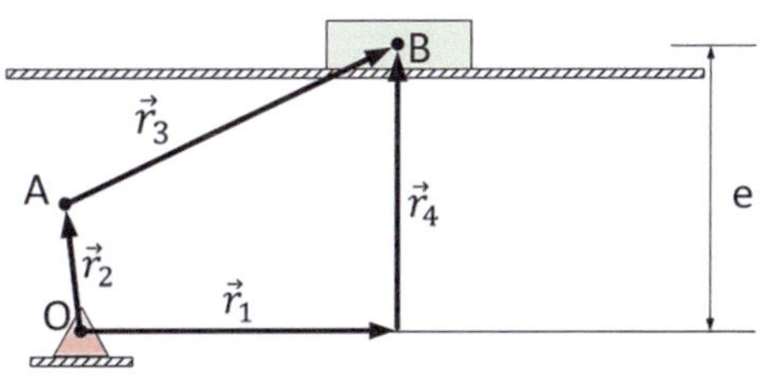

Fig. 9.13 A four-bar linkage and its corresponding vector loop ($\vec{r}_4$ is pointing up)

$$r_2 \sin \theta_2 + r_3 \sin \theta_3 = r_4 \tag{9.26}$$

Following the same procedure as in Case A, Eqs. (9.20) and (9.26) result in,

$$2r_1 r_2 \cos \theta_2 + 2r_2 r_4 \sin \theta_2 - r_2^2 + r_3^2 - r_4^2 = r_1^2 \tag{9.27}$$

or,

$$K_1 r_1 \cos \theta_2 + K_2 \sin \theta_2 + K_3 = r_1^2 \tag{9.28}$$

Equation (9.28) is identical to Eq. (9.24); however, the equations for the coefficients are slightly different as we see in Eq. (9.29):

$$K_1 = 2r_2; K_2 = 2r_2 r_4; K_3 = -r_2^2 + r_3^2 - r_4^2 \tag{9.29}$$

Comparing Eq. (9.29) with Eq. (9.25) of Case A reveals that changing the direction of $\vec{r}_4$ has resulted in a change in sign of the K_2 term.

In the four-bar linkage case of Sect. 9.3.1 there were four unknown link lengths, r_i ($i = 1, 2, 3, 4$), and the input and output variables were θ_2 and θ_4, respectively. Among the unknowns, one was a scale factor that did not change the functional relationship of the input and the output. So, there were three *real* functional variables.

In the slider-crank mechanism, there are three unknown link lengths r_i ($i = 2, 3, 4$) and the Freudenstein's Eq. (9.28) provides a relationship between the input angle θ_2 and the slider position r_1 that has three parameters, K_i. So, again the design problem has three degrees of freedom.

The three design variables r_2, r_3, and r_4 can be calculated by defining three precision points (postures) or ordered pairs ($\theta_{2,\ i}, r_{1,\ i}$) $i = 1, 2, 3$, for the slider-crank mechanism. In other words, the three design variables are calculated in such a way that the mechanism demonstrates these three required postures. Table 9.5 provides the three precision postures of a slider-crank mechanism. The aim is to calculate r_2, r_3,and r_4 in such a way that the mechanism can accurately realize the functional relationship between θ_2 and r_1 at the three postures stated in Table 9.5:

The variables $\theta_{2,\ i}$ and $r_{1,\ i}$ in Table 9.5 are defined in such a way that the first index is the variable number and the second index, *i*, is the precision point number (1 to 3). Substituting these quantities for the values in the Freudenstein's Eq. (9.28), one obtains the following three linear equations for the three unknown parameters, K_1, K_2, and K_3.

Table 9.5 Three precision points for the design of a slider-crank mechanism

Precision points for θ_2	Precision points for r_1
$\theta_{2,\ 1}$	$r_{1,\ 1}$
$\theta_{2,\ 2}$	$r_{1,\ 2}$
$\theta_{2,\ 3}$	$r_{1,\ 3}$

$$
\begin{aligned}
K_1 r_{1,1} \cos\theta_{2,1} + K_2 \sin\theta_{2,1} + K_3 &= r_{1,1}^2 \\
K_1 r_{1,2} \cos\theta_{2,2} + K_2 \sin\theta_{2,2} + K_3 &= r_{1,2}^2 \\
K_1 r_{1,3} \cos\theta_{2,3} + K_2 \sin\theta_{2,3} + K_3 &= r_{1,3}^2
\end{aligned}
\tag{9.30}
$$

Equations (9.30) can also be rewritten in the following matrix form:

$$
\begin{bmatrix} r_{1,1}\cos\theta_{2,1} & \sin\theta_{2,1} & 1 \\ r_{1,2}\cos\theta_{2,2} & \sin\theta_{2,2} & 1 \\ r_{1,3}\cos\theta_{2,3} & \sin\theta_{2,3} & 1 \end{bmatrix} \begin{Bmatrix} K_1 \\ K_2 \\ K_3 \end{Bmatrix} = \begin{Bmatrix} r_{1,1}^2 \\ r_{1,2}^2 \\ r_{1,3}^2 \end{Bmatrix} \tag{9.31}
$$

which can be easily solved for K_1, K_2, and K_3, as it was done in the case of a four-bar linkage (see Eq. (9.11)). Having calculated parameters, K_i, one can use either Eqs. (9.25) or Eqs. (9.29) for obtaining r_2, r_3, and r_4. Of course, while the calculated mechanism accurately satisfies the required functional relationship at the precision points, it can only approximately satisfy a functional relationship at other points.

The **four-bar linkage problem was scale-invariant** and the calculated link lengths could all be multiplied by the same number without any change in the functional response of the mechanism (in terms of the input and output angles). However, since in the slider-crank mechanism the output is a length, not an angle, **this mechanism is not scale-invariant**. That is, by, e.g., doubling all r_i, the response of the mechanism does not remain unchanged. Since one of the lengths that is being doubled is r_1, if it is doubled, the whole data in Table 9.5 (on which the calculation of lengths was based) would change and so do the other r_is. This is unlike the four-bar linkage problem in which multiplying the lengths of the links could be done without affecting the input and output table of precision points; since they are all angles and scaling up or down a mechanism does not change the angles! So, **in the slider-crank problem, there is no scale invariance** the way it was mentioned in the case of a four-bar linkage.

Example 9.3 Design a slider-crank mechanism (i.e., find the lengths of its links) that satisfies the precision points given in Table 9.6. Is the mechanism of the $\vec{r}_4$ pointing up or pointing down type?

Substituting the coordinates of the precision points into Eq. (9.31):

$$
\begin{bmatrix} r_{1,1}\cos\theta_{2,1} & \sin\theta_{2,1} & 1 \\ r_{1,2}\cos\theta_{2,2} & \sin\theta_{2,2} & 1 \\ r_{1,3}\cos\theta_{2,3} & \sin\theta_{2,3} & 1 \end{bmatrix} \begin{Bmatrix} K_1 \\ K_2 \\ K_3 \end{Bmatrix} = \begin{Bmatrix} r_{1,1}^2 \\ r_{1,2}^2 \\ r_{1,3}^2 \end{Bmatrix} \tag{9.31}
$$

and solving the system of equations give,

Table 9.6 Precision points for the design of a four-bar linkage

θ_2	r_1
$\theta_{2,1} = 90°$	$r_{1,1} = 4$ in
$\theta_{2,2} = 85°$	$r_{1,2} = 4.27$ in
$\theta_{2,3} = 80°$	$r_{1,3} = 4.55$ in

$$K_1 = 6.0546, K_2 = 5.3464, K_3 = 10.6536$$

Using these results, and assuming that $\vec{r}_4$ is pointing down (Case A) we apply Eq. (9.25):

$$K_1 = 2r_2, K_2 = -2r_2r_4, K_3 = -r_2^2 + r_3^2 - r_4^2 \tag{9.25}$$

and get,

$$r_2 = 3.027 \text{ in}; r_4 = -0.883 \text{ in}; r_3 = 4.538 \text{ in}$$

Since r_4 is calculated as a negative value, that does not make sense, our assumption that $\vec{r}_4$ is pointing down was incorrect. Using Eq. (9.29) for $\vec{r}_4$ pointing up (Case B):

$$K_1 = 2r_2; K_2 = 2r_2r_4; K_3 = -r_2^2 + r_3^2 - r_4^2 \tag{9.29}$$

we get:

$$r_2 = 3.027 \text{ in}; r_4 = 0.883 \text{ in}; r_3 = 4.538 \text{ in}$$

So all of the lengths of the links are obtained and the design is complete. A MATLAB code for this calculation is given below:

```
clear all
Thet21=90; Thet22=85; Thet23=80;
R11=4; R12=4.27; R13=4.55;
A=[R11*cosd(Thet21) sind(Thet21) 1;...
    R12*cosd(Thet22) sind(Thet22) 1;...
    R13*cosd(Thet23) sind(Thet23) 1]
B=[R11^2;R12^2;R13^2]
K=inv(A)*B
r2=K(1,1)/2
r4=K(2,1)/(2*r2)
r3=sqrt(K(3,1)+r2^2+r4^2)
```

References

1. Badose, L. (2022). *Design and simulation of a bridge lifting mechanism* (Capstone project report). Department of Engineering, University of Southern Maine.

2. Hannon, G. (2024). *Design and manufacture of cognate four-bar linkages* (Capstone project report). Department of Engineering, University of Southern Maine.
3. Koetsier, T., & Ceccarelli, M. (Eds.). (2012). *Explorations in the history of machines and mechanisms*. Springer.
4. Uicker, J. R., et al. (2016). *Theory of machines and mechanisms* (5th ed.). Oxford University Press.
5. Waldron, K. J., & Kinzel, G. L. (2004). *Kinematics, dynamics, and design of machinery* (2nd ed.). Wiley.

Part II
Kinetics of Machines and Mechanisms

Chapter 10
Planar Static Force Analysis

10.1 Introduction

Part II of the book is about inclusion of forces and moments in the analysis. It starts with the static equilibrium analysis of mechanisms in Chap. 10 that includes discussion of two- and three-force members as well as force analysis in spur gears.

Up to this chapter, motions in mechanisms have been analyzed with no consideration of the forces and moments that generate them (motion without forces). In this chapter, another task is performed, i.e., forces are included in the analysis, but the study is limited to cases where the applied forces and moments on a body or a mechanism are in static equilibrium (forces without motion).

In this chapter, the aim is to calculate static forces that apply on members that are in static equilibrium. Calculating the required dimensions of these members so that they can withstand the applied or induced loads is not dealt with here and it is the subject of machine design and strength of materials.

10.2 Newton's Laws

Newton's laws are our most fundamental tools for analyzing a static or a dynamic problem when forces and moments are included. **For a particle**, these laws can be expressed as follows, Meriam and Kraige [1]:

"*Law 1: A particle remains at rest or continues to move with uniform velocity (in a straight line with a constant speed) if there is no unbalance force acting on it.*

"*Law 2: The acceleration of a particle is proportional to the resultant force acting on it and is in the direction of this force.*

"*Law 3: The forces of action and reaction between interacting bodies are equal in magnitude, opposite in direction, and collinear.*"

M. Ghorashi, *Introduction to Machine Dynamics*, Mechanical Engineering Series,
https://doi.org/10.1007/978-3-031-94736-0_10

When **extended to rigid bodies** as opposed to a single particle, the **Newton's first law** can be stated as follows: If a rigid body is at rest, the sum of all forces acting on it must be zero. Furthermore, the sum of all moments of those forces about an arbitrary axis *A-A* should be zero as well. Therefore:

$$\sum \vec{F} = 0, \sum \vec{M}_{A-A} = 0 \tag{10.1}$$

Equations (10.1) are the equations of static equilibrium of the rigid body.

For a particle of mass m subjected to n forces $\vec{F}_i$, the resulting acceleration $\vec{a}$, is related to the resultant of forces by **Newton's Second Law** as follows:

$$\sum_{i=1}^{n} \vec{F}_i = m\vec{a} \tag{10.2}$$

This equation should be written using an inertial reference frame because only an inertial reference frame can detect acceleration correctly. The extension of Newton's second law to bodies of not ignorable dimensions (as opposed to particles) is discussed in Chapters 12, 13, 14 and 15.

In order to formulate **Newton's third law**, let us first show the force acting by body i on body j as $\vec{F}_{ij}$. According to Newton's third law, if such a force is applied, it is always accompanied by a force of equal magnitude, but acting in the opposite direction, applied by body j on body i, that is shown by $\vec{F}_{ji}$. Therefore, one concludes that,

$$\vec{F}_{ji} = -\vec{F}_{ij} \tag{10.3}$$

The action and reaction forces cannot be added together to produce a zero resultant! (as they are applied on different objects).

Newton's first law will be used in Chapters 10 and 11 for static analysis of mechanisms. Newton's second law is used in Chapters 12, 13, 14 and 15 where kinetics of mechanisms (i.e., forces and motion are both included) is discussed. Newton's third law is valid both in static problems and in kinetic ones. Therefore, Newton's third law will be used in all of the remaining chapters, i.e., Chapters 10, 11, 12, 13, 14 and 15.

One might wonder **why we should talk about static equilibrium in a dynamics book**. There are several reasons. The static formulation (10.1) is a special case of the dynamic formulation (10.2) and it is obtained when the mass times acceleration term in the equation of motion (10.2) is ignorable (compared to the applied forces). So, in a slow-moving mechanism, or if masses are small, after solving the dynamic problem, it may be a good idea to find the solution for the special case of ignoring the mass times acceleration term, and compare the dynamic solution with what could be obtained using a purely static solution. This comparison can be used as a means of quickly verifying at least the order of magnitude of the result of the dynamic solution.

The second reason for studying static solutions in a dynamics book is that in some applications, a so-called inertia term (minus mass times acceleration) might be small or even zero, this time due to some kinematical constraints, even though the motion is not slow and the masses are not small! In such a case, the dynamic formulation simply reduces to the static one. As an example, consider a parallelogram mechanism like the one shown in Fig. 2.7, where the coupler link can move quickly and therefore its force equation is dynamic (both mass and acceleration of mass center are non-zero). However, the moment equation is essentially static, as the angular acceleration of the coupler is identically zero (the coupler is performing curvilinear translation).

To analyze the equilibrium of a rigid body, first the **free-body diagram** (FBD) of the body is drawn and then the conditions of static equilibrium (10.1) are applied on it. When projected on the *x*, *y*, and *z* axes, these two vector equations result in six scalar equations that can be solved for six unknowns in spatial problems. In a **two-dimensional problem** in the *x*-*y* plane, the vector Eqs. (10.1) reduce to the following three scaler equations (the other three equations are automatically satisfied) that can be solved for three unknowns.

$$\sum F_x = 0, \sum F_y = 0, \sum M_z = 0 \tag{10.4}$$

If the number of unknowns in a static spatial problem is more than six, or in a static planar problem is more than three, then the unknowns cannot be calculated by static equations alone and deformation analysis is needed too. Such problems are called **statically indeterminate**. Analyzing static problems can be simplified for special cases like two-force members and three-force members.

10.3 Two-Force Members

A **two-force member** is a member that is only subjected to two forces at two different points and it is in static equilibrium. For satisfying the equilibrium conditions, the two forces at those different points should be **equal and opposite** of each other and they should also be **collinear**. Consider the illustrated four-bar linkage in Fig. 10.1 that is to be in equilibrium under the application of an input torque, T_{in}, and an output torque, T_{out}.

This mechanism has only one two-force member that is member *BC*. The FBD of this member in Fig. 10.2 illustrates the stated equilibrium condition (10.1):

$$\vec{F}_{23} + \vec{F}_{43} = 0 \rightarrow \vec{F}_{23} = -\vec{F}_{43} \tag{10.5}$$

Notice that in this case the two forces F_{23} and F_{43} are equal, opposite, and collinear with each other, and with link *BC*. If they were not collinear with this link, the sum of moments condition of equilibrium in (10.4) would not be satisfied.

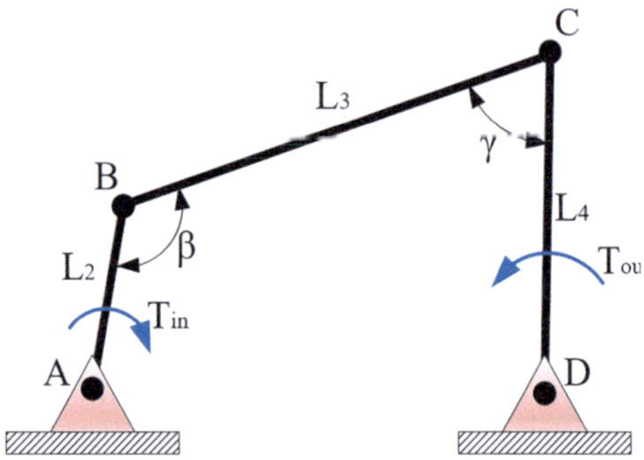

Fig. 10.1 Four-bar linkage with two-force member *BC* subjected to input and output torques

Fig. 10.2 Free body diagram of the two-force member *BC* of Fig. 10.1 in static equilibrium

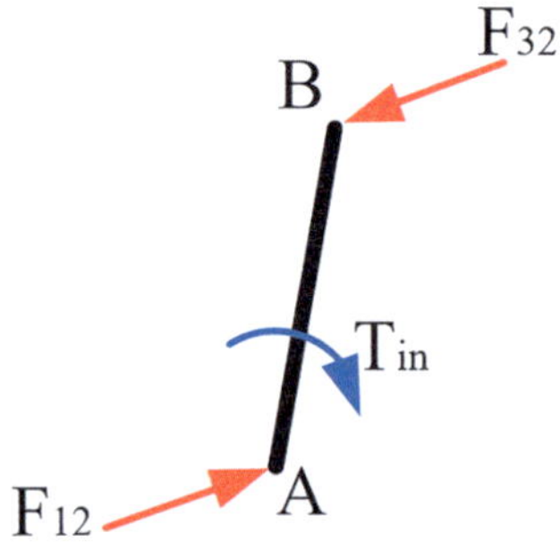

Fig. 10.3 Free body diagram of member *AB* of Fig. 10.1 in static equilibrium

Since links *AB* and *CD* are subjected to torques, together with forces, those links are not two-force members. Focusing on link *AB*, one notices that to satisfy the zero net force condition in (10.1) or (10.4), the following equation should be valid (see Fig. 10.3).

$$\vec{F}_{12} + \vec{F}_{32} = 0 \rightarrow \vec{F}_{12} = -\vec{F}_{32} \tag{10.6}$$

While these forces are parallel with each other, they are not collinear. They instead form a force couple that should balance the applied torque T_{in}. The same is true for link *CD*, whose FBD is shown in Fig. (10.4) and for which one can write:

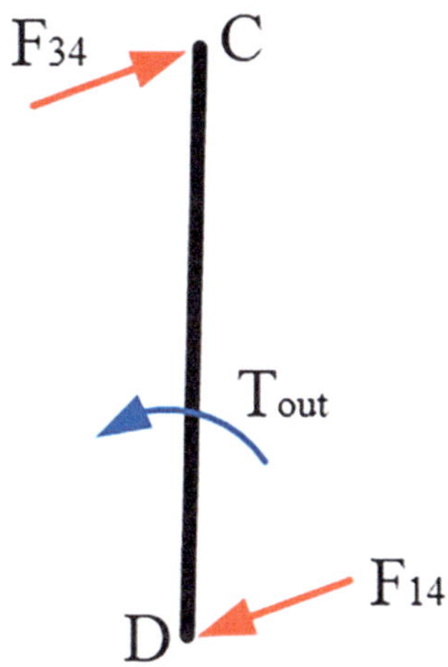

Fig. 10.4 Free body diagram of member *CD* of Fig. 10.1 in static equilibrium

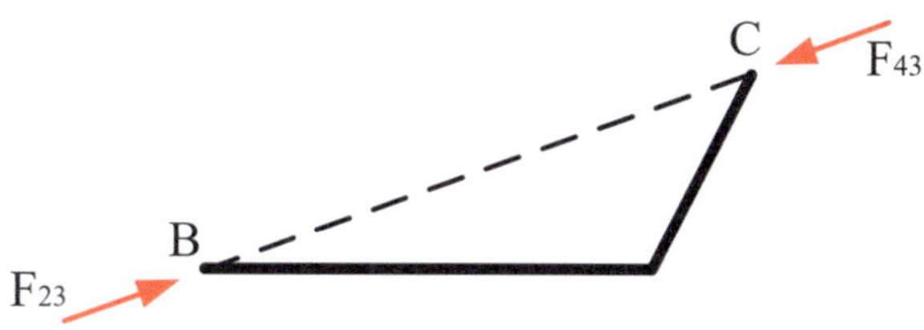

Fig. 10.5 A non-straight two-force member in static equilibrium

$$\vec{F}_{34} = -\vec{F}_{14} \tag{10.7}$$

A two-force member does not have to be a straight link. If it is a bent member like the one shown in Fig. (10.5), it would still be a two-force member and the end forces would be collinear with line *BC* connecting the two end joints of the member together.

$$\vec{F}_{23} = -\vec{F}_{43} \tag{10.8}$$

10.4 Three-Force Members

Consider Fig. 10.6 in which the piston of a slider-crank mechanism is to be in static equilibrium while being subjected to three coplanar forces (see Fig. 10.7). If two of those forces intersect at a point, for ensuring satisfaction of the equilibrium of moment equation in (10.4), the third force should also pass through the same point. Therefore, **the three forces should be concurrent**. To satisfy the equilibrium of force equations in (10.4), **the three forces should also form a triangle**. The three

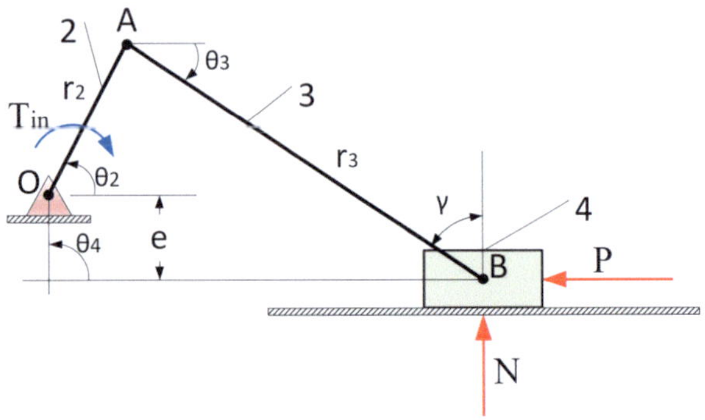

Fig. 10.6 Slider-crank mechanism in static equilibrium

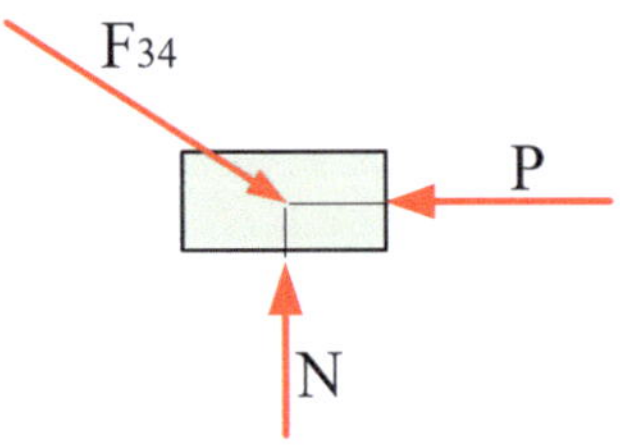

Fig. 10.7 Free body diagram of the slider in Fig. 10.6 in static equilibrium

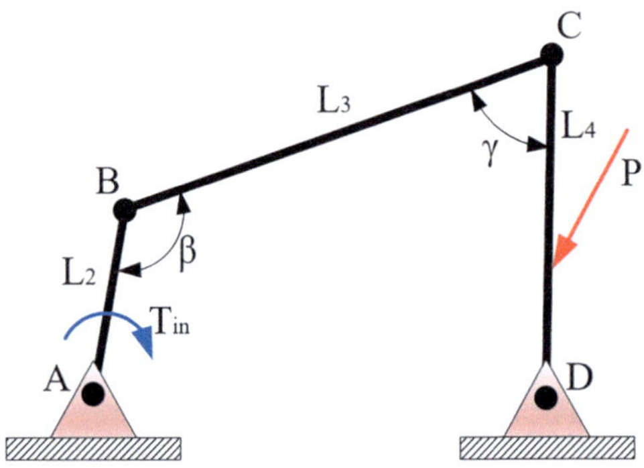

Fig. 10.8 A four-bar linkage under input torque and output force in static equilibrium

forces on the piston (see Fig. 10.7) are from the piston rod (F_{34}), the cylinder (N), and applied force (P). In the same mechanism link 3 (member AB) is a two-force member.

Example 10.1 In the mechanism shown in Fig. 10.8, that is in static equilibrium, detect a two-force member and a three-force member and draw their FBDs.

Link BC is a two-force member and link CD is a three-force member. Their FBDs are shown in Figs. 10.9 and 10.10.

Of course, having concurrent forces is only a necessary condition for equilibrium of link 4. To guarantee equilibrium, the three forces F_{34}, F_{14}, and P should also form a triangle.

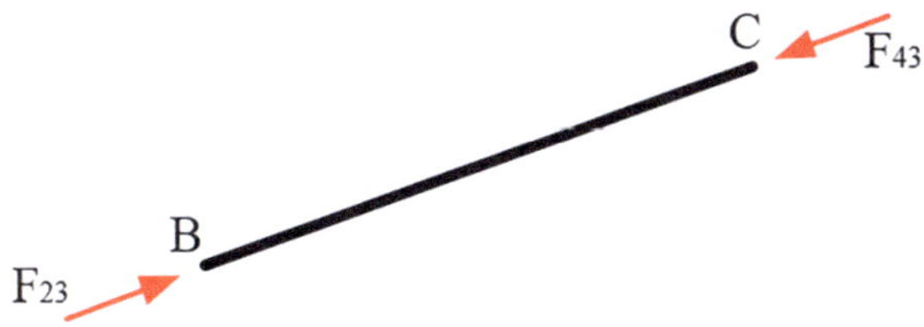

Fig. 10.9 Free body diagram of the two-force member *BC* in static equilibrium

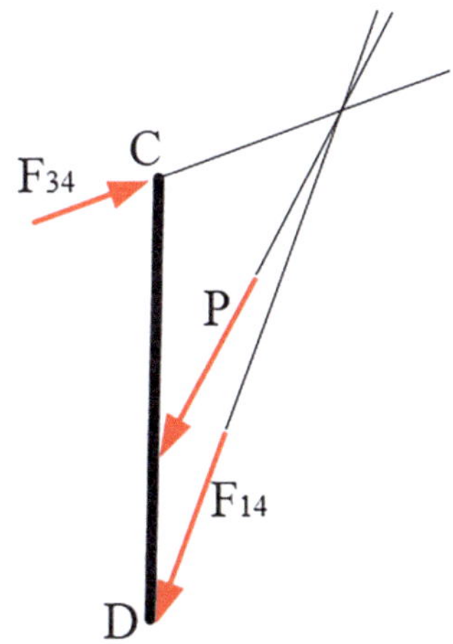

Fig. 10.10 Free body diagram of the three-force member *CD* in static equilibrium

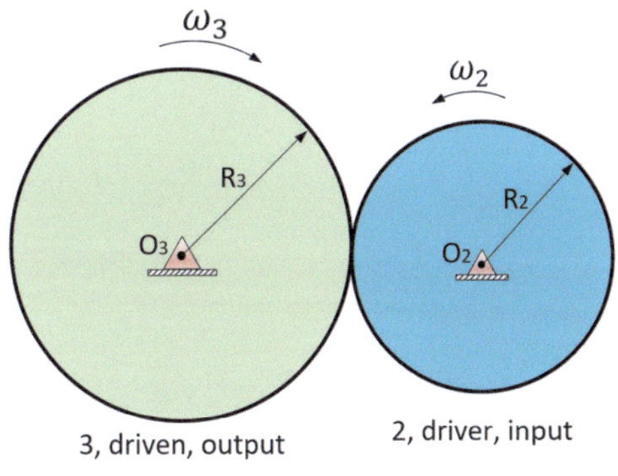

Fig. 10.11 Schematic diagram of two spur gears in mesh

10.5 Spur Gear Force Analysis

Assuming that gears have small inertia and mass, in many applications, they can be analyzed statically. In a spur gear set (see Chap. 7 and Fig. 10.11), the transmitted force between the two gears always passes through the pitch point *P* and also acts at the pressure angle φ measured from the common tangent line of the two pitch circles, as seen in Fig. 10.12.

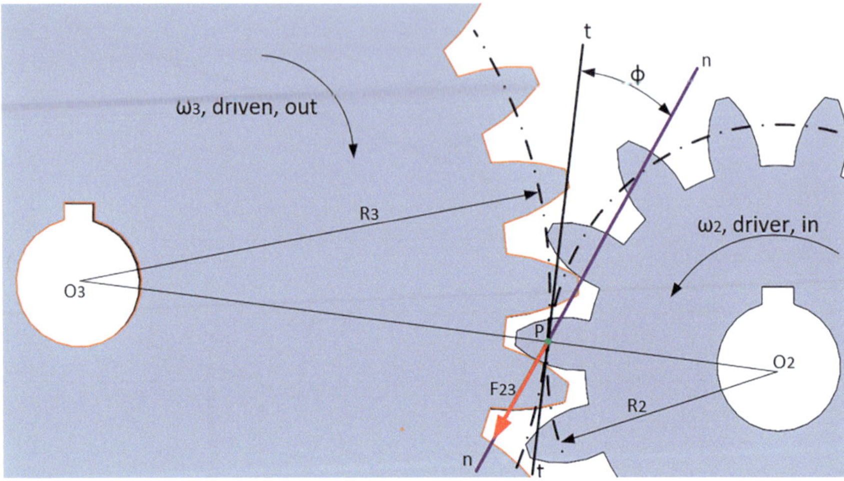

Fig. 10.12 The contact force between two gears if gear 2 is the driver and it turns CCW

Fig. 10.13 Free body diagrams of two spur gears in mesh while in static equilibrium (the input torque on gear 2 is CCW)

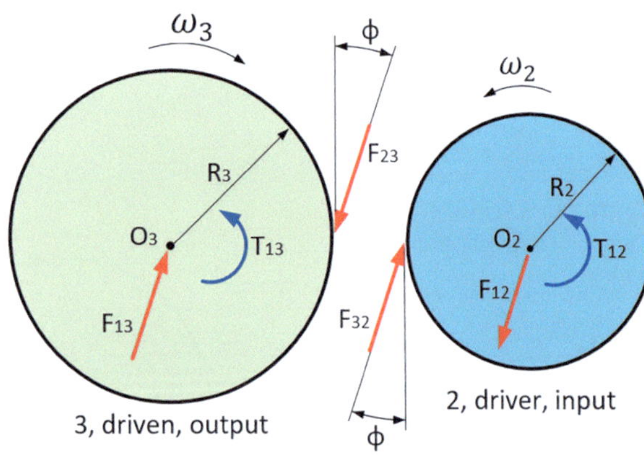

As we will soon see, in each gear set there are two feasible choices of the pressure line that satisfy the pressure angle requirement. The correct one can be found by analyzing the direction of the applied torque on the driver shaft (see Fig. 10.12). In Fig. 10.13, by analyzing the static equilibrium requirements of the two gears, we notice that in the driver gear 2 the direction of the applied torque T_{12} and that of the angular velocity ω_2 are the same. However, in the driven gear 3, the directions of the torque imposed by the load shaft on the driven gear, T_{13}, and the angular velocity of that gear, ω_3, are opposite.

While details of contact are shown in figures like 10.12, for showing the "big picture" of force and torque transmission, there is no need for drawing all of the contact details. Therefore, usually only the pitch circles of the gears are drawn, as we see in Fig. 10.13. Then, one can include all of the forces and torques on the FBD.

Gear 2 in Fig. 10.13 is receiving CCW torque T_{12} from its shaft and, as a result, it rotates in the same direction at an angular velocity ω_2. Gear 2 is in mesh with gear

3 and imposes contact force F_{23} on it that is applied at pressure angle φ with respect to the direction of the common tangent of the two pitch circles. As a result of applying F_{23} from gear 2 on 3, the reaction force F_{32} is applied by gear 3 on 2 in such a way that, according to Newton's third law one can write,

$$\vec{F}_{32} = -\vec{F}_{23} \tag{10.9}$$

The contact force F_{23} applied in the sense shown in Fig. 10.13 causes gear 3 to rotate in a CW direction.

Now, to satisfy the equilibrium of forces condition on gear 2, a force $\vec{F}_{12}$ is applied by the shaft connected to gear 2 to this gear in order to balance $\vec{F}_{32}$ applied by gear 3 on it:

$$\vec{F}_{12} = -\vec{F}_{32} \tag{10.10}$$

The two forces applied on gear 2 form a force couple. In order to balance moments on gear 2, the moment generated by this force couple should be equal and opposite to the applied torque on it (i.e., T_{12}). So writing the moment equation about O_2, i.e.,

$$\sum M_{O_2} = 0$$

gives,

$$F_{32}^t R_2 - T_{12} = 0 \tag{10.11}$$

in which F_{32}^t is the tangent component of the contact force (projected on the common tangent of the two pitch circles). This is the only component of this force that can generate torque.

$$(F_{32} \cos\varphi) R_2 - T_{12} = 0 \tag{10.12}$$

Knowing the applied torque T_{12}, one can use Eq. (10.12) to calculate contact force F_{32}. The radial component of the contact force, i.e.,

$$F_{32}^r = F_{32} \sin\varphi \tag{10.13}$$

does not generate torque and should be supported by the bearings that hold the shaft that drive gear 2.

Similar arguments can be made about gear 3. Therefore, to satisfy the equilibrium of forces condition on gear 3, a force F_{13} is applied by the shaft connected to gear 3 to this gear in order to balance F_{23} applied by gear 2 on it.

$$\vec{F}_{13} = -\vec{F}_{23} \tag{10.14}$$

The two forces applied on gear 3 form a force couple. In order to balance moments on gear 3, the moment generated by this force couple should be equal and opposite to the applied torque on it (i.e., T_{13}). So, writing the moment equation about O_3, i.e.,

$$\sum M_{O_3} = 0 \tag{10.15}$$

gives:

$$F_{23}^t R_3 - T_{13} = 0 \tag{10.16}$$

or,

$$(F_{23}\cos\varphi)R_3 - T_{13} = 0 \tag{10.17}$$

Knowing the contact force F_{23}, one can now calculate torque T_{13} that is imposed by the load (the driven element) on the shaft holding gear 3, that is transmitted to this gear.

Interestingly, considering the force equilibrium and also action-reaction forces (i.e., Newton's first and third laws) we have:

$$\vec{F}_{12} = -\vec{F}_{32} = \vec{F}_{23} = -\vec{F}_{13} \tag{10.18}$$

Therefore, all forces in Fig. 10.13 are known. The shaft forces F_{21} and F_{31} are supported by the bearings that hold the shafts that pass through points O_2 and O_3.

Finally, the equation for the input power is,

$$P_{in} = T_{12}\omega_2 \tag{10.19}$$

And the equation for the output power is:

$$P_{out} = T_{13}\omega_3 \tag{10.20}$$

If power dissipation is negligible, we have:

$$P_{in} = P_{out} \tag{10.21}$$

And therefore,

$$T_{12}\omega_2 = T_{13}\omega_3 \tag{10.22}$$

that results in gear ratio (see Chap. 7) expressible as,

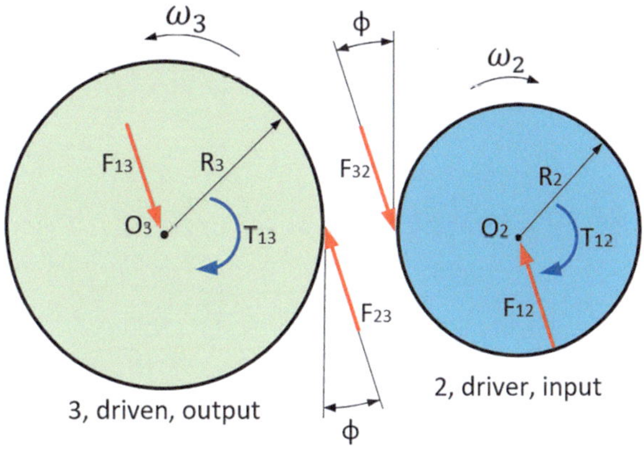

Fig. 10.14 Free body diagrams of two spur gears in mesh while in static equilibrium (the input torque on gear 2 is now CW)

$$\frac{\omega_2}{\omega_3} = \frac{T_{13}}{T_{12}} = M.A. \tag{10.23}$$

In Eq. (10.23) the mechanical advantage (M.A.) is the ratio of load torque T_{13} over effort torque T_{12}, and it will be discussed more in Chap. 11. It should be reminded that considering the no slip condition of pitch circles at the pitch point and the fundamental law of gearing we have,

$$\frac{\omega_3}{\omega_2} = \frac{R_2}{R_3} \tag{10.24}$$

Example 10.2 Redraw the FBDs of gears 2 and 3 in Fig. 10.13 if the direction of the applied torque is now CW.

The solution can be seen in Fig. 10.14 and it should be self-explanatory. This figure can be compared with Fig. 10.13 to see the effect of the change of direction of rotation of the driving gear on the direction of contact forces and the pressure angle.

In the following example, we consider adding a third gear.

Example 10.3 In Fig. 10.15, a gear train is illustrated in which all spur gears have a diametral pitch of 6 teeth/in and a pressure angle of 20°. The 10 hp. input power is given to gear 2 by its shaft, which rotates at 650 rpm. Gear 3 is an idler gear that does not transmit any power to a user connected to its shaft (i.e. $T_{13} = 0$). It is known that O_2O_3 is perpendicular to O_3O_4. The user (output or load) is connected to the shaft of gear 4. Draw the FBDs of all of the gears and calculate all of the forces and torques shown in the FBDs. For the number of gear teeth the data is: $N_2 = 22$, $N_3 = 80$, $N_4 = 60$.

The FBDs of the gears are shown in Fig. 10.16.

Gears 2 and 4 are analyzed in the same way that gears 2 and 3 were analyzed in Example 10.2. Gear 3, however, is different. Since $T_{13} = 0$, there are only three forces applied on it (i.e., the forces applied by gears 2 and 4 and a third force, F_{13} applied by its shaft) and therefore, gear 3 is a three-force member. The three forces intersect at point A and therefore,

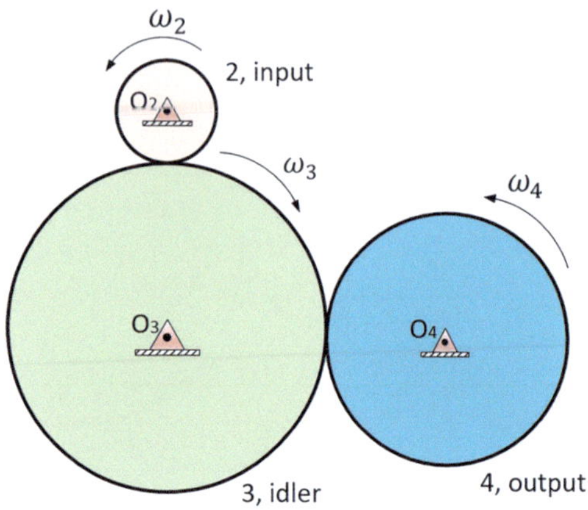

Fig. 10.15 Input and output gears are now related using an idler gear

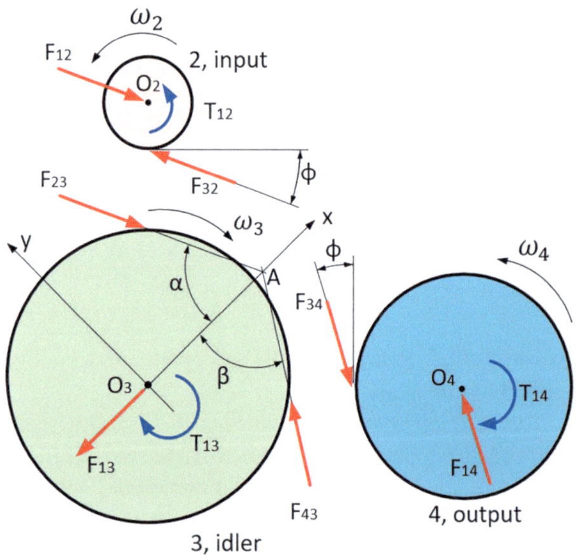

Fig. 10.16 Free body diagrams of the three spur gears in mesh shown in Fig. 10.15 while in static equilibrium

$$\sum M_A = 0 \tag{10.25}$$

Since F_{23} operates at pressure angle from the horizontal direction and F_{43} operates at pressure angle from the vertical direction, geometrical relations dictate that,

$$\alpha = 45^\circ + \varphi = 65^\circ \tag{10.26}$$

The same result is valid for β. Now, one may write the equation of equilibrium of forces in the y direction (perpendicular to $\vec{F}_{13}$, shown in Fig. 10.16) and conclude that,

$$F_{23} \sin\alpha - F_{43} \sin\beta = 0 \tag{10.27}$$

Therefore,

$$F_{23} = F_{43} \tag{10.28}$$

So the forces applied by gears 2 and 4 on gear 3 are equal. Consequently, the moment equilibrium equation about O_3 is also satisfied (as expected),

$$\sum M_{O_3} = 0 \tag{10.29}$$

because,

$$F_{23} \cos\varphi R_3 - F_{43} \cos\varphi R_3 = 0$$

and the shaft of gear 3 does not impose any moments on this gear. This gear has no effect on the speed ratio of the whole gear train either. Gear 3 only changes the direction of rotation and that is why it is called an idler gear. Idler gears are widely used to provide reverse gear in transmissions used in vehicles. Idler gears are also used where two gears cannot be directly meshed because of their distance.

One may now complete the numerical calculations. First, the radii of the pitch circles of the gears are calculated using Eq. (6.3) reminded below,

$$P = \frac{N}{2R}$$

Therefore, for gears 2, 3, and 4 one obtains,

$$R_2 = 1.833 \text{ in}, \quad R_3 = 6.666 \text{ in}, \quad R_4 = 5 \text{ in}.$$

Conversion of $\omega_2 = 650$ rpm, to rad/s results in,

$$\omega_2 = 650 \text{ rpm} = 650 \frac{\text{rev}}{\text{min}} \times \frac{2\pi \frac{\text{rad}}{\text{rev}}}{60 \frac{\text{s}}{\text{min}}} = 68.07 \text{ rad/s}$$

The input power is 10 hp. and it should be converted to lbf.in/s as follows:

$$P_{in} = 10 \text{ hp} \times \frac{550 \text{ lbf.ft/s}}{1 \text{ hp}} \times \frac{12 \text{ in}}{1 \text{ ft}} = 66{,}000 \text{ lbf.in/s}$$

So the corresponding torque applied by the shaft of gear 2 on this gear is,

$$P_{in} = T_{12}\omega_2 \rightarrow 66{,}000\ \text{lbf.in/s} = T_{12} \times 68.07\ \text{rad/s} \rightarrow T_{12} = 969.59\ \text{lbf.in}$$

Writing the moment equation about O_2 gives,

$$\sum M_{O_2} = 0 \rightarrow F^t_{32} R_2 - T_{12} = 0 \rightarrow F^t_{32} = 528.96\ \text{lbf} \tag{10.30}$$

But we know that:

$$F^t_{32} = F_{32} \cos\varphi$$

Considering the pressure angle of 20°, the contact force imposed by gear 3 on gear 2 is,

$$F_{32} = 562.91\ \text{lbf}$$

Like in Example 10.2, the considerations of static equilibrium (Newton's first law) and action-reaction forces (Newton's third law) result in,

$$F_{32} = F_{12} = F_{23} = 562.91\ \text{lbf}$$

In Eq. (10.28) we already proved that,

$$F_{23} = F_{43}$$

Therefore,

$$F_{23} = F_{43} = 562.91\ \text{lbf}$$

Finally, again the considerations of equilibrium and Newton's third law result in,

$$F_{43} = F_{34} = F_{14} = 562.91\ \text{lbf}$$

The only remaining unknown force is F_{13} that can be calculated by writing equilibrium of forces applied on gear 3 in the x direction,

$$-F_{13} + F_{23}\cos\alpha + F_{43}\cos\beta = 0$$

so,

$$F_{13} = 2 \times 562.91\ \text{lbf} \times \cos 65^{\circ} = 475.79\ \text{lbf}$$

This force is applied in the negative x direction on gear 3. The x direction itself is at 45° with respect to the centerlines O_2O_3 and O_3O_4.

The last unknown torque is T_{14}. Writing the moment equation about O_4 gives,

$$T_{14} - F_{34}\cos\varphi \times R_4 = 0 \tag{10.31}$$

Therefore,

$$T_{14} = 562.91 \text{ lbf} \times \cos 20^\circ \times 5 \text{ in} = 2644.81 \text{ lbf.in}$$

This result could also be obtained by the torque ratio (or mechanical advantage) equation. Dividing Eq. (10.31) for T_{14} by Eq. (10.30) for T_{12} gives:

$$\frac{T_{14}}{T_{12}} = \frac{R_4}{R_2} \tag{10.32}$$

So the torque ratio does not depend on the radius of the idler gear, as expected. Finally, one can calculate the angular velocities of gears 3 and 4. For gear 3, the speed ratio equation (6.10) is reminded as follows:

$$\frac{\omega_3}{\omega_2} = \frac{N_2}{N_3}$$

Therefore,

$$\omega_3 = 68.07 \text{ rad/s} \times \frac{22}{80} = 18.72 \text{ rad/s}$$

And for gear 4 we use Eq. (7.10),

$$\frac{\omega_4}{\omega_2} = \frac{N_2 N_3}{N_3 N_4} = \frac{N_2}{N_4}$$

and get,

$$\omega_4 = 68.07 \text{ rad/s} \times \frac{22}{60} = 24.96 \text{ rad/s}$$

Again it is observed that gear 3 does not change the speed of rotation of the gear train.

Reference

1. Meriam, J. L., & Kraige, L. G. (2007). *Engineering mechanics, dynamics* (6th ed.). Wiley.

Chapter 11
Mechanical Advantage

11.1 Introduction

Chapter 11 is dedicated to the definition and calculation of Mechanical Advantage (M.A.) in various machines. An inclined surface, a lever, a four-bar linkage, a slider-crank mechanism, a gear train, and a pulley system are examples of devices and machines for which mechanical advantage can be defined. The higher the mechanical advantage is, the easier a load can be overcome. In other words, mechanical advantage of a machine is a measure to express how much that machine can help us overcome (e.g., lift) a load. It is therefore defined as the ratio of the load to be overcome at the output of a machine divided by the effort to be applied at its input.

$$M.A. = \frac{\text{load at output}}{\text{effort at input}} \tag{11.1}$$

While in a gear train the mechanical advantage is a constant, in some other machines, it is not constant. For example, as we will see in this chapter, in machines that are based on four-bar linkages or slider-crank mechanisms the mechanical advantage changes during motion. It is important to make sure that M.A. does not fall below a certain lower-bound level so that the machine does not jam.

11.2 Inclined Plane

Figure 11.1 illustrates an inclined plane on which a ball of mass m is being rolled up by force F from point A to point B. In an ideal situation, if there is no energy loss, the work done by the applied effort force equates the work done on the resisting load. In other words, the product of the effort force and its corresponding displacement should equate the same product for the resisting load, i.e.,

M. Ghorashi, *Introduction to Machine Dynamics*, Mechanical Engineering Series,
https://doi.org/10.1007/978-3-031-94736-0_11

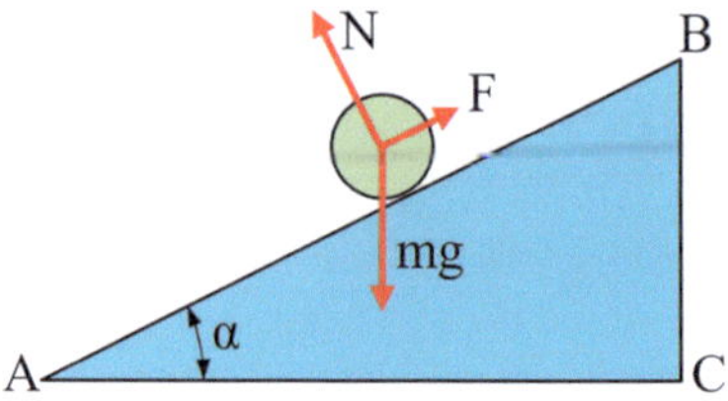

Fig. 11.1 A ball is moving up an inclined plane

$$W = F_{effort} . d_{effort} = F_{load} . d_{load} \tag{11.2}$$

It should be noted that Eq. (11.2) is essentially the same equation as the one we use for analyzing a lever and express the balance of moments about its hinge. Now, based on Eq. (11.1), one can conclude that the higher the mechanical advantage of a machine is, the less input effort would be needed for the same output load. However, as Eq. (11.2) demonstrates, if M. A. > 1 (i.e., $F_{load} > F_{effort}$), this advantage comes at a price: the input effort should be applied on a longer stroke to provide the same amount of work as the output load, i.e., $d_{effort} > d_{load}$.

For the uniform motion of the ball up the α-angle slope shown in Fig. 11.1, the equation of static equilibrium (Newton's first law) of the ball projected along the inclined surface results in:

$$F = mg \sin \alpha \tag{11.3}$$

Therefore, the mechanical advantage, or the ratio of load over effort, i.e. Eq. (11.1), results in:

$$M.A. = \frac{mg}{mg \sin \alpha} = \frac{1}{\sin \alpha} \tag{11.4}$$

We notice that for $\alpha < 90°$, the M.A. obtained from Eq. (11.4) is higher than 1. As a numerical example, for $\alpha = 30°$ one obtains, M. A. = 2, meaning that pushing a load up that inclined plane is two times easier than lifting it. However, the distance along which the effort should be applied is *AB*, that is more than distance *BC* (by the same factor as the *M. A.*), that a vertical effort force of *mg* would travel. At $\alpha = 90°$, Eq. (11.4) gives the obvious answer of: M. A. = 1 that means no improvement is achieved as there is no inclined plane at $\alpha = 90°$!

11.3 Screw Jack

Screw is an inclined plane wrapped around a cylinder. So, the screw jack shown in Fig. 11.2, is an application of the inclined plane concept. If the number of screw threads is *N*, the lead, *l* can be written as, (see Sects. 2.3.1 and 8.3):

Fig. 11.2 Inside a screw jack in operation. Public domain picture, https://www.flickr.com/photos/internetarchivebookimages/14761212352/. Accessed 12/1/2024

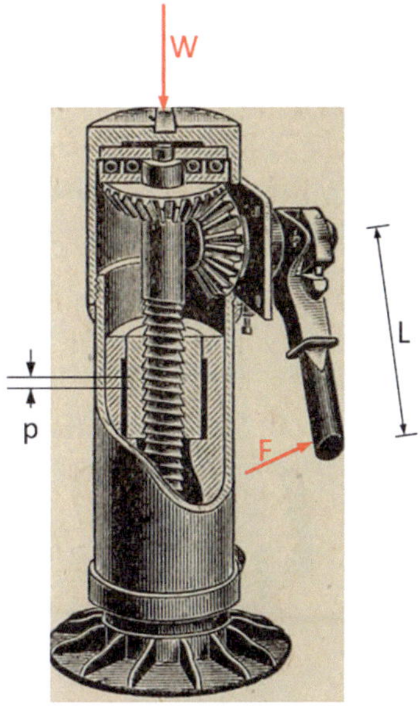

$$l = Np \tag{2.6}$$

where p is the pitch of the screw. Now, ignoring energy losses, we know that the work done by the applied force F in one revolution of the handle to raise the load W should be equal to the work done by W during its vertical motion. So, for a bevel gear ratio of one we have:

$$F(2\pi L) = Wl \tag{11.5}$$

where L is the length of the handle. Using Eqs. (11.1) and (11.5), the mechanical advantage of this screw jack can be expressed as follows:

$$M.A. = \frac{W}{F} = \frac{2\pi L}{l} \tag{11.6}$$

Example 11.1 Calculate the mechanical advantage of a single-threaded jack screw if $L = 1\text{ft}$ and $p = 1$ in:

$$M.A. = \frac{W}{F} = \frac{2\pi L}{p} = \frac{2\pi \times 1 \times 12}{1} = 75.40$$

11.4 Pulley and Belt or Sprocket and Chain

For power transmission in relatively long distances we use pulleys and belts or sprockets and chains. Figure 11.3 illustrates a 28-teeth sprocket generated by SolidWorks and Fig. 11.4 shows a pulley and belt drive that, as far as the mechanical advantage is concerned, acts similar to a sprocket and chain drive.

Considering Fig. 11.4, and assuming there is no slipping in the contact region between belt and pulleys, or that we are using timing belts to prevent such slipping, the belt speed can be written as $R_2\omega_2$ or $R_3\omega_3$ and we know that:

$$R_2\omega_2 = R_3\omega_3 \tag{11.7}$$

If member 2 is the driver pulley (or driver sprocket) and member 3 is the driven pulley (or driven sprocket), then the speed ratio is,

Fig. 11.3 A 28-teeth sprocket generated by SolidWorks

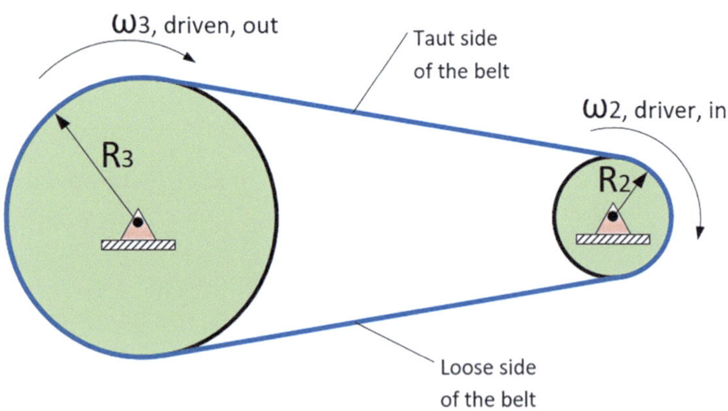

Fig. 11.4 Belt and pulley used for power transmission at long distances

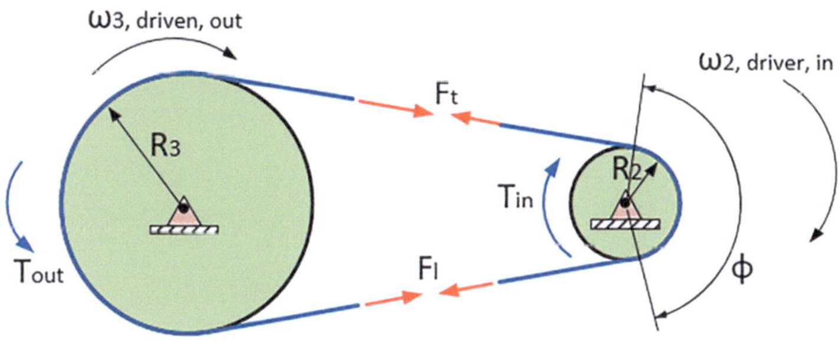

Fig. 11.5 Less input torque at high speed in this belt drive is used to support more output torque at low speed

$$\frac{\omega_{driven,out}}{\omega_{driver,in}} = \frac{\omega_3}{\omega_2} = \frac{R_2}{R_3} < 1 \tag{11.8}$$

In the belt-drive system shown in Fig. 11.5, the applied torque T_{in} on pulley 2 generates tensile force F_t in the taut side of the belt and F_l in its loose side. The two forces are not equal due to the existence of friction between the belt and the pulley. It can be shown that (if inertia forces due to rotation of the belt in contact with the pulleys are ignored) the forces F_t and F_l are related to each other by the following equation,

$$F_t = F_l e^{\mu\varphi} \tag{11.9}$$

In which μ is the coefficient of friction at the interface of the pulley and the belt. Also, angle φ is the contact angle of the belt on the pulley and it is measured in radians. It can be seen that the more this contact angle is, the larger the friction force and the difference between F_t and F_l will be. When comparing the two pulleys, we recognize that the contact angle on the smaller pulley is less than that of the larger pulley. Therefore, slipping always happens on the smaller pulley!

Considering the static equilibrium of pulley 2 under the applied torques, and ignoring inertia effects, one can write,

$$T_{in} - (F_t R_2 - F_l R_2) = 0$$

or,

$$T_{in} = (F_t - F_l) R_2 \tag{11.10}$$

So, the input power can be written as follows:

$$P_{in} = T_{in}\omega_2 = (F_t - F_l) R_2 \omega_2 \tag{11.11}$$

Similarly, considering the equilibrium of pulley 3 under the applied torques one can write,

$$T_{out} = (F_t - F_l)R_3 \tag{11.12}$$

And for the output power we have,

$$P_{out} = T_{out}\omega_3 = (F_t - F_l)R_3\omega_3 \tag{11.13}$$

By dividing Eq. (11.12) by (11.10) the mechanical advantage of the pulley system is obtained as follows:

$$M.A. = \frac{T_{out}}{T_{in}} = \frac{R_3}{R_2} > 1 \tag{11.14}$$

Recalling Eq. (11.8):

$$\frac{\omega_{driven,out}}{\omega_{driver,in}} = \frac{\omega_3}{\omega_2} = \frac{R_2}{R_3} < 1 \tag{11.8}$$

one obtains,

$$M.A. = \frac{T_{out}}{T_{in}} = \frac{\omega_{driver,in}}{\omega_{driven,out}} = \frac{\omega_2}{\omega_3} = \frac{R_3}{R_2} > 1 \tag{11.15}$$

So, in this case pulley 3 runs slower than 2, but more torque is generated in it.

Figure 11.4 could also be imagined as a schematic representation of a sprocket and chain arrangement for which Eq. (11.15) would still be valid. In a bicycle, for example, an arrangement like the one seen in Fig. 11.4 produces higher torque (needed for climbing a slope). If the driven sprocket radius is smaller than the driver one, i.e., $R_3 < R_2$ (see Fig. 11.6), then sprocket 3 would run faster than sprocket 2, but at reduced torque.

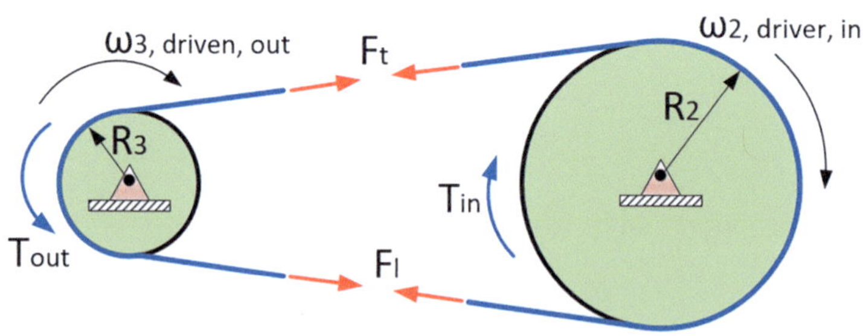

Fig. 11.6 Sprocket 3 runs faster than sprocket 2, but at reduced torque

Fig. 11.7 Three driver sprockets (chainrings) of a tandem bike (shown in the right-hand side; the smallest one has the chain on it)

Fig. 11.8 Seven driven sprockets (a seven-speed sprocket set, or cogset) on the rear wheel of a bike

$$M.A. = \frac{T_{out}}{T_{in}} = \frac{\omega_{driver,in}}{\omega_{driven,out}} = \frac{\omega_2}{\omega_3} = \frac{R_3}{R_2} < 1 \tag{11.16}$$

In a bicycle, such sprocket arrangement results in faster movement when less output torque is needed, for example, when riding in a flat trail.

A series of output sprockets in a geared bike provide the flexibility of choosing the optimal mechanical advantage during the ride. Following our discussions on Fig. 11.5, the configurations shown in Figs. 11.7 and 11.8 have maximum mechanical advantages when the driver sprocket is the smallest and the driven sprocket is the largest.

Example 11.2 Compute tables of mechanical advantages and speed ratios (i.e., the inverse of mechanical advantages) for a 21-speed bike with three driver sprockets (aka chainrings) of 28/38/48 teeth and seven driven sprockets with 12/13/15/17/20/24/28 teeth.

Since each front (driver) chainring can be used with any of the back (driven) sprocket, the total number of speed ratios (and the corresponding mechanical advantages) is $3 \times 7 = 21$. Also the number of teeth on the sprockets is proportional to the radii of sprockets. Therefore, in each equation one can replace radius with the corresponding teeth number. For example, referring to Fig. 11.6, if the 48-teeth chainring (proportional to R_2) and the 12-teeth sprocket (proportional to R_3) are selected, one obtains,

$$M.A. = \frac{T_{out}}{T_{in}} = \frac{\omega_{driver,in}}{\omega_{driven,out}} = \frac{\omega_2}{\omega_3} = \frac{R_3}{R_2} = \frac{12}{48} = 0.25$$

And the corresponding speed ratio is,

$$\frac{\omega_{driven,out}}{\omega_{driver,in}} = \frac{\omega_3}{\omega_2} = \frac{R_2}{R_3} = \frac{48}{12} = 4$$

This is the "**highest gear**," meaning it can provide the fastest speed of the bike wheels for a given speed of pedaling. The "**lowest gear**" happens when the mechanical advantage is maximum. For it to happen, the front sprocket should be the smallest of the three, i.e., the one with 28 teeth, and the rear sprocket should be the largest of the seven, i.e., the one also with 28 teeth. Therefore, the maximum mechanical advantage is,

$$M.A. = \frac{T_{out}}{T_{in}} = \frac{\omega_{driver,in}}{\omega_{driven,out}} = \frac{\omega_2}{\omega_3} = \frac{R_3}{R_2} = \frac{28}{28} = 1$$

And the corresponding speed ratio is,

$$\frac{\omega_{driven,out}}{\omega_{driver,in}} = \frac{\omega_3}{\omega_2} = \frac{R_2}{R_3} = \frac{28}{28} = 1$$

In the same way, the mechanical advantages and speed ratios for all of the 21 possibilities can be calculated. The results are listed in Tables 11.1 and 11.2, respectively.

Table 11.1 The mechanical advantages of a 21-speed bike

		Number of teeth on rear wheel sprockets						
		12	13	15	17	20	24	28
Number of teeth on chainrings	28	0.429	0.464	0.536	0.607	0.714	0.857	1.000
	38	0.316	0.342	0.395	0.447	0.526	0.632	0.737
	48	0.250	0.271	0.313	0.354	0.417	0.500	0.583

Table 11.2 The speed ratios of a 21-speed bike (number of bike wheel turns for every pedal turn)

		Number of teeth on rear wheel sprockets						
		12	13	15	17	20	24	28
Number of teeth on chainrings	28	2.331	2.155	1.866	1.647	1.401	1.167	1.000
	38	3.165	2.924	2.532	2.237	1.901	1.582	1.357
	48	4.000	3.690	3.195	2.825	2.398	2.000	1.715

11.5 Spur Gears

In spur gears, the mechanical advantage equation is similar to the one that was presented in the discussion on pulleys and sprockets. If gear 2 is the driver and gear 3 is the driven, using Eqs. (7.3) and (7.6) of Sect. 7.3, the mechanical advantage of the gear set can be written as,

$$M.A. = \frac{T_{out}}{T_{in}} = \frac{\omega_{driver,in}}{\omega_{driven,out}} = \frac{\omega_2}{\omega_3} = \frac{R_3}{R_2} = \frac{N_3}{N_2} \tag{11.17}$$

where R_2 and R_3 are the pitch circle radii of the driver and the driven gears, respectively. At a **low gear**, the output (driven) gear goes slowly, M.A. is high, and the output torque is significant too. At a **high gear**, the output gear goes faster and the output torque is small. More discussion on gears can be found in Chap. 7.

11.6 Four-Bar Linkage

Consider a four-bar linkage shown in Fig. 11.9. The mechanism is run by torque $T_{in} = T_2$ applied on link 2. The output member is link 4 that is running a load (not shown) and applies the resisting torque $T_{out} = T_4$ on link 4. The **transmission angle** $\boldsymbol{\gamma}$ is the angle between the coupler link 3 and the follower link 4. The aim is to calculate the mechanical advantage T_{out}/T_{in} at the illustrated position:

Recalling the **angular-velocity-ratio theorem** [see Eq. (4.129) of Sect. 4.7.3], and the fact that point O (aka the instantaneous center, I_{24}) should have the same velocity whether it is considered to be part of link 2 or part of link 4, one obtains,

$$\frac{\omega_{in}}{\omega_{out}} = \frac{\omega_2}{\omega_4} = \frac{I_{14}I_{24}}{I_{12}I_{24}} = \frac{OD}{OA} \tag{11.18}$$

From similarity of the right triangles ODH_4 and OAH_2 we have:

$$\frac{OD}{OA} = \frac{DH_4}{AH_2} \tag{11.19}$$

However, for triangle ABH_2 one can write:

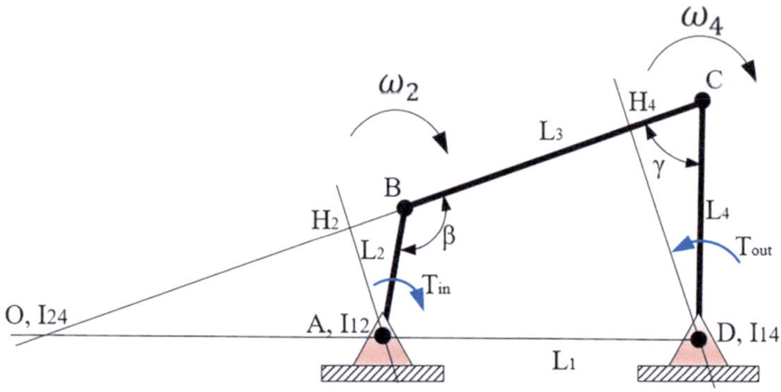

Fig. 11.9 A four-bar linkage subjected to input and output torques

$$AH_2 = AB\sin(\pi - \beta) = AB\sin\beta \tag{11.20}$$

And similarly for triangle DCH_4 we have:

$$DH_4 = DC\sin\gamma \tag{11.21}$$

Combining Eqs. (11.19) to (11.21) results in:

$$\frac{OD}{OA} = \frac{DC\sin\gamma}{AB\sin\beta} \tag{11.22}$$

And now by combining Eqs. (11.18) and (11.22) we obtain:

$$\frac{\omega_{in}}{\omega_{out}} = \frac{\omega_2}{\omega_4} = \frac{OD}{OA} = \frac{DC\sin\gamma}{AB\sin\beta} \tag{11.23}$$

During motion of the mechanism, the ratio of the angular velocities as given in Eq. (11.23) changes (since angles β and γ change).

Now recall the definition of **mechanical advantage** that is the ratio of the output (driven) torque T_{out} over the input (driver) torque T_{in},

$$M.A. = \frac{T_{out}}{T_{in}} = \frac{T_4}{T_2} \tag{11.24}$$

This ratio can be related to the ratio of angular velocities if one assumes that the mechanism is ideal (i.e., the power dissipation is negligible). In such a case, the input power, P_{in} (given by the driver link) and the output power, P_{out} (received by the driven link) are equal,

$$P_{in} = P_{out} \tag{11.25}$$

For rotational systems, power is the product of torque and angular velocity. Therefore,

$$P_{in} = T_{in}\omega_{in}; P_{out} = T_{out}\omega_{out} \tag{11.26}$$

By combining Eqs. (11.23) to (11.26) one obtains,

$$M.A. = \frac{T_{out}}{T_{in}} = \frac{\omega_{in}}{\omega_{out}} = \frac{DC \sin\gamma}{AB \sin\beta} = \frac{OD}{OA} \tag{11.27}$$

Since points A and D are fixed, according to Eq. (11.27) the value of the mechanical advantage of a four-bar linkage only depends on the location of the instantaneous center I_{24} (or point O). Clearly, during motion the location of this point changes, and therefore the mechanical advantage also changes. This conclusion is unlike what we learned about the mechanical advantages of all other mechanisms studied before in this chapter (that all remained constant values during motion).

Figure 11.10 illustrates a vise-grip at $\beta = 180°$ (compare with Fig. 11.9). In this posture, the input link 2 and the coupler link 3 are collinear and the mechanism is in its closed (locked) posture at which, based on Eq. (11.27), M. A. $= \infty$. Therefore, a small input torque can support a lot of output torque (that the object being held by the vise-grip might apply). Such a posture is widely used in locks and clamps. This posture also corresponds to a limit position of the follower link 4, as it is shown in Sect. 2.6 (see Fig. 2.51).

Furthermore, recalling the first-order kinematic coefficient θ_4' that is defined as (see Sect. 4.4):

$$\theta_4' = \frac{d\theta_4}{d\theta_2} \tag{11.28}$$

and comparing it with Eq. (11.27) one can conclude,

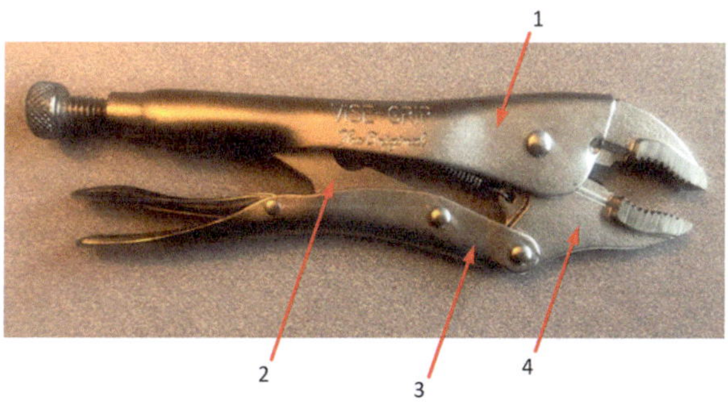

Fig. 11.10 A vise-grip in its closed (locked) posture

$$\theta_4' = \frac{d\theta_4/dt}{d\theta_2/dt} = \frac{\omega_4}{\omega_2} = \frac{\omega_{out}}{\omega_{in}} = \frac{1}{M.A.} \tag{11.29}$$

Therefore, in a four-bar linkage, the mechanical advantage is the inverse of the first-order kinematic coefficient of the output angle with respect to the input angle.

One can plot the variations of the mechanical advantage during motion in order to check whether it remains large enough during motion. As a rule of thumb, since sine of the transmission angle γ appears on the numerator of Eq. (11.27), this angle should not fall below around 45° or increase beyond 135°, otherwise the mechanism may not be able to complete its intended motion.

Example 11.3 Recalling Example 4.16, Fig. 11.11 illustrates a four-bar linkage at an instant when links 2 and 4 are both perpendicular to the frame link 1. We know that the input angular velocity is $\omega_{AB} = 2$ rad/s (CCW), $AB = 200$ mm, $BC = 400$ mm, and CD=300 mm. Calculate the mechanical advantage of the mechanism at the illustrated instant. (The input member is *AB* and the output member is *CD*.)

We recall that in this problem:

$$\omega_4 = 1.333 \text{ rad/s (CCW)}$$

Therefore, using Eq. (11.27) in the illustrated posture we have,

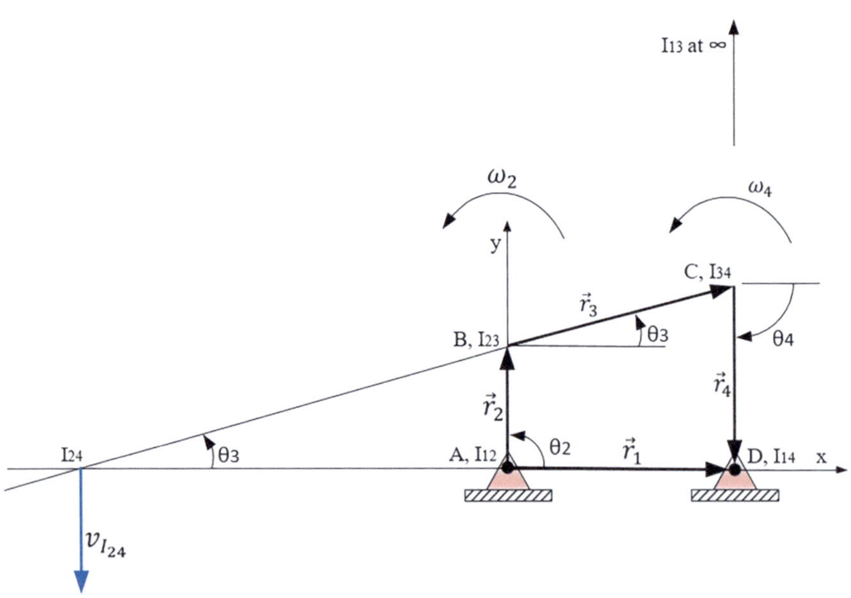

Fig. 11.11 Calculating the mechanical advantage in Example 4.16

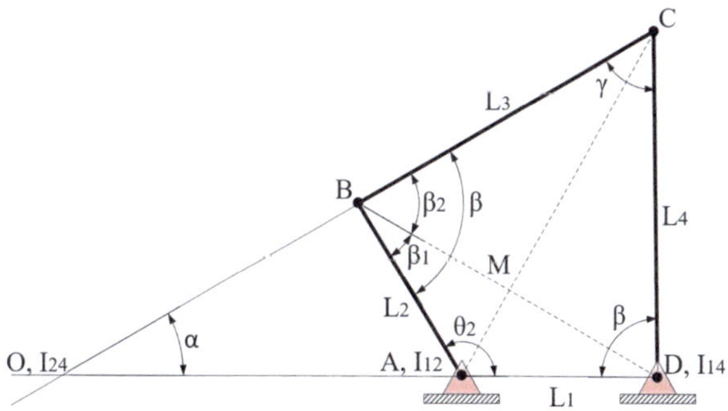

Fig. 11.12 Calculating the mechanical advantage of a four-bar linkage by analyzing its geometry

$$M.A. = \frac{T_{out}}{T_{in}} = \frac{\omega_{in}}{\omega_{out}} = \frac{2}{1.333} = 1.500$$

In the following example, we calculate the mechanical advantage of a four-bar linkage by analyzing its geometry.

Example 11.4 Consider the mechanism shown in Fig. 11.12 at $\theta_2 = 120°$. The input member is AB and the output member is CD. It is known that $AB = AD = 4$ in and $BC = CD = 7$ in.

A. Calculate M.A. by computing β and γ.
B. Calculate M.A. by locating point O.

A. **Calculate M.A. by computing β and γ.**

Recalling Eq. (11.27),

$$M.A. = \frac{T_{out}}{T_{in}} = \frac{\omega_{in}}{\omega_{out}} = \frac{DC \sin\gamma}{AB \sin\beta} = \frac{OD}{OA} \tag{11.27}$$

we need to calculate angles β and γ.

Calculating Angle β

Since the geometry of the mechanism is symmetrical about the AC line, one can conclude that,

$$\angle BAM = 0.5\theta_2 = 60°$$

So considering the right triangle BAM, we get:

$$\beta_1 = 30°$$

From analyzing triangle BAM,

$$BM = AB\cos\beta_1 = 4\cos 30\,^\circ = 3.464 \text{ in}$$

Now, studying triangle BCM results in,

$$BM = BC\cos\beta_2$$

or,

$$3.464 = 7\cos\beta_2$$

That gives, $\beta_2 = 60.340°$. Therefore we get the β angle as:

$$\beta = \beta_1 + \beta_2 = 30\,^\circ + 60.340\,^\circ = 90.340\,^\circ$$

Calculating Angle γ

To get angle γ, one can consider triangle BCM in which,

$$\beta_2 + 0.5\gamma = 90\,^\circ$$

Therefore, $\gamma = 59.320°$. Finally, the mechanical advantage is calculated by using Eq. (11.27) as follows:

$$M.A. = \frac{DC\sin\gamma}{AB\sin\beta} = \frac{7\sin 59.32}{4\sin 90.34} = 1.505$$

B. **Calculate M.A. by locating point *O***

One can obtain the same result by locating point O (aka I_{24}) and again use Eq. (11.27) for calculating the mechanical advantage. Let us first calculate angle α. For any triangle including triangle *COD* we have:

$$\alpha + \beta + \gamma = 180\,^\circ$$

Substituting the already calculated values for β and γ one obtains,

$$\alpha = 30.34\,^\circ$$

Now, to get length *OA*, we use law of sines in triangle OCD:

$$\frac{CD}{\sin\alpha} = \frac{OD}{\sin\gamma} \rightarrow \frac{7}{\sin 30.34} = \frac{OD}{\sin 59.32}$$

So, $OD = 11.918$ in and therefore, $OA = OD - AD = 7.918$ in. Consequently, Eq. (11.27) results in:

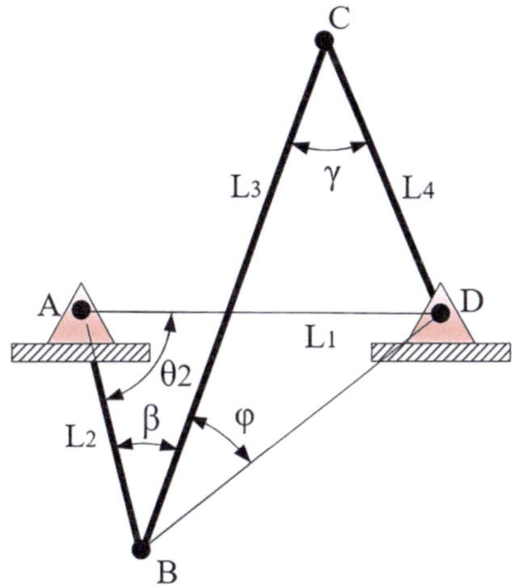

Fig. 11.13 Four-bar linkage in crossed posture

$$M.A. = \frac{OD}{OA} = \frac{11.918}{7.918} = 1.505$$

This result is in agreement with what was obtained in part A.

Since mechanical advantage is related to the torque ratio and angular velocity ratio, one can use it to calculate unknown torque or angular velocity values. The following example is based on calculation of mechanical advantage of a four-bar linkage that is in its crossed posture.

Example 11.5 Consider the mechanism shown in Fig. 11.13 when the crank angle is θ_2=80°. The input member is *AB* and the output member is *CD*. It is known that *AB* = 55 mm, *BC* = 200 mm, *CD* = 130 mm, and *AD* = 170 mm.

A. Using Grashof's law determine the type of motion of this mechanism.
B. Calculate the mechanical advantage at the illustrated instant.

A. **Using Grashof's law determine the type of motion of this mechanism.**

Recalling Grashof's law:

$$L_{min} + L_{max} < L_a + L_b \tag{2.39}$$

and using the given data we have,

$$L_{min} + L_{max} = 55 + 200 = 255 \text{ mm}$$

Also,

$$L_a + L_b = 130 + 170 = 300 \text{ mm}$$

So the Grashof's law is satisfied. Since the shortest link is beside the fixed link, this is a crank-rocker mechanism.

B. **Calculate the mechanical advantage at the illustrated instant.**

To calculate the mechanical advantage using Eq. (11.27):

$$M.A. = \frac{T_{out}}{T_{in}} = \frac{\omega_{in}}{\omega_{out}} = \frac{DC \sin \gamma}{AB \sin \beta} \tag{11.27}$$

we need to calculate the unknown angles β and γ. Here is a calculation procedure,

1. **Calculate length *BD*:** Considering triangle *ABD*, one may use law of cosines (see the Appendix for a review) to get *BD*,

$$BD^2 = AB^2 + AD^2 - 2AB.AD \cos \theta_2$$

So,

$$BD^2 = 55^2 + 170^2 - 2 \times 55 \times 170 \cos 80$$

Therefore, $BD = 169.35$ mm.

2. **Calculate angle γ:** Using law of cosines in triangle *BCD* one obtains,

$$BD^2 = BC^2 + CD^2 - 2BC.CD \cos \gamma$$

Or,

$$169.35^2 = 200^2 + 130^2 - 2 \times 200 \times 130 \cos \gamma$$

So, the transmission angle is: $\gamma = 57.13°$.

3. **Calculate angle φ:** Using law of sines in triangle *BCD* results in,

$$\frac{CD}{\sin \varphi} = \frac{BD}{\sin \gamma}$$

$$\frac{130}{\sin \varphi} = \frac{169.35}{\sin 57.13°}$$

Therefore, $\varphi = 40.15°$ or $\varphi = 139.85°$. In order to decide which one is correct, in triangle BCD law of cosines is used,

$$CD^2 = BC^2 + BD^2 - 2BC.BD\cos\varphi$$

or,

$$130^2 = 200^2 + 169.35^2 - 2 \times 200 \times 169.35\cos\varphi$$

So, $\cos\varphi = 0.764$. Since $\cos\varphi > 0$, we conclude: $\varphi < 90°$; so the correct answer is: $\varphi = 40.15°$. One can also verify that indeed, $\cos 40.15° = 0.764$.

4. **Calculate $\boldsymbol{\varphi + \beta}$:** Using law of sines in triangle ABD results in,

$$\frac{AD}{\sin(\varphi + \beta)} = \frac{BD}{\sin\theta_2}$$

or,

$$\frac{170}{\sin(\varphi + \beta)} = \frac{169.35}{\sin 80}$$

Which results in, $\varphi + \beta = 81.34°$ or $\varphi + \beta = 98.66°$. Again, to decide which answer is correct, one needs to use law of cosines in the same triangle, ABD,

$$AD^2 = AB^2 + BD^2 - 2AB.BD\cos(\varphi + \beta)$$

or,

$$170^2 = 55^2 + 169.35^2 - 2 \times 55 \times 169.35\cos(\varphi + \beta)$$

which gives,

$$\cos(\varphi + \beta) = 0.151$$

Since $\cos(\varphi + \beta) > 0$, one may conclude that $\varphi + \beta < 90°$, so the correct answer is: $\varphi + \beta = 81.34°$. One can also check to see that indeed, $\cos 81.34° = 0.151$, as expected.

5. **Calculate $\boldsymbol{\beta}$:** Having calculated $\varphi + \beta = 81.34°$, and knowing that, $\varphi = 40.15°$, one obtains,

$$\beta = 81.34° - 40.15° = 41.19°$$

6. **Calculate the M.A.:** At this step, the numerical values of all of the parameters in the M.A. Eq. (11.27) are known. Substituting the numerical values gives,

$$M.A. = \frac{T_{out}}{T_{in}} = \frac{\omega_{in}}{\omega_{out}} = \frac{130\sin 57.13^\circ}{55\sin 41.19^\circ} \rightarrow \text{M.A.} = 3.01$$

11.7 Slider-Crank Mechanism

Figure 11.14 illustrates a slider-crank mechanism in which the crank receives the input power and it is subjected to the torque T_{in} (i.e., the input torque). This torque is supposed to overcome the load applied on the slider, or F_{out}. In a can crusher, for example, T_{in} is applied by the user, and the resisting load F_{out} is the force imposed by the can while it is being crushed, see Fig. 3.46. In this case, the mechanical advantage is F_{out}/T_{in}:

$$\text{Mechanical Advantage (M.A.)} = F_{out}/T_{in} \tag{11.30}$$

The details of the geometry and force analysis of the mechanism can be seen in Fig. 11.15.

Considering the FBD of the whole mechanism, one can write,

$$\sum M_O = 0 \tag{11.31}$$

Therefore,

$$N(r_2\cos\theta_2 + r_3\cos\theta_3) - e.F_{out} - T_{in} = 0 \tag{11.32}$$

If the axial force generated in the two-force member 3 is shown by F_3, one can write the following equilibrium equations for the slider in the horizontal and the vertical directions,

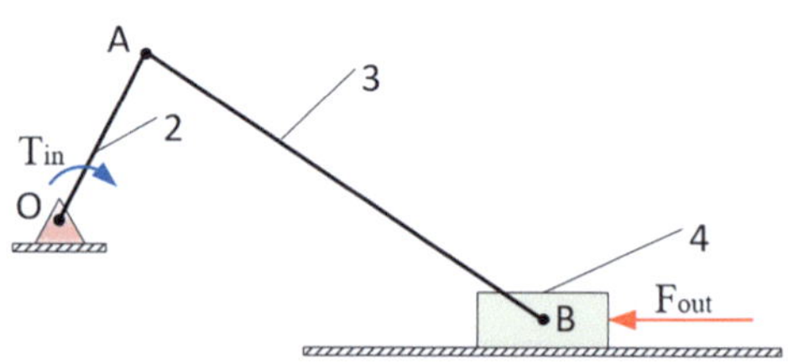

Fig. 11.14 Slider-crank mechanism with input torque applied on the crank and output force applied on the slider

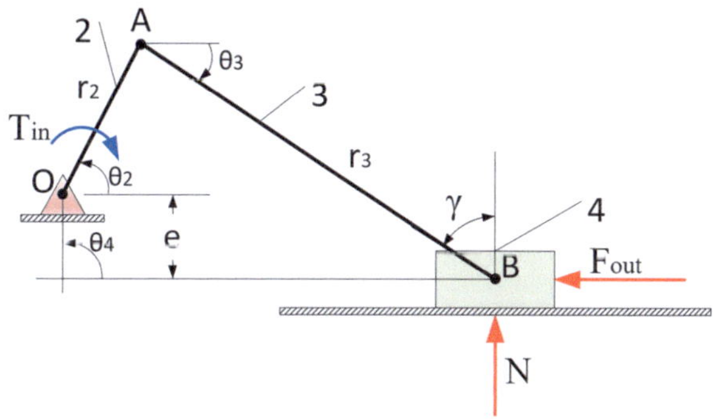

Fig. 11.15 Analyzing the mechanical advantage of a slider-crank mechanism

$$F_3 \sin \gamma = F_{out}; F_3 \cos \gamma = N \tag{11.33}$$

Dividing these equations eliminates the unknown force, F_3 and one obtains,

$$\tan \gamma = \frac{F_{out}}{N} \tag{11.34}$$

One can eliminate the unknown force N between Eqs. (11.32) and (11.34) and get,

$$\frac{F_{out}}{\tan \gamma}(r_2 \cos \theta_2 + r_3 \cos \theta_3) - e.F_{out} - T_{in} = 0$$

Therefore,

$$F_{out} = \frac{T_{in} \tan \gamma}{r_2 \cos \theta_2 + r_3 \cos \theta_3 - e \tan \gamma}$$

Finally, the mechanical advantage of a slider-crank mechanism with eccentricity is obtained as follows:

$$\frac{F_{out}}{T_{in}} = \frac{\tan \gamma}{r_2 \cos \theta_2 + r_3 \cos \theta_3 - e \tan \gamma} \tag{11.35}$$

Since during motion, angles θ_2, θ_3, and γ change, it is clear that the mechanical advantage of a slider-crank mechanism also varies. One also may notice that the higher angle γ (angle between the direction normal to the direction of motion of the slider and link 3) is, the higher the M.A. would be. That is why angle γ is called the transmission angle. It should also be pointed out that the mechanical advantage stated in **Eq. (11.35) is not unitless** (unlike all other mechanical advantages defined before).

For e = 0 and at $\gamma = 90°$ (illustrated in Fig. 11.16, also in Fig. 4.13) **the M.A. is infinity** and a lot of axial force can be supported with small input torque. This

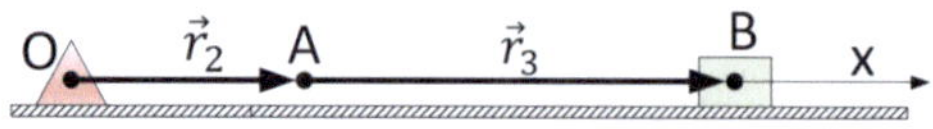

Fig. 11.16 Vector-loop diagram of a lider-crank mechanism at a position of infinite mechanical advantage (top dead center)

Fig. 11.17 Vector-loop diagram of a lider-crank mechanism at a position of zero mechanical advantage

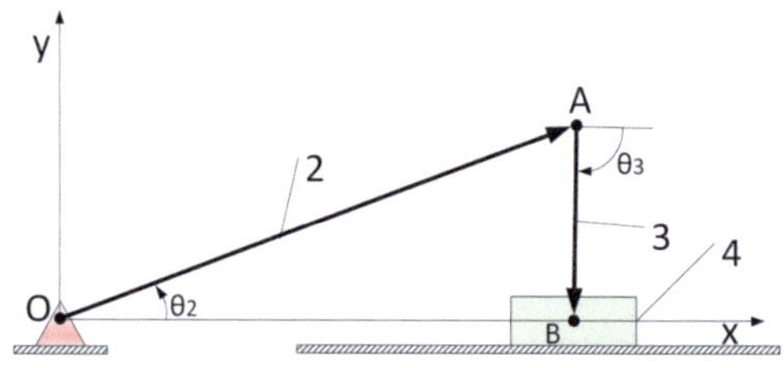

configuration is used frequently in punch presses and locking latches for obvious reasons. In punch presses a lot of force is needed at the lower end of the work stroke. However, little force is required during the remainder of the stroke. The high mechanical advantage happens at exactly when it is needed!, Sclater and Chironis [1], page 106.

At $\gamma = 0$, illustrated in Fig. 11.17, the M.A. obtained from Eq. (11.35) is zero and no output axial force on the frictionless slider can be supported when one applies input torque on the crank. The obvious reason for it is that at this instant all of the applied torque only creates a normal force that presses the slider on its horizontal guide and this force does not have any component, acting in the x direction, on the slider! (i.e., $F_{out} = 0$)

11.8 Cam and Follower

Consider Fig. 11.18 in which cam 2 is the input (driver) and follower 3 is the output (driven). Recalling the angular-velocity-ratio theorem in Sect. 4.7.3, the speed ratio can be written as follows:

$$\frac{\omega_{out}}{\omega_{in}} = \frac{\omega_3}{\omega_2} = \frac{I_{12}I_{23}}{I_{13}I_{23}} \tag{11.36}$$

where point I_{23} is the instantaneous center of 2 and 3.

Equation (11.36) shows that the speed ratio of members 2 and 3 depends only on the location of their common instantaneous center, I_{23}. In general, this ratio changes during motion. If the aim is to have a constant speed ratio, then I_{23} should be fixed. A similar logic was used in Chap. 6 for justifying that the pitch point in a gear mesh should remain fixed during motion, if one wants that the speed ratio remains constant (that, in gears, is usually the case).

If the input power is equal to the output power (i.e., power dissipations duc to friction are ignored), one can write,

$$P = T_2\omega_2 = T_3\omega_3 \tag{11.37}$$

By combining Eqs. (11.36) and (11.37), the mechanical advantage of the cam and follower mechanism can be written as follows:

$$M.A. = \frac{T_{out}}{T_{in}} = \frac{T_3}{T_2} = \frac{\omega_2}{\omega_3} = \frac{I_{13}I_{23}}{I_{12}I_{23}} \tag{11.38}$$

Equation (11.38) shows that the mechanical advantage of a cam and follower mechanism depends only on the location of their common instantaneous center, I_{23}.

Example 11.6 In the cam and follower mechanism shown in Fig. 11.18, cam 2 is the driver, and follower 3 is the driven member. Knowing that $\omega_2 = 100$ rpm, and at the illustrated instant, AP = 2 in and BP = 4 in, calculate:

A. the angular velocity ω_3 and
B. the mechanical advantage of the mechanism.

Using Eq. (11.38) we have:

$$M.A. = \frac{\omega_2}{\omega_3} = \frac{I_{13}I_{23}}{I_{12}I_{23}} = \frac{BP}{AP} \rightarrow M.A. = \frac{\omega_2}{\omega_3} = \frac{4\text{ in}}{2\text{ in}} = 2$$

So, the mechanical advantage at this instant is 2. Finally, for the angular velocity we have:

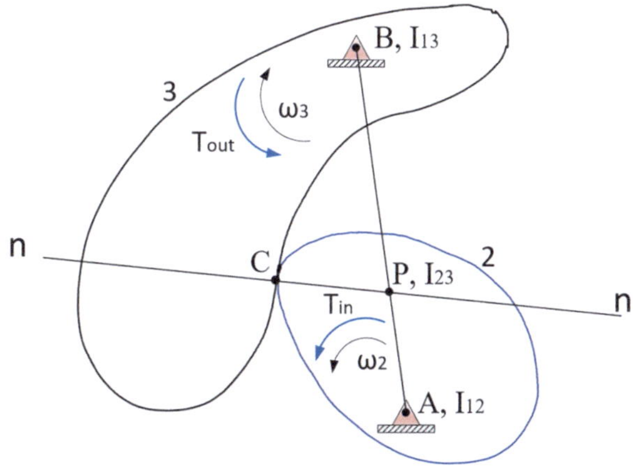

Fig. 11.18 Illustrating the angular-velocity-ratio theorem and mechanical advantage in a cam and follower mechanism

$$\frac{\omega_2}{\omega_3} = 2 \rightarrow \frac{100\ rpm}{\omega_3} = 2 \rightarrow \omega_3 = 50 \text{ rpm}$$

Reference

1. Sclater, N., & Chironis, N. P. (2007). *Mechanisms and mechanical devices* (4th ed.). McGraw-Hill.

Chapter 12
Review of Planar Dynamics

12.1 Introduction

In the chapters dedicated to kinematic analysis of mechanisms (i.e., Chaps. 2, 3, 4, 5, 6, 7, 8 and 9), attention was focused on calculating quantities like position, velocity, and acceleration, but no forces or moments were taken into consideration (*motions without forces*). In the static analysis (Chaps. 10 and 11), forces were included in the analysis, but accelerations were assumed to be zero or negligible (*forces without motions*). While statics can sometimes provide good first estimations for the solution of dynamic problems, there are many cases like resonance (in vibrations) where dynamic effects play the dominant role and cannot be ignored. In this chapter, dynamic problems in which neither accelerations nor forces are negligible are considered and the full kinetic problem is solved (*forces together with motions*).

Chapter 12 is the first one of the four chapters that are dedicated to the kinetic analysis of machines and mechanisms. In this chapter, a brief review of planar dynamics is presented and the three main solution methods, i.e., Newton's second law, impulse and momentum, and work and energy are briefly reminded. A good reference for a more comprehensive review of planar dynamics is, Meriam and Kraige [1].

12.2 Analysis by Newton's Second Law for Force and Moment

For a particle of mass m that is subjected to forces $\vec{F}_i$, $(i = 1 \text{ to } n)$, the resulting acceleration $\vec{a}$ of the particle is proportional to the resultant force and can be stated by Newton's second law as follows:

M. Ghorashi, *Introduction to Machine Dynamics*, Mechanical Engineering Series,
https://doi.org/10.1007/978-3-031-94736-0_12

$$\sum_{i=1}^{n} \vec{F}_i = m\vec{a} \tag{12.1}$$

This equation should be written by using an inertial reference frame because only an inertial reference frame can detect acceleration correctly. **For a system of particles**, if m_{tot} is the total mass of the system, $\vec{a}_G$ is the acceleration of point G (i.e., the mass center of the system), and $\vec{F}_i, (i = 1 \text{ to } n)$ are the external forces applied on the system of particles, the Newton's second law can be expressed as,

$$\sum_{i=1}^{n} \vec{F}_i = m_{tot}\vec{a}_G \tag{12.2}$$

Finally, **for a rigid body** of mass m, using the following version of Newton's second law, the acceleration of its mass center $\vec{a}_G$ is proportional to the resultant of external forces applied on it:

$$\sum_{i=1}^{n} \vec{F}_i = m\vec{a}_G \tag{12.3}$$

where $\vec{F}_i, (i = 1 \text{ to } n)$ are the external forces applied on the rigid body, as seen in Fig. 12.1.

In two-dimensional problems, the Newton's second law can be used to relate the summation of moments $\vec{M}_{G,i}, (i = 1 \text{ to } n)$ of external forces applied on a rigid body about an axis that passes through its mass center G and is perpendicular to the plane of motion, to the resulting angular acceleration of the object, α, as follows:

$$\sum_{i=1}^{n} \vec{M}_{G,i} = I_G\vec{\alpha} \tag{12.4}$$

where I_G is the mass moment of inertia of the object about an axis that passes through point G and is perpendicular to the plane of motion:

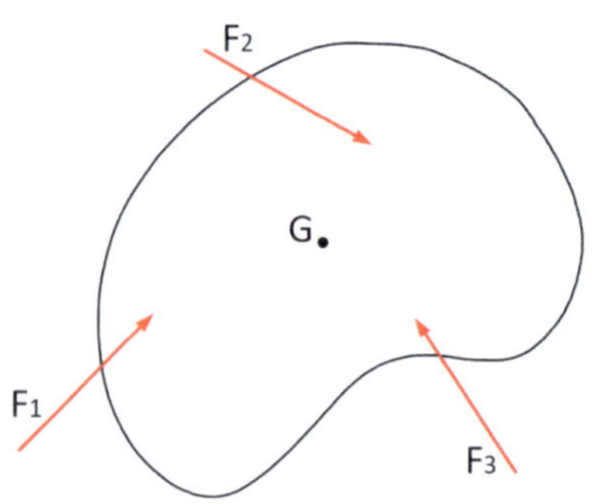

Fig. 12.1 Free body diagram of an object subjected to external forces

$$I_G = \int r^2 dm \tag{12.5}$$

In Eq. (12.5), the integral is calculated over all of the mass of the body, and distance r is measured from point G to a mass element, dm (see, e.g., Fig. 12.3). One is warned against confusing **mass moment of inertia**, discussed so far, with area moment of inertia that is frequently used in statics and strength of materials. **Area moment of inertia** is used, for example, for calculating normal bending stresses developed in a beam that is subjected to an applied bending moment M_z (see any book on strength of materials):

$$\sigma_x = -\frac{M_z y}{I_z} \tag{12.6}$$

Area moment of inertia about the z axis is defined as follows:

$$I_z = \int y^2 dA \tag{12.7}$$

In which dA is an element of area. So these quantities are completely different; however, due to the similarity of names and the similarity of notations they can be confused with each other.

Considering Eqs. (12.3) and (12.4), one can replace the system of forces shown in Fig. 12.1 and obtain the same acceleration and angular acceleration results by including only a single force $m\vec{a}_G$ acting at the center of mass and a single moment $I_G\vec{\alpha}$. The FBD of the object shown in Fig. 12.1 is therefore equivalent to what is commonly known as the kinetic diagram and is shown in Fig. 12.2. This diagram only includes the resultant elements of the stated equations of motion for force and moment.

A similar equation for two-dimensional problems can be written for the summation of moments $\vec{M}_{O,i}$, $(i = 1 \text{ to } n)$ of external forces applied on a rigid body about an

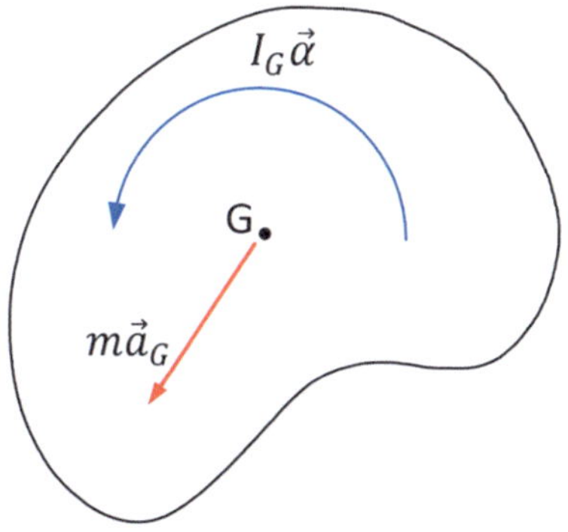

Fig. 12.2 Equivalent kinetic diagram for the FBD of a two-dimensional object shown in Fig. 12.1

axis that passes through a fixed point O and is perpendicular to the plane of motion, and the resulting angular acceleration of the object, α, as follows:

$$\sum_{i=1}^{n} \vec{M}_{O,i} = I_O \vec{\alpha} \tag{12.8}$$

in which I_O is the mass moment of inertia of the object about an axis that passes through point O and is perpendicular to the plane of motion:

$$I_O = \int r^2 dm \tag{12.9}$$

In Eq. (12.9) the integral is calculated over all of the mass of the body, and distance r is measured from point O to a mass element, dm.

In general, once the kinematic part of a problem is solved and accelerations and angular accelerations are determined, one can calculate the product of acceleration times mass and the product of angular acceleration times mass moment of inertia and get the right-hand side terms in Eqs. (12.3) and (12.4). In this way, acceleration and angular acceleration provide a strong link between kinematics and kinetics.

12.2.1 Mass Moment of Inertia and Moment Analysis

The mass moment of inertia Eqs. (12.5) and (12.9) show that moment of inertia can be changed by just **rearranging the distribution of mass in an object, without changing its total mass**. One important **application of this matter is in flywheels**. In an internal combustion engine, energy production is not a continuous function of time and it happens only when each cylinder experiences combustion. However, energy consumption in a highway at constant speed is continuous. So, an energy storage and regulating device like a flywheel is needed. A flywheel is a mechanical capacitor that absorbs kinetic energy when the supply of energy is more than demand and then gives it back when there is a shortage of supply compared to demand. In a flywheel, a high value of moment of inertia at minimal amount of mass is needed so that it is not made unnecessarily heavy (especially in transportation applications where weight reduction is a major consideration). So, we have seemingly conflicting objectives: increasing mass moment of inertia to increase its energy storage capacity, and decreasing mass to decrease the so-called dead weight. The contradiction is resolved by proper mass distribution and placing a considerable amount of mass in the outer rim. Therefore, we increase moment of inertia only by redistributing mass and for this matter we relocate a considerable part of mass to the outer edge from the inner regions where there is mass that does not contribute much to the value of the mass moment of inertia. Added mass at the outer edge of a flywheel reinforces the circumference and allows us to add teeth of a gear at the rim of a flywheel. In this way, the flywheel also serves as a gear and is used, for example, to engage with a starter gear in an internal combustion engine to transmit the starter power to the crankshaft and crank it to start the engine. (The inner hub of the flywheel is connected to the crankshaft that moves the piston rods and ultimately the pistons.)

Moments of inertia I_G and I_O are frequently expressed using **radii of gyration** k_G and k_O about G and O, respectively, and are defined as follows:

$$I_o = k_O^2 m \tag{12.10}$$

$$I_G = k_G^2 m \tag{12.11}$$

Therefore, the radii of gyration k_G or k_O are distances from G or O, respectively, that if the whole mass of the body is located at those points, then the resulting moment of inertia will be equal to I_G or I_O, respectively. While mass m does not need an axis for its definition, talking about a moment of inertia or radius of gyration without providing reference to the axis about which they are calculated is meaningless.

Example 12.1 Consider a homogeneous disk of mass m, mass density ρ, thickness t, and radius R that is shown in Fig. 12.3.

A. Derive the equation of mass moment of inertia of the disk about its mass center, I_G (i.e., about an axis that passes through point G and is perpendicular to the plane of motion).
B. Derive the equation of radius of gyration of the disk about the same axis, k_G.

A. **Derive the equation of mass moment of inertia I_G about its mass center.**

Recalling Eq. (12.5):

$$I_G = \int r^2 dm \tag{12.5}$$

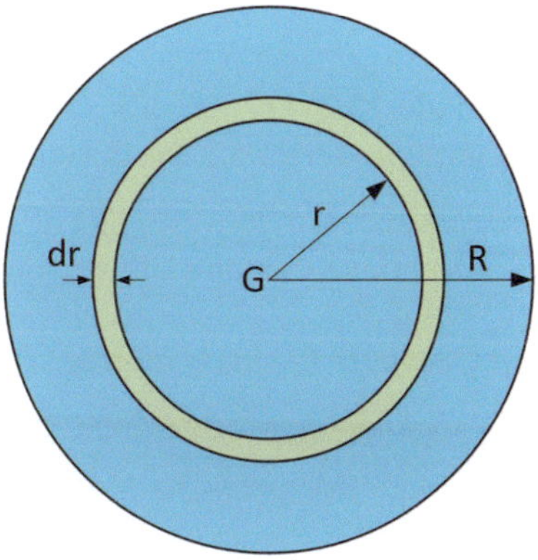

Fig. 12.3 Calculating the mass moment of inertia of a uniform disk

and by using the shown element dr, its corresponding ring element area, dA, and shell element volume, dV, Eq. (12.5) can be rewritten as follows:

$$I_G = \int r^2 dm = \int r^2 (\rho dV) = \int r^2 \rho t dA = \int_0^R r^2 \rho t (2\pi r dr) = \frac{2\pi \rho t}{4} R^4 \quad (12.12)$$

But for the mass density we have,

$$\rho = \frac{m}{V} = \frac{m}{\pi R^2 t} \quad (12.13)$$

Substitution of ρ into Eq. (12.12) gives,

$$I_G = \frac{1}{2} m R^2 \quad (12.14)$$

B. **Derive the equation of radius of gyration of the disk, k_G.**

Combining Eq. (12.11),

$$I_G = k_G^2 m \quad (12.11)$$

and Eq. (12.14) one obtains:

$$\frac{1}{2} m R^2 = k_G^2 m$$

Therefore, the radius of gyration of a uniform disk with radius R about its mass center is obtained as follows:

$$k_G = \frac{R}{\sqrt{2}} \quad (12.15)$$

We notice that k_G is a geometrical parameter and does not depend on mass of the disk.

Example 12.2 For the disk considered in Example 12.1, assume the following parameters: weight 0.2 lbf, thickness 0.5 in, and radius 2 in.

A. Calculate the mass moment of inertia of the disk about its mass center.
B. Determine its radius of gyration about the mass center.

A. **Calculate the mass moment of inertia of the disk about its mass center.**

To get the mass of a 0.2 lbf disk one can divide its weight by the gravitational acceleration, i.e., 32.2 ft/s^2. Then convert ft to in and get,

$$m = \frac{0.2\ \text{lbf}}{32.2\ \text{ft/s}^2} \times \frac{1\ \text{ft}}{12\ \text{in}} = \frac{0.2\ \text{lbf}}{386.4\ \text{in/s}^2} = 5.18 \times 10^{-4}\ \frac{\text{lbf}}{\text{in/s}^2}$$

Substituting the calculated value of m into Eq. (12.14) gives:

$$I_G - \frac{1}{2} mR^2 = \frac{1}{2} \times 5.18 \times 10^{-4} \frac{\text{lbf}}{\text{in/s}^2} \times (2\text{in})^2 = 1.035\ \text{lbf.in.s}^2$$

B. **Determine its radius of gyration about the mass center.**

Using Eq. (12.15) we have:

$$k_G = \frac{R}{\sqrt{2}} = \frac{2\ \text{in}}{\sqrt{2}} = 1.414\ \text{in}$$

Example 12.3 A ball of mass m and radius R is thrown on the horizontal surface of a table with the horizontal velocity v_0 of its mass center and with no initial angular velocity, see Fig. 12.4. If the kinetic coefficient of friction is μ, answer the following questions:

A. How long does it take for the ball to finish sliding and start its pure rolling motion (t_r)?
B. What are the velocity of the mass center and the angular velocity of the ball (v_r and ω_r, respectively) when the pure rolling motion starts?

A. **How long does it take for the ball to finish slipping and start its pure rolling motion (t_r)?**

Figure 12.5 illustrates the FBD of the ball after touching the surface of the table and during its combined rolling and sliding motion.

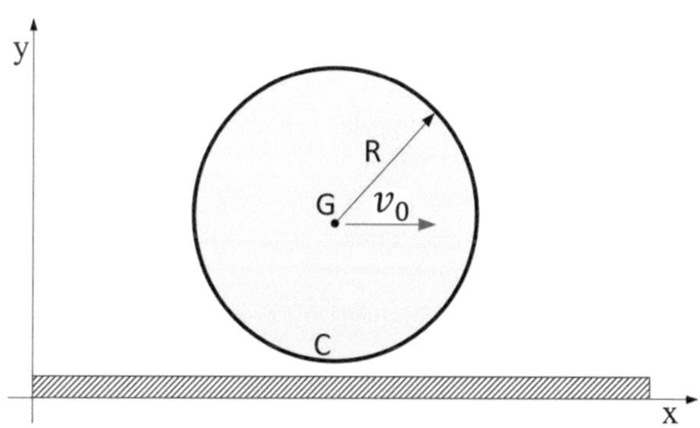

Fig. 12.4 A ball is thrown on a horizontal surface with initial horizontal velocity v_0 of its mass center

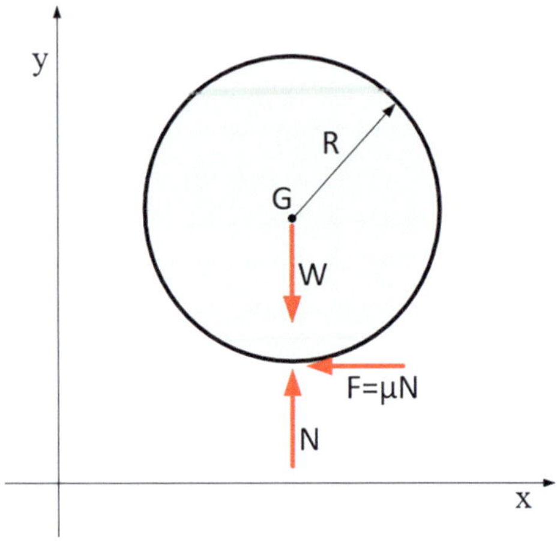

Fig. 12.5 Free body diagram of the released ball while rolling and sliding on the table

Using Newton's second law, i.e., Eq. (12.3), in the horizontal x direction we have:

$$-F = -\mu N = ma_G \tag{12.16}$$

In the vertical y direction, there is no acceleration. Therefore we have:

$$N = W = mg \tag{12.17}$$

Also, from the moment equation (12.4) we have:

$$FR = I_G \alpha \tag{12.18}$$

Therefore,

$$a_G = \frac{-F}{m}, \alpha = \frac{FR}{I_G} \tag{12.19}$$

Since a_G is negative, the linear velocity of the center of mass is decreasing, but since α is positive, the angular velocity of the ball is increasing. Sliding will end and pure rolling will start when,

$$v_r = R\omega_r \tag{12.20}$$

Recall kinematic equations for constant acceleration,

$$v_r = v_0 + a_G t \tag{12.21}$$

$$\omega_r = \alpha t \tag{12.22}$$

Sliding will end and pure rolling will start at t_r when condition (12.20) is satisfied. By substituting Eqs. (12.19), (12.21), and (12.22) into (12.20) we get:

$$v_0 - \frac{F}{m}t_r = R\left(\frac{FR}{I_G}\right)t_r \tag{12.23}$$

or,

$$\left(\frac{F}{m} + \frac{FR^2}{I_G}\right)t_r = v_0 \tag{12.24}$$

Combining Eqs. (12.16) and (12.17) gives,

$$F = \mu N = \mu mg \tag{12.25}$$

By substituting Eq. (12.25) into Eq. (12.24) we get,

$$\left(\mu g + \frac{\mu mgR^2}{I_G}\right)t_r = v_0$$

So the time t_r between touch down and when pure rolling starts to happen is obtained as follows:

$$t_r = \frac{v_0}{\mu g\left(1 + \frac{mR^2}{I_G}\right)} \tag{12.26}$$

For a uniform spherical ball of mass m and radius R we have,

$$I_G = \frac{2}{5}mR^2 \tag{12.27}$$

Substitution of Eq. (12.27) into (12.26) gives:

$$t_r = \frac{2v_0}{7\mu g} \tag{12.28}$$

B. **What are the velocity of the mass center and the angular velocity of the ball (v_r and ω_r, respectively) when the pure rolling motion starts?**

Using Eqs. (12.16), (12.17), (12.21), (12.22), and (12.28), the velocity of the mass center of the ball and its angular velocity when sliding ends are obtained as follows:

$$v_r = v_0 - \frac{F}{m} \times \frac{2v_0}{7\mu g} = v_0 - \frac{2v_0}{7} = \frac{5}{7} v_0 \tag{12.29}$$

$$\omega_r = \alpha t_r = \frac{FR}{I_G} \times \frac{2v_0}{7\mu g} = \frac{\mu mgR}{\frac{2}{5} mR^2} \times \frac{2v_0}{7\mu g} = \frac{5}{7} \frac{v_0}{R} \tag{12.30}$$

We see that Eqs. (12.29) and (12.30) satisfy the pure rolling condition (12.20):

$$v_r = R\omega_r \tag{12.20}$$

So, before sliding ends, the linear velocity of the center of mass decrease and correspondingly the angular velocity of the ball increase. Changes happen until the values of these quantities satisfy the pure rolling condition. After this instant, no sliding happens and therefore no sliding friction operates on the ball. Consequently, the resultant force in the horizontal direction is zero and the same is true for the resultant moment of forces about the mass center. So, we conclude that after t_r the velocity of the center of mass and the angular velocity of the ball no longer change. (In reality, however, other types of dissipation, especially rolling friction do decrease the speed of motion of the ball and eventually stop it.)

12.2.2 Parallel-Axis Theorem and Its Application

If the mass moment of inertia of an object of mass m about an axis that passes through its mass center G is I_G and one would like to calculate the mass moment of inertia of this object, I_A about a parallel axis passing through point A at distance d from the axis that passes through point G, the two values are related by,

$$I_A = I_G + md^2 \tag{12.31}$$

Example 12.4 Consider a uniform slender rod of mass m and length l, as shown in Fig. 12.6. It is known that the mass moment of inertia of this rod about axis x-x (that passes through its mass center) is,

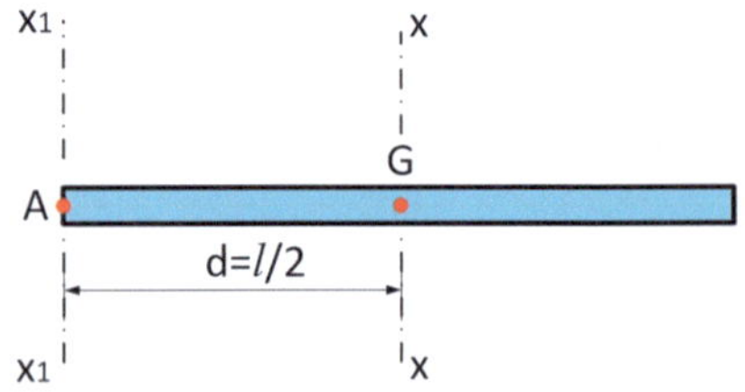

Fig. 12.6 Using the parallel-axis theorem to calculate the mass moment of inertia of a rod about an axis that passes through one end of it and is perpendicular to the rod

$$I_G = \frac{1}{12} m l^2 \tag{12.32}$$

Calculate the mass moment of inertia I_A about axis $x_1 - x_1$ that passes through one end of the rod.

In this case, $d = l/2$. Therefore, Eq. (12.31) results in:

$$I_A = I_G + md^2 = \frac{1}{12} m l^2 + m(l/2)^2 = \frac{1}{3} m l^2 \tag{12.33}$$

Example 12.5 The slender and uniform rod shown in Fig. 12.7 has 3 kg mass and is 2 m long. Calculate its mass moment of inertia and radius of gyration about axes that pass through any of the following points: *G*, *B*, and *A*, and are perpendicular to the rod.

Recalling Eq. (12.32) we have,

$$I_G = \frac{1}{12} m l^2 = \frac{1}{12} \times 3\ \text{kg} \times (2\text{m})^2 = 1\ \text{kg.m}^2$$

And for point *A*, using Eq. (12.33) one obtains,

$$I_A = \frac{1}{3} m l^2 = \frac{1}{3} \times 3\ \text{kg} \times (2\text{m})^2 = 4\ \text{kg.m}^2$$

Finally for point *B*, using the parallel-axis theorem (12.31), we have

$$I_B = I_G + m\left(\frac{l}{4}\right)^2 = 1\ \text{kg.m}^2 + \frac{1}{16} \times 3\ \text{kg} \times (2\text{m})^2 = 1.75\ \text{kg.m}^2$$

The corresponding radii of gyration are obtained using the definition given in equations like (12.10) and (12.11), i.e.,

$$I_G = k_G^2 m, I_A = k_A^2 m, I_B = k_B^2 m$$

Using the available data we get,

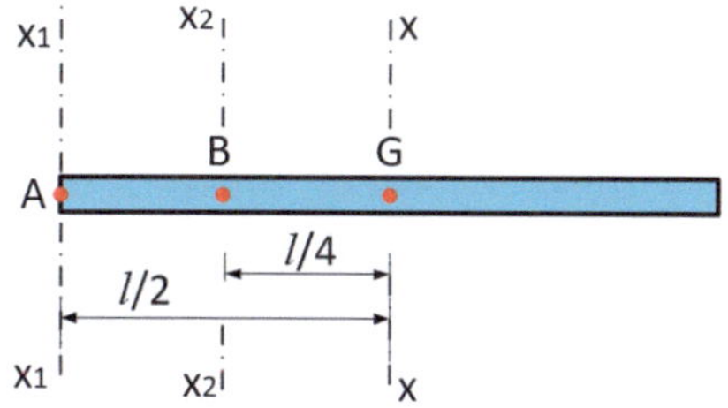

Fig. 12.7 Calculating radii of gyration of a rod about parallel axes that pass through various points along the rod

$$k_G = 0.577 \text{ m}, k_A = 1.154 \text{ m}, k_B = 0.764 \text{ m}$$

These results reveal that the further the measurement point for calculating mass moment of inertia is from the mass center, the higher the mass moment of inertia and the corresponding radius of gyration would be.

12.2.3 D'Alembert's Principle

Solving static problems is much easier than solving dynamic problems. There is a way by which a dynamic problem can be converted to an equivalent static problem, and in this way, be solved much easier. The method is based on what is called D'Alembert's principle.

Recall Newton's second law for a rigid body of mass m and acceleration of mass center $\vec{a}_G$:

$$\sum_{i=1}^{n} \vec{F}_i = m\vec{a}_G \tag{12.3}$$

Similarly, for moments we have:

$$\sum_{i=1}^{n} \vec{M}_{G,i} = I_G \vec{\alpha} \tag{12.4}$$

$$\sum_{i=1}^{n} \vec{M}_{O,i} = I_O \vec{\alpha} \tag{12.8}$$

One may rewrite these equations as follows:

$$\sum_{i=1}^{n} \vec{F}_i + \left(-m\vec{a}_G\right) = 0 \tag{12.34}$$

$$\sum_{i=1}^{n} \vec{M}_{G,i} + \left(-I_G \vec{\alpha}\right) = 0 \tag{12.35}$$

$$\sum_{i=1}^{n} \vec{M}_{O,i} + \left(-I_O \vec{\alpha}\right) = 0 \tag{12.36}$$

and consider them as equations of equilibrium, but this time, they are called **dynamic equilibrium equations**! In this way, we can state that a dynamic problem is converted to an equivalent static one if to the real forces and moments that exist in a problem, the so-called ***inertia force*** $-m\vec{a}_G$ and ***inertia moment*** $-I_G\vec{\alpha}$ or $-$

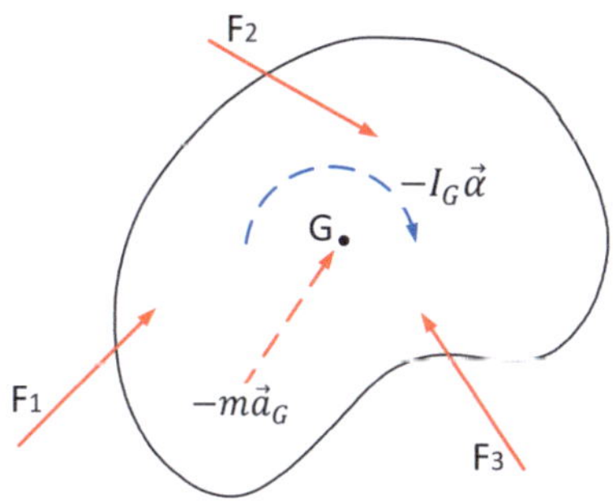

Fig. 12.8 The usual free body diagram (with solid arrows) is augmented with inertia forces and inertia moments (dashed arrows). It now displays an object in dynamic equilibrium

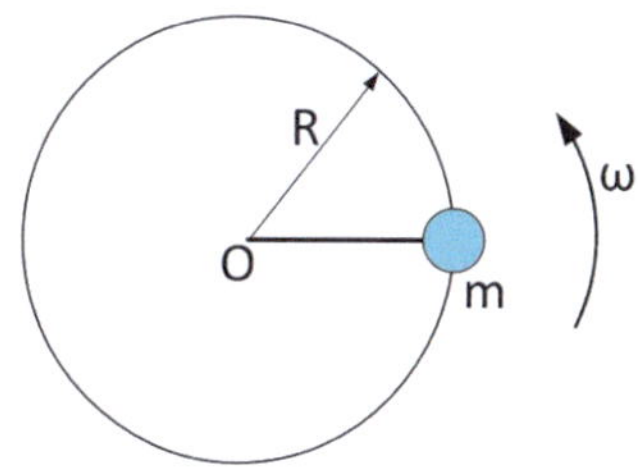

Fig. 12.9 Massless rod is hinged at point O and at the tip it is attached to a particle. The assembly is rotating about a vertical axis perpendicular to the plane of motion with constant angular velocity

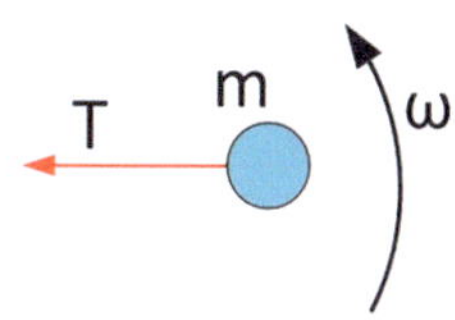

Fig. 12.10 Free body diagram of the tip mass in Fig. 12.9

$I_O\vec{\alpha}$ are added (see Fig. 12.8). The inertia force and inertia moment are shown by dashed arrows to differentiate them from the real forces and moments that are applied to the object. In other words, by adding inertia forces and inertia moments to a dynamic problem, it is converted to an equivalent static one that usually can be solved much faster and easier. This equivalency is the basis of the D'Alembert's principle.

Example 12.6 Consider a massless rod of length R that is hinged at point O and at the tip it is attached to a particle of mass m, as shown in Fig. 12.9. The assembly is rotating about a vertical axis (perpendicular to the plane of motion) at a constant angular velocity ω. Draw the FBD, the kinetic diagram, and the equivalent dynamic equilibrium diagram (obtained by the application of the D'Alembert's principle to the tip mass, m). Calculate the generated tensile force T in the rod if $m = 2$ kg, $R = 3$ m, and $\omega = 50$ rad/s.

The original FBD on the plane of motion (see Fig. 12.10) only includes the rod tensile force T that creates the centripetal acceleration of the tip mass. The kinetic diagram is shown in Fig. 12.11 and it only includes the mass times centripetal acceleration part (i.e., $-mr\omega^2$). Finally, by applying the inertia force (in this case,

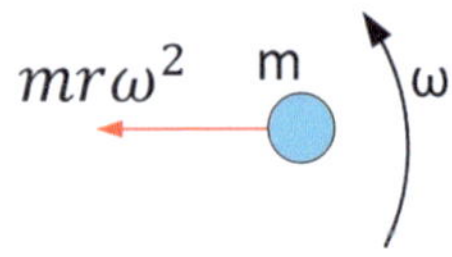

Fig. 12.11 Kinetic diagram (see Fig. 12.2) showing the centripetal force

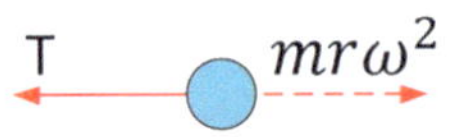

Fig. 12.12 Dynamic equilibrium of the tensile force T and the centrifugal (inertia) force using D'Alember's principle

it is also called the centrifugal force) $mr\omega^2$, and using D'Alembert's principle, we have the dynamic equilibrium shown in Fig. 12.12.

Considering the dynamic equilibrium of forces shown in Fig. 12.12 we have:

$$\sum \vec{F} = 0 \rightarrow -T + mr\omega^2 = 0$$

Therefore, the following relation between the tensile force T developed in the rod and the inertia force (centrifugal force) $mr\omega^2$ exists:

$$T = mr\omega^2 \tag{12.37}$$

By substituting the given numerical values for the parameters one obtains,

$$T = 2\ \text{kg} \times 3\ \text{m}\ (50\ \text{rad/s})^2 = 15\ \text{kN}$$

12.2.4 *Center of Percussion*

The center of percussion of a rod, bat, or a racquet that is held as a hinge support at one end is a point where a transversal force or impact does not generate any force at the hinge support (or the holding hand). That is why in various sports this point is also called the **sweet spot**. We would like to derive an equation for the location of this sweet spot.

Example 12.7 Consider an object that is being held from one of its ends in the vertical plane, as shown in Fig. 12.13. A. Find a general equation for the location of the center of percussion (sweet spot). B. Calculate the result for a uniform slender rod of mass m and length l.

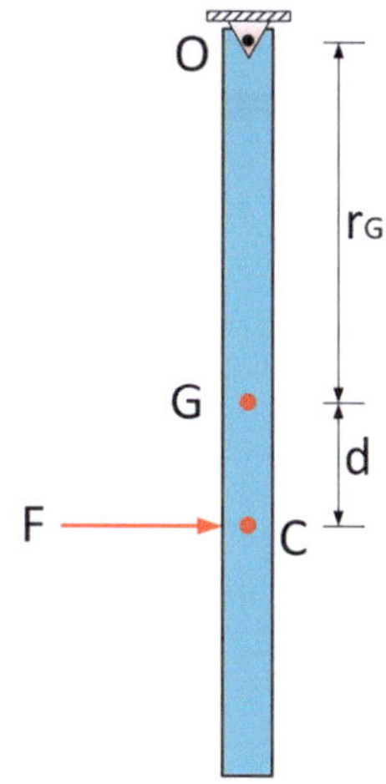

Fig. 12.13 Locating the center of percussion of a hinged rod constrained to move in the vertical plane

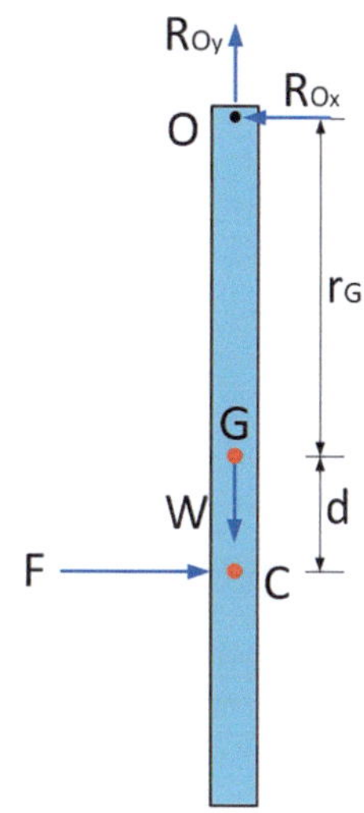

Fig. 12.14 Free body diagram of the hinged rod shown in Fig. 12.13

A. **Find a general equation for the location of the center of percussion (sweet spot).**

The FBD of the part can be drawn as shown in Fig. 12.14,

The Newton's second law in the x direction can then be written as follows:

$$\sum F_x = ma_{G_x}$$

And it results in,

$$F - R_{O_x} = mr_G\alpha \tag{12.38}$$

The Newton's second law in the y direction can be written as follows:

$$\sum F_y = ma_{G_y}$$

which results in,

$$R_{O_y} - W = mr_G\omega^2 \tag{12.39}$$

If from an impact analysis the angular velocity is known, it can be substituted into Eq. (12.39) and R_{O_y} be calculated. Since initially (i.e., in the vertical position) R_{O_y} does not generate any moment about point G, its value does not affect the rest of this analysis.

The equation of motion for moments about point G, i.e., Eq. (12.4) is reminded as follows:

$$\sum_{i=1}^{n} \vec{M}_{G,i} = I_G\,\vec{\alpha}$$

Using Fig. 12.14, the moment equation reduces to the following:

$$Fd + R_{O_x} r_G = I_G\alpha \tag{12.40}$$

This equation can be combined with the Newton's second law equation in the x direction,

$$F - R_{O_x} = mr_G\alpha \tag{12.41}$$

To satisfy the center of percussion condition, hinge reaction forces should be zero. Therefore,

$$R_{O_x} = 0 \tag{12.42}$$

Substituting Eq. (12.42) into Eqs. (12.40) and (12.41) results in,

$$Fd = I_G\alpha; F = mr_G\alpha \tag{12.43}$$

Dividing the two equations in (12.43) gives,

$$d = \frac{I_G}{mr_G} \tag{12.44}$$

where d is the distance between the center of percussion and the mass center (measured along the extension of OG).

B. **Calculate the result for a uniform slender rod of mass m and length l.**

For a uniform and slender rod of mass m and length l, the following is known,

$$I_G = \frac{1}{12} ml^2, r_G = \frac{l}{2} \tag{12.45}$$

Substituting Eqs. (12.45) into (12.44) gives,

$$d = \frac{l}{6} \tag{12.46}$$

Example 12.8 As an application of the D'Alembert's principle, consider Fig. 12.15 in which a conical pendulum consisting of a massless rod of length l is hinged at point O and has an attached tip mass m. The rod and the attached mass revolve at an angular velocity ω about a vertical axis. Calculate the steady-state angle θ of the pendulum as a function of ω and other system parameters.

Figure 12.15 shows that mass m is moving on a circular path with radius r. Therefore, the centripetal acceleration of this mass is $-r\omega^2$ and, like in Example 12.6, the inertia force or the centrifugal force applied on this mass is $mr\omega^2$. Figure 12.16 illustrates the FBD of the tip mass and its dynamic equilibrium using D'Alembert's principle. Therefore, adding inertia forces has converted the dynamic three-dimensional problem to a static two-dimensional one.

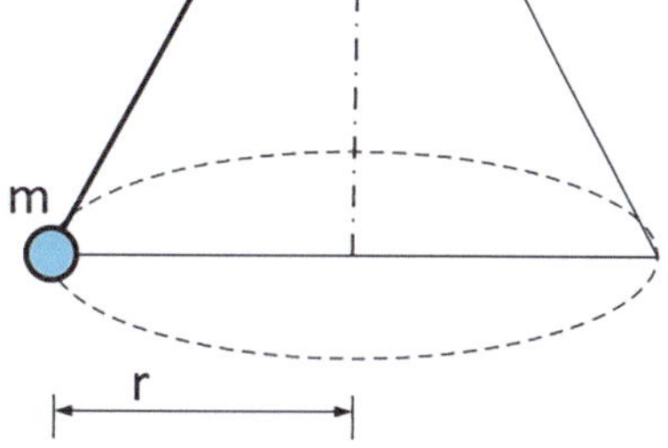

Fig. 12.15 Conical pendulum consisting of a massless rod hinged at point O and attached to tip mass m

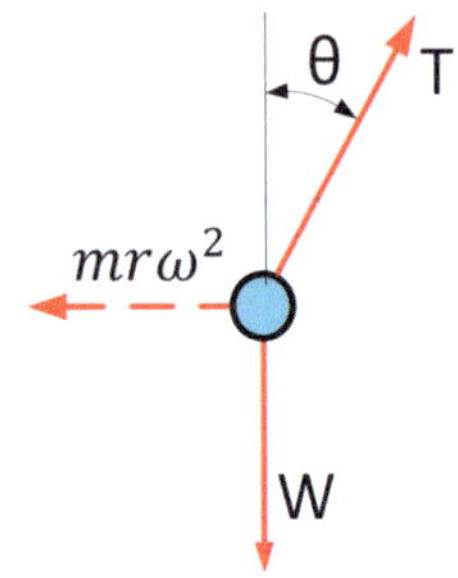

Fig. 12.16 Free body diagram of the tip mass and its related inertia force for the problem shown in Fig. 12.15

The equations of dynamic equilibrium are,

$$\sum F_x = 0 \rightarrow T\sin\theta - mr\omega^2 = 0 \tag{12.47}$$

$$\sum F_y = 0 \rightarrow T\cos\theta - W = 0 \tag{12.48}$$

Noticing that,

$$r = l\sin\theta;\, W = mg \tag{12.49}$$

Equations (12.47) and (12.48) reduce to,

$$T\sin\theta = ml\omega^2\sin\theta \tag{12.50}$$

$$T\cos\theta = mg \tag{12.51}$$

For any value of the angular velocity ω one trivial root of Eq. (12.50) is $\sin\theta = 0$ with the following results,

$$\theta = 0;\, 180\,° \tag{12.52}$$

In which case Eq. (12.51) gives,

$$T = mg \tag{12.53}$$

If $\sin\theta \neq 0$, Eq. (12.50) results in,

$$T = ml\omega^2 \tag{12.54}$$

By combining Eqs. (12.51) and (12.54) we get the following non-trivial solution,

$$\cos\theta = \frac{g}{l\omega^2} \tag{12.55}$$

Equation (12.55) can only be valid if it results in $\cos\theta \le 1$. It can be seen that as the angular velocity increases, cosine decreases, and the angle θ increases (as it is intuitively expected). We notice that the response angle of the pendulum does not depend on the mass of the tip object. The $\cos\theta \le 1$ condition results in,

$$\cos\theta \le 1 \rightarrow \omega^2 \ge \frac{g}{l} \tag{12.56}$$

Furthermore, the asymptotic limits of the ranges of θ and ω values are as follows:

$$\omega = \omega^* = \sqrt{\frac{g}{l}}; \theta = 0 \tag{12.57}$$

$$\omega \text{ is very large}; \theta = 90\,^\circ \tag{12.58}$$

At small values of ω, condition (12.56) is not satisfied. So, the other root, i.e., Eq. (12.52) is the correct answer. Therefore, we have:

$$\omega^2 \le \frac{g}{l} \rightarrow \theta = 0; 180\,^\circ \tag{12.59}$$

So, at this low-velocity range, the trivial solution $\theta = 0$ is the only solution that is stable. At the high-velocity range of $\omega^2 \ge \frac{g}{l}$, still θ=0 is a solution, but it is now unstable. These arguments are summarized in Fig. 12.17 where the stable solutions are in thick continuous lines and the unstable solutions are in thick dashed lines. The concept of stability of steady-state solutions is discussed in Chap. 15.

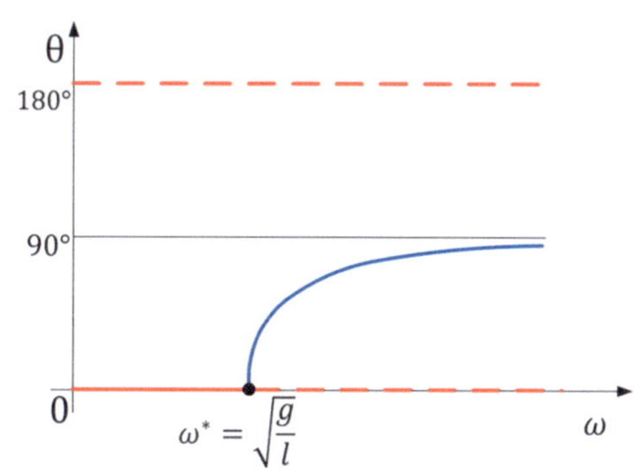

Fig. 12.17 Steady-state angle θ as a function of ω in Example 12.8

Example 12.9 Referring to the mechanism shown in Fig. 12.15, plot the response of a 1-m-long conical pendulum as its angular velocity increases from zero to 15 rad/s.

In this case, the ω^* value expressed in Eq. (12.57) can be calculated as follows:

$$\omega^* = \sqrt{\frac{g}{l}} = \sqrt{9.806/1} = 3.103 \text{ rad/s}$$

So the stable solution at $\omega \leq \omega^*$ is, $\theta = 0$. For $\omega \geq 3.103$ rad/s, the solution is given by Eq. (12.55):

$$\cos\theta = \frac{g}{l\omega^2} \tag{12.55}$$

For example,

$$\omega = 5 \text{ rad/s} \rightarrow \theta = 66.91\,^\circ$$

$$\omega = 15 \text{ rad/s} \rightarrow \theta = 87.50\,^\circ$$

The response, plotted by MATLAB, is shown in Fig. 12.18:

We notice that the solution given in Fig. 12.18 follows the same pattern as we saw in Fig. 12.17.

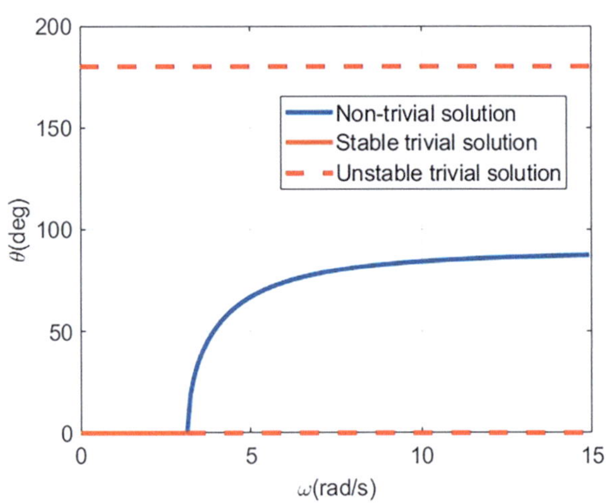

Fig. 12.18 Steady-state angle θ as a function of ω in Example 12.9

12.2.5 Equivalent Mass Moment of Inertia in Parallel Shafts

Figure 12.19 illustrates an electric motor that is running a rigid shaft 2 which in turn drives a gear set that reduces speed and increases torque and drives a parallel rigid shaft 3 which in turn drives a flywheel. The aim is to model an equivalent system that is only based on shaft 2; i.e., it only has the input and the output attributed to system 2 in it. In this way, we would like to transfer all of the elements that rotate with shaft 3, to equivalent elements that now rotate with shaft 2.

One can now draw the FBDs of the left-hand view of the two gears as shown in Fig. 12.20 (see Section 10.5).

The equation of motion of the gear 2 sub-assembly (that includes gear 2, its shaft, and motor parts that rotate with this shaft) using Eq. (12.8) can be written as follows:

$$\sum M_{O_2} = I_{O_2}\alpha_2 \tag{12.60}$$

Since shaft 2 is rigid, all elements connected to it rotate at the same angular displacement, angular velocity, and angular acceleration (θ_2, ω_2, and α_2, respectively). Therefore, if the total amount of the moments of inertia of all of the elements connected to shaft 2 about an axis passing through O_2 and perpendicular to the face of gear 2 is I_2, one obtains the following equation of motion for the gear 2 sub-assembly (not just gear 2). Notice that the CCW direction in the left-hand view is considered to be the positive direction.

$$T_m - R_2 F_{32}\cos\varphi = I_2\alpha_2 \tag{12.61}$$

In Eq. (12.61), T_m is the torque applied to the shaft 2 sub-assembly by the motor (and it is not necessarily equal to torque T_2 directly applied on gear 2 and shown on the FBD because of the inertia of other elements on sub-assembly 2). Notice that only the torque applied by the motor and the torque applied by the force exerted by gear 3 are the external torques applied on the gear 2 sub-assembly. (For this sub-assembly, torque T_2 is just an internal moment.)

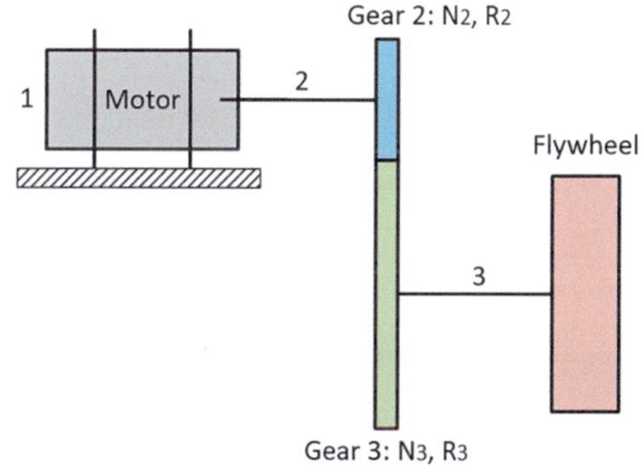

Fig. 12.19 A motor runs a flywheel by a gear set and parallel shafts

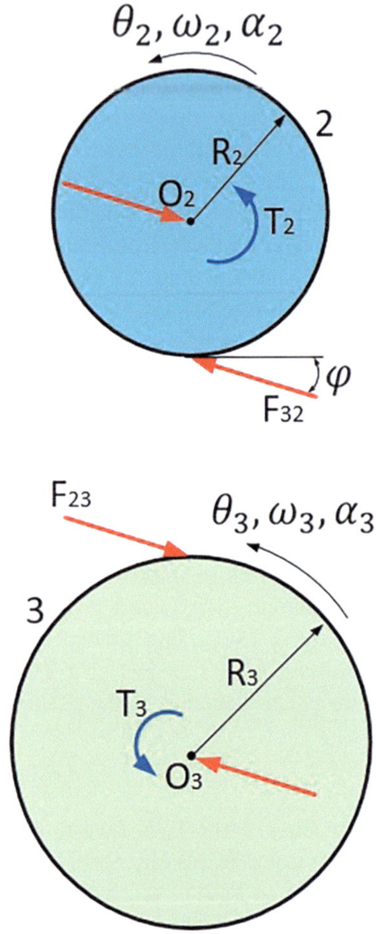

Fig. 12.20 Free body diagrams of the gears of the system shown in Fig. 12.19

Similarly, the equation of motion of the gear 3 sub-assembly (that includes gear 3, its shaft, and the attached flywheel) using Eq. (12.8) can be written as,

$$\sum M_{O_3} = I_{O_3}\alpha_3 \tag{12.62}$$

Again, since shaft 3 is rigid, all elements connected to it rotate at the same angular displacement, angular velocity, and angular acceleration (θ_3, ω_3, and α_3, respectively). Therefore, looking at Fig. 12.20, if the total amount of the moments of inertia of all of the elements connected to shaft 3 about an axis passing through O_3 and perpendicular to the face of gear 3 is I_3, one obtains,

$$-F_{23}\cos\varphi R_3 = I_3\alpha_3 \tag{12.63}$$

Notice that only the torque applied by the force exerted by gear 2 is an external torque applied on the gear 3 sub-assembly. (For this sub-assembly, torque T_3 directly applied on gear 3 and shown on the FBD is just an internal moment.) Equation (12.63) results in,

$$F_{23} = \frac{-I_3\alpha_3}{\cos\varphi R_3} \tag{12.64}$$

Since F_{32} and F_{23} are action and reaction forces, they have the same magnitudes, but act in opposite directions. Therefore,

$$\vec{F}_{23} = -\vec{F}_{32} \text{ so} : F_{23} = F_{32} \tag{12.65}$$

Therefore, Eq. (12.64) can be rewritten as follows:

$$F_{32} = \frac{-I_3\alpha_3}{\cos\varphi R_3} \tag{12.66}$$

Substituting Eq. (12.66) into Eq. (12.61) results in,

$$T_m = I_2\alpha_2 - \frac{R_2}{R_3}I_3\alpha_3 \tag{12.67}$$

For gears 2 and 3, the angular accelerations ratio is the same as the velocity ratio, the angular displacement ratio, and the inverse of the radii ratio with opposite sign. So we have:

$$\frac{\alpha_3}{\alpha_2} = \frac{\omega_3}{\omega_2} = \frac{\theta_3}{\theta_2} = -\frac{R_2}{R_3} = -\frac{N_2}{N_3} \tag{12.68}$$

The negative sign is due to the external contact of gears that causes them to always turn in opposite directions of each other (see Sect. 3.7, for example). Therefore, Eq. (12.67) reduces to,

$$T_m = I_2\alpha_2 - \frac{R_2}{R_3}I_3\left(-\frac{R_2}{R_3}\alpha_2\right) \rightarrow T_m = \left[I_2 + \left(\frac{R_2}{R_3}\right)^2 I_3\right]\alpha_2 \tag{12.69}$$

Equation (12.69) shows the way the moment of inertia of sub-assembly 3 should be modified in order to be added with the moment of inertia of sub-assembly 2. Therefore, **the equivalent moment of inertia of the whole system transferred to shaft 2, $I_{eq\ on\ 2}$**, that has input on shaft 2 (i.e., T_m) and also output on shaft 2 (i.e., α_2) is obtained as,

$$I_{eq\ on\ 2} = I_2 + \left(\frac{R_2}{R_3}\right)^2 I_3 \tag{12.70}$$

Alternatively, we may call 2 as the destination for the transfer of moments of inertias and 3 as the branch for this transfer and write,

$$I_{eq\ on\ destination} = I_{destination} + \left(\frac{R_{destination}}{R_{branch}}\right)^2 I_{branch} \tag{12.71}$$

The following numerical example illustrates these concepts.

Example 12.10 In the system shown in Fig. 12.19, the total mass moment of inertia of the parts on sub-system 2 (i.e., the contributing motor parts, shaft 2, and gear 2) is 0.1 lbf.in.s^2 and the total mass moment of inertia of parts on sub-system 3 (i.e., shaft 3, gear 3, and flywheel) is 5 lbf.in.s^2. If the motor is started by a torque of 100 lbf.in that operates in the CCW direction on the sub-system 2 in the left-hand view, calculate the resulting acceleration of the flywheel. We know that $N_2 = 20$ and $N_3 = 60$.

We can generate the equivalent system on shaft 2 using Eq. (12.70),

$$I_{eq\ on\ 2} = I_2 + \left(\frac{R_2}{R_3}\right)^2 I_3 \tag{12.70}$$

or equivalently,

$$I_{eq\ on\ 2} = I_2 + \left(\frac{N_2}{N_3}\right)^2 I_3 \tag{12.72}$$

Therefore,

$$I_{eq\ on\ 2} = 0.1 + \left(\frac{20}{60}\right)^2 \times 5 = 0.6556\ \text{lbf.in.}s^2$$

Now, substituting this result into Eq. (12.69), i.e.,

$$T_m = \left[I_2 + \left(\frac{R_2}{R_3}\right)^2 I_3\right]\alpha_2 \tag{12.69}$$

and using the available data gives,

$$100\ \text{lbf.in} = 0.6556\ \text{lbf.in}.s^2\alpha_2$$

or,

$$\alpha_2 = 152.53\ \text{rad/s}^2(\text{CCW, in the left} - \text{hand view})$$

But this is only the acceleration of system 2. In order to get the acceleration of the flywheel one needs to get α_3. So, we need to use Eq. (12.68),

$$\frac{\alpha_3}{\alpha_2} = \frac{\omega_3}{\omega_2} = \frac{\theta_3}{\theta_2} = -\frac{R_2}{R_3} = -\frac{N_2}{N_3} \tag{12.68}$$

and get,

$$\alpha_3 = \alpha_2\left(-\frac{N_2}{N_3}\right) = 152.53\frac{\text{rad}}{\text{s}^2} \times \left(-\frac{20}{60}\right) =$$
$$-50.74\ \text{rad/s}^2\ (\text{CW, in the left} - \text{hand view})$$

This is the acceleration of the flywheel right at the start of the engine.

The above discussion can be generalized to more complex systems of parallel shafts. First, consider Fig. 12.21 in which power is given to gear 2 and then it is distributed between gears 3 and 4 that both are in direct contact with gear 2. In this case, one can say that **the parallel shafts system is in parallel**! (Because power is distributed between two other parallel shafts directly from shaft 2.)

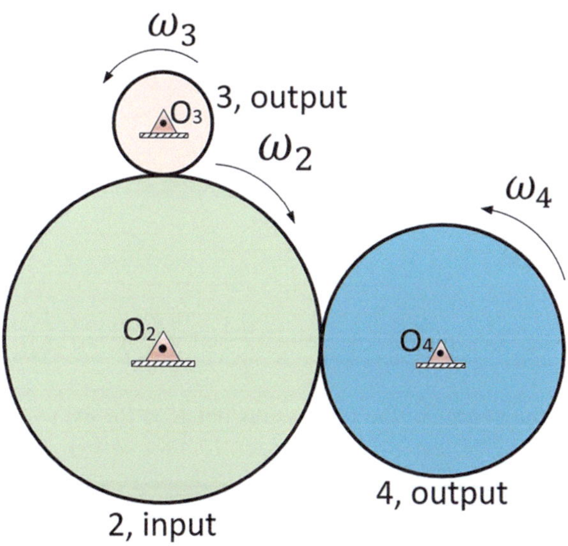

Fig. 12.21 A system of parallel shafts in parallel

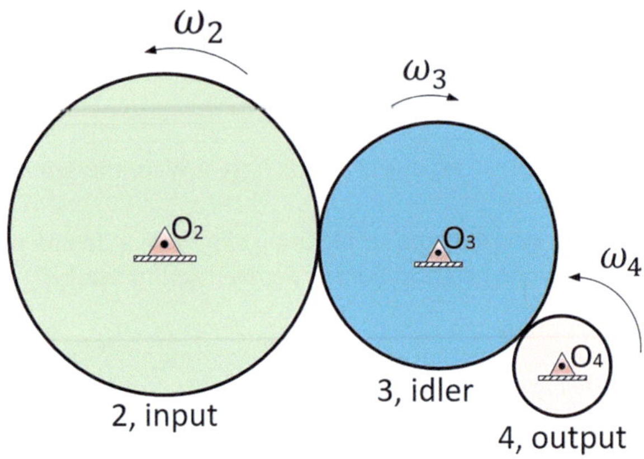

Fig. 12.22 A system of parallel shafts in series

The equivalent inertia transferred to the input system 2 can be obtained by successive use of Eq. (12.70),

$$I_{eq\ on\ 2} = I_2 + \left(\frac{R_2}{R_3}\right)^2 I_3 + \left(\frac{R_2}{R_4}\right)^2 I_4 \qquad (12.73)$$

Now, consider Fig. 12.22 where power is given to gear 2 and it transmits it first to gear 3 and then gear 3 transmits it to gear 4. So there is no direct contact between gear 2 and gear 4. In this case, one can say that **the parallel shafts system is in series**! (Because power is distributed to two other parallel shafts but in a sequential manner: first to 3 and then to 4.)

The equivalent inertia transferred to the input system 2 is calculated by using Eq. (12.70). The transfer from shaft 3 to 2 is exactly like in Eq. (12.70). The transfer from shaft 4 to 2 happens in two stages: first transfer from shaft 4 to 3 followed by a transfer from shaft 3 to 2. Therefore we obtain:

$$I_{eq\ on\ 2} = I_2 + \left(\frac{R_2}{R_3}\right)^2 I_3 + \left(\frac{R_2}{R_3}\right)^2 \left(\frac{R_3}{R_4}\right)^2 I_4$$

or,

$$I_{eq\ on\ 2} = I_2 + \left(\frac{R_2}{R_3}\right)^2 I_3 + \left(\frac{R_2}{R_4}\right)^2 I_4 \qquad (12.74)$$

Interestingly, the final results obtained for the two cases shown in Eqs. (12.73) and (12.74) are identical. In Sect. 12.4 the concept of equivalent inertia is also used for analyzing kinetic energy.

12.3 Analysis by the Impulse and Momentum Method

The impulse and momentum equation can be obtained by integrating the Newton's second law with respect to time. Recall Newton's second law for a rigid body,

$$\sum_{i=1}^{n} \vec{F}_i = m\vec{a}_G \tag{12.3}$$

which can be rewritten as,

$$\sum_{i=1}^{n} \vec{F}_i = \frac{d\left(m\vec{v}_G\right)}{dt} \tag{12.75}$$

So, the sum of the applied forces on an object at an instant is equal to the time derivative of the linear momentum of that object. Now, one may integrate both sides of Eq. (12.75) in a time span of interest (let us say from t_0 to t_f) to get the following result:

$$\sum_{i=1}^{n} \vec{F}_i dt = d\left(m\vec{v}_G\right) \rightarrow \int_{t_0}^{t_f} \sum_{i=1}^{n} \vec{F}_i dt = \int d\left(m\vec{v}_G\right)$$

Finally,

$$\int_{t_0}^{t_f} \sum_{i=1}^{n} \vec{F}_i dt = m\vec{v}_G\left(t_f\right) - m\vec{v}_G(t_0) \tag{12.76}$$

The left-hand side is the sum of the linear impulses of all external forces applied on the object and the right-hand side is the change in its linear momentum, both measured in the same time interval.

Similarly, recalling the sum of moments equations for a rigid body about mass center G and fixed point O, i.e.,

$$\sum_{i=1}^{n} \vec{M}_{G,i} = I_G\vec{\alpha} \tag{12.4}$$

and,

$$\sum_{i=1}^{n} \vec{M}_{O,i} = I_O\vec{\alpha} \tag{12.8}$$

they can be rewritten as,

$$\sum_{i=1}^{n} \vec{M}_{G,i} = \frac{d\left(I_G \vec{\omega}\right)}{dt} \tag{12.77}$$

and,

$$\sum_{i=1}^{n} \vec{M}_{O,i} = \frac{d\left(I_O \vec{\omega}\right)}{dt} \tag{12.78}$$

So, the sum of the applied moments about G or O on an object at an instant is equal to the time derivative of the angular momentum of that object about G or O. Integration of both sides of Eqs. (12.77) and (12.78) with respect to time in a time span of interest (let us say from t_0 to t_f) results in,

$$\int_{t_0}^{t_f} \sum_{i=1}^{n} \vec{M}_{G_i} dt = I_G \vec{\omega}\left(t_f\right) - I_G \vec{\omega}\left(t_0\right) \tag{12.79}$$

and,

$$\int_{t_0}^{t_f} \sum_{i=1}^{n} \vec{M}_{O_i} dt = I_O \vec{\omega}\left(t_f\right) - I_O \vec{\omega}\left(t_0\right) \tag{12.80}$$

The left-hand sides of Eqs. (12.79) and (12.80) are the sum of angular impulses of all external moments applied on the object (about G or O, respectively) and the right-hand sides are the changes in the angular momenta of the object about the same points, all measured in the same time interval.

If the net linear or the net angular impulse in a time period is zero, the right-hand side of the corresponding Eq. (12.76), (12.79) or (12.80) will be zero too. Therefore, in the mentioned time period that particular momentum will be conserved.

Example 12.11 One would like to hit a billiard ball of mass m (shown in Fig. 12.23) horizontally in such a way that the ball would only roll without slipping and therefore no friction force is generated at its point of contact with the table. Calculate the necessary vertical distance, d, between the line of action of this force and the mass center of the ball.

Starting with the kinematics of the problem, since the motion of the ball is rolling without slipping, the contact point C is the instantaneous center of zero velocity of the ball. Furthermore, considering the CW direction of rolling of the ball, ω is negative and the velocity of the mass center of the ball can be written as follows:

Fig. 12.23 Horizontal impact on a billiard ball

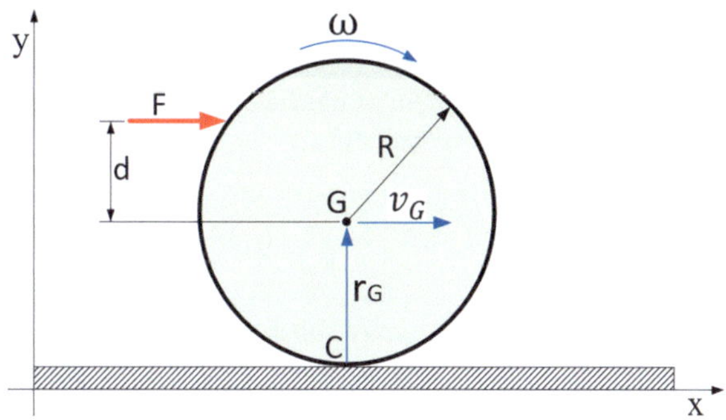

$$\vec{v}_G = \vec{\omega} \times \vec{r}_G = \omega\hat{k} \times r_G\hat{j} = -r_G\omega\hat{i} \tag{12.81}$$

Now, we use the kinetic equations of impulse and momentum:

$$\int_{t_0}^{t_f} \sum_{i=1}^{n} \vec{F}_i dt = m\vec{v}_G(t_f) - m\vec{v}_G(t_0) \tag{12.76}$$

$$\int_{t_0}^{t_f} \sum_{i=1}^{n} \vec{M}_{G_i} dt = I_G\vec{\omega}(t_f) - I_G\vec{\omega}(t_0) \tag{12.79}$$

If the duration of application of the linear impulse is Δt, Eqs. (12.76) and (12.79) result in,

$$F.\Delta t = mv_G \tag{12.82}$$

$$-Fd.\Delta t = I_G\omega \tag{12.83}$$

where v_G and ω are the final linear velocity of the mass center of the ball and the final angular velocity of the ball after the impact, respectively. Eqs. (12.81), (12.82), and (12.83) are three equations from kinematics and kinetics with three unknowns, d, v_G, and ω. Dividing the obtained kinetics Eqs. (12.82) and (12.83) gives,

$$d = \frac{-I_G\omega}{mv_G} \tag{12.84}$$

Substituting v_G from the kinematics Eq. (12.81) into (12.84) results in,

$$d = \frac{I_G}{mr_G} \tag{12.85}$$

Interestingly, Eq. (12.85) is the same as Eq. (12.44) for the center of percussion obtained using Newton's second law! And the obtained location is indeed the center of percussion of the billiard ball. The pure rolling condition that necessitates zero friction force at the contact point, imposes the same zero force condition as we had for the center of percussion (see Sect. 12.2.4).

Now, the mass moment of inertia of a uniform spherical ball of mass m and radius R about an axis that passes through its mass center is,

$$I_G = \frac{2}{5} m R^2 \tag{12.86}$$

Noticing $r_G = R$, one may substitute Eq. (12.86) into Eq. (12.85) and obtain the correct location of the center of percussion, d^*as follows:

$$d^* = \frac{2}{5} R \tag{12.87}$$

Finally, considering the distance of the applied force from the center of mass of the ball as a variable, d, and studying Eqs. (12.82) and (12.83), one concludes that while v_G does not depend on d, ω does, and we have:

$$\begin{cases} \text{if } d < d^*; \text{the ball rotates slower than the pure rolling case} \\ \text{if } d > d^*; \text{the ball rotates faster than the pure rolling case} \end{cases} \tag{12.88}$$

12.4 Application of the Work and Energy Method

The work and energy equation can be obtained by integrating Newton's second law with respect to displacement. As an example, consider a particle of mass m subjected to forces $\vec{F}_i (i = 1 \text{ to } n)$. From Newton's second law we have,

$$\sum_{i=1}^{n} \vec{F}_i = m \vec{a} \tag{12.89}$$

If the particle is displaced from its initial position $\vec{r}_1$ to its final position $\vec{r}_2$, considering $d\vec{r}$ to be tangent to the path of motion of the particle, one can dot product both sides of Eq. (12.89) by $d\vec{r}$ and integrate,

$$\int_{\vec{r}_1}^{\vec{r}_2} \sum_{i=1}^{n} \vec{F}_i . \mathrm{d}\vec{r} = \int_{\vec{r}_1}^{\vec{r}_2} m \vec{a} . \mathrm{d}\vec{r} \tag{12.90}$$

The left-hand side is the total work done by all external forces during this displacement and it is shown by U_{1-2}. In the right-hand side, since $\mathrm{d}\vec{r}$ is tangent to the path of motion, the non-zero part of the dot product will only have the tangential acceleration, a_t and we get,

$$U_{1-2} = \int_{s_1}^{s_2} ma_t ds \tag{12.91}$$

in which, ds is the magnitude of $d\vec{r}$. But, from the definition of tangential acceleration we have,

$$a_t ds = vdv \tag{12.92}$$

Therefore, Eq. (12.91) reduces to the following,

$$U_{1-2} = \int_{s_1}^{s_2} mvdv \rightarrow U_{1-2} = \frac{1}{2}m\left(v_2^2 - v_1^2\right) \tag{12.93}$$

By defining,

$$T = \frac{1}{2}mv^2 \tag{12.94}$$

as the kinetic energy of a particle of mass m and moving at velocity v, Eq. (12.93) can be interpreted as,

$$U_{1-2} = T_2 - T_1 \tag{12.95}$$

This is the **Work and Kinetic Energy Principle** for a particle. In words, it states that the work done by all external forces on a particle between its two positions $\vec{r}_1$ and $\vec{r}_2$ is equal to the change in the kinetic energy of that particle between those two positions (also shown by ΔT).

The work done by all forces, U_{1-2}, can be broken down into the work done by gravitational force (that is the negative of the change in the gravitational potential energy, ΔV_g), the work done by elastic forces (that is negative of the change in the elastic potential energy, ΔV_e) and the work done by non-conservative forces, shown by U'_{1-2}. Therefore, Eq. (12.95) could be expanded as,

$$\left(-\Delta V_g\right) + \left(-\Delta V_e\right) + U'_{1-2} = \Delta T$$

And finally we get,

$$U'_{1-2} = \Delta T + \Delta V_g + \Delta V_e \tag{12.96}$$

Equation (12.96) is also valid for rigid bodies. For a rigid body, together with the work done by forces, that was presented in Eq. (12.90), we need to define work done by moments. A moment $\vec{M}$ shown in Fig. 12.24, is the resultant of a force couple $\vec{F}$ in such a way that we have:

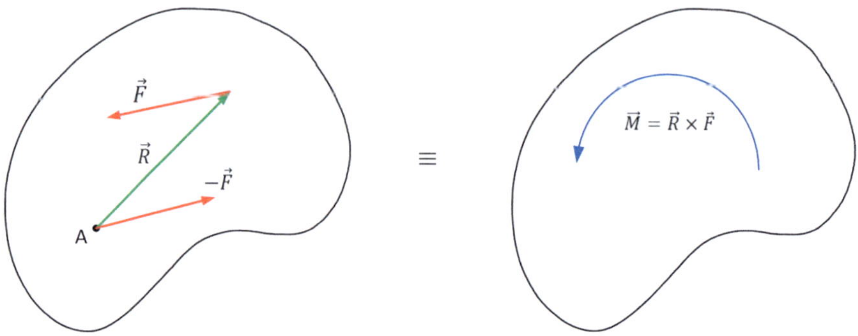

Fig. 12.24 The identity between a moment and a force couple having the same resultant moment

$$\vec{M} = \vec{R} \times \vec{F} \tag{12.97}$$

The corresponding work during an angular displacement from position θ_1 to θ_2 of the rigid body can be expressed as,

$$\int_{\theta_1}^{\theta_2} M d\theta \tag{12.98}$$

Also, the kinetic energy of a rigid body of mass m and mass moment of inertia about the mass center I_G in planar motion can be expressed as,

$$T = \frac{1}{2} m v_G^2 + \frac{1}{2} I_G \omega^2 \tag{12.99}$$

where v_G is the velocity of its mass center and ω is the angular velocity of the object. If the object rotates about a fixed point O, Eq. (12.99) reduces to,

$$T = \frac{1}{2} I_O \omega^2 \tag{12.100}$$

The same form of equation is valid if a point C is found to be the instantaneous center of zero velocity of the object,

$$T = \frac{1}{2} I_C \omega^2 \tag{12.101}$$

Recalling,

$$U'_{1-2} = \Delta T + \Delta V_g + \Delta V_e \tag{12.96}$$

in case there are no non-conservative forces operating on the object, i.e.,

$$U'_{1-2} = 0$$

Equation (12.96) reduces to:

$$\Delta T + \Delta V_g + \Delta V_e = 0 \tag{12.102}$$

or,

$$\Delta\left(T + V_g + V_e\right) = 0 \tag{12.103}$$

Naming,

$$E = T + V_g + V_e \tag{12.104}$$

as the total mechanical energy, by reviewing Eqs. (12.103) and (12.104) one can conclude that if no non-conservative forces exist, then: $\Delta E = 0$. So during motion the total mechanical energy, *E*, of the object remains constant. **This is the law of conservation of energy.**

Power is the time rate of performing work. So it is the derivative of work with respect to time. For a force $\vec{F}$ applied on an object that during motion of the body its point of application on the body has been displaced by $\overrightarrow{dr}$, the differential amount of work is,

$$dU = \vec{F}.\mathrm{d}\vec{r} \tag{12.105}$$

So, the corresponding power of force $\vec{F}$ during time interval *dt* is,

$$P = \frac{dU}{dt} = \vec{F}.\frac{\mathrm{d}\vec{r}}{dt} = \vec{F}.\vec{v} \tag{12.106}$$

In a similar way, in planar motion for a moment *M* applied on an object during its rotation by $d\theta$ the differential amount of work is,

$$dU = Md\theta \tag{12.107}$$

So, the corresponding power of moment *M* during time interval *dt* is,

$$P = \frac{dU}{dt} = M\frac{\mathrm{d}\theta}{dt} = M\omega \tag{12.108}$$

Clearly, the total power is the summation of the one associated with forces and the one associated with moments. Adding Eqs. (12.106) and (12.108), one obtains:

$$P = \vec{F}.\vec{v} + M\omega \tag{12.109}$$

Example 12.12 Consider the machine shown in Fig. 12.19 that was discussed in Sect. 12.2.5. In this machine a motor is running a flywheel through a speed reduction gear set. The total mass moment of inertia of the parts on sub-system 2 (i.e., the contributing motor parts, shaft 2, and gear 2) is 0.1 lbf.in.s^2 and the total mass moment of inertia of parts on sub-system 3 (i.e., shaft 3, gear 3, and flywheel) is 5 lbf.in.s^2. Calculate the necessary constant motor power (in hp) so that starting from rest and after 3 s of operation the angular velocity of shaft 2 reaches 3000 rpm. Neglect any friction losses. Finally, the gear teeth numbers are as follows: $N_2 = 20$ and $N_3 = 60$.

In Sect. 12.2.5 for the system shown in Fig. 12.19, an equivalent rotational system was created in which all of the inertia in sub-system 3 was transferred to sub-system 2 (where both system input and output referred to sub-system 2). In the transferred system, all elements rotating with shaft 3 were converted to equivalent elements that rotate with shaft 2. The equivalent mass moment of inertia of the system on shaft 2 was calculated as,

$$I_{eq\ on\ 2} = I_2 + \left(\frac{R_2}{R_3}\right)^2 I_3 \tag{12.70}$$

or equivalently,

$$I_{eq\ on\ 2} = I_2 + \left(\frac{N_2}{N_3}\right)^2 I_3 \tag{12.72}$$

Therefore,

$$I_{eq\ on\ 2} = 0.1 + \left(\frac{20}{60}\right)^2 \times 5 = 0.6556\ \text{lbf.in.}s^2 \tag{12.110}$$

Since the input power is P and it is constant, after time t the total amount of energy given to the system is Pt. Using Eq. (12.100), the resulting change in the total kinetic energy of the system is,

$$\Delta T = \frac{1}{2} I_{eq\ on\ 2} \times \left(\omega_{2f}^2 - \omega_{2i}^2\right) \tag{12.111}$$

in which ω_{2i} and ω_{2f} are the initial and the final angular velocities of shaft 2, respectively. Since energy losses in the system are negligible, the energy given to the system is equal to the change in its kinetic energy. The final angular velocity is,

$$\omega_{2f} = \frac{3000 \text{ rev}}{\text{min}} \times \frac{2\pi \text{ rad/rev}}{60 \text{ s/ min}} = 314.16 \text{ rad/s} \tag{12.112}$$

Therefore, the input energy can be related to the resulting kinetic energy as follows:

$$P.t = \frac{1}{2} \times 0.6556 \text{ lbf.in.}s^2 \left[(314.16)^2 - 0\right] \tag{12.113}$$

or,

$$P.t = 32352.71 \text{ lbf.in} \rightarrow \text{P} \times 3\text{s} = 32352.56 \text{ lbf.in}$$

Finally we get,

$$P = 10784.2 \text{ lbf.}\frac{\text{in}}{\text{s}} = 898.7 \text{ lbf.}\frac{\text{ft}}{\text{s}} = 898.7 \text{ lbf.ft/s} \times \frac{1 \text{ hp}}{550 \text{ lbf.ft/s}}$$
$$= 1.63 \text{ hp} \tag{12.114}$$

So, the required power of the electric motor is 1.63 hp.

Reference

1. Meriam, J. L., & Kraige, L. G. (2007). *Engineering mechanics, dynamics* (6th ed.). Wiley.

Chapter 13
Inverse Planar Dynamics

13.1 Introduction

In forward dynamics problems, the applied forces on a machine or a mechanism are given and we want to calculate the resulting motion. In Chap. 13, the inverse planar dynamic problems are introduced in which the **desired motion of a machine or mechanism is known and we are looking for compatible forces and moments** that can generate that motion. **The inverse dynamics analysis is necessary for choosing suitable motors and actuators when the required motion is known.** The connecting link between motion, forces, and moments is Newton's second law and through finding the acceleration terms.

The calculated dynamic forces and moments that are needed to generate the desired motion are compared with the corresponding static ones and used as a means for verification of the obtained results. It is shown how static solutions can provide approximations for dynamic solutions in cases where inertia effects are not significant (low mass times acceleration). Finally, application of matrix algebra for solving systems of equations of static equilibrium (in statics) or equations of dynamic equilibrium (in dynamics) is illustrated.

13.2 General Solution Algorithm of Inverse Planar Dynamics

As it was stated in Chap. 10, if in a dynamic problem the generated inertia forces were negligible, (compared to the applied forces) a static analysis of the problem would provide sufficiently accurate results. Otherwise, a complete dynamic analysis is required in which applied forces and inertia forces are all taken into consideration. In what follows, first the steps needed for a static force analysis of a problem are presented. Then the kinematic analysis of the same problem is reviewed, accelerations are

M. Ghorashi, *Introduction to Machine Dynamics*, Mechanical Engineering Series,
https://doi.org/10.1007/978-3-031-94736-0_13

calculated and used to provide connections between kinematics and kinetics. Accelerations are also used to provide inertia forces and inertia moments. Finally, the equations of dynamic equilibrium are written and solved for the unknown dynamic forces.

The complete solution algorithm for solving an inverse dynamics problem is given below:

Part A: Statics.

A1. Draw the FBDs of all parts in static equilibrium.
A2. Write the corresponding static equations of equilibrium for each object.
A3. Solve the static equations of equilibrium and calculate the unknown static forces.

Part B: Kinematics.

B1. Draw the vector-loop diagrams.
B2. Write the vector-loop eqs.
B3. Solve the position problem.
B4. Solve the velocity problem.
B5. Solve the acceleration problem and get the acceleration of all mass centers and angular accelerations of all rigid bodies.

Part C: Kinetics.

C1. Calculate the inertia forces and the inertia moments.
C2. Draw FBDs of all parts; showing dynamic forces.
C3. Write the corresponding equations of motion for each object.
C4. Solve the equations of motion and calculate the unknown dynamic forces.
C5. Compare the calculated dynamic forces with the corresponding static values as a means for verification of the results.

In what follows, a few problems that illustrate the application of the above-mentioned algorithm are discussed. Additional examples for use of this method can be found in Stanišić [1].

13.2.1 Application Problem 1

The following example includes part A (static analysis) of a problem.

Example 13.1 In the mechanism shown in Fig. 13.1, sliders 2 and 4 are connected by link AB (whose length is L) using revolute joints. The mechanism is on a horizontal plane. Slider 4 is pushed in the $-y$ direction by force $P = 100$ N. If the mechanism is supposed to maintain the illustrated posture at $\theta = 45°$, calculate the necessary force R that should be applied on slider 2.

A kinematic problem related to this mechanism was discussed in Chap. 5 (see Example 5.9). Here we are dealing with static analysis. Since at $\theta = 45°$ there is complete symmetry in the mechanism, one expects to get,

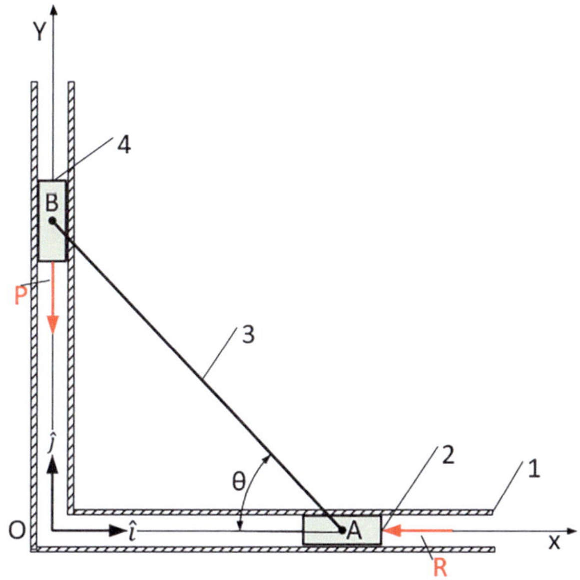

Fig. 13.1 Double-slider mechanism subjected to two forces

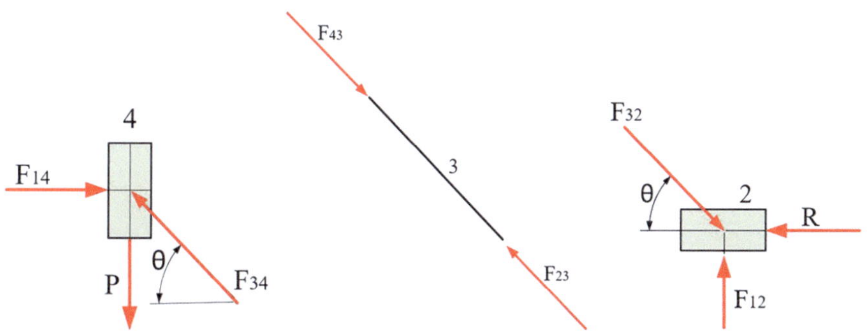

Fig. 13.2 Static free body diagrams of members of the mechanism shown in Fig. 13.1

$$R = P = 100 \text{ N} \tag{13.1}$$

This expectation can now be verified. The FBDs shown in Fig. 13.2 make it clear that sliders 2 and 4 are three-force members and link 3 is a two-force member. (In other words, the forces acting on sliders 2 and 4 form triangles and the forces acting on link 3 are collinear.)

Considering the FBD of slider 4 and its corresponding static equilibrium equations one obtains,

$$F_{34} \sin 45^\circ = P = 100 \text{ N}$$

Therefore,

$$F_{34} = 141.42 \text{ N} \tag{13.2}$$

Application of the equilibrium equation of link 3 and Newton's third law for slider 4 and link 3 gives,

$$F_{23} = F_{43} = F_{34} = 141.42 \text{ N} \tag{13.3}$$

Newton's third law for slider 2 and link 3 results in,

$$F_{32} = F_{23} = 141.42 \text{ N} \tag{13.4}$$

Note that action and reaction forces act in opposite directions:

$$\vec{F}_{34} = -\vec{F}_{43}, \vec{F}_{23} = -\vec{F}_{32} \tag{13.5}$$

This matter has already been taken into account in the FBDs. Equations (13.3) and (13.4) are about the magnitudes of these forces (that are equal). Finally, considering the equilibrium of forces on slider 2 we have:

$$R = F_{32} \cos 45° = 100 \text{ N} \tag{13.6}$$

Therefore, our expectation, i.e., Eq. (13.1) has been verified. Of course, the presented results here are only the static ones.

Considering Examples 13.1 and 5.9 as well as the three-part solution of an inverse dynamics problem given in Sect. 13.2, one notices that parts *A* (statics) and *B* (kinematics) of the solution have already been completed. Now, to calculate the necessary force *R* that should be applied on slider 2 so that the mechanism moves exactly in a way that is expressed in the kinematics problem (i.e., Example 5.9), one needs to complete the kinetics part of the analysis (i.e., part *C*). The details of this analysis are given in Example 13.2.

Example 13.2 In the mechanism shown in Fig. 13.3, sliders 2 and 4 are connected by link *AB* (whose length is $L = 10$ cm) using revolute joints. The mechanism is on a horizontal plane. Slider 4 is pushed in the $-y$ direction by force $P = 100$ N and the aim is to apply force *R* on slider 2 so that (as a result of application of *P* and *R*) slider 2 moves with a constant velocity $v_A = 2$ cm/s to the right. Calculate the necessary force *R* that should be applied on slider 2 when $\theta = 45°$. We know that: $m_2 = m_4 = 0.2$ kg, and $m_3 = 0.3$ kg.

Sliders 2 and 4 only translate, so their moments of inertia will not enter into the equations of motion. Rod 3, however, simultaneously translates and rotates so we need to calculate its mass moment of inertia about its mass center:

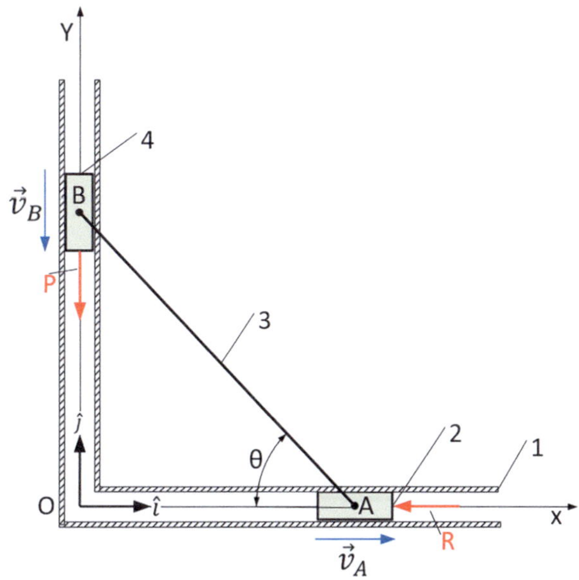

Fig. 13.3 Dynamics of the double-slider mechanism subjected to two forces

$$I_{G3} = I_C = \frac{1}{12} m_3 AB^2 = \frac{1}{12} \times 0.3 \times 0.10^2 = 2.50 \times 10^{-4}\ \text{kg.m}^2 \tag{13.7}$$

We have already completed part A (statics) of this problem in Example 13.1 and got the static force R necessary for maintaining static equilibrium given in Eq. (13.6):

$$R = 100\ \text{N} \tag{13.6}$$

We also completed most of part B (kinematics) of this problem in Example 5.9 and got the following results when slider 2 moves with a constant velocity $v_A = 2$ cm/s to the right ($a_A = 0$, see Fig. 13.4).

$$\begin{aligned} \omega_3 &= 0.283\ \text{rad/s (CCW)};\ v_B = -2.00\ \text{cm/s} \\ \alpha_3 &= 0.080\ \text{rad/s}^2\ \text{(CCW)};\ a_B = -1.132\ \text{cm/s}^2 \end{aligned} \tag{13.8}$$

So the angular accelerations of all members are known (angular accelerations of the sliders are identically zero as their motion is translational). The accelerations of the mass centers of the sliders are known as well. Therefore, in order to complete the last segment of the kinematic analysis (i.e., *B*5), we need to calculate the acceleration of the mass center of link 3, i.e., point *C*. One can use the relative motion method for calculating the acceleration of point *C*. By placing the origin of the moving coordinate system at *A* and attaching this coordinate system to link 3 one can use Eq. (5.16) for calculating the acceleration of point *C*,

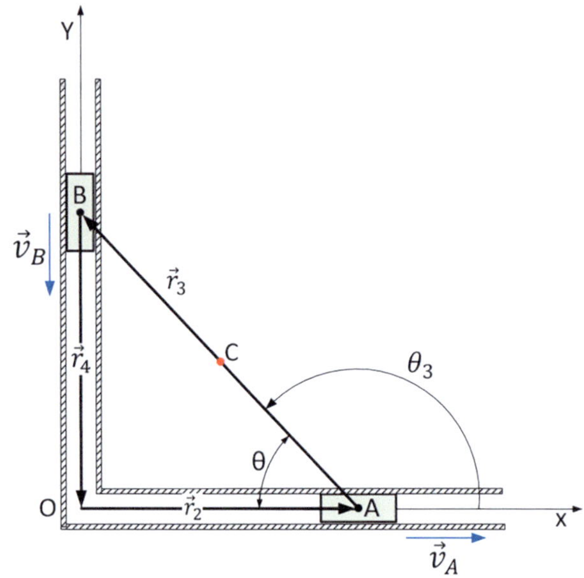

Fig. 13.4 Kinematics of the double-slider mechanism

$$\vec{a}_C = \vec{a}_A + \vec{\alpha} \times \vec{r} + \vec{\omega} \times \left(\vec{\omega} \times \vec{r}\right) \tag{13.9}$$

After recognizing vectors $\vec{r}$, $\vec{\omega}$, and $\vec{\alpha}$ in the present problem, Eq. (13.9) can be written as follows:

$$\vec{a}_C = \vec{a}_A + \vec{\alpha}_3 \times \overrightarrow{AC} + \vec{\omega}_3 \times \left(\vec{\omega}_3 \times \overrightarrow{AC}\right) \tag{13.10}$$

The unit vector in the $\overrightarrow{AC}$ direction can be introduced as,

$$\hat{e}_{\overrightarrow{AC}} = \left(-\cos\theta\hat{i} + \sin\theta\hat{j}\right)$$

By substituting the available data into Eq. (13.10), it can be rewritten as follows:

$$\vec{a}_C = 0 + 0.080\hat{k} \times 0.05\left(-\cos\theta\hat{i} + \sin\theta\hat{j}\right) + 0.05 \times 0.283^2\left(\cos\theta\hat{i} - \sin\theta\hat{j}\right)$$

Calculating this expression at $\theta = 45°$ results in,

$$\vec{a}_C = 3.15 \times 10^{-6}\hat{i} - 5.66 \times 10^{-3}\hat{j}\ \mathrm{m/s^2} \tag{13.11}$$

The x component of $\vec{a}_C$ is essentially a round-off error. (Its accurate value is zero.) Furthermore, since point C is half way between A and B, its acceleration should be half of the acceleration at point B, and in fact it is!

Now that all of the acceleration components have been calculated, analyzing part B of the solution (i.e., kinematics) is complete and one can start to work on the kinetics part, i.e., part *C*.

C1. Calculate the inertia forces and the inertia moments.

For slider 2, $a_A = 0$, $\alpha_2 = 0$ therefore the inertia force and inertia moment terms are,

$$-m_2 a_{G2} = -m_2 a_A = 0;\ -I_{G2}\alpha_2 = 0 \tag{13.12}$$

For rod 3,

$$\vec{a}_C = 3.15 \times 10^{-6}\hat{i} - 5.66 \times 10^{-3}\hat{j}\ \text{m/s}^2; \alpha_3 = 0.080\ \text{rad}/s^2\ (\text{CCW}) \tag{13.13}$$

Therefore, the inertia force and the inertia moment terms are,

$$\begin{aligned} -m_3\vec{a}_{G3} &= -m_3\vec{a}_C = -9.45 \times 10^{-7}\hat{i} + 1.698 \times 10^{-3}\hat{j}\ \text{N} \\ -I_{G3}\vec{\alpha}_3 &= -I_C\vec{\alpha}_3 = -2.50 \times 10^{-4}\ \text{kg.m}^2 \times 0.080\hat{k}\ \text{rad}/s^2 = -2 \times 10^{-5}\hat{k}\ \text{N.m} \end{aligned} \tag{13.14}$$

For slider 4, using Eq. (13.8) we have:

$$\vec{a}_{G4} = \vec{a}_B = -1.132 \times 10^{-2}\hat{j}\ \text{m/s}^2; \alpha_4 = 0 \tag{13.15}$$

Therefore, the inertia force and the inertia moment terms associated with slider 4 are,

$$-m_4\vec{a}_{G4} = -m_4\vec{a}_B = -0.2\ \text{kg} \times -1.132 \times 10^{-2}\hat{j}\ \text{m/s}^2 = 2.264 \times 10^{-3}\hat{j}\ \text{N} \tag{13.16}$$

$$-I_{G2}\alpha_2 = 0 \tag{13.17}$$

C2. Draw FBDs of all parts showing dynamic forces.

In order to differentiate dynamic forces from the previously calculated static ones we add a prime to their names. Also, since member 3 is now subjected to the inertia force and the inertia moment (13.14), it is no longer a two-force member. So F'_{23} and F'_{43} are no longer necessarily equal in magnitude,

$$F'_{23} \neq F'_{43} \tag{13.18}$$

However, Newton's third law is always valid so,

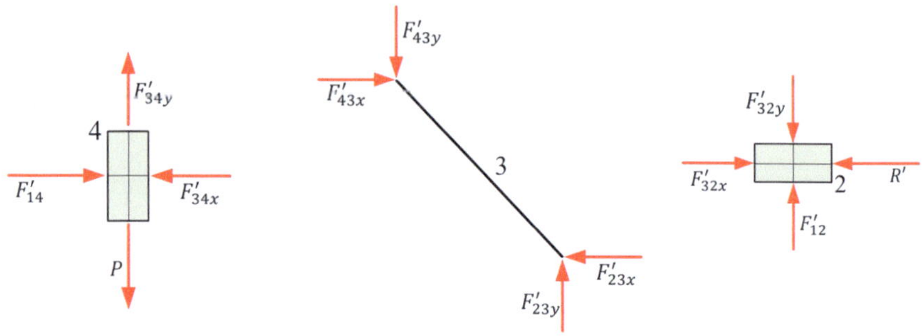

Fig. 13.5 Dynamic FBDs of the members of the double-slider mechanism

$$\vec{F}'_{34} = -\vec{F}'_{43}; \vec{F}'_{23} = -\vec{F}'_{32} \tag{13.19}$$

Since link 3 is no longer a two-force member, the angles of operation of the four forces mentioned in Eq. (13.19) are no longer the same as the inclination angle of link 3 (i.e., θ, in Fig. 13.4) and they are unknown. That is why in the dynamic FBDs, these forces are shown by their x and y components that both are unknown. The dynamic FBDs are shown in Fig. 13.5.

C3. Write the corresponding equations of motion for each object.

Now, for each member, the dynamic force and the dynamic moment equations of motion are written. For the sliders, the moment equations are trivial ($0 = 0$). The right-hand sides of the force and moment equations are occupied by the negative of the inertia forces and the inertia moments that were calculated in part *C1*. So, we have:

For slider 4:

$$F'_{14} - F'_{34x} = 0 \tag{13.20}$$

$$F'_{34y} - P = m_4 a_{G4y} = m_4 a_{By} = -2.264 \times 10^{-3} \tag{13.21}$$

For rod 3:

The force equations are:

$$F'_{43x} - F'_{23x} = m_3 a_{G3x} = m_3 a_{Cx} = 9.45 \times 10^{-7} \tag{13.22}$$

$$F'_{23y} - F'_{43y} = m_3 a_{G3y} = m_3 a_{Cy} = -1.698 \times 10^{-3} \tag{13.23}$$

And the moment equation about point C can be written by noticing that at $\theta = 45°$ all forces that are acting on the rod have the same moment arm about C.

$$\left(F'_{23y} + F'_{43y} - F'_{23x} - F'_{43x}\right) \times \frac{L}{2} \cos 45° = I_{G3}\alpha_3 = 2 \times 10^{-5}$$

or, recalling that $L = 10$ cm, we get,

$$\left(F'_{23y}+F'_{43y}-F'_{23x}-F'_{43x}\right)\times\frac{0.1}{2}\cos 45^{\circ}=I_{G3}\alpha_3=2\times 10^{-5}$$

finally,

$$F'_{23y}+F'_{43y}-F'_{23x}-F'_{43x}=5.6569\times 10^{-4} \tag{13.24}$$

For slider 2:

$$F'_{32x}-R'=m_2a_{G2x}=0 \tag{13.25}$$

$$F'_{12}-F'_{32y}=0 \tag{13.26}$$

Using Newton's third law we have:

$$F'_{23x}=F'_{32x}, F'_{23y}=F'_{32y}, F'_{34x}=F'_{43x}, F'_{34y}=F'_{43y} \tag{13.27}$$

Notice that the fact action-reaction forces act in opposite directions and we have:

$$\vec{F}'_{34}=-\vec{F}'_{43}, \vec{F}'_{23}=-\vec{F}'_{32} \tag{13.28}$$

has already been taken into account in the FBDs shown in Fig. 13.5 and Eqs. (13.27) are only about the magnitudes of these forces (that are equal).

The seven derived dynamic equations can be expressed in terms of the following seven unknowns:

$$\left\{F'_{14}, F'_{34x}, F'_{34y}, F'_{23x}, F'_{23y}, F'_{12}, R'\right\} \tag{13.29}$$

After doing so, the dynamic equations can be written as follows:

$$F'_{14}-F'_{34x}=0 \tag{13.30}$$

$$F'_{34y}-P=-2.264\times 10^{-3} \tag{13.31}$$

$$F'_{34x}-F'_{23x}=9.45\times 10^{-7} \tag{13.32}$$

$$F'_{23y}-F'_{34y}=-1.698\times 10^{-3} \tag{13.33}$$

$$F'_{23y}+F'_{34y}-F'_{23x}-F'_{34x}=5.6569\times 10^{-4} \tag{13.34}$$

$$F'_{23x}-R'=0 \tag{13.35}$$

$$F'_{12}-F'_{23y}=0 \tag{13.36}$$

C4. Solve the equations of motion and calculate the unknown dynamic forces.

One can write the system of the obtained seven equations in seven unknowns in the following matrix form:

$$AX = B \tag{13.37}$$

In which, X and B are the vector of unknowns and the vector of known right-hand side values, respectively, and they are as follows:

$$X^t = \left\{ F'_{14}, F'_{34x}, F'_{34y}, F'_{23x}, F'_{23y}, F'_{12}, R' \right\} \tag{13.38}$$

$$B^t = \left\{ 0, P - 2.264 \times 10^{-3}, 9.45 \times 10^{-7}, -1.698 \times 10^{-3}, 5.6569 \times 10^{-4}, 0, 0 \right\} \tag{13.39}$$

Matrix A is the coefficient matrix which can be easily populated by analyzing the coefficients of the derived seven Eqs. (13.30) to (13.36). The whole matrix equation can be written as follows:

$$\begin{bmatrix} F'_{14} & F'_{34x} & F'_{34y} & F'_{23x} & F'_{23y} & F'_{12} & R' \\ - & - & - & - & - & - & - \\ 1 & -1 & 0 & 0 & 0 & 0 & 0 \\ 0 & 0 & 1 & 0 & 0 & 0 & 0 \\ 0 & 1 & 0 & -1 & 0 & 0 & 0 \\ 0 & 0 & -1 & 0 & 1 & 0 & 0 \\ 0 & -1 & 1 & -1 & 1 & 0 & 0 \\ 0 & 0 & 0 & 1 & 0 & 0 & -1 \\ 0 & 0 & 0 & 0 & -1 & 1 & 0 \end{bmatrix} \begin{Bmatrix} F'_{14} \\ F'_{34x} \\ F'_{34y} \\ F'_{23x} \\ F'_{23y} \\ F'_{12} \\ R' \end{Bmatrix} = \begin{Bmatrix} 0 \\ P - 2.264 \times 10^{-3} \\ 9.45 \times 10^{-7} \\ -1.698 \times 10^{-3} \\ 5.6569 \times 10^{-4} \\ 0 \\ 0 \end{Bmatrix} \tag{13.40}$$

The following MATLAB script solves the mentioned system of equations:

```
P=100;
A=[1 -1 0 0 0 0 0 ;
  0 0 1 0 0 0 0;
  0 1 0 -1 0 0 0 ;
  0 0 -1 0 1 0 0 ;
  0 -1 1 -1 1 0 0 ;
  0 0 0 1 0 0 -1;
  0 0 0 0 -1 1 0];
B=[0;
   P-0.002264;
```

Table 13.1 Comparison of the static and the dynamic solutions of force *R*, for various choices of the applied load *P* (forces are in N)

P	R_{Static}	$R_{Dynamic}$	$[(R_{Static}-R_{Dynamic})/R_{Dynamic}] \times 100$
100	100	99.9966	0.0034%
10	10	9.9966	0.034%
1	1	0.9966	0.34%
0.1	0.1	0.0966	3.52%
0.01	0.01	0.0066	51.43%

```
   9.45e-7;
   -1.6983e-3;
   5.6569e-4;
   0;
   0];
X=linsolve(A,B)
```

Recalling, $P = 100$ N, the code results in:

$$F'_{14}=99.9966, F'_{34x}=99.9966, F'_{34y}=99.9977, \tag{13.41}$$

$$F'_{23x}=99.9966, F'_{23y}=99.9960 \tag{13.42}$$

$$F'_{12}=99.9960, R'=99.9966 \tag{13.43}$$

Since all of the calculated forces are positive, the directions that were assumed for all of the unknown forces on the FBDs in Fig. 13.5 were correct.

C5. Compare the calculated dynamic forces with the corresponding static values as a means for verification of the results.

We notice that the calculated value of the dynamic force R' in Eq. (13.43) is very close to its corresponding static value of Eq. (13.1). The reason for this closeness is that the inertia effects calculated in part *C1* were all small and in the range of 0.001 N. Therefore, one would expect that the static solution should be able to provide a good approximation for the dynamic problem.

If the magnitude of the applied force gets closer to the calculated inertia forces, the difference between the calculated static and dynamic values of *R* increases. To verify this matter, we keep the expected kinematics given in the problem statement unchanged and just gradually reduce the applied load *P* and keep recording the resulting static and dynamic values of the required force *R* that should be applied so that the expected response is realized. The results of this study are given in Table 13.1 and Fig. 13.6.

We see that by decreasing the applied load *P* and making it closer and closer to the orders of magnitude of the existing inertia effects, the dynamic values of *R* get further and further away from their corresponding static value, as expected. Therefore, only at high values of the applied load compared to inertia effects can a static analysis provide sufficiently accurate results for a dynamic problem.

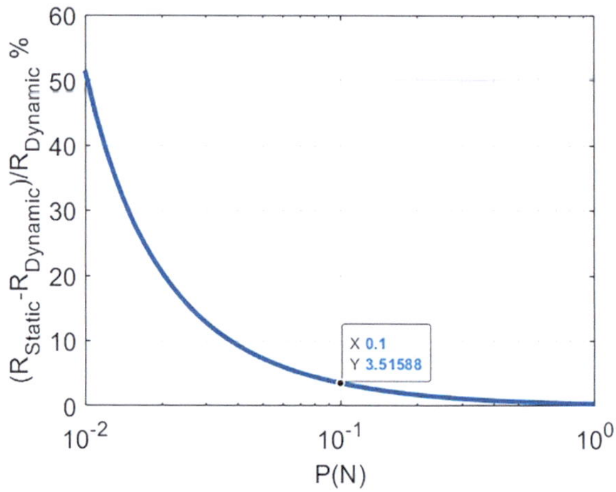

Fig. 13.6 Differences between the calculated dynamic values of force *R* and their corresponding static values in Example 13.2 and for various choices of the applied load *P*

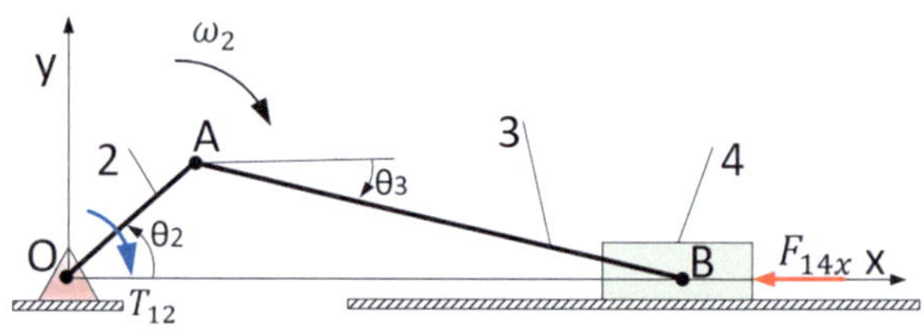

Fig. 13.7 Slider-crank mechanism under applied torque and force

13.2.2 Application Problem 2

Example 13.3 Consider the slider-crank mechanism shown in Fig. 13.7 with $r_2 =$ 75 mm, $r_3 =$ 200 mm, and with no eccentricity ($e = 0$). Note that the crank starts its motion at $\theta_2 = 0$ and $\theta_3 = 0$ (i.e., the top dead center) and rotates at constant speed of 2000 rpm (CW). When the crank angle is: $\theta_2 = 40\ ^\circ$, the crank is subjected to $T_{12} = 1000$ N.m torque that operates in the clockwise direction. The aim is to apply enough force F_{14x} on the piston so that crank 2 continues to rotate at constant speed of 2000 rpm (CW) as it passes the $\theta_2 = 40^\circ$ angle (corresponding to $\theta_3 = -13.95^\circ$). The masses of the members are: $m_2 = 0.3$ kg, $m_3 = 0.4$ kg, $m_4 = 0.2$ kg.

The algorithm for solving the complete inverse dynamics problem is reminded below. In short, we will first solve the static problem for calculating the necessary F_{14x} (as a first estimation to the solution for the dynamic problem). Then, we will solve the dynamic problem; and finally compare the two sets of results.

Part A: Statics.

A1. Draw the FBDs of all parts in static equilibrium.
A2. Write the corresponding static equations of equilibrium for each object.
A3. Solve the static equations of motion and calculate the unknown static forces.

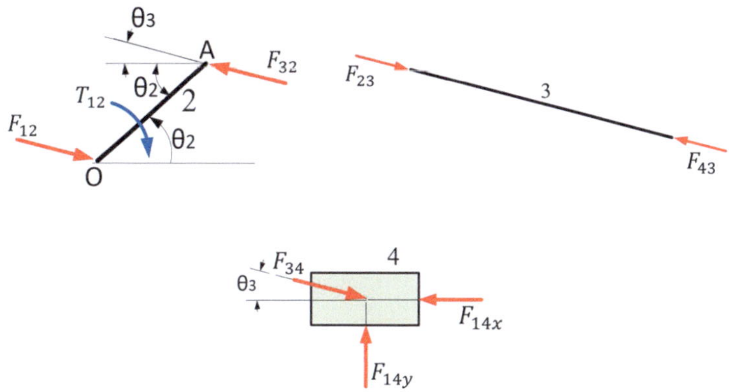

Fig. 13.8 Static free body diagrams of members of the mechanism shown in Fig. 13.7

Figure 13.8 includes the static free body diagrams of members of the mechanism. Using these FBDs, and as a first approximation for solving the dynamic problem, we would like to calculate the necessary force F_{14x} for balancing T_{12} statically. Therefore, the summation of the applied moments on part 2 about O should be zero and we get:

$$-T_{12}\hat{k} + \overrightarrow{OA} \times \vec{F}_{32} = 0 \tag{13.44}$$

or,

$$-T_{12} + OA.F_{32}.\sin(\theta_2 + |\theta_3|) = 0$$

where $OA.\sin(\theta_2 + |\theta_3|)$ is the moment arm of F_{32} about point O. Substituting the available data results in:

$$-T_{12} + 0.075 \times F_{32}.\sin(40^\circ + 13.95^\circ) = 0$$

Therefore, we get:

$$F_{32} = 16.491 T_{12}\ \mathrm{N} \tag{13.45}$$

Considering Newton's third law between members 2 and 3 and members 3 and 4, and also the fact that in the static case, link 3 is a two-force member, one obtains,

$$F_{32} = F_{23} = F_{43} = F_{34} = 16.491 T_{12}\ \mathrm{N} \tag{13.46}$$

Finally, equilibrium of forces acting on slider 4 result in,

$$F_{14x} = F_{34}\cos\theta_3 = 16.491T_{12} \times \cos 13.95\,^{\circ}\,\text{N} = 16.005T_{12}\ \text{N} \tag{13.47}$$

Also we have,

$$F_{14y} + F_{34}\sin\theta_3 = 0$$

or,

$$F_{14y} = -16.491T_{12} \times \sin(-\,13.95\,^{\circ}\,)\text{N} = 3.976T_{12}\ \text{N} \tag{13.48}$$

Recalling $T_{12} = 1000$ N.m, Eqs. (13.47) and (13.48) result in:

$$F_{14x} = 16{,}005\ \text{N};\ F_{14y} = 3{,}976\ \text{N} \tag{13.49}$$

Part B: Kinematics.

B1. Draw the vector-loop diagrams.
B2. Write the vector-loop eqs.
B3. Solve the position problem.
B4. Solve the velocity problem.
B5. Solve the acceleration problem and get the acceleration of all mass centers and angular accelerations of all rigid bodies.

Considering the vector-loop diagram shown in Fig. 13.9, the corresponding vector-loop equation can be written as follows:

$$\vec{r}_2 + \vec{r}_3 = \vec{r}_4 \tag{13.50}$$

The projected components of Eq. (13.50) in the direction of $\vec{r}_1$ and the direction perpendicular to $\vec{r}_1$ are as follows:

$$r_2\cos\theta_2 + r_3\cos\theta_3 = r_4 \tag{13.51}$$

$$r_2\sin\theta_2 + r_3\sin\theta_3 = 0 \tag{13.52}$$

If the available data for r_2, r_3 and θ_2 are substituted into Eq. (13.52) we get: $\sin\theta_3 = -\,0.241$ or, as it was given before,

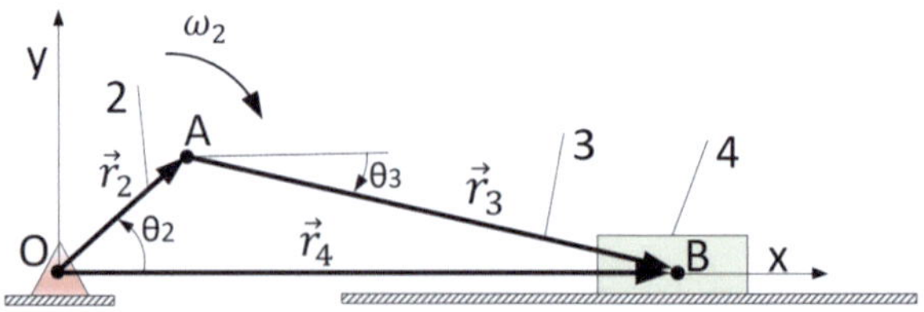

Fig. 13.9 Kinematic analysis of the slider-crank mechanism in Example 13.3

$$\theta_3 = -13.95\,^\circ \tag{13.53}$$

Similarly, Eq. (13.51) results in:

$$r_4 = r_2 \cos\theta_2 + r_3 \cos\theta_3 = 251.55 \text{ mm} \tag{13.54}$$

Furthermore, converting the angular velocity of link 2 to rad/s gives,

$$\omega_2 = -2000 \text{ rpm} = -2000 \text{ rev/ min} \times \frac{2\pi \text{ rad/rev}}{60 \text{ s/ min}} = -209.44 \text{ rad/s (CW)} \tag{13.55}$$

Substituting the above-mentioned results into Eq. (4.28):

$$\omega_3 = -\frac{r_2\omega_2 \cos\theta_2}{r_3 \cos\theta_3};\, v_B = \frac{r_2\omega_2 \sin(\theta_3 - \theta_2)}{\cos\theta_3} \tag{4.28}$$

gives,

$$\omega_3 = 61.99 \text{ rad/s (CCW)};\, v_B = 13.086 \text{ m/s} \tag{13.56}$$

The positive value of ω_3 makes sense at it means at $\theta_2 = 40°$ link 3 is moving in the CCW direction which can be verified by examining the posture of the mechanism at $\theta_2 = 40°$. The same is true for the sign of v_B that means motion of the slider to the right, which can also be verified by inspection.

Now, to complete the *B5* section of the solution algorithm, we recall that for the slider-crank mechanism the acceleration equations are as follows:

$$\begin{bmatrix} r_3 \sin\theta_3 & 1 \\ -r_3 \cos\theta_3 & 0 \end{bmatrix} \begin{Bmatrix} \alpha_3 \\ a_B \end{Bmatrix} = \begin{Bmatrix} -r_2\alpha_2 \sin\theta_2 - r_2\omega_2^2 \cos\theta_2 - r_3\omega_3^2 \cos\theta_3 \\ r_2\alpha_2 \cos\theta_2 - r_2\omega_2^2 \sin\theta_2 - r_3\omega_3^2 \sin\theta_3 \end{Bmatrix} \tag{5.9}$$

where the coefficient matrix developed in this matrix equation is the Jacobian matrix:

$$J = \left[r_3 \sin\, \theta_3 1 - r_3 \cos\theta_3 0 \right] \rightarrow |J| = r_3 \cos\theta_3 \tag{13.57}$$

Using Cramer's rule, the solution can be written as,

$$\alpha_3 = \frac{\begin{vmatrix} -r_2\alpha_2 \sin\theta_2 - r_2\omega_2^2 \cos\theta_2 - r_3\omega_3^2 \cos\theta_3 & 1 \\ r_2\alpha_2 \cos\theta_2 - r_2\omega_2^2 \sin\theta_2 - r_3\omega_3^2 \sin\theta_3 & 0 \end{vmatrix}}{|J|} \tag{13.58}$$

$$a_B = \frac{\begin{vmatrix} r_3 \sin\theta_3 & -r_2\alpha_2 \sin\theta_2 - r_2\omega_2^2 \cos\theta_2 - r_3\omega_3^2 \cos\theta_3 \\ -r_3 \cos\theta_3 & r_2\alpha_2 \cos\theta_2 - r_2\omega_2^2 \sin\theta_2 - r_3\omega_3^2 \sin\theta_3 \end{vmatrix}}{|J|} \tag{13.59}$$

By substitution of the available numerical values into Eqs. (13.58) and (13.59) we get:

$$\alpha_3 = \frac{\begin{vmatrix} -75 \times (-209.44)^2 \cos 40° - 200 \times (61.99)^2 \cos(-13.95°) & 1 \\ -75 \times (-209.44)^2 \sin 40° - 200 \times (61.99)^2 \sin(-13.95°) & 0 \end{vmatrix}}{200 \times \cos(-13.95°)}$$

$$= \frac{\begin{vmatrix} -3266081.7 & 1 \\ -1929417.6 & 0 \end{vmatrix}}{194.10}$$

$$\alpha_3 = +9940.33 \text{ rad/s}^2 \tag{13.60}$$

and:

$$a_B = \frac{\begin{vmatrix} 200 \times \sin(-13.95°) & -75 \times (-209.44)^2 \cos 40° - 200 \times (61.99)^2 \cos(-13.95°) \\ -200 \times \cos(-13.95°) & -75 \times (-209.44)^2 \sin 40° - 200 \times (61.99)^2 \sin(-13.95°) \end{vmatrix}}{200 \times \cos(-13.95°)}$$

$$a_B = \frac{\begin{vmatrix} -48.22 & -3266081.7 \\ -194.10 & -1929417.6 \end{vmatrix}}{194.10} = -2786759.1 \text{ mm/s}^2 = -2786.76 \text{ m/s}^2$$

$$a_B = -2786.76 \text{ m/s}^2 \tag{13.61}$$

Therefore, now the angular accelerations of all members are known:

$$\alpha_2 = 0; \alpha_3 = +9940.33 \text{ rad/s}^2; \alpha_4 = 0 \tag{13.62}$$

In order to calculate accelerations of centers of masses we consider Fig. 13.10:
For calculating $\vec{a}_{G2}$ we start with:

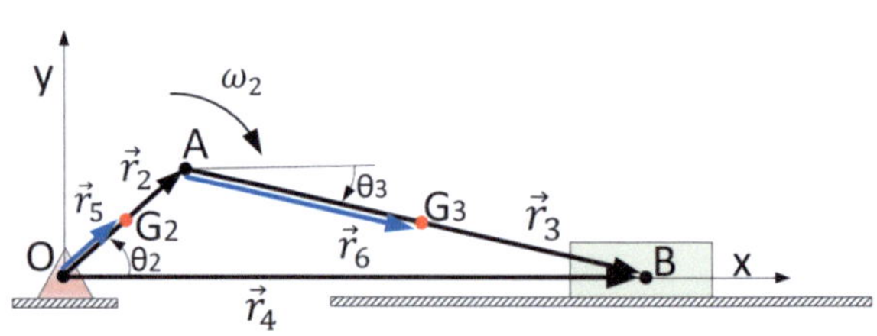

Fig. 13.10 Calculating accelerations of mass centers in Example 13.3

$$\vec{a}_{G2} = \frac{d^2 \vec{r}_5}{dt^2} \tag{13.63}$$

Referring to Fig. 13.10 we have:

$$\vec{r}_5 - \frac{1}{2}\vec{r}_2 = \frac{1}{2} \times 0.075\left(\cos\theta_2\hat{i} + \sin\theta_2\hat{j}\right) \tag{13.64}$$

so,

$$\vec{v}_{G2} = \frac{d\vec{r}_5}{dt} = \frac{1}{2} \times 0.075\left(-\sin\theta_2\hat{i} + \cos\theta_2\hat{j}\right)\omega_2 \tag{13.65}$$

And recalling $\alpha_2 = 0$, we get:

$$\vec{a}_{G2} = \frac{d^2\vec{r}_5}{dt^2} = \frac{-1}{2} \times 0.075\left(\cos\theta_2\hat{i} + \sin\theta_2\hat{j}\right)\omega_2^2 \tag{13.66}$$

Now, substituting $\theta_2 = 40°$ and Eq. (13.55):

$$\omega_2 = -209.44 \text{ rad/s (CW)} \tag{13.55}$$

into Eq. (13.66) we get:

$$\vec{a}_{G2} = -1644.94\left(\cos 40°\,\hat{i} + \sin 40°\,\hat{j}\right) \text{ m/s}^2 \tag{13.67}$$

It is clear that this acceleration is centripetal (toward the hinge point O), as expected. Equation (13.67) can be rewritten as follows:

$$\vec{a}_{G2} = -1260.10\hat{i} - 1057.35\hat{j} \text{ m/s}^2 \tag{13.68}$$

For calculating $\vec{a}_{G3}$ we start with:

$$\vec{a}_{G3} = \frac{d^2\vec{r}_{G3}}{dt^2} \tag{13.69}$$

Referring to Fig. 13.10 we have:

$$\vec{r}_{G3} = \vec{r}_2 + \vec{r}_6 = \vec{r}_2 + \frac{1}{2}\vec{r}_3 = 0.075\left(\cos\theta_2\hat{i} + \sin\theta_2\hat{j}\right) + \frac{1}{2} \times 0.2\left(\cos\theta_3\hat{i} + \sin\theta_3\hat{j}\right) \tag{13.70}$$

Taking a derivative with respect to time gives:

$$\vec{v}_{G3}=\frac{d\vec{r}_{G3}}{dt}=0.075\left(-\sin\theta_2\hat{i}+\cos\theta_2\hat{j}\right)\omega_2+0.1\left(-\sin\theta_3\hat{i}+\cos\theta_3\hat{j}\right)\omega_3 \quad (13.71)$$

A second derivative gives:

$$\vec{a}_{G3}=\frac{d^2\vec{r}_{G3}}{dt^2}=\frac{d\vec{v}_{G3}}{dt}$$

or,

$$\vec{a}_{G3}=-0.075\left(\cos\theta_2\hat{i}+\sin\theta_2\hat{j}\right)\omega_2^2-0.1\left(\cos\theta_3\hat{i}+\sin\theta_3\hat{j}\right)\omega_3^2 + 0.1\left(-\sin\theta_3\hat{i}+\cos\theta_3\hat{j}\right)\alpha_3 \quad (13.72)$$

Substituting the available data listed below:

$$\begin{gathered}\theta_2=40^{\circ};\omega_2=-209.44\ \text{rad/s (CW)};\alpha_2=0\\ \theta_3=-13.95^{\circ};\omega_3=61.99\ \text{rad/s (CCW)};\alpha_3=9940.33\ \text{rad/s}^2\ \text{(CCW)}\end{gathered} \quad (13.73)$$

into Eq. (13.72) gives:

$$\vec{a}_{G3}=-2653.50\hat{i}-1057.34\hat{j}\ \text{m/s}^2 \quad (13.74)$$

Finally, **for $\vec{a}_{G4}$** it was shown before that,

$$\vec{a}_{G4}=\vec{a}_B=-2786.76\,\hat{i}\ \text{m/s}^2 \quad (13.61)$$

***Part C: Kinetics*.**

C1. Calculate the inertia forces and the inertia moments.
C2. Draw FBDs of all parts showing dynamic forces.
C3. Write the corresponding equations of motion for each object.
C4. Solve the equations of motion and calculate the unknown dynamic forces.
C5. Compare the calculated dynamic forces with the corresponding static values as a means for verification of the results.

Having already calculated the accelerations of all of the mass centers, and the angular accelerations of all of the members of the mechanism, we can now calculate the inertia forces and inertia moments:

For Link 2 Using Eq. (13.68) we have:

$$-m_2\vec{a}_{G2} = -0.3\text{ kg}\times\left(-1260.10\hat{i} - 1057.35\hat{j}\right)\text{ m/s}^2 = 378.03\,\hat{i} + 317.21\hat{j}\text{ N} \tag{13.75}$$

Also since $\alpha_2 = 0$:

$$-I_{G2}\alpha_2 = 0 \tag{13.76}$$

For Link 3 Using Eq. (13.74) we have:

$$-m_3\vec{a}_{G3} = -0.4\text{ kg}\times\left(-2653.50\hat{i} - 1057.34\,\hat{j}\right)\text{ m/s}^2 = 1061.4\hat{i} + 422.94\hat{j}\text{ N} \tag{13.77}$$

For calculating the inertia moment, $-I_{G3}\alpha_3$ we first calculate the mass moment of inertia:

$$I_{G3} = \frac{1}{12}m_3r_3^2 = \frac{1}{12}\times 0.4\text{ kg}\times 0.2^2\text{ m}^2 = 1.33\times 10^{-3}\text{ kg.m}^2 \tag{13.78}$$

Therefore, using Eqs. (13.60) and (13.78), the inertia moment corresponding to link 3 is:

$$-I_{G3}\vec{\alpha}_3 = -1.33\times 10^{-3}\text{ kg.m}^2\times 9940.33\hat{k}\text{ rad/s}^2 \rightarrow -I_{G3}\vec{\alpha}_3 = -13.22\,\hat{k}\text{ N.m} \tag{13.79}$$

Finally, for the Slider Using Eq. (13.61) we have:

$$-m_4\vec{a}_{G4} = -0.2\text{ kg}\times\left(-2786.76\,\hat{i}\text{ m/s}^2\right) = 557.35\hat{i}\text{ N} \tag{13.80}$$

Since the slider only performs reciprocating motion and does not rotate, $\alpha_4 = 0$, and we get:

$$-I_{G4}\times\alpha_4 = 0 \tag{13.81}$$

C2. Draw FBDs of all parts showing dynamic forces.

Figure 13.11 illustrates the FBDs of all of the parts with their associated dynamic forces. Like in Example 13.2, in order to distinguish dynamic forces from the previously calculated static ones, we add a prime to their names. Also, since member 3 is under inertia force and inertia moment it is no longer a two-force member. So,

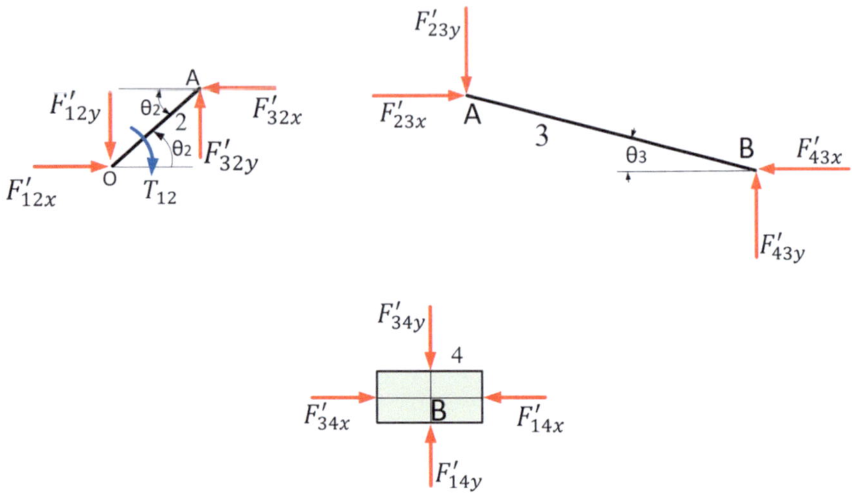

Fig. 13.11 Dynamic free body diagrams of members of the mechanism in Example 13.3

F'_{23} and F'_{43} are no longer necessarily equal in magnitude. However, Newton's third law is always valid and we have,

$$\vec{F}'_{34} = -\vec{F}'_{43}, \vec{F}'_{23} = -\vec{F}'_{32} \tag{13.82}$$

Since link 3 is no longer a two-force member, the angles of operation of the four forces mentioned in Eq. (13.82) are no longer the same as the inclination angle of link 3 (i.e., θ_3) and therefore, they are unknown. In the dynamic FBDs, these forces are shown by their x and y components and both of these components are unknown. On the FBD of link 2, unlike the static case, forces $\vec{F}'_{12}$ and $\vec{F}'_{32}$ are not parallel, equal and operating in opposite directions. Also, on slider 4 the acting forces no longer form a triangle.

C3. Write the corresponding equations of motion for each object.

Now, for each member, the force and moment equations of motion are written. For the slider, the moment equation is trivial ($0 = 0$). The right hand of the force and moment equations are the negative of the inertia forces and inertia moments that were calculated in step *C2*. So, we have:

For Link 2
The force and moment equations are as follows:

$$F'_{12x} - F'_{32x} = m_2 a_{G2x} = -378.03 \tag{13.83}$$

$$F'_{32y} - F'_{12y} = m_2 a_{G2y} = -317.21 \tag{13.84}$$

$$-T_{12}\hat{k} + \overrightarrow{G_2O} \times \vec{F}'_{12} + \overrightarrow{G_2A} \times \vec{F}'_{32} = I_{G2}\alpha_2 = 0 \tag{13.85}$$

Notice that G_2 and G_3 are the centers of masses of links 2 and 3, respectively (see Fig. 13.10). Also we have, $\overrightarrow{G_2A} = \vec{r}_5$ therefore, Eq. (13.85) results in:

$$-T_{12}\hat{k} + \frac{1}{2}OA\left(-\cos\theta_2\hat{i} - \sin\theta_2\hat{j}\right) \times \left(F'_{12x}\hat{i} - F'_{12y}\hat{j}\right) + \frac{1}{2}OA\left(\cos\theta_2\hat{i} + \sin\theta_2\hat{j}\right)$$
$$\times\left(-F'_{32x}\hat{i} + F'_{32y}\hat{j}\right) = 0$$

Substituting $\theta_2 = 40°$, $OA = 75$ mm, and performing the cross products result in,

$$-T_{12} + 0.0287F'_{12y} + 0.0241F'_{12x} + 0.0287F'_{32y} + 0.0241F'_{32x} = 0 \tag{13.86}$$

In the static case we could calculate F_{32} only by writing the equilibrium of moments equation for member 2 (since in that case direction of F_{32} was known, as member 3 was a two-force member). In the dynamic case, however, no unknown could be calculated before the whole system is modeled since the direction of F_{32} is no longer known (as link 3 is no longer a two-force member). Mathematically, we notice that in the dynamic case, for link 2, we have three equations for 4 unknowns: $F'_{12x}, F'_{12y}, F'_{32x}$, and F'_{32y}. So additional equations from other members are needed.

For Link 3

The force equations are:

$$F'_{23x} - F'_{43x} = m_3 a_{G3x} = -1061.4 \tag{13.87}$$

$$-F'_{23y} + F'_{43y} = m_3 a_{G3y} = m_3 a_{Cy} = -422.94 \tag{13.88}$$

And the moment equation about point G_3 can be written as follows:

$$\overrightarrow{G_3A} \times \vec{F}'_{23} + \overrightarrow{G_3B} \times \vec{F}'_{43} = I_{G3}\vec{\alpha}_3$$

Note that $\overrightarrow{G_3B} = \vec{r}_6$ in Fig. 13.10. Expanding each term results in:

$$\frac{1}{2}AB\left(-\cos\theta_3\hat{i} - \sin\theta_3\hat{j}\right) \times \left(F'_{23x}\hat{i} - F'_{23y}\hat{j}\right) + \frac{1}{2}AB\left(\cos\theta_3\hat{i} + \sin\theta_3\hat{j}\right)$$
$$\times\left(-F'_{43x}\hat{i} + F'_{43y}\hat{j}\right) = I_{G3}\vec{\alpha}_3$$

By using Eqs. (13.62) and (13.78), substituting $\theta_3 = -13.95°$ and $AB = 200$ mm, and after performing the cross products we get,

$$0.0971F'_{23y} - 0.0241F'_{23x} + 0.0971F'_{43y} - 0.0241F'_{43x} = 13.22 \tag{13.89}$$

The signs in Eq. (13.89) show that the forces in the y direction generate positive (CCW) moments whereas the forces in the x direction generate negative (CW) moments, as it is also clearly seen in the Fig. 13.11.

We notice that in the dynamic case, for link 3, we have three equations for 4 unknowns: $F'_{23x}, F'_{23y}, F'_{43x}$, and F'_{43y}. So additional equations from other members are needed.

For Slider 4

The force equations can be written as follows:

$$F'_{34x} - F'_{14x} = m_4 a_{G4x} = m_4 a_B; F'_{14y} - F'_{34y} = 0 \tag{13.90}$$

Recalling Eq. (13.80) we get:

$$F'_{34x} - F'_{14x} = -557.35; F'_{14y} - F'_{34y} = 0 \tag{13.91}$$

Finally, since the slider is in translation, the moment equation is automatically satisfied.

We notice that in the dynamic case, for slider 4, we have two equations for four unknowns: $F'_{14x}, F'_{14y}, F'_{34x}$, and F'_{34y}. So again additional equations from other members are needed.

Recalling the relations from Newton's third law we have:

$$F'_{23x} = F'_{32x}; F'_{23y} = F'_{32y}; F'_{34x} = F'_{43x}; F'_{34y} = F'_{43y} \tag{13.92}$$

Notice that the fact action–reaction forces act in opposite directions and we have:

$$\vec{F}'_{34} = -\vec{F}'_{43}; \vec{F}'_{23} = -\vec{F}'_{32} \tag{13.93}$$

has already been taken into account in the FBDs shown in Fig. 13.11 and Eqs. (13.92) are only about the magnitudes of these forces (that are equal). Looking at Fig. 13.11 one notices that there are eight independent unknowns in this problem that are listed below:

$$\left\{ F'_{12x}, F'_{12y}, F'_{23x}, F'_{23y}, F'_{34x}, F'_{34y}, F'_{14x}, F'_{14y} \right\} \tag{13.94}$$

Using Eqs. (13.83), (13.84), (13.86), (13.87), (13.88), (13.89), and (13.91) one can write eight dynamics equations that are expressed in terms of the eight unknowns listed in (13.94).

For Link 2

$$F'_{12x} - F'_{23x} = -378.03 \tag{13.95}$$

$$F'_{23y} - F'_{12y} = -317.21 \tag{13.96}$$

$$0.0287F'_{12y} + 0.0241F'_{12x} + 0.0287F'_{23y} + 0.0241F'_{23x} = T_{12} \tag{13.97}$$

For Link 3:

$$F'_{23x} - F'_{34x} = -1061.4 \tag{13.98}$$

$$-F'_{23y} + F'_{34y} = -422.94 \tag{13.99}$$

$$0.0971F'_{23y} - 0.0241F'_{23x} + 0.0971F'_{34y} - 0.0241F'_{34x} = 13.22 \tag{13.100}$$

For Slider 4

$$F'_{34x} - F'_{14x} = -557.35 \tag{13.101}$$

$$F'_{14y} - F'_{34y} = 0 \tag{13.102}$$

C4. Solve the equations of motion and calculate the unknown dynamic forces.

One can write the system of eight equations in eight unknowns in the following matrix form:

$$AX = B \tag{13.103}$$

In which, X and B are the vector of unknowns and the vector of known right-hand side values, respectively, and they are as follows:

$$X^t = \left\{ F'_{12x}, F'_{12y}, F'_{23x}, F'_{23y}, F'_{34x}, F'_{34y}, F'_{14x}, F'_{14y} \right\} \tag{13.104}$$

$$B^t = \{-378.03, -317.21, T_{12}, -1061.4, -422.94, 13.22, -557.35, 0\} \tag{13.105}$$

Matrix A is the coefficient matrix which can be easily populated by analyzing the coefficients of the derived eight equations. The whole matrix equation can be expanded as,

$$\begin{bmatrix} F'_{12x} & F'_{12y} & F'_{23x} & F'_{23y} & F'_{34x} & F'_{34y} & F'_{14x} & F'_{14y} \\ - & - & - & - & - & - & - & - \\ 1 & 0 & -1 & 0 & 0 & 0 & 0 & 0 \\ 0 & -1 & 0 & 1 & 0 & 0 & 0 & 0 \\ .0241 & .0287 & .0241 & .0287 & 0 & 0 & 0 & 0 \\ 0 & 0 & 1 & 0 & -1 & 0 & 0 & 0 \\ 0 & 0 & 0 & -1 & 0 & 1 & 0 & 0 \\ 0 & 0 & -0.0241 & .0971 & -.0241 & .0971 & 0 & 0 \\ 0 & 0 & 0 & 0 & 1 & 0 & -1 & 0 \\ 0 & 0 & 0 & 0 & 0 & -1 & 0 & 1 \end{bmatrix} \begin{Bmatrix} F'_{12x} \\ F'_{12y} \\ F'_{23x} \\ F'_{23y} \\ F'_{34x} \\ F'_{34y} \\ F'_{14x} \\ F'_{14y} \end{Bmatrix} = \begin{Bmatrix} -378.03 \\ -317.21 \\ T_{12} \\ -1061.4 \\ -422.94 \\ 13.22 \\ -557.35 \\ 0 \end{Bmatrix}$$

The following MATLAB script solves the mentioned system of equations: Notice that the negative sign of T_{12} was already taken into account in the derivations.

```
T12=1000;

A=[1 0 -1 0 0 0 0 0 ; ...
   0 -1 0 1 0 0 0 0 ; ...
   0.0241 0.0287 0.0241 0.0287 0 0 0 0 ; ...
   0 0 1 0 -1 0 0 0; ...
   0 0 0 -1 0 1 0 0 ; ...
   0 0 -0.0241 0.0971 -0.0241 0.0971 0 0;
   0 0 0 0 1 0 -1 0;
   0 0 0 0 0 -1 0 1]

B=[-378.03;
    -317.21;
    T12;
    -1061.4;
    -422.94;
    13.22;
    -557.35;
    0]

X=linsolve(A,B)
```

Therefore, we get:

$$\begin{aligned} &F'_{12x} = 15{,}258\ \text{N}, F'_{12y} = 4{,}609\ \text{N}, F'_{23x} = 15{,}636\ \text{N}, F'_{23y} = 4{,}292\ \text{N} \\ &F'_{34x} = 16{,}697\ \text{N}, F'_{34y} = 3{,}869\ \text{N}, F'_{14x} = 17{,}255\ \text{N}, F'_{14y} = 3{,}869\ \text{N} \end{aligned} \tag{13.106}$$

Since all of the calculated forces are positive, the directions that were assumed for all of the unknown forces in the FBDs given in Fig. 13.11 were correct.

C5. Compare the calculated dynamic forces with the corresponding static values as a means for verification of the results.

The static solution was obtained in Eq. (13.47) as:

$$F_{14x} = 16.005 T_{12}\ \text{N} \tag{13.47}$$

At an applied torque of $T_{12} = 1000$ N.m, Eq. (13.47) provides the following value for the required static load to be applied on the slider for maintaining static equilibrium:

$$\text{Static}: F_{14x} = 16{,}005\ \text{N} \tag{13.107}$$

The calculated dynamic load is listed in Eq. (13.106) and it is as follows:

$$\text{Dynamic}: F'_{14x} = 17{,}255\ \text{N} \tag{13.108}$$

We notice that the dynamic force is 7.24% higher than the corresponding static one. Therefore, the static solution provides a fairly good approximation for this dynamic problem.

One can also compare the generated forces at points A and B and explore the extent of departure of member 3 from being a two-force member in this dynamic problem. According to the results listed in Eq. (13.106), while at point A the induced forces are:

$$F'_{23x} = 15{,}636\ \text{N};\ F'_{23y} = 4{,}292\ \text{N}$$

At point B we have:

$$F'_{43x} = 16{,}697\ \text{N};\ F'_{43y} = 3{,}869\ \text{N}$$

So, while the corresponding forces at points A and B are close, they are no longer equal to each other. This is in contrast with the way these forces were in the static problem when AB was a simple two-force member.

Finally, we would like to verify that if the applied torque T_{12} has an order of magnitude that is closer to that of the existing inertia torques (recall Eq. (13.79) in which $I_{G3}\alpha_3 = 13.22$ N. m), the dynamic effects will be more significant and the differences between the calculated static and dynamic values of F_{14x} will increase. On the other hand, if the differences in the orders of magnitude of the applied torque T_{12} and that of the existing inertia torques increase, the dynamic results will be even closer to the static ones.

To verify the mentioned points, we keep the expected kinematics of the problem unchanged (see the problem statement) and just gradually change the applied torque.

Table 13.2 Differences between the calculated dynamic values of force F_{14x} and its corresponding static values (both in N) in Example 13.3 for various choices of the input torque (in N.m)

T_{12}	F_{Static}	$F_{Dynamic}$	$[(F_{Dynamic}-F_{Static})/F_{Static}]\times 100$
10,000	160,050	161,380	0.82%
2,000	32,010	33,268	3.78%
1,000	16,005	17,255	7.24%
500	8,002.5	9,247.7	13.46%
100	1,600.5	2,842.2	43.69%

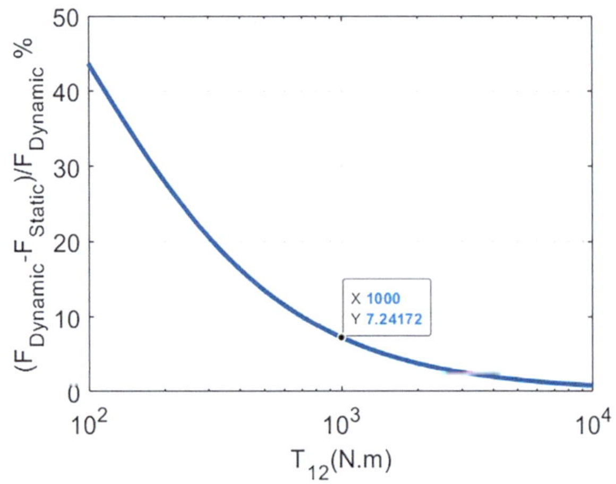

Fig. 13.12 Differences between the calculated dynamic values of force F_{14x} and its corresponding static values in Example 13.3 and for various choices of the input torque

For each value of the applied torque, the resulting static and dynamic values of the required force F_{14x} that should be applied so that the expected response of the system is realized are recorded. The results of this study are given in Table 13.2 and Fig. 13.12. In Table 13.2, the previously obtained results for the baseline case of $T_{12} = 1000$ N.m are highlighted.

We see that by decreasing the applied torque T_{12} and choosing it to be closer and closer to the inertia effect's order of magnitude, the difference between the dynamic value of F_{14x} and its corresponding static value increases significantly. In such cases, providing a static approximation for an inherently dynamic problem is incorrect.

On the other hand, by increasing the applied load even further, we notice that the existing inertia effect of $I_{G3}\alpha_3 = 13.22$ N.m would be negligible and the error resulting from ignoring dynamic effects reduces to only 0.82%, at $T_{12} = 10{,}000$ N.m. Therefore, at high values of the applied loading and small values of mass times acceleration, the static analysis can successfully provide a quick and sufficiently accurate result.

Reference

1. Stanišić, M. L. (2015). *Mechanisms and machines: Kinematics, dynamics, and synthesis*. Cengage Learning.

Chapter 14
Static and Dynamic Unbalance and Balancing

14.1 Introduction

In many applications that include rotating or translating objects severe vibrations can occur. Such applications include vehicle wheels, internal combustion engines, reciprocating compressors, fans, electric motors, and steam and gas turbines. At high speeds, or speeds that are close to the natural frequencies of vibrations of a machine, significant levels of vibrations can be experienced. You just need to remember the last time that you drove a car in which the wheels were not properly balanced! Severe vibrations can also result in noise, cyclic loading, and structural fatigue. In order to prevent such problems, we should keep vibrations at an acceptable level by balancing.

Chapter 14 covers the static and dynamic balancing of short and long rotors. These are important topics and find applications in rotating machinery and especially at high speeds. This chapter also includes discussions on calculating shaking forces in machines that have reciprocating members like internal combustion engines and reciprocating compressors.

14.2 Single-Plane Balancing of Short Rotors

An important cause of vibrations in machinery is that a rotating or a translating member is not properly balanced. Even seemingly small amounts of unbalance at certain speeds of operation (aka critical speeds) can generate considerable levels of vibrations. A necessary condition for having a balanced rotor (i.e., one in which only negligible amounts of forces and moments are generated during operation) is that its center of mass be located on the axis of rotation. For perfect balance, the mass center of each thin transversal slice of a rotor should be on the axis of revolution. In such a situation no dynamic forces are created when the rotor is rotated at any angular

M. Ghorashi, *Introduction to Machine Dynamics*, Mechanical Engineering Series,
https://doi.org/10.1007/978-3-031-94736-0_14

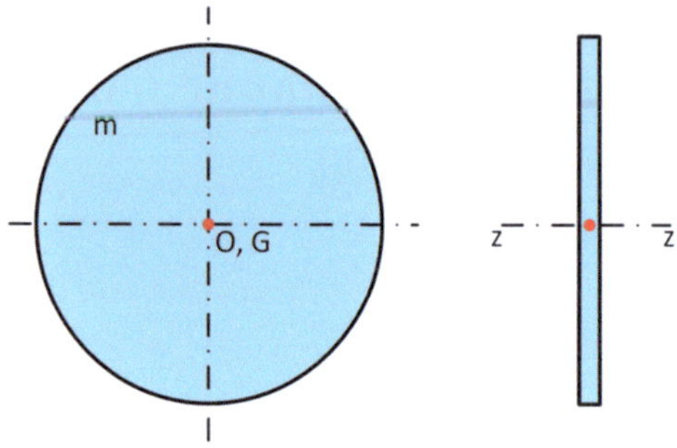

Fig. 14.1 Statically and dynamically balanced thin disk in rotations about the z-z axis

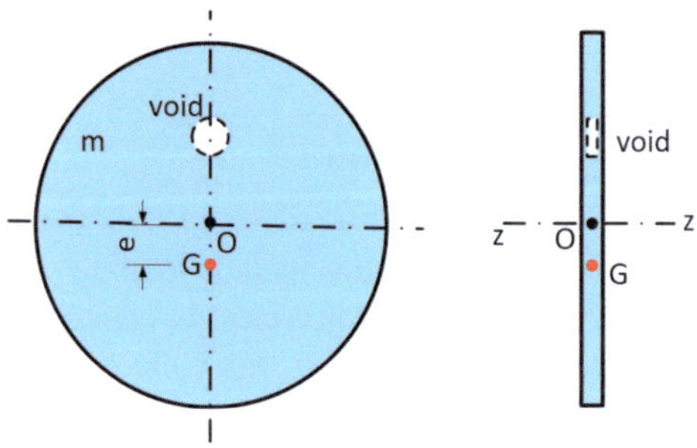

Fig. 14.2 A thin disk with an internal void has a center of mass that is not located on its axis of rotation z-z and has this *preferred static equilibrium position*

velocity. It is important to note that even if the geometry of a rotor might seem to be completely symmetrical with respect to its axis of rotation, the part might have, e.g., internal voids due to manufacturing methods like casting. Such voids can cause the mass center of the object to be eccentric from the axis of rotation. If the amount of eccentricity of the mass center of a rotor of mass m is e, during rotation at an angular velocity ω, an unbalanced inertia force of $me\omega^2$ (that is centrifugal) is generated that causes vibrations and dynamic reaction forces at the supporting bearings.

Consider a uniform thin disk of mass m shown in Fig. 14.1. The disk can rotate about the z-z axis that intersects the disk at point O (that is also the geometric center of the disk). If this disk has uniform mass distribution, its mass center G will also be located on the axis of rotation and its distance from point O (or the amount of eccentricity) will be zero. As a result, the disk has neutral stability; meaning that if we rotate it about the z-z axis by any angle θ, it will stay at that position with no oscillations. **The disk has no preferred static equilibrium position. Such a disk is said to be statically balanced.** If this disk is now rotated at an angular velocity ω about the z-z axis, since the eccentricity of its mass center with respect to the rotation axis is zero, the rotation of this disk does not generate any centrifugal force. Therefore, this disk is also said to be **dynamically balanced**.

Now, consider Fig. 14.2 in which a similar thin disk of mass m can rotate about the z-z axis that intersects the disk at point O that is also the geometric center of the disk. But this time the disk has a void inside it that causes the center of mass to be shifted in the opposite side of the void by an amount e, that is called the eccentricity of the mass center. If the disk is now rotated about the z-z axis by an angle θ and then released, it will oscillate a bit (depending on the amount of friction in the bearings)

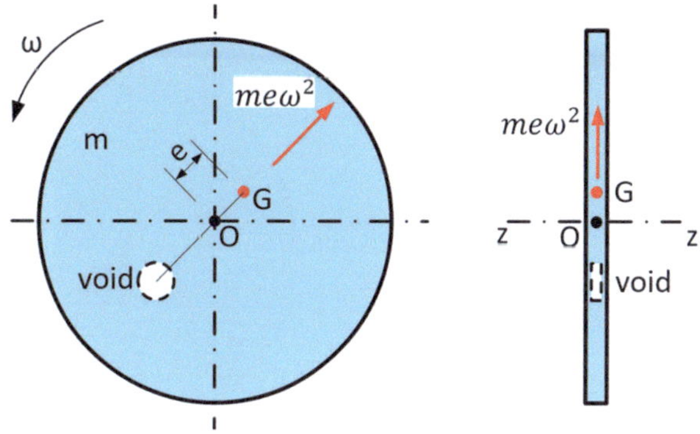

Fig. 14.3 The same unbalanced disk as shown in Fig. 14.2, but now it is rotating at an angular velocity ω

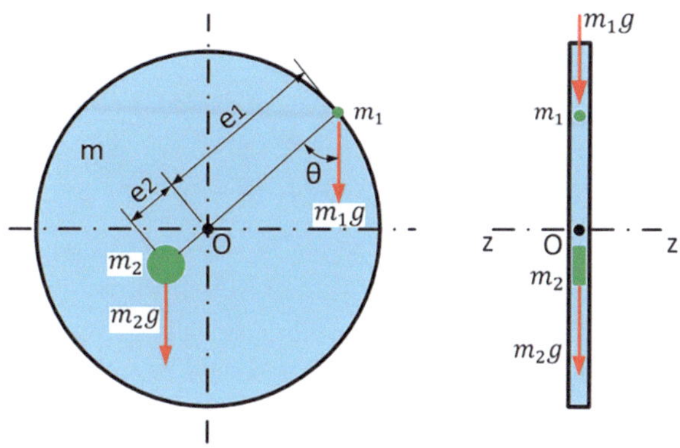

Fig. 14.4 Static balance of a thin rotor

and finally comes to rest at the location shown in Fig. 14.2; in which the center of mass is at the lowest possible point (i.e., line OG is vertical). **Since in this case there is a preferred static equilibrium position, the disk is said to be statically unbalanced.**

If the disk shown in Fig. 14.2 is now rotated about the z-z axis at an angular velocity ω, as shown in Fig. 14.3, since the eccentricity of the mass center with respect to the rotation axis is not zero, the rotation of this disk generates a rotating centrifugal force $F = me\omega^2$. This force, which rotates with the disk, should be supported by the bearings that hold its shaft (not shown in the figure). Because of generation of this unbalanced force during rotation, this disk is also said to be **dynamically unbalanced**.

Having a void that has been generated by manufacturing methods (like casting) is not the only way by which a rotor can become unbalanced. For example, if a blade of a perfectly balanced fan breaks, then the broken fan is unbalanced and it starts to become noisy and make vibrations. One solution, of course, is to break another blade that is located opposite to the already broken one and in this way re-balance the fan!

Now, we see why in balancing terminology **the term *m.e* is named the amount of unbalance**. In Fig. 14.4, for balancing a rotor that has an added eccentric mass m_1 at radius e_1, another mass m_2 (aka counterbalance or correction mass) at radius e_2 and 180° out of phase with respect to the first one is placed in such a way that the

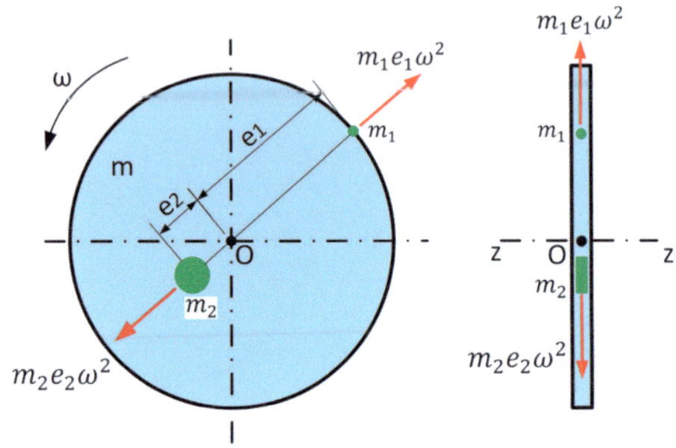

Fig. 14.5 Dynamic balance of a thin rotor

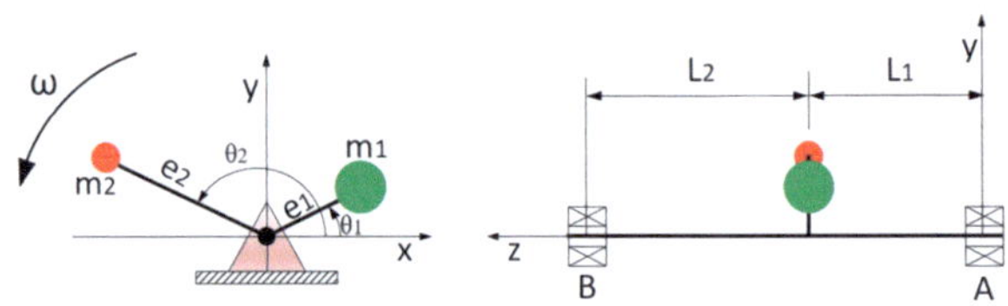

Fig. 14.6 A shaft with two masses attached to it in one plane

resulting mass center of the thin disk falls on the axis of rotation. Therefore, the following condition for balancing should be satisfied.

$$m_1e_1 = m_2e_2 \tag{14.1}$$

If Eq. (14.1) is satisfied, then the unbalance amounts would be equal and therefore their moments (generated by gravity) will cancel each other at any arbitrary angle θ:

$$m_1ge_1 \sin\theta = m_2ge_2 \sin\theta \tag{14.2}$$

For thin rotors, as we see in Fig. 14.5, once static balancing is accomplished, the rotor will be dynamically balanced too because the corresponding centrifugal forces would be equal as well:

$$m_1e_1\omega^2 = m_2e_2\omega^2 \tag{14.3}$$

For balancing an unbalanced rotor, one should always decide about the amount of the mass to be added (or taken out), the radius at which it should happen, and its angular position. In practice, balancing machines can be used to determine these unknowns. In this way, the thin rotor is balanced statically and dynamically in one plane (single-plane balancing) and for a thin object, usually that is all we need for proper balancing.

Example 14.1 Figure 14.6 illustrates a light, angle-shaped link to which masses m_1 and m_2 are attached. The corresponding radii are e_1 and e_2. The link is connected to a

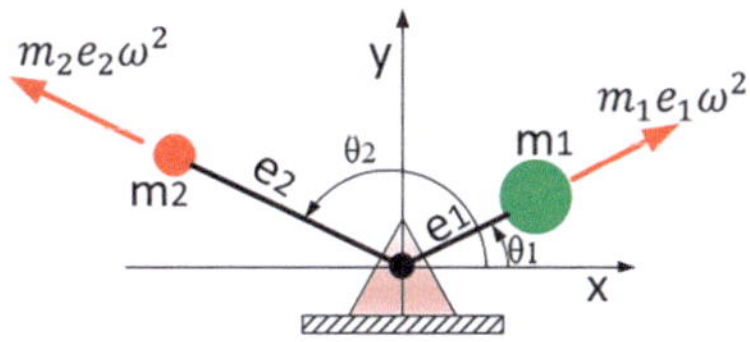

Fig. 14.7 Inertia forces apply on the rotor shown in Fig. 14.6 while in rotation

shaft and the shaft is supported by two bearings at A and B. The shaft can be modeled as a simply-supported rod. We know that: $m_1 = 200$ g, $m_2 = 100$ g, $e_1 = 200$ mm, $e_2 = 400$ mm, $L_1 = 1$ m, $L_2 = 1.5$ m, $\theta_1 = 25°$, $\theta_2 = 155°$.

A. If the shaft rotates at 300 rpm, calculate the resulting reaction forces that are generated in the bearings at A and B.
B. Calculate the amount of the correction mass m_c and its angular position so that no unbalanced force is generated during rotation of the shaft if the correction mass is to be added at the radius of $e_c = 300$ mm.

Clearly, the rotor is not statically balanced as its mass center is not located on the axis of rotation. For the same reason, the rotor is not dynamically balanced either. Therefore, for balancing it, we need to add a proper correction mass at a suitable radius and a suitable angular position.

A. **If the shaft rotates at 300 rpm, calculate the resulting reaction forces that are generated in the bearings at A and B.**

First, we express the angular velocity of the rotor in rad/s,

$$\omega = 300 \text{ rpm} = 300 \text{ rev/ min} \times \frac{2\pi \frac{\text{rad}}{\text{rev}}}{60 \frac{\text{s}}{\text{min}}} = 31.42 \text{ rad/s}$$

Now, the two centrifugal forces (aka inertia forces) shown in Fig. 14.7 are calculated as follows:

$$m_1 e_1 \omega^2 = 0.2\ kg \times 0.2\ m \times (31.42 \text{ rad/s})^2 = 39.49 \text{ N}$$

$$m_2 e_2 \omega^2 = 0.1\ kg \times 0.4\ m \times (31.42 \text{ rad/s})^2 = 39.49 \text{ N}$$

Using the coordinate system x, y, z that rotates with the shaft and is shown in Figs. 14.6 and 14.7, one can write the calculated inertia forces in vector form as follows:

$$\vec{F}_1 = 39.49\left(\cos\theta_1\hat{i} + \sin\theta_1\hat{j}\right) \text{ N}$$

$$\vec{F}_2 = 39.49\left(\cos\theta_2\hat{i} + \sin\theta_2\hat{j}\right) \text{ N}$$

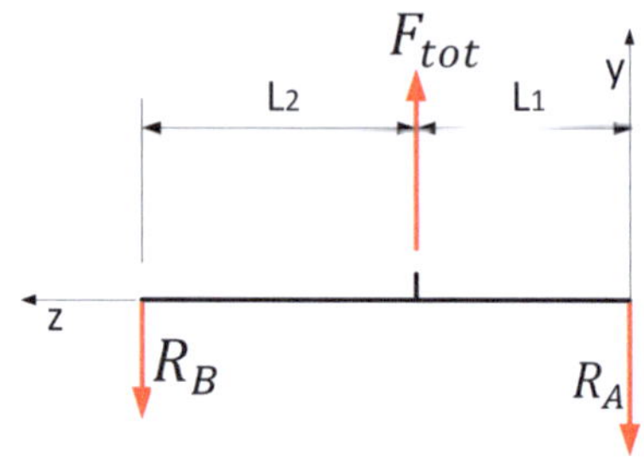

Fig. 14.8 Free body diagram of the rotor shown in Fig. 14.7 with its resultant inertia force and the reaction forces

where, $\vec{F}_1$ and $\vec{F}_2$ are the inertia forces associated with the rotation of m_1 and m_2, respectively. Substituting the given values for θ_1 and θ_2 we get these inertia forces that act on the *x*-*y* plane,

$$\vec{F}_1 = 35.79\hat{i} + 16.69\hat{j}\ \mathrm{N}$$

$$\vec{F}_2 = -35.79\hat{i} + 16.69\hat{j}\ \mathrm{N}$$

Therefore, the total inertia force is:

$$\vec{F}_{tot} = \vec{F}_1 + \vec{F}_2 = 33.38\hat{j}\ \mathrm{N} \tag{14.4}$$

As expected, this force operates in the *y* direction (see Fig. 14.8). In fact, the two inertia forces are symmetrical with respect to the *y* axis, so when they are added, their *x* components cancel out. Since we applied inertia forces to the rotor, using D'Alembert's principle (see Sect. 12.2.3) the problem has now been converted to the dynamic equilibrium of a simply-supported beam subjected to a transversal force, $\vec{F}_{tot}$. Using the equilibrium conditions of the shaft, we can calculate the unknown reaction forces R_A and R_B as follows:

$$\begin{gathered} R_A + R_B = F_{tot} \\ R_A = \frac{L_2}{L_1 + L_2} F_{tot};\ R_B = \frac{L_1}{L_1 + L_2} F_{tot} \end{gathered} \tag{14.5}$$

Therefore,

$$R_A = \frac{1.5}{1 + 1.5} \times 33.38 = 20.03\ \mathrm{N},\ R_B = 13.35\ \mathrm{N}$$

These reaction forces act on the *y*-*z* plane that contains F_{tot}, (i.e., the resultant of inertia forces of the two masses m_1 and m_2), as well as the shaft. Therefore, as the masses rotate about the shaft, these reaction forces rotate as well (and so does $\vec{F}_{tot}$). Even though the calculated reaction forces R_A and R_B have constant magnitudes, they rotate with the shaft; so their horizontal and vertical components in a stationary coordinate system X, Y, Z can be stated as follows: at bearing A:

$$R_{AX} = R_A \cos \omega t; R_{AY} = R_A \sin \omega t \tag{14.6}$$

In Eq. (14.6), it is assumed that at $t = 0$ the moving coordinate axis y (along which the resultant inertia force and the bearing reaction forces operate) is collinear with the stationary coordinate axis X. By substituting the available data one obtains,

$$R_{AX} = 20.03 \cos(31.42\text{t})\ \text{N}; R_{AY} = 20.03 \sin(31.42\text{t})\ \text{N}$$

Similarly at bearing B we have:

$$R_{BX} = R_B \cos \omega t; R_{BY} = R_B \sin \omega t \tag{14.7}$$

which results in,

$$R_{BX} = 13.35 \cos(31.42\text{t})\ \text{N}; R_{BY} = 13.35 \sin(31.42\text{t})\text{N}$$

The stress effect of the unbalanced force on the shaft is not time dependent, as the shaft itself rotates with the unbalanced force! (No relative motion exists.) Therefore, a point on the shaft that is subjected to tensile stress remains so during motion. However, the bearings are subjected to cyclic loading and fatigue.

B. **Calculate the amount of the correction mass m_c and its angular position so that no unbalanced force is generated during rotation of the shaft.**

Method 1 The correction mass m_c should be chosen in such a way that after it is added to the rotor, the center of mass of the rotor G, coincides with the axis of rotation of the shaft. As a result, the net eccentricity of the existing masses reduce to zero and the rotor will be statically balanced:

$$\vec{e}_G = 0 \tag{14.8}$$

Combining Eq. (14.8) and the equation for the position of mass center in a system of particles we get:

$$\vec{e}_G = \frac{\sum m_i \vec{e}_i}{\sum m_i} \rightarrow \sum m_i \vec{e}_i = 0 \tag{14.9}$$

or,

$$m_1 \vec{e}_1 + m_2 \vec{e}_2 + m_c \vec{e}_c = 0 \tag{14.10}$$

Using the rotating coordinate system x, y, Eq. (14.10) can be expanded as follows,

$$m_1e_1\left(\cos\theta_1\hat{i}+\sin\theta_1\hat{j}\right)+m_2e_2\left(\cos\theta_2\hat{i}+\sin\theta_2\hat{j}\right)+m_ce_c\left(\cos\theta_c\hat{i}+\sin\theta_c\hat{j}\right)=0 \tag{14.11}$$

Substituting the available data and $e_c = 300$ mm, into Eq. (14.11) results in:

$$0.2\times0.2\left(\cos 25^\circ\hat{i}+\sin 25^\circ\hat{j}\right)+0.1\times0.4\left(\cos 155^\circ\hat{i}+\sin 155^\circ\hat{j}\right)$$
$$+0.3m_c\left(\cos\theta_c\hat{i}+\sin\theta_c\hat{j}\right)=0$$

This vector equation can be broken into two scalar equations for the two unknowns m_c and θ_c:

$$0.3m_c\cos\theta_c=0; 0.3m_c\sin\theta_c=-0.0338 \tag{14.12}$$

Since $\cos\theta_c = 0$, one concludes that,

$$\theta_c=90^\circ \text{ or } 270^\circ$$

Since $\sin\theta_c < 0$, the correct angular position of the correction mass is:

$$\theta_c=270^\circ$$

This result is in harmony with our intuitive expectations and what we see in Eq. (14.4):

$$\vec{F}_{tot}=\vec{F}_1+\vec{F}_2=33.38\hat{j}\,\text{N} \tag{14.4}$$

Since the total inertia force $\vec{F}_{tot}$ is pointing up, in the y direction, it was indeed expected that the balancing mass be placed in the $-\vec{F}_{tot}$ direction which is the 270° angular position. The correction mass, m_c, is now calculated using the following equation that can be obtained from Eq. (14.12) at $\theta_c = 270°$,

$$0.3m_c=0.0338$$

Therefore we get:

$$m_c=112.67\text{ g} \tag{14.13}$$

So the suitable correction mass m_c, shown in Fig. 14.9, and located at the 300 mm radius is 112.67 g and it should be placed at 270°.

Method 2 The same result could be obtained by first calculating the location of the mass center of m_1 and m_2 (call it $\vec{e}_{tot}$):

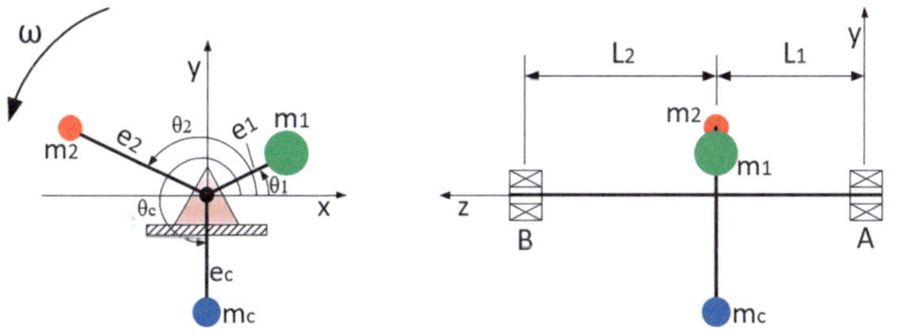

Fig. 14.9 Correction mass m_c is added to balance the originally unbalance rotor shown in Fig. 14.7

$$\vec{e}_{tot} = \frac{m_1 \vec{e}_1 + m_2 \vec{e}_2}{m_1 + m_2} \tag{14.14}$$

Substituting the available data in Eq. (14.14) results in:

$$\vec{e}_{tot} = \frac{0.2\times0.2\left(\cos25°\hat{i} + \sin25°\hat{j}\right) + 0.1\times0.4\left(\cos155°\hat{i} + \sin155°\hat{j}\right)}{0.2+0.1} = 0.11270\hat{j}m$$

Then, we place the correction mass in the $-\vec{e}_{tot}$ direction (in this case, $-\hat{j}$). The correction mass should also be located at the requested $e_c = 300$ mm radius from the axis of rotation. So, its location is fully known and only its magnitude is still unknown. To balance the inertia force due to the rotation of two masses m_1 and m_2 that is applied at their mass center, $\vec{e}_{tot}$, we should have:

$$(m_1 + m_2)e_{tot}\omega^2 = m_c e_c \omega^2$$

The ω^2 term is canceled from both sides (meaning that the result is valid for all angular velocities!). By substituting the available data we get:

$$(0.2 + 0.1) \times 0.11270 = m_c \times 0.3$$

And finally:

$$m_c = 112.70 \text{ g} \tag{14.15}$$

This result is within round-off error from Eq. (14.13) in which $m_c = 112.67$ g. So the suitable correction mass m_c, shown in Fig. 14.9, and located at the 300 mm radius is 112.70 g and it should be placed at 270°. Furthermore, we conclude that the total unbalance of the original rotor is:

$$m_c e_c = 33{,}810 \text{ g.mm} \tag{14.16}$$

Fig. 14.10 Balancing masses on a car wheel ring

As we saw in Example 14.1, usually the **correction mass radius e_c, is dictated by the particular balancing application** and it is considered as a given value. For example, for balancing a car wheel it is impractical to add correction masses to the wheel at any place other than the rim of the ring (see Fig. 14.10). Our design variables for balancing (as we saw in Example 14.1) are the amount of mass to be added and its corresponding angular position.

In single-plane balancing of a thin rotor, the condition for balancing was shown to be that the center of mass should be located on the axis of rotation:

$$\sum m_i \vec{e}_i = 0 \tag{14.9}$$

For dynamic balancing of a thin rotor, the summation of all inertia forces should be equal to zero:

$$\sum m_i \vec{e}_i \omega^2 = 0 \tag{14.17}$$

As a result, no dynamic bearing reaction forces would be generated. In Eq. (14.17), the angular velocity term can be factored out to give:

$$\omega^2 \sum m_i \vec{e}_i = 0$$

that again results in Eq. (14.9). So, in thin rotors, when the center of mass is located on the axis of rotation the summation of inertia forces will be zero and the rotor is both statically and dynamically balanced.

14.3 Long Rotor Unbalance vs Short Rotor Unbalance

The decision as to which rotor can be classified as long, and which one as short (related to the number of correction planes needed for adding correction masses for balancing), is made based on the length-to-diameter ratio and the angular velocity of operation. In short rotors, the inertia moments created by unbalance and inertia forces are small (since the moment arms are small). But even short rotors may need two-plane balancing if they operate at high speeds. In case of a long rotor, the inertia forces have considerable moment arms and therefore can create significant inertia moments that should also be balanced. An example is a crankshaft of an internal combustion engine shown in Fig. 14.11 that is supposed to rotate at a few thousands of revolution per minute. The counterweights shown in Fig. 14.11 are used to balance this crankshaft.

An unbalance rotor can become balanced by strategically adding perfectly calculated masses (by methods like screwing bolts) or by drilling holes at certain radii and angular positions. In the rest of this chapter, we will see how those calculations are carried out. Figure 14.12 illustrates that same idea in which bolts of different masses can be screwed at various radii and angular positions in order to help balance an otherwise unbalance rotor.

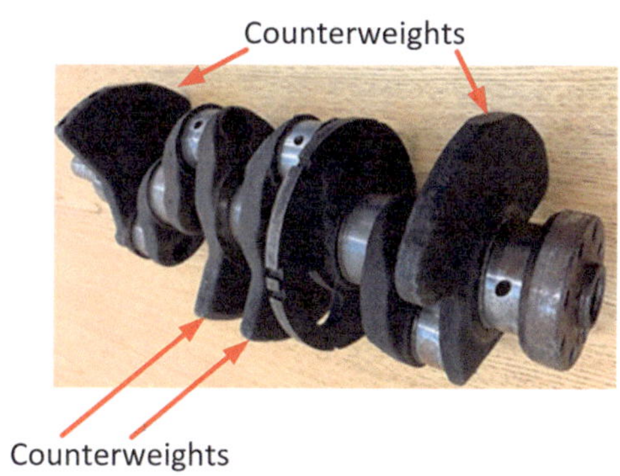

Fig. 14.11 Crankshaft of a 1995 Saturn SL1 and its counterweights as an example of the need to balance a long rotor

Fig. 14.12 Machinery Fault Simulator by SpectraQuest, VA

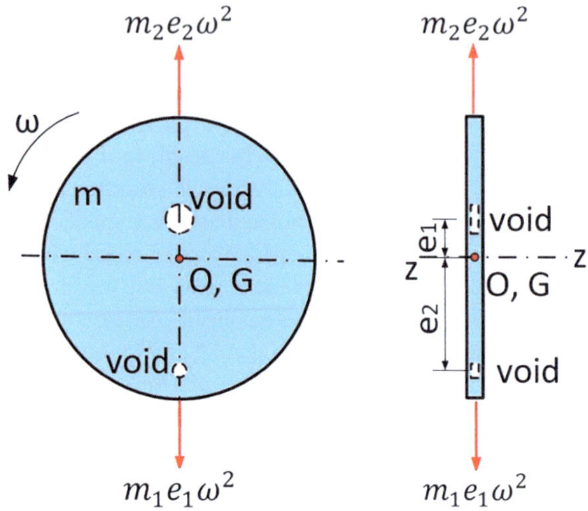

Fig. 14.13 Statically balanced short rotor is now dynamically balanced as well

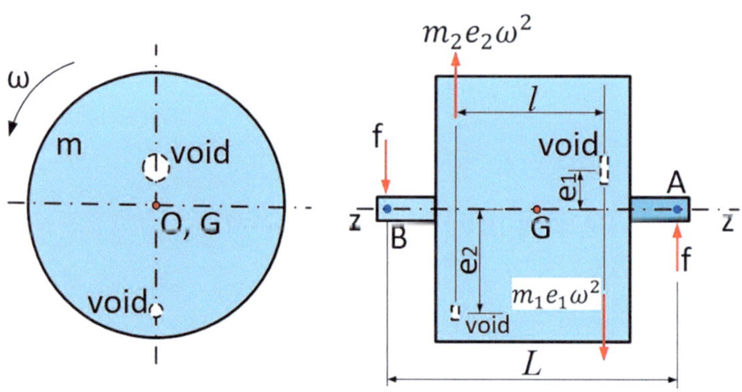

Fig. 14.14 This statically balanced long rotor is not dynamically balanced. (forces are shown only in one view)

Consider a short rotor and a long rotor shown in Figs. 14.13 and 14.14, respectively. It is seen that both rotors have two internal casting voids that are 180° apart. Let us assume that the missing masses for voids located at e_1 and e_2 are m_1 and m_2, respectively. If we have,

$$m_1e_1 = m_2e_2 \tag{14.18}$$

in both the short and the long rotors the center of mass will be located on the axis of rotation and both of the rotors will be statically balanced. Satisfying Eq. (14.18) guarantees the dynamic balancing of the short rotor shown in Fig. 14.13. However, as we see in Fig. 14.14 for the long rotor, the moment arm of the inertia forces is not negligible and therefore while we still have,

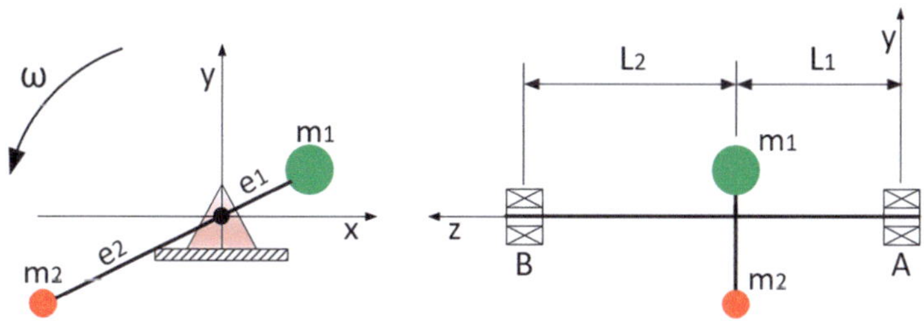

Fig. 14.15 Model of a short rotor discussed in Example 14.2

$$F = m_1 e_1 \omega^2 = m_2 e_2 \omega^2 \tag{14.19}$$

and the inertia forces are balanced, the two forces form a couple that can have a significant moment arm, l. The resulting moment of this inertia force couple is:

$$M = Fl \tag{14.20}$$

where F is defined in Eq. (14.19). This moment is balanced by an equal moment,

$$M = fL \tag{14.21}$$

that is applied on the rotor shaft by the shaft bearings located at A and B (with an axial distance L from each other). Each of these bearings is subjected to force f. Consequently, in the case of a long rotor, even if the inertia forces cancel each other out, their moment effect is not zero and can induce significant stresses on the shaft, rotor, and the bearings. Therefore, while the rotor shown in Fig. 14.14 is statically balanced, it is not dynamically balanced. Such a rotor cannot be balanced by adding or removing masses only in one plane. As we will see in Sections 14.4 and 14.5, a long rotor must be balanced in two planes. This is exactly what is done for balancing wheels in cars.

Example 14.2 Consider the short and long rotors shown in Figs. 14.15 and 14.16, respectively, for which we have the following parameters:

$m_1 = 100$ g, $m_2 = 50$ g, $e_1 = 50$ mm, $e_2 = 100$ mm, $L_1 = 2$ m, $L_2 = 2$ m, $L_3 = 2$ m, $\omega = 300$ rad/s.

A. Investigate to see if either case is statically or dynamically balanced.
B. In case any one of the rotors is dynamically unbalance, calculate the unbalance force or moment and the bearing reactions.

A. **Investigate to see if either case is statically or dynamically balanced.**

We first analyze the short rotor. The requirement for it to be statically and dynamically balanced is that the following equation is satisfied:

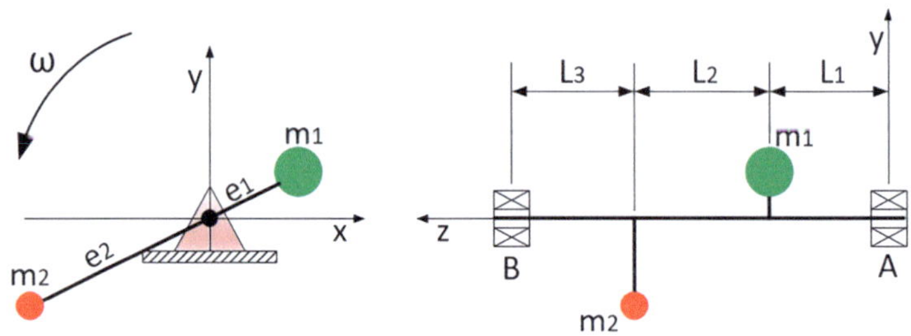

Fig. 14.16 Model of a long rotor discussed in Example 14.2

$$m_1 e_1 = m_2 e_2 \tag{14.1}$$

Substitution of the available data in Eq. (14.1) reveals that this equation is satisfied. So the short rotor shown in Fig. 14.15 is statically and also dynamically balanced and its dynamic bearing reaction forces are zero.

Now, we consider the long rotor shown in Fig. 14.16. The front view of this rotor is identical to the front view of the short rotor given in Fig. 14.15. Exploring the right view of the long rotor in Fig. 14.16 shows that, in this case, the two masses are not located at the same axial position along the rotor. However, in this case too Eq. (14.1) is satisfied; so the long rotor shown in Fig. 14.16 is statically balanced.

As to the dynamic balancing of the long rotor shown in Fig. 14.16, the net inertia force is also zero and indeed,

$$F = m_1 e_1 \omega^2 = m_2 e_2 \omega^2$$

where

$$F = 0.100 \text{ kg} \times 0.050 \text{ m} \times (300 \text{ rad/s})^2 = 450 \text{ N}$$

B. **In case any one of the rotors is dynamically unbalance, calculate the unbalance force or moment and the bearing reactions.**

The net inertia moment is not zero and we have:

$$M = FL_2 = 450 \text{ N} \times 2 \text{ m} = 900 \text{ N.m}$$

Considering the symmetry of the problem, the bearing reaction forces, f at the two bearings are equal, act in opposite directions (similar to the case shown in Fig. 14.14) and we have:

$$f \times (L_1 + L_2 + L_3) = 900 \text{ N.m}$$

So, the dynamic reaction forces at the bearings are:

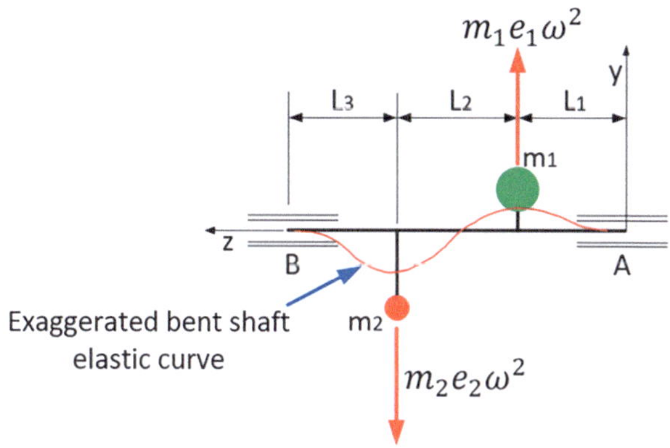

Fig. 14.17 Exaggerated deformation of the shaft shown in Fig. 14.16 during rotation due to the unbalanced inertia moment applied on the shaft

$$f = 900 \text{ N.m}/6\text{m} = 150 \text{ N}$$

We notice that while this rotor is statically balanced, it is dynamically unbalanced and its shaft will deform under application of the bending moment generated by inertia forces, as seen in Fig. 14.17:

Before proceeding with the two-plane balancing of long rotors, let us answer two important questions: Is it possible that a dynamically balanced rotor be statically unbalanced? No! If e is the eccentricity of the mass center of the rotor from the axis of rotation, in a dynamically balanced rotor we have, $me\omega^2 = 0$, that results in: $e = 0$. Therefore, the mass center of the rotor is on the axis of rotation and the rotor is also statically balanced.

The follow-up question is as follows: Is it possible that a statically balanced rotor be dynamically unbalanced? Yes! In a long rotor we just saw that it is completely possible. **So, if a long rotor is statically balanced, one cannot rule that it is also dynamically balanced. However, if it is dynamically balanced, it is definitely statically balanced as well.**

In other words, we can conclude that for short rotors static balancing is enough for ensuring dynamic balancing. For long rotors, static balancing does not guarantee dynamic balancing. However, the converse is true: **If a rotor is dynamically balanced (i.e., balanced for any ω), it is also statically balanced (i.e., balanced for $\omega = 0$).**

In the discussions on balancing so far, we considered a few special cases in which a short rotor with unbalanced masses located in the same plane could be successfully balanced by adding or taking away a mass in that same plane (see Example 14.1). To this end, only a single vector force equation was used that included the needed correction mass and its angular position as the two unknowns.

We also considered cases in which unbalanced masses were located in two planes (see Fig. 14.16) and the balancing masses could be added in those same planes. In what follows, balancing of rotors will be discussed in a more general way. We will consider long rotors whose unbalance masses are distributed in several planes; or they are continuously distributed along the axis of the rotor. **We will see that a rotor can always be balanced and provide zero reaction forces on the bearings if proper masses are added on two planes that are determined suitable for adding or removing masses in a particular application.** The two suitable balancing masses and their corresponding two angular positions (a total of four unknowns) can then be calculated by solving the dynamic equilibrium force and moment equations (a total of four scalar equations).

14.4 Two-Plane Balancing of Long Rotors with Lumped Masses

For balancing long rotors, both the force and the moment equations should be implemented as the moment arms of inertia forces are no longer small. Usually, two correction planes have already been identified as suitable and feasible locations on which balancing masses are to be added or taken away from (see Fig. 14.18). On each plane, the radial positions of the balancing masses are usually predetermined as there are limited positions one can easily add or take away material from the rotor. (See Fig. 14.10 that shows a balanced car wheel with a suitable balancing plane and a suitable balancing radius.)

Having decided about the longitudinal coordinates of the correction masses (i.e., the location of the correction planes) and the radial locations of the correction masses on the two correction planes, the only remaining unknowns are the amounts of correction masses and their angular positions. So there are two unknowns on each correction plane and a total of four unknowns. We use D'Alembert's principle and write dynamic equilibrium equations of moments about convenient points on each

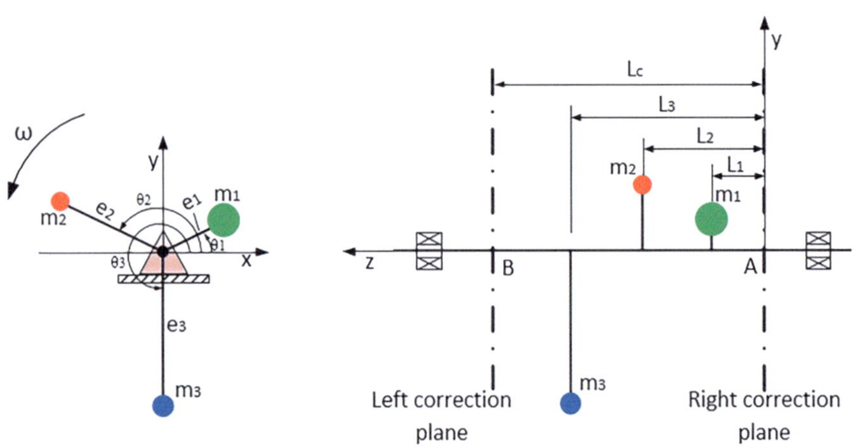

Fig. 14.18 A long rotor and the locations of its masses with respect to the right correction plane (chosen for installing or taking off correction masses)

one of the two correction planes. These two vector equations can be converted to four scalar equations, if their component equations are considered. Therefore, there are a total of four dynamic equilibrium equations that can be solved for the four unknowns mentioned before.

In order to see details of the mentioned calculations, consider the rotor shown in Fig. 14.18. The figure illustrates a long rotor with a few discrete eccentric masses that is to be balanced both statically and dynamically by adding two correction masses on two correction planes. We name the correction masses to be added on the left and right correction planes as m_l and m_r, respectively. The left and right correction masses are to be located at e_l and e_r distances from the rotation axis and at angular positions, θ_l and θ_r, respectively (both measured from the rotating x axis shown in Fig. 14.18). When the rotor is perfectly balanced, zero dynamic reaction forces are generated at the bearings of the shaft (that is our aim). By application of the D'Alembert's principle, we see that the only forces acting on the FBD of the shaft are the inertia forces due to the rotor masses and the inertia forces due to the added correction masses. Therefore, the dynamic equilibrium equation for inertia forces can be written as follows:

$$\sum m_i \vec{e}_i \omega^2 = 0 \tag{14.22}$$

or,

$$\sum m_i \vec{e}_i = 0 \tag{14.23}$$

which is the same as equation Eq. (14.9) that we obtained for balancing short rotors. Expansion of this summation results in,

$$m_1 \vec{e}_1 + m_2 \vec{e}_2 + m_3 \vec{e}_3 + m_l \vec{e}_l + m_r \vec{e}_r = 0 \tag{14.24}$$

The vector Eq. (14.24) can be projected in the x and y directions and provide two scalar equations. Since in practice the proper radial eccentricities e_l and e_r are considered as given values, Eq. (14.24) should be solved for the remaining four unknowns, i.e., balancing masses m_l and m_r, as well as their corresponding angular positions, θ_l and θ_r (included in $\vec{e}_l$ and $\vec{e}_r$).

In order to complete the solution, one needs to write two more scalar equations that relate the listed unknowns. These equations come from the dynamic equilibrium of the moments of the inertia forces about a convenient point. Considering inertia forces and assuming reaction forces at the bearings to be zero (that is our aim), one may write an equation for the dynamic equilibrium of moments of inertia forces about point A on the right correction plane (see Fig. 14.18). This equation has mass and the angular position of the correction mass located on the left correction plane. The reason for this choice is because the moments of the inertia forces acting on the right correction plane about point A is zero so no more unknowns related to those forces enter the resulting moment equation. Therefore, the summation of inertia moments about point A, i.e.,

$$\sum \vec{M}_A = 0 \tag{14.25}$$

results in:

$$\sum L_i m_i \vec{e}_i \omega^2 = 0 \tag{14.26}$$

where L_i is the moment arm or the axial distance of the i-th inertia force from point A. Notice that we do not need to write the cross product calculations to get the moment equation (14.26). If we do write moment as a complete cross product, the outcome will be the same as Eq. (14.26), but each vector will be rotated by 90°, due to the cross product. So the relative positions of the vectors will not change and the result will be the same!

By taking out the ω^2 term in Eq. (14.26) we get:

$$\sum L_i m_i \vec{e}_i = 0 \tag{14.27}$$

Expansion of the summation in Eq. (14.27) results in,

$$L_1 m_1 \vec{e}_1 + L_2 m_2 \vec{e}_2 + L_3 m_3 \vec{e}_3 + L_c m_l \vec{e}_l = 0 \tag{14.28}$$

Equation (14.28) can be projected in the x and y directions and provide two scalar equations. In practice, the proper radial eccentricities e_l and e_r (where the correction masses are to be located) are considered as given. Therefore, these two scalar equations should be solved for two unknowns, i.e., balancing mass m_l and its angular position, θ_l (included in $\vec{e}_l$).

Since moment is a cross product, the signs used for the terms in Eq. (14.28) are not always positive and they depend on the axial location of the contributing inertia force along the z axis. If any mass had been located in the right-hand side of the right plane shown in Fig. 14.18, its contribution in Eq. (14.28) should have had a negative sign. In order to guarantee correct sign for each term, we place the y axis on the right plane and use coordinate z instead of L_i for writing Eq. (14.28). In this way, the mentioned sign change automatically takes place. So the moment equation about point A on the right correction plane can be rewritten as follows:

$$z_1 m_1 \vec{e}_1 + z_2 m_2 \vec{e}_2 + z_3 m_3 \vec{e}_3 + z_l m_l \vec{e}_l = 0 \tag{14.29}$$

in which z_i's are the z coordinates of each of the masses and z_l is the z coordinate of the left correction plane. In order to complete the solution one may either use the balance of inertia forces Eq. (14.24) or solve a separate balance of moments equation; this time about point B on the left correction plane (see Fig. 14.19). Either

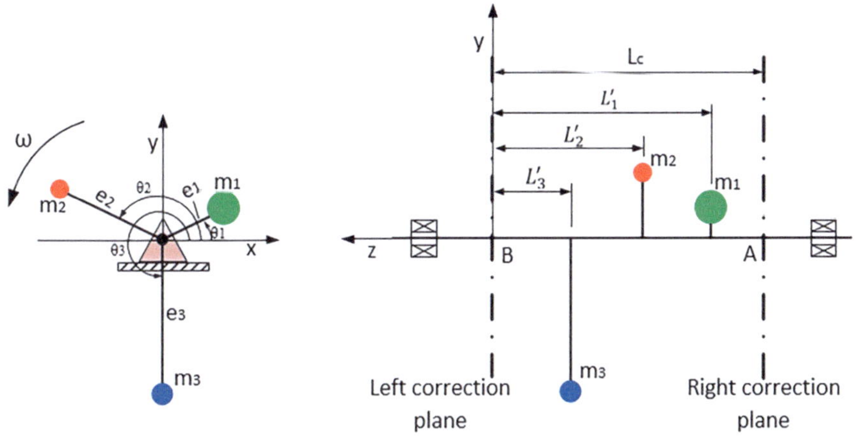

Fig. 14.19 A long rotor and the locations of its masses with respect to the left correction plane (chosen for installing or taking off correction masses)

equation has the remaining two unknowns in it; i.e., the right correction mass m_r and its angular position, θ_r and can be solved as a system of two scalar equations for two unknowns.

As to the moment of inertia forces about point B on the left correction plane we have:

$$\sum \vec{M}_B = 0 \tag{14.30}$$

In a way similar to what we did before for deriving Eq. (14.27) about the right correction plane, one can show that:

$$\sum L'_i m_i \vec{e}_i = 0 \tag{14.31}$$

where, as we see in Fig. 14.19, the moment arm L'_i is now measured from the left correction plane.

Expansion of this summation results in,

$$L'_1 m_1 \vec{e}_1 + L'_2 m_2 \vec{e}_2 + L'_3 m_3 \vec{e}_3 + L_c m_r \vec{e}_r = 0 \tag{14.32}$$

Equation (14.32) can be projected in the x and y directions and provide two scalar equations. These two equations should be solved for the remaining two unknowns, i.e., balancing mass m_r and its angular position, θ_r (included in $\vec{e}_r$).

Again, since moment is a cross product, the signs used for the terms in Eq. (14.32) are not always positive and they depend on the axial location of the contributing inertia force along the z axis. If any mass had been located in the left-hand side of the left plane in Fig. 14.19, its contribution in Eq. (14.32) should have had a negative

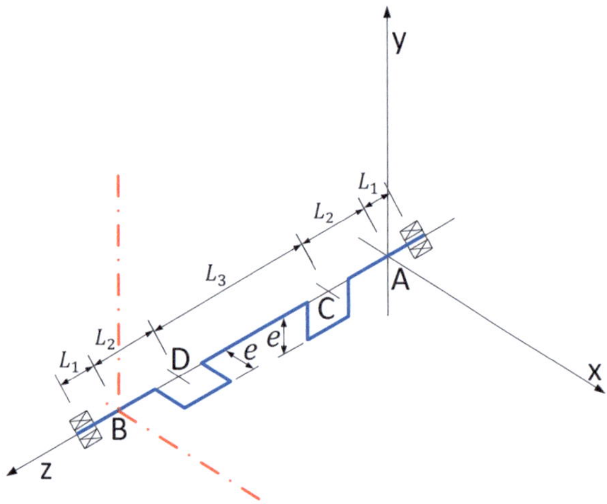

Fig. 14.20 A crankshaft with two throws is supported by two bearings

sign. In order to guarantee correct sign for each term, we place the y axis on the left plane and use coordinate z instead of L_i' for writing Eq. (14.32). In this way, the mentioned sign change automatically takes place. So the moment equation about point B on the left correction plane can be rewritten as follows:

$$z_1 m_1 \vec{e}_1 + z_2 m_2 \vec{e}_2 + z_3 m_3 \vec{e}_3 + z_r m_r \vec{e}_r = 0 \tag{14.33}$$

in which the z_i's are the z coordinates of each of the masses and z_r is the z coordinate of the right correction plane.

The two moment equations (14.29) and (14.33), each converts to a system of two scalar equations for two unknowns [instead of writing the force Eq. (14.23) and a moment equation]. These two sets of two scalar equations for two unknowns (one set for each correction plane) can be solved for m_l and m_r as well as their corresponding angular positions, θ_l and θ_r.

In the force and moment equations discussed above, no reaction forces on the shafts at the bearing locations were taken into account; as they are to vanish when suitable correction masses are implemented. Furthermore, the obtained equations for the correction masses and their angular positions do not depend on the angular velocity of the shaft, ω. Therefore, if a rotor is balanced at a certain angular velocity, it will still be balanced if the angular velocity changes.

Example 14.3 Figure 14.20 illustrates a crankshaft with two throws that is supported by two bearings at its ends. The crankshaft rotates about the z axis and we would like to calculate balancing (correction) masses and their angular positions with respect to the coordinate system x, y, z, that rotates with the crankshaft. The correction masses are to be placed on the right and left correction planes that pass through points A and B, respectively, and are at L_1 distances from the bearings. It is desirable to place the correction masses at the tip of two light links (not shown) with

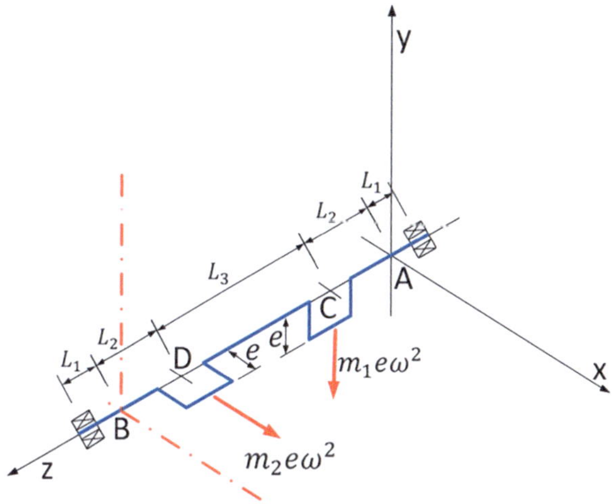

Fig. 14.21 Inertia forces on the crankshaft shown in Fig. 14.20 and finding balancing masses on the left correction plane

$e_l = e_r = 4$ in, that rotate with the shaft on the left and right correction planes. Each crank throw can be modeled as a 5 lbm mass that is eccentric by $e = 2.5$ in from the axis of rotation, z. The rest of the data is as follows:

$$L_1 = 4 \text{ in}, L_2 = 6 \text{ in}, L_3 = 15 \text{ in}$$

We are looking for the dynamically balanced condition at which no reaction forces are created at the bearings (see Fig. 14.21). We start by writing the moment equation about point A on the right correction plane in order **to get the mass and its angular position on the left correction plane** (that passes through B). Recalling Eq. (14.29):

$$z_1 m_1 \vec{e}_1 + z_2 m_2 \vec{e}_2 + z_3 m_3 \vec{e}_3 + z_l m_l \vec{e}_l = 0 \tag{14.29}$$

for this case we get,

$$L_2 m_1 e\left(-\hat{j}\right) + (L_2 + L_3) m_2 e\hat{i} + (2L_2 + L_3) m_l \vec{e}_l = 0$$

or,

$$6\,\text{in} \times 5\,\text{lbm} \times 2.5\,\text{in}\left(-\hat{j}\right) + (6\text{in} + 15\text{in}) \times 5\,\text{lbm} \times 2.5\,\text{in}\left(\hat{i}\right) + (2 \times 6\,\text{in} + 15\,\text{in}) m_l \vec{e}_l = 0$$

$$m_l \vec{e}_l = -9.722\,\hat{i}\ \text{lbm.in} + 2.778\,\hat{j}\ \text{lbm.in} \tag{14.34}$$

One may consider,

$$\vec{e}_l = e_l \widehat{n}_l = 4\text{in}\ \widehat{n}_l \tag{14.35}$$

in which, $\widehat{n}_l$ is the unit vector determining the direction of the correction mass on the left correction plane. Therefore, we have:

$$m_l \widehat{n}_l = -2.431\widehat{i}\ \text{lbm} + 0.695\widehat{j}\ \text{lbm} = 2.528\ \text{lbm}\left(-0.961\widehat{i} + 0.275\widehat{j}\right) \tag{14.36}$$

and finally we get:

$$m_l = 2.528\ \text{lbm} \tag{14.37}$$

and,

$$\tan\theta_l = \frac{0.275}{-0.961} \rightarrow \theta_l = -15.969^\circ; 164.030^\circ \tag{14.38}$$

Since the x component of $\widehat{n}_l$ is negative and its y component is positive, θ_l is in the second quadrant and the correct answer is:

$$\theta_l = 164.030^\circ \tag{14.39}$$

For determining the **balancing mass and its angular position on the right correction plane** we place the coordinate system on the left correction plane and write moments about point B on the left correction plane (see Fig. 14.22). In this

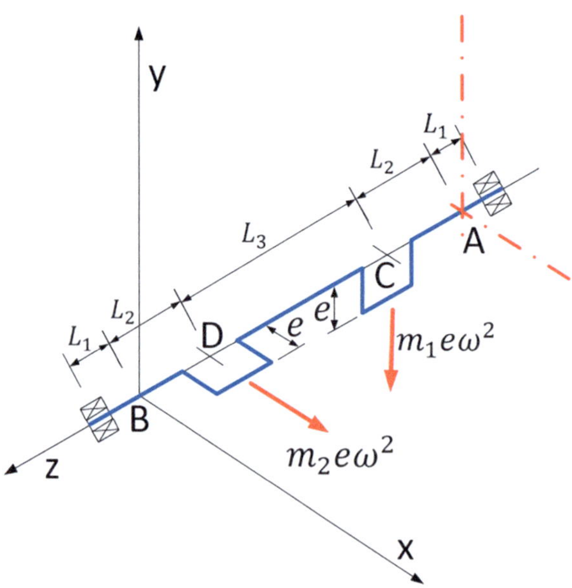

Fig. 14.22 Inertia forces on the crankshaft shown in Fig. 14.20 and finding balancing masses on the right correction plane

case, the correction mass we just calculated on the left correction plane has zero moment arm and does not enter the formulation. Recalling Eq. (14.33):

$$z_1 m_1 \vec{e}_1 + z_2 m_2 \vec{e}_2 + z_3 m_3 \vec{e}_3 + z_r m_r \vec{e}_r = 0 \tag{14.33}$$

for this case we have,

$$-(L_2 + L_3) m_1 e\left(-\hat{j}\right) + (-L_2) m_2 e \hat{i} + (-2L_2 - L_3) m_r \vec{e}_r = 0 \tag{14.40}$$

so,

$$-(6\text{in} + 15\text{in}) \times 5\ \text{lbm} \times 2.5\ \text{in}\left(-\hat{j}\right) + (-6\ \text{in}) \times 5\ \text{lbm} \times 2.5\ \text{in}\left(\hat{i}\right)$$
$$+ (-2 \times 6\ \text{in} - 15\ \text{in}) m_r \vec{e}_r = 0$$

or,

$$m_r \vec{e}_r = 9.722\,\hat{j}\ \text{lbm.in} - 2.778\,\hat{i}\ \text{lbm.in} \tag{14.41}$$

One may consider,

$$\vec{e}_r = e_r \hat{n}_r = 4\text{in}\ \hat{n}_r \tag{14.42}$$

In which, $\hat{n}_r$ is the unit vector determining the direction of the correction mass on the right correction plane. Substituting Eq. (14.42) into Eq. (14.41) gives,

$$m_r \hat{n}_r = -0.695\hat{i}\ \text{lbm} + 2.431\hat{j}\ \text{lbm} = 2.528\ \text{lbm}\left(-0.275\hat{i} + 0.962\hat{j}\right) \tag{14.43}$$

Therefore,

$$m_r = 2.528\ \text{lbm} \tag{14.44}$$

Also,

$$\tan\theta_r = \frac{0.962}{-0.275} \rightarrow \theta_r = -74.047\,°;\ 105.953\,° \tag{14.45}$$

Since the x component of $\hat{n}_r$ is negative and its y component is positive, θ_r is in the second quadrant and the correct answer is:

$$\theta_r = 105.953\,° \tag{14.46}$$

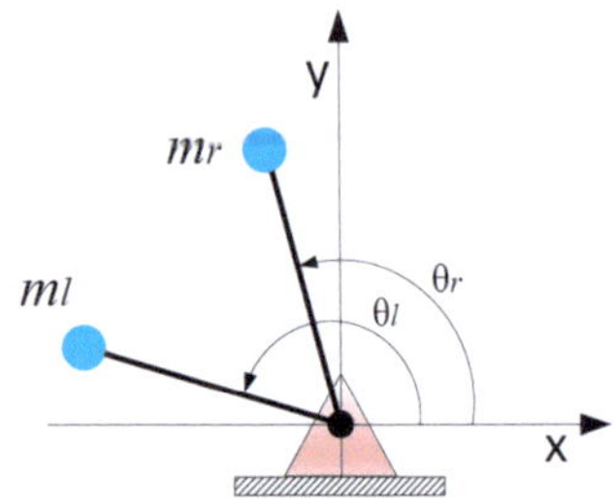

Fig. 14.23 Side view of the correction masses added to the crankshaft of Fig. 14.20, showing their radial and angular positions

Finally, we check to see if the calculated correction masses and angular positions satisfy the balance of inertia forces equation. (note: $m_3=0$)

$$m_1 \vec{e}_1 + m_2 \vec{e}_2 + m_l \vec{e}_l + m_r \vec{e}_r = 0 \tag{14.24}$$

Substituting the known quantities into Eq. (14.24) results in:

$$5\times 2.5\left(-\hat{j}\right) + 5\times 2.5\left(\hat{i}\right) + \left(-9.722\,\hat{i} + 2.778\,\hat{j}\right) + \left(9.722\,\hat{j} - 2.778\,\hat{i}\right) = 0$$

$$-12.5\hat{j} + 12.5\hat{i} - 12.5\hat{i} + 12.5\hat{j} = 0$$

So, the equation of dynamic equilibrium of inertia forces is satisfied, as expected. Therefore, the suitable correction masses and their angular positions are given by Eqs. (14.37), (14.39), (14.44), and (14.46) which are repeated below:

$$\begin{aligned} m_l &= 2.528 \text{ lbm}; \theta_l = 164.030\,^\circ \\ m_r &= 2.528 \text{ lbm}; \theta_r = 105.953\,^\circ \end{aligned} \tag{14.47}$$

and shown in Fig. 14.23.

14.5 Two-Plane Balancing of Long Rotors with Distributed Mass

So far we considered balancing lumped-parameter rotors where locations of the unbalanced segments were known in advance. Now, we study the more practical case of an actual rotor modeled as a distributed mass shown in Fig. 14.24 where points a and b are the locations of the two bearings that support the shaft of the rotor 2. During the operation of this unbalanced rotor, it has been detected that forces $\vec{F}_{a21}$ and $\vec{F}_{b21}$ are imposed by the rotor shaft on the bearings. The reaction forces $\vec{F}_{a12}$ and $\vec{F}_{b12}$ are imposed by the bearings on the rotor shaft (equal and opposite to $\vec{F}_{a21}$ and $\vec{F}_{b21}$, respectively, imposed by the shaft on the bearings; when correction

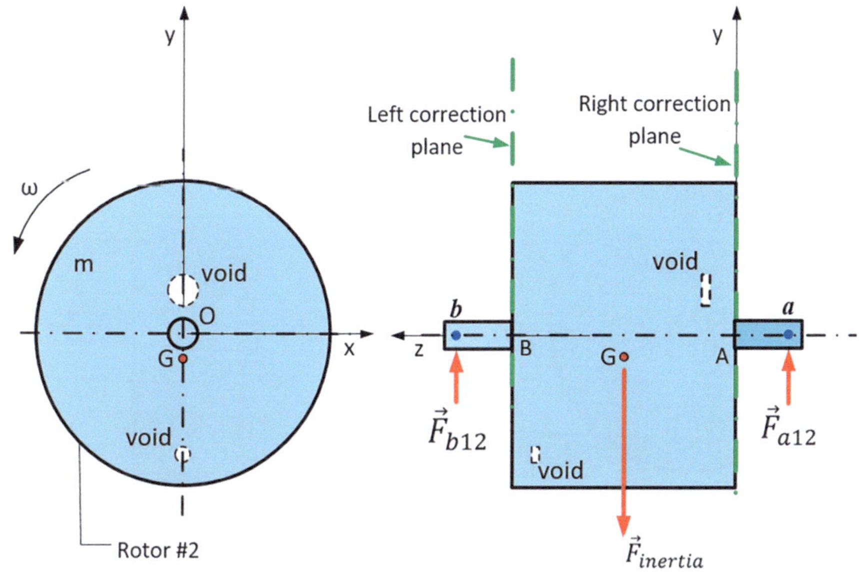

Fig. 14.24 Free body diagram of an unbalanced long rotor including the inertia force and the reaction forces applied by the bearings on the shaft (without correction masses, forces are shown only on one view)

masses are absent). The aim is to **calculate suitable balancing masses, their radial positions, and their angular positions on each of the two correction planes** so that the resulting rotor with the added (or taken out) balancing masses does not generate unbalanced dynamic forces on the bearings (see Fig. 14.25).

We have two correction planes, but the total number of variables in this case are six (three variables for correction masses, i.e., the suitable balancing mass value, the proper radial position, and the correct angular position, on each one of the two correction planes). However, as we noticed in Sect. 14.4, by using D'Alembert's principle, we can write only two moment equations or one moment and one force equation. When the two vector equations are projected on the two coordinates on the cross-section of the rotor (i.e., x and y), we get four scalar equations that can be used for solving not more than four unknowns. In practice, usually not only the suitable planes on which balancing masses can be added (or taken out) can relatively easily be decided, but also their practical radial positions are dictated by the application. So the only remaining unknowns are the amount of mass and the angular position of the correction mass on each of the two correction planes. Therefore, there are a total of four remaining unknowns that can be calculated by using the four mentioned scalar equations. This method, in which balancing masses are calculated on the two planes by analyzing the created unbalanced forces, is called the **Mechanical Compensation method of balancing**.

For distributed mass systems, one cannot write simple inertia force and inertia moment equations for the existing unbalance in a rotor. Instead, we would have to use the reaction forces at the bearings that are generated when the correction masses

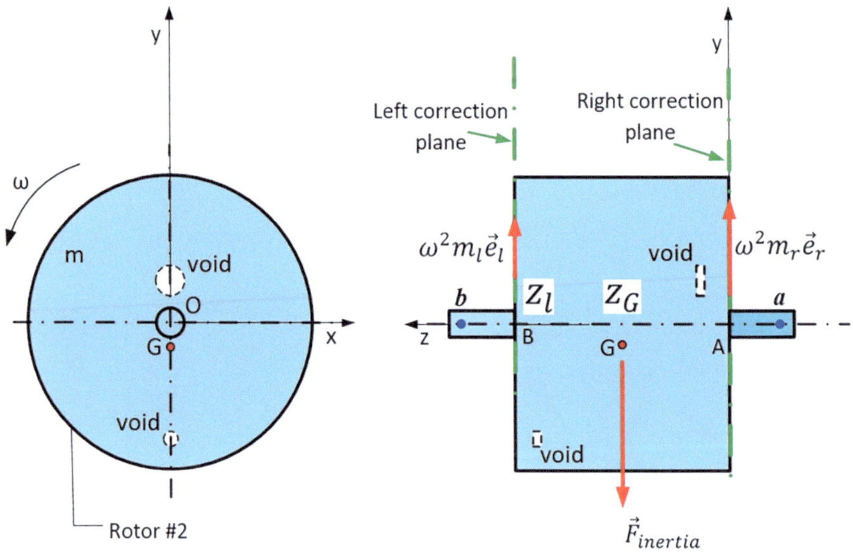

Fig. 14.25 Free body diagram showing the dynamic equilibrium of the now balanced rotor (no dynamic bearing forces exist) including the inertia forces of the correction masses (forces are shown only on one view)

are absent (in order to gage the amount of unbalance in the rotor). Fig. 14.24 illustrates the original FBD of the shaft (i.e., before the correction masses are added) using D'Alembert's principle, that shows its dynamic equilibrium under the rotor inertia force $\vec{F}_{inertia}$ and bearing forces.

The corresponding force equilibrium equation is,

$$\vec{F}_{a12} + \vec{F}_{b12} + \vec{F}_{inertia} = 0 \tag{14.48}$$

Referring to Fig. 14.24 and by using moment arms z_a, z_b, and z_G for the right bearing, left bearing, and mass center, respectively, the moment equilibrium equation about point A on the right correction plane (before any correction masses are added) is:

$$z_a \vec{F}_{a12} + z_b \vec{F}_{b12} + z_G \vec{F}_{inertia} = 0 \tag{14.49}$$

Next, we use equations of dynamic equilibrium considering the inertia forces and moments due to the correction masses added (or taken out) on the correction planes so that they can impose forces to balance $\vec{F}_{inertia}$ and moments to balance $z_G \vec{F}_{inertia}$ (see Fig. 14.25). The suitable correction masses to be added on the left and right correction planes are named as m_l and m_r, respectively. The left and right correction masses are to be located at e_l and e_r distances from the rotation axis and at angular positions, θ_l and θ_r, respectively. When the suitable correction masses are added, it is expected that the reaction forces at the bearings vanish and the dynamic equilibrium FBD looks like what we see in Fig. 14.25.

So, we have the following dynamic force and moment equilibrium equations. (Note that when the moment equation is written, e.g., about point A on the right correction plane, z is measured from that plane and thus, $z_r = 0$.)

$$\vec{F}_{inertia} + \omega^2 m_l \vec{e}_l + \omega^2 m_r \vec{e}_r = 0 \tag{14.50}$$

$$z_G \vec{F}_{inertia} + \omega^2 z_l m_l \vec{e}_l + \omega^2 z_r m_r \vec{e}_r = 0 \tag{14.51}$$

One can now eliminate the unknown $\vec{F}_{inertia}$ and $z_G \vec{F}_{inertia}$ terms among Eqs. (14.48) to (14.51) and get,

$$-\vec{F}_{a12} - \vec{F}_{b12} + \omega^2 m_l \vec{e}_l + \omega^2 m_r \vec{e}_r = 0 \tag{14.52}$$

$$-z_a \vec{F}_{a12} - z_b \vec{F}_{b12} + \omega^2 z_l m_l \vec{e}_l + \omega^2 z_r m_r \vec{e}_r = 0 \tag{14.53}$$

Since in practice we can usually measure the reaction forces applied *by the rotor on the bearing* and frame, i.e., $\vec{F}_{a21}$ and $\vec{F}_{b21}$, we get the final suitable form of the force and moment equations for this case of a long rotor as follows:

$$\vec{F}_{a21} + \vec{F}_{b21} + \omega^2 m_l \vec{e}_l + \omega^2 m_r \vec{e}_r = 0 \tag{14.54}$$

$$z_a \vec{F}_{a21} + z_b \vec{F}_{b21} + \omega^2 z_l m_l \vec{e}_l + \omega^2 z_r m_r \vec{e}_r = 0 \tag{14.55}$$

These are two vector equations that can be converted to four scalar equations. These four equations can be solved for four unknowns (that are the two values of the correction masses and their two angular positions) in such a way that they completely balance and remove the previously measured reaction forces at the bearings.

One can also simplify Eqs. (14.54) and (14.55) for the special case of a short rotor. In the case of a short rotor (that has small z values) operating at low speeds equation (14.55) essentially reduces to 0=0. Having fewer number of equations means that for short rotors we can have (or we need) only one correction mass. Therefore, we will have only two scalar equations with two unknowns, i.e., the angle and the mass amount of the single correction mass that is needed for balancing.

Example 14.4 Testing a rotor at the speed of 100 rad/s has revealed rotating forces (phasors) imposed by the rotor on its bearings and the rotor frame, i.e., $\vec{F}_{a21}$ and $\vec{F}_{b21}$. In Fig. 14.26, the corresponding reaction forces applied on the rotor, i.e., $\vec{F}_{a12}$ and $\vec{F}_{b12}$ are shown:

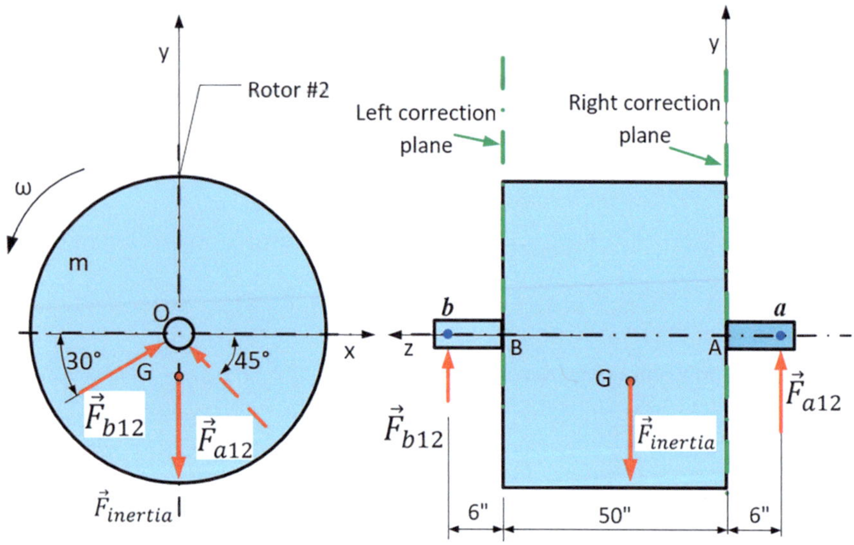

Fig. 14.26 Recorded dynamic reaction forces on an unbalanced rotor

$$\vec{F}_{a12} = 50\angle 135\,°\,\text{lbf},\ \vec{F}_{b12} = 30\angle 30\,°\,\text{lbf}$$

We would like to balance this rotor (i.e., reduce the induced bearing reaction forces to zero) by adding material to it or removing material from it on the shown left and right correction planes. The correction masses are all to be placed at 3 in radii from the rotation axis.

Calculate the amounts and the angular positions of the two correction masses with respect to the shown rotating coordinate system (x and y) in the following cases:

A. Masses are to be added to the correction planes.
B. Masses are to be removed from the correction planes.

A. **Masses are to be added to the correction planes.**

First, we need to calculate $\vec{F}_{a21}$ and $\vec{F}_{b21}$ that are:

$$\vec{F}_{a21} = -\vec{F}_{a12} = 50\angle -45\,°\,\text{lbf};\ \vec{F}_{b21} = -\vec{F}_{b12} = 30\angle 210\,°\,\text{lbf} \tag{14.56}$$

Recalling the moment equilibrium Eq. (14.55):

$$z_a\vec{F}_{a21} + z_b\vec{F}_{b21} + \omega^2 z_l m_l\vec{e}_l + \omega^2 z_r m_r\vec{e}_r = 0 \tag{14.55}$$

if it is written about the right correction plane we have:

$$z_a = -6 \text{ in}, z_b = 56 \text{ in}, z_l = 50 \text{ in}, z_r = 0 \tag{14.57}$$

Substituting Eqs. (14.56) and (14.57) into (14.55) results in:

$$(-6 \text{ in} \times 50\angle -45° \text{ lbf}) + (56 \text{ in} \times 30\angle 210° \text{ lbf}) + \left(100^2 \times 50m_l \times 3\angle\theta_l\right) = 0$$

or,

$$-300 \text{ lbf.in}\left[\cos(-45°)\widehat{i} + \sin(-45°)\widehat{j}\right]$$
$$+ 1680 \text{ lbf.in}\left[\cos(210°)\widehat{i} + \sin(210°)\widehat{j}\right] + 1.5 \times 10^6 m_l\angle\theta_l \text{ in}^2/\text{s}^2 = 0$$

in which,

$$\angle\theta_l = \widehat{n}_l \tag{14.58}$$

is the unit vector that determines the direction of the installed correction mass on the left correction plane. Simplification of the above equation gives:

$$-1667.05\hat{i} \text{ lbf.in} - 627.87\hat{j} \text{ lbf.in} + 1.5 \times 10^6 m_l\hat{n}_l \text{ in}^2/\text{s}^2 = 0$$

or,

$$m_l\widehat{n}_l = 1.111 \times 10^{-3}\widehat{i} + 0.418 \times 10^{-3}\widehat{j} \text{ lbf.s}^2/\text{in} \tag{14.59}$$

We can separate the mass and the unit vector portions by normalizing the $m_l\widehat{n}_l$ vector:

$$m_l\widehat{n}_l = 1.187 \times 10^{-3} \times \left[\frac{1}{1.187 \times 10^{-3}}\left(1.11 \times 10^{-3}\widehat{i} + 0.418 \times 10^{-3}\widehat{j}\right)\right] \text{lbf.s}^2/\text{in} \tag{14.60}$$

Now, since 1 lbm weighs 1 lbf, we can make the following unit conversions to get:

$$m_l = 1.187 \times 10^{-3} \text{ lbf.s}^2/\text{in} = 1.187 \times 10^{-3} \text{ lbf.s}^2/\text{in} \times \frac{32.2 \text{ lbm}\frac{\text{ft}}{\text{s}^2} \times 12\frac{\text{in}}{\text{ft}}}{1 \text{ lbf}}$$

Therefore, the necessary correction mass on the left correction plane is:

$$m_l = 0.459 \text{ lbm} \tag{14.61}$$

Recalling Eq. (14.60) we conclude that:

$$\widehat{n}_l = 0.936\widehat{i} + 0.352\widehat{j}$$

Therefore, the angular position of the left correction mass is:

$$\theta_l = \tan^{-1}\left(\frac{0.352}{0.936}\right) = 20.61\,^\circ \tag{14.62}$$

Now, to get **the correction mass and its angular position on the right correction plane** we can use either the dynamic equilibrium of forces Eq. (14.54):

$$\vec{F}_{a21} + \vec{F}_{b21} + \omega^2 m_l \vec{e}_l + \omega^2 m_r \vec{e}_r = 0 \tag{14.54}$$

or the equation of equilibrium of moments about the left correction plane:

$$z_a \vec{F}_{a21} + z_b \vec{F}_{b21} + \omega^2 z_l m_l \vec{e}_l + \omega^2 z_r m_r \vec{e}_r = 0 \tag{14.55}$$

where,

$$z_a = -56 \text{ in}, z_b = 6 \text{ in}, z_l = 0, z_r = -50 \text{ in} \tag{14.63}$$

Let us first use the force equilibrium Eq. (14.54). Substituting the known data, it converts to the following equation:

$$(50\angle -45\,^\circ \text{lbf}) + (30\angle 210\,^\circ \text{lbf}) + \left(100^2 \times m_l \times 3\angle\theta_l\right) + \left(100^2 \times m_r \times 3\angle\theta_r\right) = 0$$

or,

$$50 \text{ lbf}\left[\cos(-45\,^\circ)\widehat{i} + \sin(-45\,^\circ)\widehat{j}\right] + 30 \text{ lbf}\left[\cos(210\,^\circ)\widehat{i} + \sin(210\,^\circ)\widehat{j}\right]$$
$$+ \left(3 \times 10^4 \times m_l\angle\theta_l\right) + \left(3 \times 10^4 \times m_r\angle\theta_r\right) = 0$$

We can name the unit vectors defining the angular positions of the left and right correction masses as follows:

$$\angle\theta_l = \widehat{n}_l, \angle\theta_r = \widehat{n}_r \tag{14.64}$$

Simplification of the above equation and using Eq. (14.59) gives:

$$9.375\widehat{i} \text{ lbf} - 50.355\widehat{j} \text{ lbf} + 3 \times 10^4 \frac{\text{in}}{\text{s}^2}\left(1.111 \times 10^{-3}\widehat{i} + 0.418 \times 10^{-3}\widehat{j}\right)\text{lbf}\frac{\text{s}^2}{\text{in}} + 3$$
$$\times 10^4 \times m_r \widehat{n}_r = 0$$

or,

$$42.705\,\widehat{i}\,\text{lbf} - 37.815\,\widehat{j}\,\text{lbf} + 3 \times 10^4 \times m_r \widehat{n}_r \frac{\text{in}}{\text{s}^2} = 0$$

Finally, we get:

$$m_r \widehat{n}_r = -1.424 \times 10^{-3} \widehat{i} + 1.261 \times 10^{-3} \widehat{j}\,\text{lbf.s}^2/\text{in} \tag{14.65}$$

We can separate the mass and the unit vector portions by normalizing the $m_r \widehat{n}_r$ vector:

$$m_r \widehat{n}_r = 1.902 \times 10^{-3} \times \left[\frac{1}{1.902 \times 10^{-3}} \left(-1.424 \times 10^{-3} \widehat{i} + 1.261 \times 10^{-3} \widehat{j} \right) \right] \text{lbf.s}^2/\text{in}$$

So,

$$m_r = 1.902 \times 10^{-3} \text{lbf.s}^2/\text{in} = 1.902 \times 10^{-3}\ \text{lbf.s}^2/\text{in} \times \frac{32.2\ \text{lbm.ft/s}^2 \times 12\ \text{in/ft}}{1\ \text{lbf}}$$

Therefore, the necessary correction mass on the right correction plane is obtained as:

$$m_r = 0.735\ \text{lbm} \tag{14.66}$$

also,

$$\widehat{n}_r = -0.749 \widehat{i} + 0.663 \widehat{j}$$

Based on which the angular position of the right correction mass is:

$$\theta_r = \tan^{-1}\left(\frac{0.663}{-0.749} \right) = 138.485^\circ$$

Notice that using a calculator one might have got

$$\theta_r = \tan^{-1}\left(\frac{0.663}{-0.749} \right) = -41.515^\circ$$

However, since the x component of the unit vector is negative it cannot be located in the fourth quadrant and the correct solution is:

$$\theta_r = \tan^{-1}\left(\frac{0.663}{-0.749} \right) = 138.485^\circ \tag{14.67}$$

So, the masses and the angular positions to be added in the 3in radii on the left and right correction planes are given by Eqs. (14.61), (14.62), (14.66), and (14.67) that are summarized below and shown in Fig. 14.27:

$$\begin{aligned} m_l &= 0.459 \text{ lbm}, \theta_l = 20.61\,^\circ \\ m_r &= 0.735 \text{ lbm}, \theta_r = 138.485\,^\circ \end{aligned} \tag{14.68}$$

One can verify that the obtained results satisfy the equation of equilibrium of moments about the left correction plane as well, i.e.,

$$z_a \vec{F}_{a21} + z_b \vec{F}_{b21} + \omega^2 z_l m_l \vec{e}_l + \omega^2 z_r m_r \vec{e}_r = 0 \tag{14.55}$$

where,

$$z_a = -56 \text{ in}, z_b = 6 \text{ in}, z_l = 0, z_r = -50 \text{ in} \tag{14.63}$$

Substituting Eq. (14.63) and other provided data into the moment equation (14.55) gives:

$$\begin{aligned} &-56 \text{ in} \times 50 \text{ lbf}\left[\cos(-45\,^\circ)\hat{i} + \sin(-45\,^\circ)\hat{j}\right] + 6 \text{ in} \\ &\times 30 \text{ lbf}\left[\cos(210\,^\circ)\hat{i} + \sin(210\,^\circ)\hat{j}\right] + 0 + \left(100\frac{\text{rad}}{\text{s}}\right)^2 \times (-50 \text{ in}) m_r \vec{e}_r = 0 \end{aligned}$$

or,

$$-2135.78\hat{i} \text{ lbf.in} + 1889.90\hat{j} \text{ lbf.in} - 0.5 \times 10^6 \times 3 m_r \hat{n}_r \frac{\text{in}^2}{\text{s}^2} = 0$$

and finally,

$$m_r \hat{n}_r = -1.424 \times 10^{-3}\hat{i} + 1.261 \times 10^{-3}\hat{j} \text{ lbf.s}^2/\text{in}$$

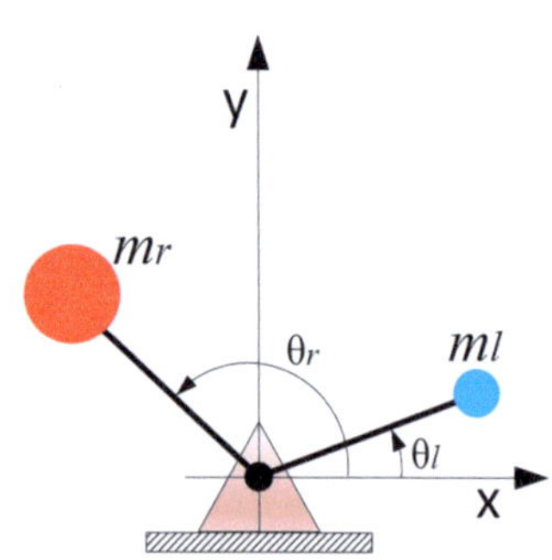

Fig. 14.27 Correction masses needed to be added for balancing the rotor shown in Fig. 14.26 (Illustrating Eq. (14.68) in a side view)

This is the same as Eq. (14.65) that was obtained before using the equilibrium of forces equation. So the results are verified.

B. **Masses are to be removed from the correction planes.**

Now, the aim is to calculate the masses and the angular positions of the materials that should be **removed** from the left and the right correction planes for dynamic balancing. To this end, we just need to choose angular positions that are opposite to those calculated in part A (i.e., they have 180° phase shift compared to those of part A). Therefore, recalling the solution given in Eq. (14.68) and shown in Fig. 14.27 the solution to part B, i.e., the amount and location of the materials to be removed from the rotor is simply:

$$\begin{aligned} m_l &= 0.459 \text{ lbm}; \theta_l = 200.61\,^\circ \\ m_r &= 0.735 \text{ lbm}; \theta_r = -41.52\,^\circ \end{aligned} \tag{14.69}$$

14.6 Lumped-Mass Model of Slider-Crank Mechanism

In the slider-crank mechanism shown in Fig. 14.28, link 2 performs pure rotation with constant angular velocity ω_2 and slider 4 undergoes pure translation. The connecting rod 3 has a combined rotation and translation motion. The aim here is to generate a lumped-mass model for this mechanism that includes masses that only translate or only rotate. By creating this model, calculation of inertia forces and the resulting shaking forces (due to combined translation and rotation) can be easily accomplished.

Points A and B are the crank pin and the wrist pin, respectively, and they are the points at which all of the rotational and the translational masses are to be placed, in that same order.

First, focusing on the crank, we would like to place mass m_A at point A in such a way that it generates the same inertia force as the distributed crank mass m_2 that has a center of mass at G_2, see Fig. 14.29. Therefore, the equivalent mass at A should satisfy the following condition:

$$m_A r_2 \omega_2^2 = m_2 r_5 \omega_2^2 \tag{14.70}$$

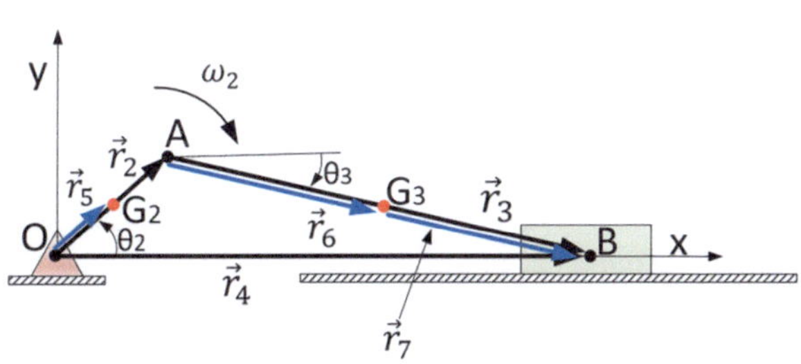

Fig. 14.28 Slider-crank mechanism and the mass centers of its members

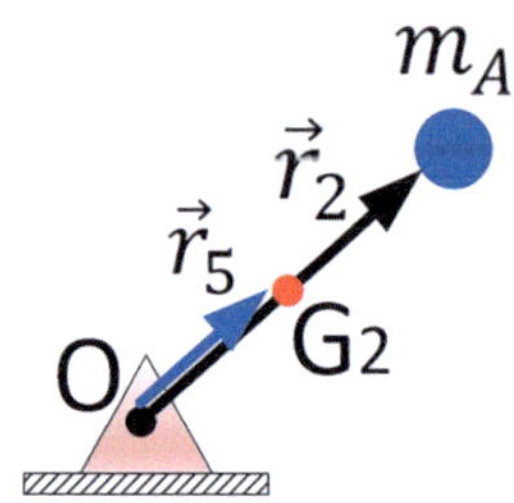

Fig. 14.29 Focus on motion of the crank in Fig. 14.28 to find equivalent mass at point *A*

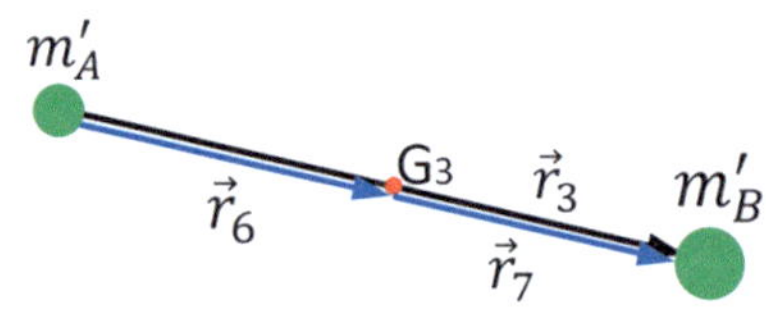

Fig. 14.30 Focus on motion of the connecting rod in Fig. 14.28 to find equivalent masses at points *A* and *B*

Therefore, the lumped mass at crank pin *A* that generates the same amount of inertia force as the distributed mass of crank 2 is:

$$m_A = \frac{r_5}{r_2} m_2 \tag{14.71}$$

Similarly, we would like to replace the connecting rod 3 with two lumped masses m'_A and m'_B that are connected with a massless rod of length r_3, see Fig. 14.30. We obtain the unknown lumped masses m'_A and m'_B by using two conditions. First, the total mass is to remain unchanged:

$$m'_A + m'_B = m_3 \tag{14.72}$$

Second, the center of mass of the new link 3 is to remain at G_3 so the moment of m'_A about *B* should be equal to the moment of m_3 about this point:

$$m'_A(r_6 + r_7) = m_3 r_7 \rightarrow m'_A = \frac{r_7}{r_3} m_3 \tag{14.73}$$

Therefore, Eqs. (14.72) and (14.73) give:

$$m'_B = \frac{r_6}{r_3} m_3 \tag{14.74}$$

This modeling is not perfect. While the total mass and the mass center of the link are unchanged, we have no control on the resulting moment of inertia of the simulated link 3. Therefore, while the inertia forces should be correct, the inertia moment can be inaccurate. Since the simulated masses m'_A and m'_B are placed at the two ends of the simulated connecting rod, the radius of gyration has increased and therefore, the resulting mass moment of inertia of the simulated piston rod is somewhat higher than the actual one.

14.7 Balancing of Shaking Forces in Slider-Crank Mechanisms

Having calculated m_A, m'_A, and m'_B in Sect. 14.6, now the simulated lumped-mass model of a slider-crank mechanism has two masses $m_{A,\ eq}$ at point A (rotational), and $m_{B,\ eq}$ at point B (translational) that can be defined as follows:

$$m_{A,eq} = m_A + m'_A \tag{14.75}$$

$$m_{B,eq} = m_4 + m'_B \tag{14.76}$$

The acceleration at point A is:

$$\vec{a}_A = -\vec{r}_2\omega_2^2 \tag{14.77}$$

And it was previously shown that the acceleration of the slider can be approximated by:

$$a_B \approx -r_2\omega_2^2(\cos\theta_2 + \lambda\cos 2\theta_2); \lambda = r_2/r_3 \tag{5.2}$$

Therefore, the total inertia force associated with the motion of the slider-crank mechanism can be written as follows:

$$\vec{F}_{inertia} = -m_{A,eq}\vec{a}_A + \left(-m_{B,eq}\vec{a}_B\right) \tag{14.78}$$

Substituting Eqs. (14.77) and (5.2) into Eq. (14.78) results in:

$$\vec{F}_{inertia} = m_{A,eq}\left(r_2\omega_2^2\right)\left(\cos\theta_2\hat{i} + \sin\theta_2\hat{j}\right) + m_{B,eq}r_2\omega_2^2[\cos\theta_2 + \lambda\cos 2\theta_2]\hat{i}$$

Therefore, **the total inertia force associated with the motion of the slider-crank mechanism** is:

$$\vec{F}_{inertia} = r_2\omega_2^2\left[\left(m_{A,eq} + m_{B,eq}\right)\cos\theta_2 + \lambda m_{B,eq}\cos 2\theta_2\right]\hat{i} + r_2\omega_2^2\left(m_{A,eq}\sin\theta_2\right)\hat{j} \tag{14.79}$$

Finally, let us **add a counterbalance**, as shown in Fig. 14.31, to the crank in order to reduce the calculated inertia (shaking) force. If the counterbalance mass is m_{cb} and it is added at radius r_{cb}, then its equivalent value at point A would be obtained using Eq. (14.71):

Fig. 14.31 Cutaway of an internal combustion engine. 1. Engine body, 2. Crankshaft, 3. Connecting rod, and 4. Piston

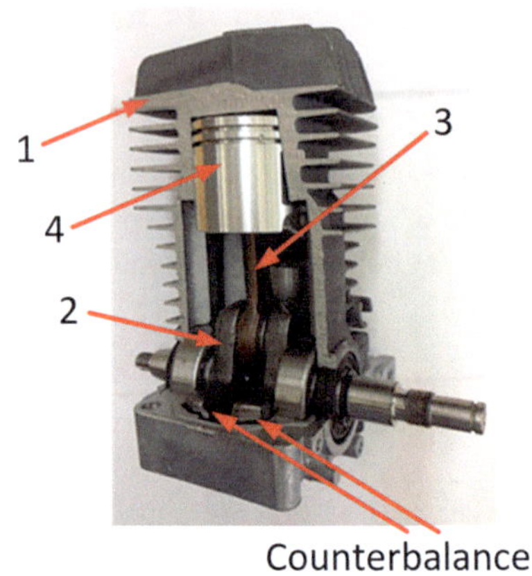

$$m_A = \frac{r_5}{r_2} m_2 \tag{14.71}$$

Therefore we have,

$$m_{cb,eq} = \frac{r_{cb}}{r_2} m_{cb} \tag{14.80}$$

The effect of counterbalance is reduction of the effective mass at point A, i.e., $m_{A,\ eq}$. Therefore, the inertia force Eq. (14.79) is modified as follows:

$$\begin{aligned}\vec{F}_{inertia} = {} & r_2\omega_2^2\left[\left(m_{A,eq} + m_{B,eq} - m_{cb,eq}\right)\cos\theta_2 + \lambda m_{B,eq}\cos 2\theta_2\right]\hat{i} \\ & + r_2\omega_2^2\left(m_{A,eq} - m_{cb,eq}\right)\sin\theta_2\hat{j}\end{aligned} \tag{14.81}$$

Example 14.5 A slider-crank mechanism shown in Fig. 14.28 has the following parameters:

$$m_2 = 4\ \text{kg}; m_3 = 4\ \text{kg}; m_4 = 3\ \text{kg}$$
$$r_2 = 15\ \text{cm}; r_3 = 60\ \text{cm}$$

Considering that the speed of rotation of the crank is 1000 rpm (CCW), plot the total inertia force due to the motion of the mechanism. The crank and the connecting rod are assumed to be uniform bars with their mass centers located at their centroids.

Recalling Eq. (14.71) we have:

$$m_A = \frac{r_5}{r_2} m_2 \rightarrow m_A = \frac{1}{2} m_2 = 2\ \text{kg}$$

Furthermore, using Eqs. (14.73) and (14.74) one obtains:

$$m'_A = \frac{r_7}{r_3} m_3 \rightarrow m'_A = \frac{1}{2} m_3 = 2 \text{ kg}$$

$$m'_B = \frac{r_6}{r_3} m_3 \rightarrow m'_B = \frac{1}{2} m_3 = 2 \text{ kg}$$

Therefore, using Eqs. (14.75) and (14.76) we get:

$$m_{A,eq} = m_A + m'_A = 4 \text{ kg}$$

$$m_{B,eq} = m_4 + m'_B = 5 \text{ kg}$$

Now recalling Eq. (5.2),

$$\lambda = \frac{r_2}{r_3} = 0.25$$

and,

$$\omega_2 = 1000 \text{ rpm} = 1000 \text{ rev/ min} \times \frac{2\pi \frac{\text{rad}}{\text{rev}}}{60 \frac{\text{s}}{\text{min}}} = 104.72 \text{ rad/s (CCW)}$$

the inertia (shaking) force Eq. (14.81)

$$\vec{F}_{inertia} = r_2\omega_2^2\left[\left(m_{A,eq} + m_{B,eq} - m_{cb,eq}\right)\cos\theta_2 + \lambda m_{B,eq}\cos 2\theta_2\right]\hat{i} + r_2\omega_2^2\left(m_{A,eq} - m_{cb,eq}\right)\sin\theta_2\hat{j} \tag{14.81}$$

results in:

$$\vec{F}_{inertia} = 0.15\text{m} \times (104.72 \text{ rad/s})^2[(4 \text{ kg} + 5 \text{ kg})\cos\theta_2 + 0.25 \times 5\text{kg}\cos 2\theta_2]\hat{i} + 0.15\text{m} \times (104.72 \text{ rad/s})^2(4 \text{ kg})\sin\theta_2\hat{j}$$

or,

$$\vec{F}_{inertia} = 14804.5\cos\theta_2\hat{i} + 2056.2\cos 2\theta_2\hat{i} + 6579.8\sin\theta_2\hat{j} \text{ N} \tag{14.82}$$

Here is a MATLAB code for plotting the magnitude of $\vec{F}_{inertia}$, given by Eq. (14.82) for one rotation cycle of the crank, (for $m_{cb,\ eq} = 0$).

```
clear all
r2=0.15;
Omeg2=104.72;
Maeq=4;
```

```
Mbeq=5;
Lambd=0.25;
Mcb=0;
  for i=1:1:361
  x(i)=(i-1)*pi/180;
  F_inertia_x(i)=r2*Omeg2^2*((Maeq+Mbeq-Mcb)*cos(x(i))
+Lambd*Mbeq*cos(2*x(i)));
  F_inertia_y(i)=r2*Omeg2^2*(Maeq-Mcb)*sin(x(i));
  F_inertia(i)=sqrt((F_inertia_x(i))^2+(F_inertia_y(i))^2);
  end
xdeg=x*180/pi;
figure
plot(xdeg,F_inertia,'LineWidth',3)
xlabel('\theta_2(deg)')
ylabel('F_i_n_e_r_t_i_a(N)')
grid on
figure
plot(F_inertia_x,F_inertia_y,'LineWidth',3)
xlabel('F_i_n_e_r_t_i_a_,_x(N)')
ylabel('F_i_n_e_r_t_i_a_,_y(N)')
grid on
```

With no counterbalance, the output of this code is shown in Fig. 14.32. The closed loop in Fig. 14.33 illustrates the way the x and y components of $\vec{F}_{inertia}$ change in one cycle.

Next, we **add an equivalent counterbalance** of:

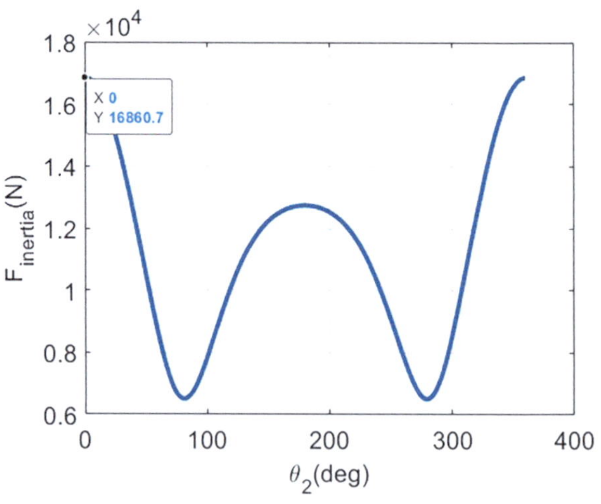

Fig. 14.32 Total inertia force using Eq. (14.82), with no counterbalance

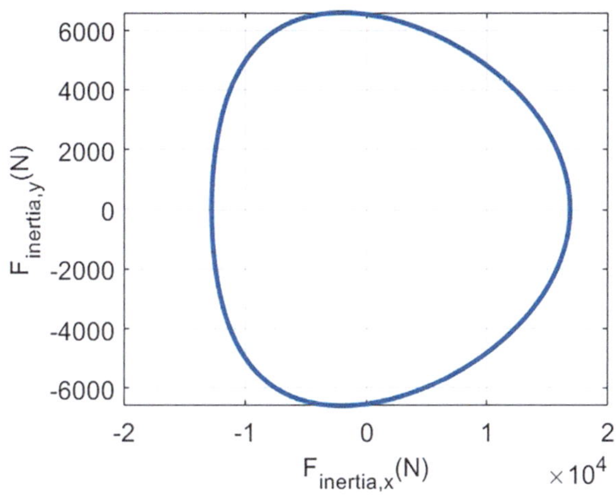

Fig. 14.33 The x and y components of the total inertia force shown in Fig. 14.32

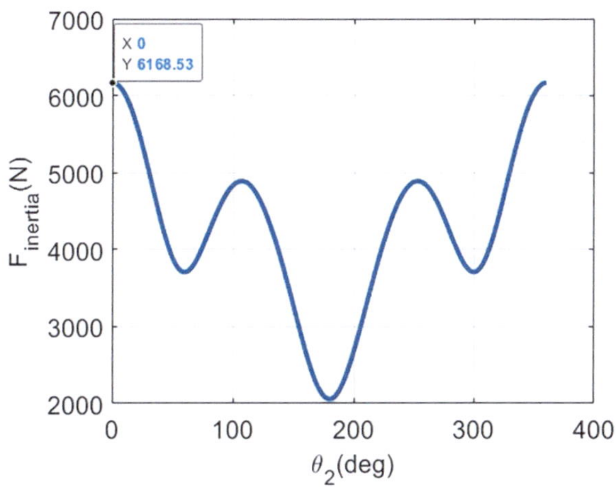

Fig. 14.34 Total inertia force if a counterbalance of $m_{cb,\ eq} = 6.5$ kg is added to the crank in Example 14.5

$$m_{cb,eq} = m_{A,eq} + \frac{1}{2} m_{B,eq} = 6.5 \text{ kg} \tag{14.83}$$

at radius r_2 and plot the magnitude of the revised $\vec{F}_{inertia}$ obtained from Eq. (14.81) in one cycle. The result is shown in Fig. 14.34. The corresponding x and y components of $\vec{F}_{inertia}$ with the mentioned counterbalance mass are shown in Fig. 14.35.

Figure 14.36 provides a comparison between Figs. 14.32 and 14.34 in which it is clear that the counterbalance has been able to reduce the maximum magnitude of $\vec{F}_{inertia}$ from 16.86 kN to 6.17 kN (a 63.4% reduction). This matter shows the importance of using counterbalance in slider-crank applications as we see in

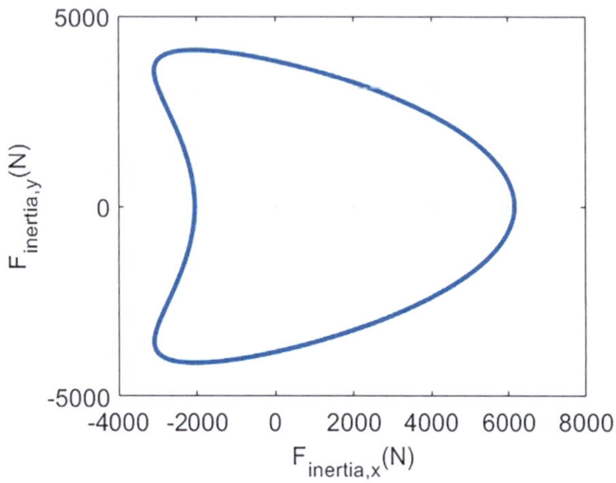

Fig. 14.35 The x and y components of the total inertia force shown in Fig. 14.34

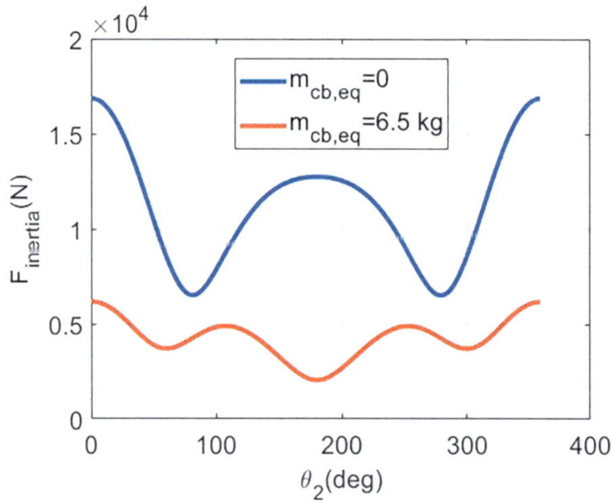

Fig. 14.36 Comparison of the total inertia forces without and with counterbalance in Figs. 14.32 and 14.34, respectively

Fig. 14.31. The corresponding x and y components of the total inertia forces are compared in Fig. 14.37.

Finally, we run the provided MATLAB code for the following range of counter-balance values:

$$m_{cb,eq} = m_{A,eq} \text{ to } m_{cb,eq} = m_{A,eq} + m_{B,eq} \tag{14.84}$$

and at each value of $m_{cb,\ eq}$ we obtain the maximum magnitude of $\vec{F}_{inertia}$ and plot that value versus the chosen $m_{cb,\ eq}$. In this way, Fig. 14.38 is obtained. Using this figure, one can find **the optimal $m_{cb,eq}$ value that corresponds to the one for which the maximum magnitude of $\vec{F}_{inertia}$ is minimum** (this is a minimax problem) as seen in Fig. 14.38.

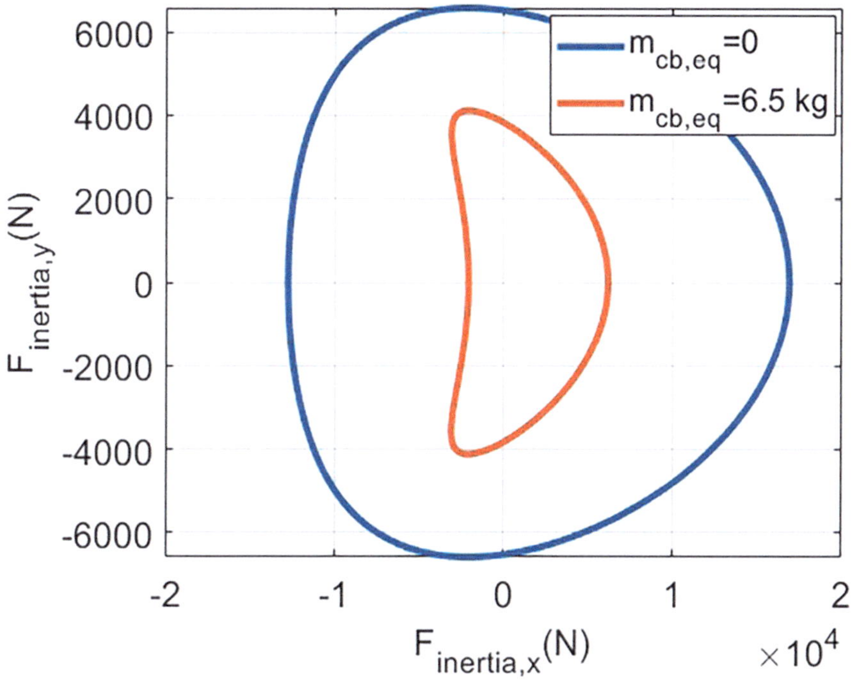

Fig. 14.37 Plot of the *x* and *y* components of the total inertia forces shown in Fig. 14.36

Fig. 14.38 Variation of the maximum total inertia force as a function of the chosen counterbalance and locating *the minimum of the maximum intertia force*

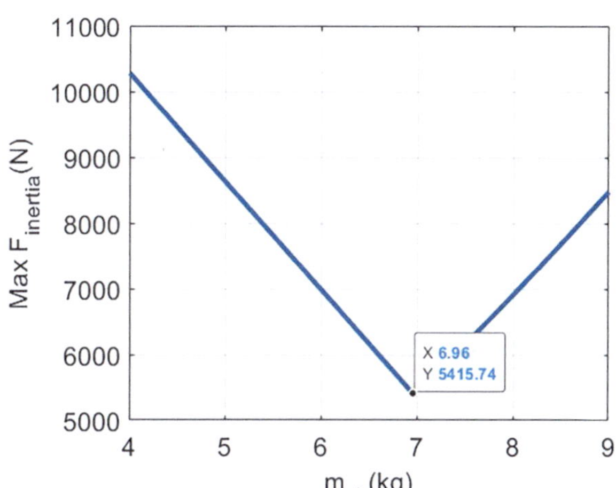

The minimax solution in this problem is $m_{cb,\ eq} = 6.96$ kg that is a bit away from the usually recommended value given below: (see Waldron and Kinzel [1], page 648)

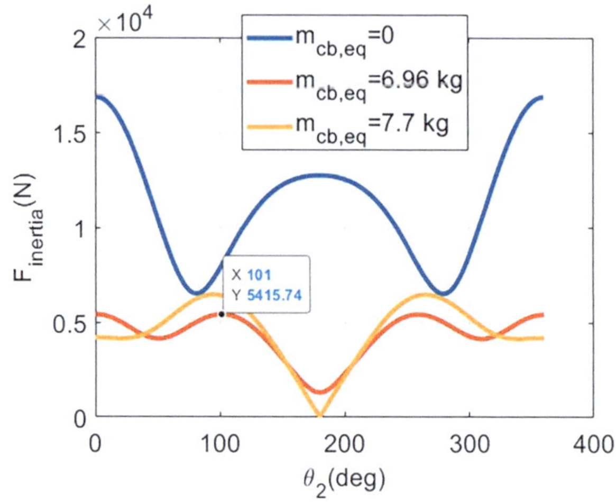

Fig. 14.39 Comparison of the total inertia forces obtained for different values of counterbalance masses

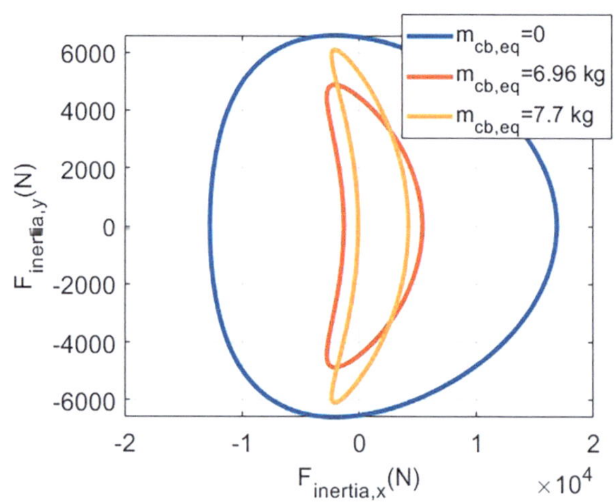

Fig. 14.40 Plot of the *x* and *y* components of the total inertia forces shown in Fig. 14.39

$$m_{cb,eq} = m_{A,eq} + \left(\frac{2}{3}\right) m_{B,eq} = 4 + \left(\frac{2}{3}\right) \times 5 = 7.333 \text{ kg} \tag{14.85}$$

Figures 14.39 and 14.40 illustrate the magnitude of $\vec{F}_{inertia}$ and variations of its x and y components, respectively, for the following values of the counterbalance mass:

$$m_{cb,eq} = 0; 6.96 \text{ kg}; 7.70 \text{ kg} \tag{14.86}$$

Once again it is clear that $m_{cb,\ eq} = 6.96$ kg provides the minimum of the maximum $\vec{F}_{inertia}$ values among the counterbalance masses listed in Eq. (14.86). The reduction of the maximum magnitude of $\vec{F}_{inertia}$ from 16.86 kN to 5.415 kN (a 67.88% reduction) shows the importance of using the optimal counterbalance in slider-crank applications.

Reference

1. Waldron, K. J., & Kinzel, G. L. (2004). *Kinematics, dynamics, and design of machinery* (2nd ed.). Wiley.

Chapter 15
Three-Dimensional Dynamics

15.1 Introduction

Almost all of the mechanisms and machines discussed so far were examples of planar problems. There are, however, many important machines and mechanisms with motions that are not planar. Gyroscopes and all machines that use gyroscopic effect for navigation can be listed as examples. Chapter 15 covers the basic principles of three-dimensional dynamics, including Euler's equations of motion and equations for parallel-plane motion. A discussion about gyroscope and the gyroscopic effect concludes this chapter.

15.2 Mass Moment of Inertia and Angular Momentum

The Newton's second law for forces applied to a rigid body with mass center G, i.e.,

$$\sum \vec{F} = m\vec{a}_G \tag{15.1}$$

can be easily extended to the third dimension and we have:

$$\sum F_x = ma_{G_x}, \sum F_y = ma_{G_y}, \sum F_z = ma_{G_z} \tag{15.2}$$

However, finding the analog equation for moments in three-dimensional dynamics is a bit more complicated. First of all, recall that in planar dynamics (see Chap. 12) for summation of external moments applied on an object about its mass center G we have:

M. Ghorashi, *Introduction to Machine Dynamics*, Mechanical Engineering Series,
https://doi.org/10.1007/978-3-031-94736-0_15

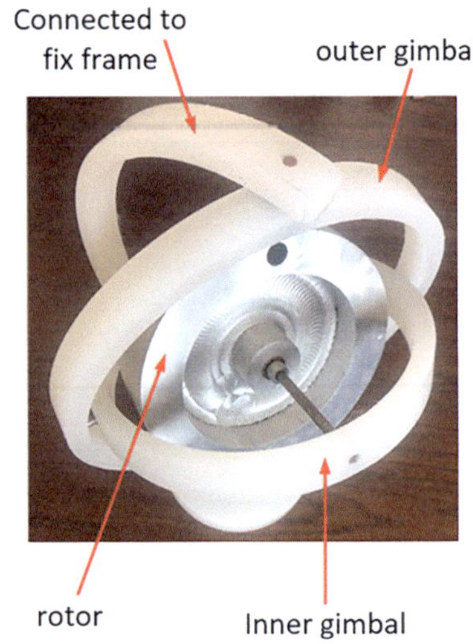

Fig. 15.1 Model of a gyroscope. (USM term project)

$$\sum \vec{M}_G = I_G \vec{\alpha} \tag{15.3}$$

where I_G is the mass moment of inertia of the object about an axis that passes through point G and is perpendicular to the plane of motion. Also, α is the angular acceleration of the object. A similar equation for two-dimensional problems can be written for moments about a fixed point O about which the object rotates,

$$\sum \vec{M}_O = I_O \vec{\alpha} \tag{15.4}$$

where I_O is the mass moment of inertia of the object about an axis that passes through point O and is perpendicular to the plane of motion.

Equations (15.3) and (15.4) are limited to planar problems and they cannot be used in three-dimensional cases like the gyroscope model shown in Fig. 15.1. However, we can still state that the resultants of applied moments about G or O are equal to the time derivative of angular momentum about the same points, respectively:

$$\sum \vec{M}_G = \dot{\vec{H}}_G \rightarrow \sum \vec{M}_O = \dot{\vec{H}}_O \tag{15.5}$$

We are going to see how these equations can result in other moment equations that are convenient to use in spatial cases. In what follows, some basic information on this matter is provided. For more detailed treatment of the subject, the reader is referred to references like, Meriam and Kraige [1].

Let us first start with the fact that in spatial problems, the whole inertia matrix of the object should be considered (and not just one term) for mass moment of inertia (as we do in planar cases). Using an x, y, z, coordinate system for calculating mass moments of inertia, they can be defined as follows:

$$I_{xx} = \int (y^2 + z^2)dm, I_{yy} = \int (x^2 + z^2)dm, I_{zz} = \int (x^2 + y^2)dm \tag{15.6}$$

If in a particular direction, say z, an object is thin, then Eq. (15.6) can be approximated as:

$$I_{xx} = \int y^2 dm, I_{yy} = \int x^2 dm \tag{15.7}$$

Therefore, in this special case, for I_{zz} we have:

$$I_{zz} = \int (x^2 + y^2)dm = I_{xx} + I_{yy} \tag{15.8}$$

Finally, in spatial problems, products of inertia can also enter into our calculations. Using the same x, y, z, coordinate system, they are defined as follows:

$$I_{xy} = I_{yx} = \int xydm, I_{xz} = I_{zx} = \int xzdm, I_{yz} = I_{zy} = \int yzdm \tag{15.9}$$

Now the complete mass moment of inertia matrix of the object and in the chosen x, y, z, coordinate system can be written as follows:

$$I = \begin{bmatrix} I_{xx} & -I_{xy} & -I_{xz} \\ -I_{yx} & I_{yy} & -I_{yz} \\ -I_{zx} & -I_{zy} & I_{zz} \end{bmatrix} \tag{15.10}$$

For any object, one may rotate the chosen coordinate system (while keeping its origin fixed) in such a way that for a specific choice of orientation of axes (i.e., principal directions of inertia), the off-diagonal elements (i.e., the products of inertia) vanish and we get, Meriam and Kraige [1]:

$$I = \begin{bmatrix} I_{xx} & 0 & 0 \\ 0 & I_{yy} & 0 \\ 0 & 0 & I_{zz} \end{bmatrix} \tag{15.11}$$

The elements of the mass moment of inertia matrix in this case are called the principal mass moments of inertia. One of these elements is the largest and another one is the smallest mass moment of inertia among all moment of inertia values about axes that can pass through that same point of origin.

Consider an object with mass moment of inertia matrix, I, for calculating which an x, y, z coordinates system with origin at its mass center G or fixed point O has been used. The coordinate system is assumed to be attached to the object and the object is performing angular velocities ω_x, ω_y, ω_z. Therefore, the angular momentum of the object about the origin of the coordinate system can be written as:

$$\vec{H} = \begin{Bmatrix} H_x \\ H_y \\ H_z \end{Bmatrix} = \begin{bmatrix} I_{xx} & -I_{xy} & -I_{xz} \\ -I_{yx} & I_{yy} & -I_{yz} \\ -I_{zx} & -I_{zy} & I_{zz} \end{bmatrix} \begin{Bmatrix} \omega_x \\ \omega_y \\ \omega_z \end{Bmatrix} \tag{15.12}$$

So the components of the angular momentum vector are as follows:

$$\begin{cases} H_x = I_{xx}\omega_x - I_{xy}\omega_y - I_{xz}\omega_z \\ H_y = -I_{yx}\omega_x + I_{yy}\omega_y - I_{yz}\omega_z \\ H_z = -I_{zx}\omega_x - I_{zy}\omega_y + I_{zz}\omega_z \end{cases} \tag{15.13}$$

15.3 Euler's Equations of Motion

From the impulse and momentum equation of a rigid body in spatial motion (about mass center G or fixed point O), we know that the resultant moment applied on a rigid body about either of these two points equals to the time derivative of the corresponding angular momentum:

$$\sum \vec{M} = \frac{d\vec{H}}{dt} \tag{15.14}$$

in which,

$$\vec{H} = H_x\hat{i} + H_y\hat{j} + H_z\hat{k} \tag{15.15}$$

Since the coordinate system x, y, z is attached to the object, its angular velocity vector is the same as the angular velocity vector of the rigid body itself. Using Eq. (15.14), taking a time derivative of $\vec{H}$ given in Eq. (15.15), and noticing that time derivatives of unit vectors are calculated as:

$$\frac{d\hat{i}}{dt} = \vec{\omega} \times \hat{i},\ \frac{d\hat{j}}{dt} = \vec{\omega} \times \hat{j},\ \frac{d\hat{k}}{dt} = \vec{\omega} \times \hat{k} \tag{15.16}$$

one obtains:

$$\begin{cases} \sum M_x = \dot{H}_x - H_y\omega_z + H_z\omega_y \\ \sum M_y = \dot{H}_y - H_z\omega_x + H_x\omega_z \\ \sum M_z = \dot{H}_z - H_x\omega_y + H_y\omega_x \end{cases} \tag{15.17}$$

Now, let us choose a coordinate system that while still attached to the rigid body, it is now aligned with its principal directions of inertia. In this case, the angular momentum terms in Eq. (15.13) reduce to:

$$H_x = I_{xx}\omega_x; H_y = I_{yy}\omega_y; H_z = I_{zz}\omega_z \tag{15.18}$$

Substituting Eq. (15.18) into the moment equations (15.17) results in:

$$\begin{cases} \sum M_x = I_{xx}\dot{\omega}_x - (I_{yy} - I_{zz})\omega_y\omega_z \\ \sum M_y = I_{yy}\dot{\omega}_y - (I_{zz} - I_{xx})\omega_z\omega_x \\ \sum M_z = I_{zz}\dot{\omega}_z - (I_{xx} - I_{yy})\omega_x\omega_y \end{cases} \tag{15.19}$$

These are Euler's equations of motion and they are the three-dimensional analogs of the two-dimensional Eqs. (15.3) and (15.4):

$$\sum \vec{M}_G = I_G\vec{\alpha} \tag{15.3}$$

$$\sum \vec{M}_O = I_O\vec{\alpha} \tag{15.4}$$

For using Euler's equations, one should notice that the moments are written about either the mass center G, or a fixed point O, the coordinates system should have its origin located at mass center G or fixed point O, and the coordinate system x, y, z should be attached to the rigid body and its axes be aligned with the principal directions of inertia of the rigid body.

For two-dimensional cases with motion in the x-y plane, we have

$$\omega_x = 0; \omega_y = 0 \tag{15.20}$$

Therefore, we get:

$$\sum M_x = 0; \sum M_y = 0; \sum M_z = I_{zz}\dot{\omega}_z \tag{15.21}$$

which are the expected two-dimensional moment equations. In what follows, we reconsider the conical pendulum problem that was discussed before for a particle (i.e., Example 12.8). This time we analyze it as a rigid body pendulum in three-dimensional motion.

15.3.1 Application Problem

Consider Fig. 15.2 in which a conical pendulum consisting of a uniform bar of length l and mass m is hinged at point O. This bar is connected to a vertical shaft (not shown) that rotates at angular velocity ω. Calculate the steady-state deviation angle θ of the bar with respect to the vertical direction using the following methods; and finally assess the stability of the steady-state solutions.

A. D'Alembert's principle and inertia forces.
B. Euler's equations for three-dimensional motion.
C. Assess the stability of the steady-state solutions.

A. **D'Alembert's principle and inertia forces.**

By applying inertia force (in this case, centrifugal force) on the rotating bar and according to D'Alembert's principle, the bar will be in dynamic equilibrium. The centrifugal force, unlike the gravitational force, is not uniformly distributed along the bar. It is proportional to the radius of rotation at each point and this radius changes linearly from zero at pivot point O to R at the tip of the bar. Considering this variation, one cannot assume that the total centrifugal force on the bar acts at its center of mass. Therefore, the centrifugal force action should be modeled at the differential element level. Figure 15.3 shows an element dm of the bar that is located at position x from the pivot. This element is at distance r from the axis of rotation; when the bar has reached dynamic equilibrium and steady-state deviation angle θ, due to the application of its weight, the force applied at the hinge, and the centrifugal force.

The dynamic equilibrium condition for moments about the hinge point O can be written as follows:

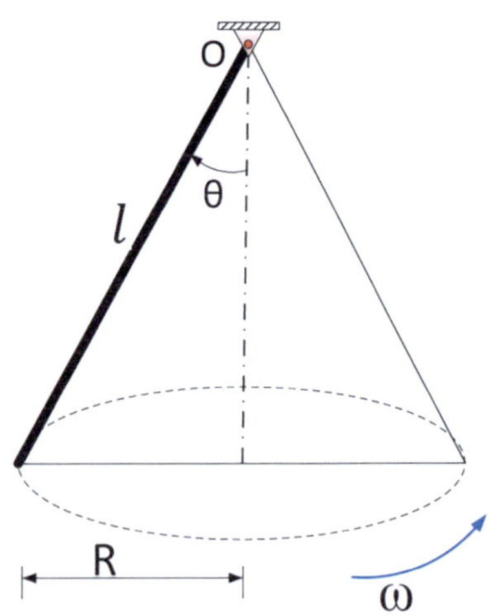

Fig. 15.2 Conical pendulum consisting of a hinged bar connected to a vertical shaft (not shown) that rotates at angular velocity ω

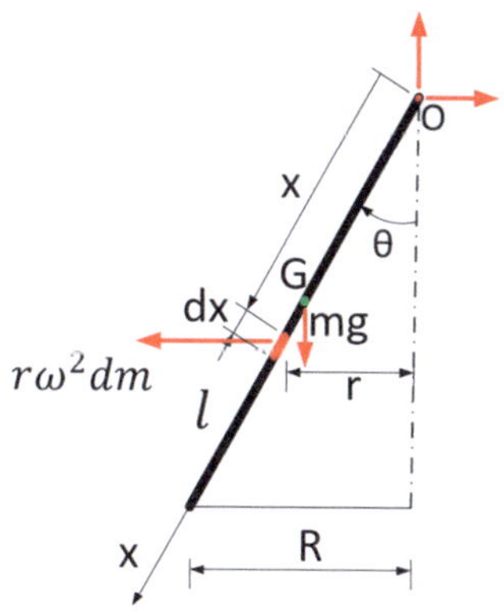

Fig. 15.3 Analyzing a differential element of the rotating rod shown in Fig. 15.2

$$\sum \vec{M}_O = 0 \tag{15.22}$$

The moment of the differential centrifugal force acting on an element dx about point O is:

$$\left(r\omega^2 dm\right) x \cos\theta$$

Since $r = x \sin\theta$ and $dm = \rho dx$ we get:

$$\omega^2 x^2 \sin\theta \cos\theta (\rho dx)$$

or,

$$\frac{1}{2}\omega^2 x^2 \sin 2\theta \rho dx$$

in which, ρ is the mass per unit length of the bar:

$$\rho = m/l$$

Therefore, the equilibrium of moments of forces gives:

$$\sum M_O = \int_0^l \frac{1}{2}\omega^2 x^2 \sin 2\theta \rho dx - mg\left(\frac{l}{2}\right) \sin\theta = 0 \tag{15.23}$$

or,

$$\left(\frac{2}{3}\omega^2 l \cos\theta - g\right) \sin\theta = 0 \tag{15.24}$$

One root of Eq. (15.24) is the trivial solution:

$$\sin\theta = 0 \tag{15.25}$$

which gives the following trivial solutions,

$$\theta = 0, \pi \tag{15.26}$$

The other root is:

$$\frac{2}{3}\omega^2 l\cos\theta - g = 0$$

or,

$$\cos\theta = \frac{3g}{2l\omega^2} \tag{15.27}$$

This is the non-trivial solution for the tilt angle of the bar rotating at angular velocity, ω. This solution is valid only if:

$$\frac{3g}{2l\omega^2} < 1 \tag{15.28}$$

or when,

$$\omega^2 > \frac{3g}{2l} \tag{15.29}$$

In other words, at low angular velocities, the trivial solutions (15.26) is the answer. As the rotational speed increases beyond,

$$\omega_{min} = \sqrt{\frac{3g}{2l}} \tag{15.30}$$

The trivial solution becomes unstable and the non-trivial solution takes over. This is the minimum value of ω for which θ becomes non-zero. By increasing ω from ω_{min} given in Eq. (15.30), θ increases from 0 and then for very high rotational speeds approaches to 90°.

B. **Use Euler's equations to calculate the steady-state deviation angle θ.**

Now we try to solve the same problem using Euler's equations. In order to do that, the coordinate system should be placed at *G* or *O* and be aligned with the principal directions of inertia. Placing the origin at *O* is preferred as it excludes the reaction forces that operate at *O* from Euler equations. (Their moments about O are zero.) We also reorient the coordinate system, as seen in Fig. 15.4, so that it is aligned with the principal directions of inertia.

Recalling Euler's Eqs. (15.19),

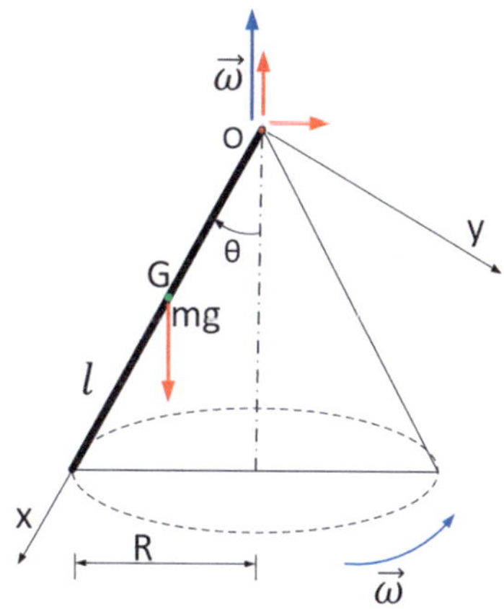

Fig. 15.4 Free body diagram of the rod rotating about a vertical axis

$$\begin{cases} \sum M_x = I_{xx}\dot{\omega}_x - (I_{yy} - I_{zz})\omega_y\omega_z \\ \sum M_y = I_{yy}\dot{\omega}_y - (I_{zz} - I_{xx})\omega_z\omega_x \\ \sum M_z = I_{zz}\dot{\omega}_z - (I_{xx} - I_{yy})\omega_x\omega_y \end{cases} \tag{15.19}$$

The only force that generates moment about O is the weight of the bar and its moment is in the z direction:

$$\sum M_z = mg(l/2)\sin\theta \tag{15.31}$$

As to the angular velocity components, considering the angular velocity vector one obtains,

$$\omega_x = -\omega\cos\theta;\, \omega_y = -\omega\sin\theta \tag{15.32}$$

Substituting these values in the only non-trivial Euler's equation results in:

$$mg\left(\frac{l}{2}\right)\sin\theta = -(I_{xx} - I_{yy})\omega_x\omega_y \tag{15.33}$$

in this case,

$$I_{xx} \approx 0, I_{yy} = I_{zz} = \frac{1}{3}ml^2 \tag{15.34}$$

Therefore we get:

$$mg\left(\frac{l}{2}\right)\sin\theta = -\left(0 - \frac{1}{3}ml^2\right)(-\omega\cos\theta)(-\omega\sin\theta)$$

So, the steady-state solution θ^* satisfies,

$$\left(\frac{2}{3}\omega^2 l \cos\theta^* - g\right)\sin\theta^* = 0 \tag{15.35}$$

This is the same as Eq. (15.24) and therefore the same conclusions follow.

C. **Assess the stability of the steady-state solutions.**

The stability of the results could be confirmed by the perturbations method in which the dynamic solution is written as:

$$\theta = \theta^* + \widehat{\theta} \tag{15.36}$$

where θ^* is the calculated steady-state solution and $\widehat{\theta}$ is a small perturbation from the steady-state. We would like to see if such perturbations decay or amplify over time. Recalling the dynamic (not the steady-state) version of Euler's equation in the z direction:

$$\sum M_z = I_{zz}\dot{\omega}_z - \left(I_{xx} - I_{yy}\right)\omega_x\omega_y \tag{15.19}$$

where,

$$\omega_x = -\omega\cos\theta;\ \omega_y = -\omega\sin\theta;\ \omega_z = -\dot{\theta} \tag{15.37}$$

and substituting the known values of the variables in Eq. (15.19) we have,

$$mg\left(\frac{l}{2}\right)\sin\left(\theta^* + \widehat{\theta}\right) = I_{zz}\left(-\dot{\widehat{\theta}}\right) - \left(0 - \frac{1}{3}ml^2\right)\left[-\omega\cos\left(\theta^* + \widehat{\theta}\right)\right]\left[-\omega\sin\left(\theta^* + \widehat{\theta}\right)\right]$$

or after simplification:

$$\dot{\hat{\theta}} - \frac{1}{2}\omega^2\sin\left(2(\theta^* + \hat{\theta})\right) + \frac{3g}{2l}\sin\left(\theta^* + \hat{\theta}\right) = 0$$

Considering the trivial steady-state solution: $\theta^* = 0$, and for small values of perturbation, $\widehat{\theta}$, the above equation reduces to the following first-order differential equation:

$$\dot{\widehat{\theta}} + \left(\frac{3g}{2l} - \omega^2\right)\widehat{\theta} = 0 \tag{15.38}$$

For:

$$\omega < \sqrt{\frac{3g}{2l}} \tag{15.39}$$

the perturbations in Eq. (15.38) decay with time, and therefore, the dynamic solution given in Eq. (15.36) converges to the steady-state solution. So, in this region the trivial steady-state solution is stable. However, for:

$$\omega > \sqrt{\frac{3g}{2l}} \tag{15.40}$$

The perturbations in Eq. (15.38) amplify with time, so in this range the $\theta^* = 0$ solution is unstable. Repeating the same procedure for the non-trivial steady-state solution, i.e.,

$$\cos\theta^* = \frac{3g}{2l\omega^2} \tag{15.27}$$

after some mathematical manipulations we get:

$$\ddot{\hat{\theta}} - \omega^2\left[\left(\frac{3g}{2l\omega^2}\right)^2 - 1\right]\hat{\theta} = 0 \tag{15.41}$$

So, for stability of the non-trivial θ^* we should have:

$$1 - \left(\frac{3g}{2l\omega^2}\right)^2 > 0 \tag{15.42}$$

or,

$$\omega > \sqrt{\frac{3g}{2l}} \tag{15.43}$$

Finally, in a similar way, it can be shown that the trivial $\theta^* = \pi$ solution is always unstable (as expected). Table 15.1 provides a summary of the obtained results.

Example 15.1 Plot the response of a 1-m-long conical pendulum as its angular velocity increases from zero to 15 rad/s.

In this case the ω_{min} given by Eq. (15.30) would be as follows:

$$\omega_{min} = \sqrt{\frac{3g}{2l}} = \sqrt{(3\times 9.806)/(2\times 1)} = 3.835 \text{ rad/s}$$

So the stable solution at $\omega \le \omega_{min}$ is $\theta = 0$. For $\omega \ge 3.835$ rad/s, the solution is,

Table 15.1 Steady-state solutions of the 15.3.1 Application Problem and their stability domains

Steady-state solution	$\omega < \sqrt{\frac{3g}{2l}}$	$\omega > \sqrt{\frac{3g}{2l}}$
$\theta^* = 0$	Stable	Unstable
$\theta^* = 180°$	Unstable	Unstable
$\theta^* = \cos^{-1}\left(\frac{3g}{2l\omega^2}\right)$	N/A	Stable

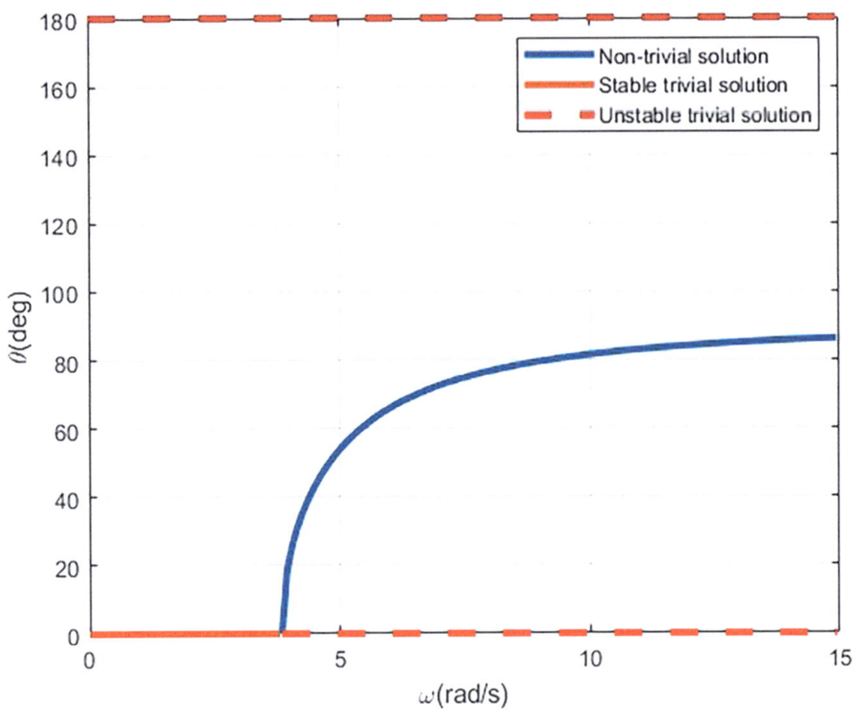

Fig. 15.5 Steady-state solutions of Example 15.1 and their stability domains

$$\cos\theta = \frac{3g}{2l\omega^2} \tag{15.27}$$

For example, at $\omega = 5$ rad/s, the result is $\theta = 53.96°$ and at $\omega = 15$ rad/s one obtains, $\theta = 86.25°$. The response, plotted by MATLAB, is shown in Fig. 15.5.

15.4 Parallel-Plane Motion

The parallel-plane motion happens frequently in rotating shafts where all particles in a part move on parallel planes. We attach the rotating coordinate system to the part to be studied, and orient it in such a way that the z coordinate is on the shaft axis, and coordinates x and y form a plane that all particles in the part rotate parallel to it (see, e.g., Fig. 15.6). Therefore,

$$\omega_x = 0, \omega_y = 0, \vec{\omega} = \omega_z \widehat{k} \tag{15.44}$$

Substituting Eq. (15.44) into Eq. (15.13):

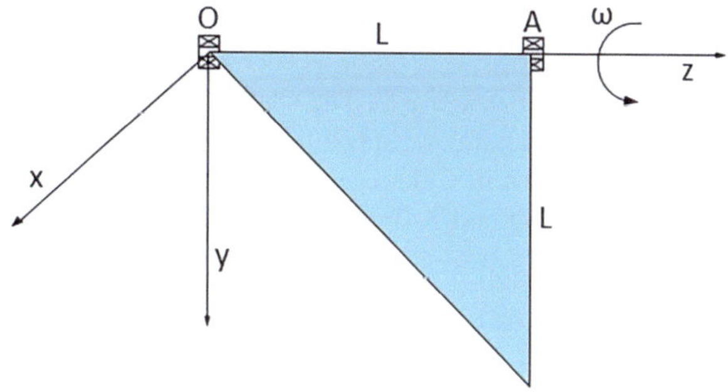

Fig. 15.6 Thin plate attached to a shaft and rotating about the z axis at constant angular velocity ω

$$\begin{cases} H_x = I_{xx}\omega_x - I_{xy}\omega_y - I_{xz}\omega_z \\ H_y = -I_{yx}\omega_x + I_{yy}\omega_y - I_{yz}\omega_z \\ H_z = -I_{zx}\omega_x - I_{zy}\omega_y + I_{zz}\omega_z \end{cases} \tag{15.13}$$

one obtains:

$$H_x = -I_{xz}\omega_z, H_y = -I_{yz}\omega_z, H_z = +I_{zz}\omega_z \tag{15.45}$$

So,

$$\vec{H} = -I_{xz}\omega_z\hat{i} - I_{yz}\omega_z\hat{j} + I_{zz}\omega_z\hat{k} \tag{15.46}$$

Now we take a time derivative of Eq. (15.46), use Eq. (15.14):

$$\sum \vec{M} = \frac{d\vec{H}}{dt} \tag{15.14}$$

and:

$$\frac{d\hat{i}}{dt} = \vec{\omega} \times \hat{i}, \frac{d\hat{j}}{dt} = \vec{\omega} \times \hat{j}, \frac{d\hat{k}}{dt} = \vec{\omega} \times \hat{k} \tag{15.16}$$

to get:

$$\sum \vec{M} = \left(-I_{xz}\dot{\omega}_z + I_{yz}\omega_z^2\right)\hat{i} + \left(-I_{yz}\dot{\omega}_z - I_{xz}\omega_z^2\right)\hat{j} + I_{zz}\dot{\omega}_z\hat{k} \tag{15.47}$$

Equation (15.47) is valid about G or about any origin of the coordinate system on the rotating shaft. The moment equation should be solved together with the following force equations:

$$\sum F_x = ma_{G_x}, \sum F_y = ma_{G_y}, \sum F_z = 0 \tag{15.48}$$

Example 15.2 A part that can be modeled as a thin right isosceles triangular plate of mass m is attached to a shaft, as seen in Fig. 15.6, and the assembly is rotating at constant angular velocity ω. The shaft is supported by two bearings and the coordinates are chosen with origin O on the left bearing and plane y-z on the triangular plate and rotating with it. Calculate the dynamic forces imposed by the bearings on the shaft at points O and A. Ignore the weight of the plate compared to the generated dynamic forces.

Recall the parallel-plane motion Eq. (15.47):

$$\sum \vec{M} = \left(-I_{xz}\dot{\omega}_z + I_{yz}\omega_z^2\right)\hat{i} + \left(-I_{yz}\dot{\omega}_z - I_{xz}\omega_z^2\right)\hat{j} + I_{zz}\dot{\omega}_z\hat{k} \tag{15.47}$$

Since the angular velocity is constant we have:

$$\dot{\omega}_z = 0$$

Therefore Eq. (15.47) reduces to:

$$\sum M_x = I_{yz}\omega_z^2,\ \sum M_y = -I_{xz}\omega_z^2,\ \sum M_z = 0 \tag{15.49}$$

Since the plate is thin, we can assume that:

$$I_{xz} = 0 \tag{15.50}$$

So, considering Eq. (15.49) we get:

$$\sum M_y = 0 \tag{15.51}$$

Therefore, the only non-trivial moment equation in (15.49) is:

$$\sum M_x = I_{yz}\omega_z^2 \tag{15.52}$$

in which I_{yz} is not ignorable and we have:

$$I_{yz} = \int yz dm = \iint yz(\rho dy dz) = \rho \int_0^L z\left\{\int_0^z y dy\right\} dz \tag{15.53}$$

where ρ is the mass density of the plate per unit area:

$$\rho = \frac{2m}{L^2}$$

Therefore, Eq. (15.53) reduces to the following:

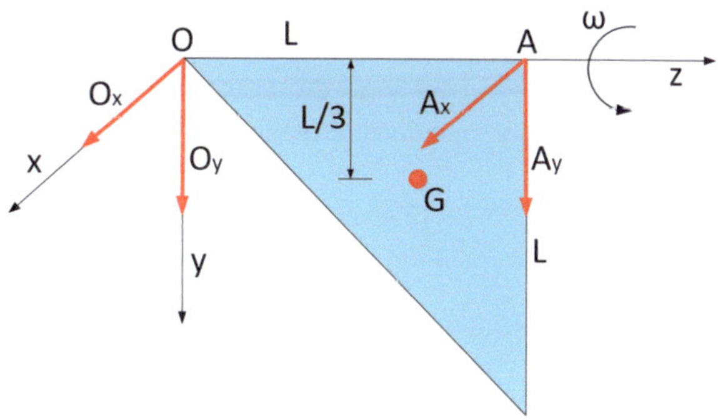

Fig. 15.7 Free body diagram of the plate shown in Fig. 15.6, illustrating the dynamic forces acting on it

$$I_{yz} = \frac{2m}{L^2} \times \frac{L^4}{8} = \frac{1}{4} mL^2 \tag{15.54}$$

Substituting Eq. (15.54) into the non-trivial Eq. (15.52) results in,

$$\sum M_{Ox} = \frac{1}{4} mL^2 \omega_z^2 \tag{15.55}$$

Considering the FBD of the plate shown in Fig. 15.7, the moment of forces about point O and in the x direction is:

$$\sum M_{Ox} = -A_y L \tag{15.56}$$

Equating Eqs. (15.55) and (15.56) results in,

$$A_y = -\frac{1}{4} mL\omega_z^2 \tag{15.57}$$

To get O_y, we use the Newton's second law equation in the y direction:

$$\sum F_y = ma_{G_y} \tag{15.48}$$

That gives,

$$A_y + O_y = -m\left(\frac{L}{3}\right)\omega_z^2 \tag{15.58}$$

Equations (15.57) and (15.58) result in:

$$O_y = -\frac{1}{12} mL\omega_z^2 \tag{15.59}$$

Since the other equations of motion have reduced to just $0 = 0$, we conclude that:

$$A_x = 0, O_x = 0 \tag{15.60}$$

15.5 Gyroscopic Effect

15.5.1 *The Concept of Gyroscopic Effect*

Conservation of angular momentum allows spinning objects to keep their direction of motion. This phenomenon has been utilized in many applications. Bicycles, spinning tops, and gyroscopes (that are used in navigation of ships and airplanes) are all stabilized by conservation of angular momentum.

Consider Fig. 15.8 where a spinning object, rotating about the x axis at constant spin rate ω_s, is subjected to moment, M, in a direction perpendicular to the spin direction, say the y axis. In this case, instead of turning the rotor in the direction of the applied moment, i.e., in the y direction, we would see that the applied moment causes the rotor to turn about the third axis (i.e., z) in what is called the steady precession motion at rate ω_p. So, precession takes place about the third axis that is perpendicular to both the spin axis and the applied moment axis. This phenomenon is called the gyroscopic effect and it happens (in a somewhat different manner) in, e.g., motion of a spinning top, where spin and the moment due to weight generate the familiar precession motion of the top. This effect cannot be explained by two-dimensional formulations.

The converse of what was just described is true as well, i.e., if a spinning object is forced to perform the precession motion in a direction that is perpendicular to the spinning direction, a moment will be exerted on the object in a direction that is perpendicular to both the spin and the precession directions. This is called forced precession and it is experienced during turns in many applications like cars,

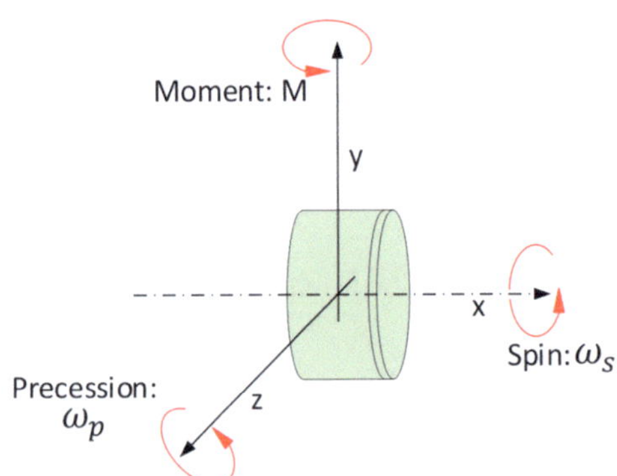

Fig. 15.8 A rotor spinning about the x axis, simultaneously subjected to a moment about the y axis, experiences rotation about the z axis

airplanes, and ships. In all of these cases, spin and precession motions are the inputs and moment is the output. The moment that is developed during forced precession can impose additional forces on the bearings; therefore, if those forces are significant, they should be taken into account in designing components and bearings. We see that in all of these cases spin is the fundamental input. **If spin accompanies an applied moment, precession is generated. If spin accompanies precession, moment is generated.**

A gyroscope rotor like the model shown in Fig. 15.1 is able to maintain its axis of rotation, since it is free from external moments. An instrument with this property is ideal for pilots and ship captains as they always need to know the orientation of their ship or airplane with respect to the true vertical direction. However, they cannot use a level for finding the horizontal and vertical directions during motion. The trick for the operation of a gyroscope lies on its use of a Cardan suspension system that has two gimbals that protect the gyroscope rotor from external moments. The axis of the rotor in connected to the inner gimbal and the inner gimbal is connected to the outer gimbal at two points. The outer gimbal is then connected to a frame that is connected to the ship or airplane. The existence of the inner gimbal protects the gyroscope rotor from external moments that could be applied by ship or plane movements. Since a gyroscope is free from external moments, it is the ideal tool for being used as a gyro compass, or artificial horizon. It is also a major part of an autopilot system. A gyroscope is used as a stabilizing device as well. For example, it is used in ships to counteract rolling in waves and prevent instability and capsizing of ships. Finally, a gyro compass has an added advantage that it is not affected by the existence of metallic parts and electronic equipment. The spin rate in a gyroscope rotor can be as large as 25,000 rpm.

15.5.2 Formulating the Gyroscopic Effect

The most common type of gyroscopic effect is the steady state one that happens when both the spin and the precession motions happen at constant rates. Considering Fig. 15.9, we assume the rotor is spinning at a large rate, ω_s. If the mass moment of inertia of the rotor about the spin axis is I, the angular momentum of the rotor could be written as:

$$\vec{H} = I\vec{\omega}_s \tag{15.61}$$

Now consider application of moment $\vec{M}$ that is perpendicular to $\vec{\omega}_s$ and $\vec{H}$. Since,

$$\vec{M} = \frac{d\vec{H}}{dt} \tag{15.14}$$

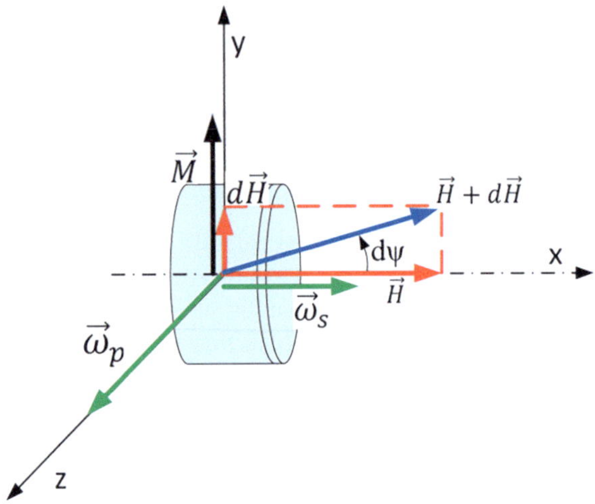

Fig. 15.9 Application of moment $\vec{M}$ perpendicular to the direction of the angular momentum $\vec{H}$ changes it and creates the precession motion $\vec{\omega}_p$

The change of angular momentum caused by application of the moment would be:

$$d\vec{H} = \vec{M}.dt \quad (15.62)$$

Therefore, the change of angular momentum is in the $\vec{M}$ direction and it is perpendicular to the original angular momentum $\vec{H}$. The new $\vec{H}$ is therefore at an angle dψ with respect to the old $\vec{H}$. However, since the incremental change $d\vec{H}$ is perpendicular to the original $\vec{H}$ the magnitude of angular momentum does not change and the $\vec{H}$ vector keeps rotating as shown. The result is the precession motion $\vec{\omega}_p$ in the direction that is perpendicular to both $\vec{\omega}_s$ and $\vec{M}$.

So, the spin angular velocity $\vec{\omega}_s$ generates angular momentum $\vec{H}$. The applied moment $\vec{M}$ generates $d\vec{H}$. And $d\vec{H}$ is the cause of the precession angular velocity, $\vec{\omega}_p$. For a quantitative relation, considering Fig. 15.9 we have:

$$dH = H.d\psi \quad (15.63)$$

Recalling,

$$d\vec{H} = \vec{M}dt, \vec{H} = I\vec{\omega}_s \quad (15.64)$$

We have:

$$M.dt = I\omega_s.d\psi \quad (15.65)$$

So, the precession angular velocity is obtained as,

$$\omega_p = \frac{d\psi}{dt} = \frac{M}{I\omega_s} \tag{15.66}$$

Considering the directions of $\vec{\omega}_s$, $\vec{M}$, and $\vec{\omega}_p$, and Eq. (15.66), a vector equation among these quantities can be stated as follows,

$$\vec{M} - \vec{\omega}_p \times I\vec{\omega}_s \tag{15.67}$$

It can be seen that the relation is linear, so, e.g., doubling moment M, results in a double value of the precession rate, ω_p.

Example 15.3 A turbojet aircraft is undergoing a yaw motion of $R = 5000$ m radius while traveling at a speed of 1000 km/h. The aircraft has a 150 kg rotor with 300 mm radius of gyration about its mass center and its spin velocity is 20,000 rpm. Calculate the gyroscopic moment that this forced precession motion will induce on the airframe.

$$I = k^2 m = (0.3 \text{ m})^2 \times 150 \text{ kg} = 13.5 \text{ kg.m}^2$$

$$\omega_s = 20{,}000 \text{ rpm} = 20{,}000 \text{ rev/ min} \times \frac{2\pi \text{ rad/rev}}{60 \text{ s/ min}} = 2094.4 \text{ rad/s}$$

And for ω_p we have:

$$v = \omega_p R \tag{15.68}$$

Therefore,

$$\omega_p = \frac{v}{R} = \frac{1000km}{h} \times \frac{1000m}{1\ km} \times \frac{1\ h}{3600\ s} \times \frac{1}{5000\ m} = 0.0556 \text{ rad/s}$$

Recalling,

$$\vec{M} = \vec{\omega}_p \times I\vec{\omega}_s \tag{15.67}$$

one obtains:

$$M = 0.0556 \text{ rad/s} \times 13.5 \text{ kg.m}^2 \times 2094.4 \text{ rad/s} = 1572.1 \text{ N.m}$$

Reference

1. Meriam, J. L., & Kraige, L. G. (2007). *Engineering mechanics, dynamics* (6th ed.). Wiley.

Appendix

For geometrical solution of mechanism problems, including the geometrical calculation of the mechanical advantage in a four-bar linkage, one frequently uses law of sines and law of cosines for triangles. We now briefly review these laws. In all of the materials that follow, we consider a triangle that has sides a, b, and c that are opposite to angles A, B, and C, as seen in Fig. A.1:

In Fig. A.1, R is the radius of the circumscribing circle of triangle ABC (with its center located at the intersection point of the perpendicular bisectors of this triangle that are concurrent and meet at point O_1). Also, r is the radius of the inscribed circle of triangle ABC (with its center located at the intersection point of the angle bisectors of this triangle that are concurrent and meet at point O_2).

Law of Cosines

Using the mentioned terms, one can prove that,

$$\begin{aligned} a^2 &= b^2 + c^2 - 2bc\cos A \\ b^2 &= a^2 + c^2 - 2ac\cos B \\ c^2 &= a^2 + b^2 - 2ab\cos C \end{aligned} \tag{A.1}$$

It is reminded that in every triangle,

$$\widehat{A} + \widehat{B} + \widehat{C} = 180^\circ \tag{A.2}$$

For the special case of $\widehat{A} = 90^\circ$, the first equation of Eqs. (A.1) reduces to the familiar Pythagorean theorem statement of,

M. Ghorashi, *Introduction to Machine Dynamics*, Mechanical Engineering Series,
https://doi.org/10.1007/978-3-031-94736-0

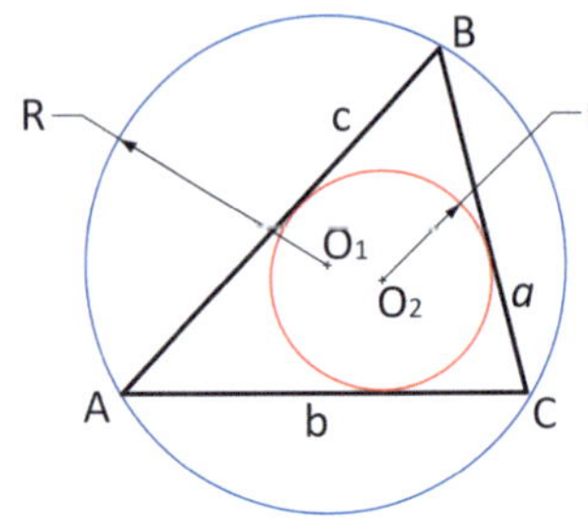

Fig. A.1 Triangle *ABC* with its circumscribing circle (radius *R*) and inscribed circle (radius *r*)

$$a^2 = b^2 + c^2 \tag{A.3}$$

Law of Sines

Using the mentioned terms, one can prove that,

$$\frac{a}{\sin A} = \frac{b}{\sin B} = \frac{c}{\sin C} = 2R \tag{A.4}$$

in which, *R*, the radius of the circumscribing circle of triangle *ABC*, can be calculated using *S*, the area of this triangle, or, *p* that is half its perimeter,

$$R = \frac{abc}{4S} \tag{A.5}$$

$$S = \sqrt{p(p-a)(p-b)(p-c)} \tag{A.6}$$

$$p = \frac{1}{2}(a+b+c) \tag{A.7}$$

The radius of the inscribed circle of triangle *ABC* is simply,

$$r = \frac{S}{p} \tag{A.8}$$

In a particular problem, choosing which one of these laws to use (law of sines or law of cosines) depends on the nature of the available data. The following examples clarify the matter. First, consider an example in which only one angle is initially known,

Example A.1 In triangle *ABC*, it is known that: $a = 200$ mm, $c = 100$ mm, and $\widehat{B} = 80°$. Calculate the following three unknowns: $\widehat{A}$, *b*, and $\widehat{C}$.

Using the law of cosines given in Eqs. (A.1) provides the solution for *b* in one step:

$$b^2 = a^2 + c^2 - 2ac\cos B$$

Since all of the parameters in the right side of the equation are known we get,

$$b = 207.49 \text{ mm}$$

One can now proceed to use law of sines, Eq. (A.4):

$$\frac{b}{\sin B} = \frac{c}{\sin C}$$

to find the remaining angles $\widehat{A}$ and $\widehat{C}$. Substituting the known parameter values gives,

$$\frac{207.5}{\sin 80} = \frac{100}{\sin C} \rightarrow \sin C = 0.475$$

The two possible solutions are,

$$\widehat{C} = 28.33\,^\circ \text{ and } \widehat{C} = 180\,^\circ - 28.33\,^\circ = 151.67\,^\circ$$

The only acceptable solution is $\widehat{C} = 28.33\,^\circ$ since the other one results in,

$$\widehat{B} + \widehat{C} = 231.67\,^\circ$$

That clearly violates Eq. (A.2):

$$\widehat{A} + \widehat{B} + \widehat{C} = 180\,^\circ \tag{A.2}$$

The final unknown is now calculated using Eq. (A.2) to give,

$$\widehat{A} = 180\,^\circ - 80\,^\circ - 28.33\,^\circ = 71.67\,^\circ$$

Had one started by using law of sines (A.4)

$$\frac{a}{\sin A} = \frac{b}{\sin B} = \frac{c}{\sin C} = 2R \tag{A.4}$$

after substituting the known quantities one would have gotten,

$$\frac{200}{\sin A} = \frac{b}{\sin 80} = \frac{100}{\sin C}$$

This result is equivalent to the following two independent equations,

$$\frac{200}{\sin A} = \frac{b}{\sin 80} \tag{A.9}$$

$$\frac{b}{\sin 80} = \frac{100}{\sin C} \tag{A.10}$$

Furthermore, every triangle satisfies condition (A.2) between its interior angles,

$$\widehat{A} + \widehat{B} + \widehat{C} = 180\,^\circ \tag{A.2}$$

In the present case, condition (A.2) reduces to the following equation,

$$\widehat{A} + 80\,^\circ + \widehat{C} = 180\,^\circ \tag{A.11}$$

We now have a system of three Eqs. (A.9), (A.10), and (A.11) for the three unknowns, $\widehat{A}$, $\widehat{C}$, and b. This system is not easily solvable! Therefore, starting the solution with using the law of cosines was a correct decision.

In the next example, we consider the same triangle, but assume a different set of given values (with more known angles). In such a case, using law of sines would be advantageous.

Example A.2 In triangle *ABC*, it is known that: a = 200 mm, $\widehat{B} = 80\,^\circ$, and $\widehat{C} = 28.33\,^\circ$. Calculate the following three unknowns: $\widehat{A}$, b, and c.

Finding the only unknown angle requires the use of condition (A.2)

$$\widehat{A} + \widehat{B} + \widehat{C} = 180\,^\circ \tag{A.2}$$

which gives,

$$\widehat{A} = 180\,^\circ - 80\,^\circ - 28.33\,^\circ = 71.67\,^\circ$$

From this point on, if one tries to use the law of cosines (A.1),

$$\begin{aligned} a^2 &= b^2 + c^2 - 2bc \cos A \\ b^2 &= a^2 + c^2 - 2ac \cos B \\ c^2 &= a^2 + b^2 - 2ab \cos C \end{aligned} \tag{A.1}$$

the following difficult system of equations is obtained,

$$b^2 = (200)^2 + c^2 - 2 \times 200c \cos 80$$

$$c^2 = (200)^2 + b^2 - 2 \times 200b \cos 28.33$$

which results in,

$$b^2 = 40{,}000 + c^2 - 69.46c$$

$$c^2 = 40{,}000 + b^2 - 352.09b$$

This is a system of two coupled quadratic equations in terms of b and c that is challenging to solve.

Instead, for finding b and c, one may use the law of sines (A.4).

$$\frac{a}{\sin A} = \frac{b}{\sin B} = \frac{c}{\sin C} = 2R \tag{A.4}$$

The first two segments give:

$$\frac{200}{\sin 71.67^\circ} = \frac{b}{\sin 80}$$

or,

$$b = 207.49 \text{ mm}$$

Finally, c can be calculated using the next two segments of the law of sines,

$$\frac{207.49}{\sin 80} = \frac{c}{\sin 28.33^\circ} \rightarrow c = 99.98 \text{ mm}$$

These results are essentially the same as the ones obtained in the previous example, with the exception of some round-off error.

Index

M. Ghorashi, *Introduction to Machine Dynamics*, Mechanical Engineering Series,
https://doi.org/10.1007/978-3-031-94736-0

MIX
Papier aus verantwortungsvollen Quellen
Paper from responsible sources
FSC® C105338

If you have any concerns about our products,
you can contact us on
ProductSafety@springernature.com

In case Publisher is established outside the EU,
the EU authorized representative is:
Springer Nature Customer Service Center GmbH
Europaplatz 3, 69115 Heidelberg, Germany

Printed by Libri Plureos GmbH
in Hamburg, Germany